AVIAN MOLECULAR EVOLUTION AND SYSTEMATICS

AVIAN MOLECULAR EVOLUTION AND SYSTEMATICS

Edited by

DAVID P. MINDELL

Museum of Zoology & Department of Biology
University of Michigan
Ann Arbor, Michigan

ACADEMIC PRESS

SAN DIEGO LONDON BOSTON NEW YORK SYDNEY TOKYO TORONTO

Cover illustration by Jonathan Higgins.

This book is printed on acid-free paper. ♾

Academic Press
a division of Harcourt Brace & Company
525 B Street, Suite 1900, San Diego, California 92101-4495, USA
http://www.apnet.com

Academic Press Limited
24-28 Oval Road, London NW1 7DX, UK
http://www.hbuk.co.uk/ap/

Library of Congress Cataloging-in-Publication Data

Avian molecular evolution and systematics / edited by David P. Mindell.
p. cm.
Includes bibliographical references and index.
ISBN 0-12-498315-4 (alk. paper)
1. Birds--Molecular genetics. 2. Birds--Evolution. 3. Birds--Morphology. I. Mindel, David P.
QL696.5.A95 1997
572.8'3818--dc21 97-6621
CIP

PRINTED IN THE UNITED STATES OF AMERICA
97 98 99 00 01 02 QW 9 8 7 6 5 4 3 2 1

Contents

PART I

Molecular Sequences and Evolutionary History in Birds

5 *Phylogeny and Evolution of 12S rDNA in Gruiformes (Aves)*

Peter Houde, Alan Cooper, Elizabeth Leslie, Allan E. Strand, and Gabriel A. Montaño

6 *Phylogeny of the Pelacaniformes: Molecular Systematics of a Privative Group*

Douglas Siegel-Causey

7 *The Phylogeny of Ratite Birds: Resolving Conflicts between Molecular and Morphological Data Sets*

Krista Lee, Julie Feinstein, and Joel Cracraft

8 *Phylogenetic Relationships among and within Select Avian Orders Based on Mitochondrial DNA*

David P. Mindell, Michael D. Sorenson, Christopher J. Huddleston, Hector C. Miranda, Jr., Alec Knight, Steven J. Sawchuk, and Tamaki Yuri

PART II

Applying Phylogeny and Population Genetics to Broader Issues

9 *Relevance of Microevolutionary Processes to Higher Level Molecular Systematics*

Scott V. Edwards

10 *Phylogeny in Studies of Bird Ecology, Behavior, and Morphology*

Frederick H. Sheldon and Linda A. Whittingham

11 *Phylogeographic Studies of North American Birds*

Robert M. Zink

12 *The Speciation of South American and African Birds in Montane Regions*

Michael S. Roy, José Maria Cardoso da Silva, Peter Arctander, Jaime García-Moreno, and Jon Fjeldså

13 *Studies of Avian Ancient DNA: From Jurassic Park to Modern Island Extinctions*

Alan Cooper

Contributors

Numbers in parentheses indicate the pages on which the authors' contributions begin.

Peter Arctander (325)
Zoological Institute
University of Copenhagen
DK 2100 Denmark

Allan J. Baker (51)
Centre for Biodiversity and Conservation Biology
Royal Ontario Museum
Toronto, Ontario M5S 2C6, Canada

José Maria Cardoso da Silva (325)
Zoological Museum
University of Copenhagen
DK 2100 Denmark

Alan Cooper (121, 345)
Institute of Molecular Medicine
Oxford University
Oxford
OX2 6Q5, United Kingdom

Joel Cracraft (173)
Department of Ornithology
American Museum of Natural History
New York, New York 10024

Victor R. DeFilippis (83)
Department of Biological Sciences
Wayne State University
Detroit, Michigan 48202

Scott V. Edwards (251)
Department of Zoology and Burke Museum
University of Washington
Seattle, Washington 98195

Julie Feinstein (173)
Department of Ornithology
American Museum of Natural History
New York, New York 10024

Jon Fjeldså (325)
Zoological Institute
University of Copenhagen
DK 2100 Denmark

Jaime García-Moreno (325)
Zoological Institute
University of Copenhagen
DK 2100 Denmark

Peter Houde (121)
Department of Biology
New Mexico State University
Las Cruces, New Mexico 88003

Christopher J. Huddleston (213)
Laboratory of Molecular Systematics
National Museum of Natural History
Smithsonian Institution
Washington, DC 20560

Alec Knight (213)
Department of Biology
Sul Ross State University
Alpine, Texas 79832

Krista Lee (173)
Museum of Natural Science
Louisiana State University
Baton Rouge, Louisiana 70803

Elizabeth Leslie (121)
Neuroscience Laboratory
University of Michigan
Ann Arbor, Michigan 48109

H. Dawn Marshall (51)
Department of Zoology
University of Toronto
Toronto, Ontario M5S 3G5, Canada

David B. McDonald (29)
Department of Zoology and Physiology
University of Wyoming
Laramie, Wyoming 82071

David P. Mindell (213)
Department of Biology and Museum of Zoology
University of Michigan
Ann Arbor, Michigan 48109

Hector C. Miranda, Jr. (213)
Department of Biological Sciences
University of Cincinnati
Cincinnati, Ohio 45221

Gabriel A. Montaño (121)
Department of Biology
New Mexico State University
Las Cruces, New Mexico 88003

William S. Moore (83)
Department of Biological Sciences
Wayne State University
Detroit, Michigan 48202

Wayne K. Potts (29)
Department of Biology
University of Utah
Salt Lake City, Utah 84112

Thomas W. Quinn (3)
Department of Biological Sciences
University of Denver
Denver, Colorado 80208

Michael S. Roy (325)
Zoological Institute
University of Copenhagen
DK 2100 Denmark

Steven J. Sawchuk (213)
School of Veterinary Medicine
Ohio State University
Columbus, Ohio 43210

Frederick H. Sheldon (279)
Museum of Natural Science
Louisiana State University
Baton Rouge, Louisiana 70803

Douglas Siegel-Causey (159)
School of Biological Sciences
University of Nebraska
Lincoln, Nebraska 68588

Michael D. Sorenson (213)
Museum of Zoology
University of Michigan
Ann Arbor, Michigan 48109

Allan E. Strand (121)
Department of Biology
New Mexico State University
Las Cruces, New Mexico 88003

Linda A. Whittingham (279)
Museum of Natural Science
Louisiana State University
Baton Rouge, Louisiana 70803

Tamaki Yuri (213)
Department of Biology and Museum of Zoology
University of Michigan
Ann Arbor, Michigan 48109

Robert M. Zink (301)
J. F. Bell Museum
University of Minnesota
St. Paul, Minnesota 55108

Preface

The objective of this book is to synthesize and explore current theory and practice of molecular evolution and systematics focusing on birds. Chapters are written by active practitioners discussing current controversies, demonstrating methods, reviewing new findings, and assessing directions for future research.

As indicated by the title, *Avian Molecular Evolution and Systematics*, studies of organismal phylogeny and of evolution at the molecular level have become closely linked. This stems from widespread use of molecular characters that are believed to be homologs (similar due to common descent) in phylogenetic inference and from reciprocal use of phylogenetic trees in studying the evolution and hypotheses of homology for the characters themselves. The mutually informing nature of organismal and molecular evolution (Fig. 1) is general, applying to all of life and across taxonomic levels. This generality bodes well for the field, signifying a shared research agenda for many evolutionary biologists, in using analyses of molecular evolution to inform phylogenetic analyses and vice versa. Indeed, a great deal has been learned about the evolution of both organisms and molecular sequences during the past decade in this fashion.

However, all change at the molecular level is not so impervious to convergence, natural selection, varying functional constraints, or chance events as to provide a linear measure of the passage of time or of relatedness of taxa. It is increasingly clear that evolution of some molecular characters can be as quirky and unique to individual lineages as that seen for some phenotypic characters. The need to reconcile evolutionary inferences across numerous molecular and phenotypic analyses presents biologists with both challenges and opportunities. For example, assessing discord between molecular and morphological analyses challenges biologists to provide well-corroborated phylogenies based on both molecular and nonmolecular characters, and provides the opportunity to learn about the different constraints on change for different data sets. Assessing discord between gene trees and species trees

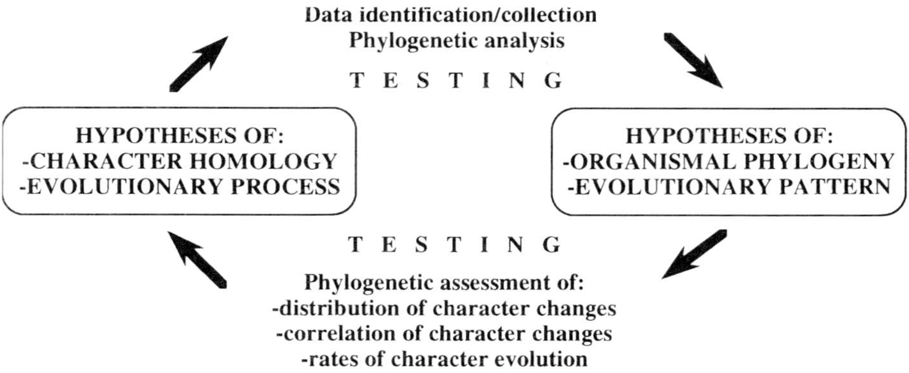

FIGURE 1 Diagram indicating the mutually informing nature of hypotheses of organismal phylogeny and processes of molecular evolution.

provides the same challenge, as well as the opportunity to learn about potential character divergence within and among populations prior to speciation. A degree of uncoupling for evolutionary change across data sets can complicate recovery of the actual phylogenetic pattern; however, this reflects differences in evolutionary processes operating at different levels of organization and on different character sets, and an understanding of both process and pattern is key in a comprehensive view of evolution (Fig. 1).

Birds have long been a source of insight into the workings of nature. This may be attributed, at least in part, to birds being both conspicuous and widespread, attracting numerous field observers. This has in turn yielded detailed understanding of many facets of avian behavior, distribution, and ecology. As an indication of the historical role of ornithology, observations of birds figure prominently in Aristotle's 4th-century BCE *History of Animals*, in which he lays the groundwork for the eventual organization of biology into physiology, morphology, systematics, embryology, ethology, and ecology. In the 13th century, Frederick II was an ardent natural historian of birds and falconer and pioneered a return to direct observations in seeking explanations for the natural world, when nearly all others sought explanations in the revealed word of the churches. Darwin's observations of finches on the Galapagos Islands were vital in development of thought on evolution by means of natural selection. In recent times, birds have been central to work in many fields relevant to evolutionary biology, including studies of species formation and species definition; comparative morphology, physiology, and endocrinology; studies of mating systems and reproductive strategies; roles of kinship in evolution; population and community ecology; effects of environmental change and conservation strategies for populations and species; and the roles of vicariance and dispersal in biogeography. As a consequence, researchers undertaking new molecular studies of birds have available a wealth of background material on comparative avian biology to provide context and inform interpretation of the molecular findings. Thus, there is good reason to be optimistic about the prospects for further insight into evolutionary patterns and processes to be gained from the continued study of birds.

This volume had its beginnings in the symposium "Avian Molecular Evolution" held at the 113th stated meeting of the American Ornithologist' Union, August 15–19, 1995 in Cincinnati, Ohio. Six of the chapter authors (Edwards, McDonald, Mindell, Quinn, Sheldon, Zink) gave presentations in the symposium, and subsequently, seven others were invited to contribute chapters along with their collaborators. As organizer and editor of this book, I have attempted to seek some balance in coverage of topics, taxa, and taxonomic levels. However, this volume is not intended to be encyclopedic in its coverage, being restricted by the current interests and expertise of the various authors. It is my hope that this volume will stimulate inquiry and promote understanding among researchers and students working in this area, as well as provide specialists and nonspecialists alike with useful topic reviews and empirical examples.

The chapters are divided into two sections: (I) Molecular Sequences and Evolutionary History in Birds and (II) Applying Phylogeny and Population Genetics to Broader Issues. Authors focusing on the evolution and utility of molecular markers consider current understanding of the evolution of the mitochondrial genome in birds and closely related vertebrates (Quinn), the inherent difficulties and applications for nuclear DNA microsatellites (McDonald and Potts), the applicability of mitochondrial control region sequences to studies of population structure (Baker and Marshall), and the range of taxonomic resolution for mitochondrial cytochrome *b* (Moore and DeFilippis). Investigators consider methodological issues in systematics and variously present new data and analyses of phylogeny for select species or populations of Charadriiformes, Apodiformes, and Passeriformes (Baker and Marshall, Edwards, Sheldon and Whittingham, Zink); for species of Piciformes (Moore and DeFilippis), Gruiformes (Houde *et al.*), Pelecaniformes (Siegel-Causey), and ratites (Lee *et al*; Cooper); and for Falconiformes, Strigiformes, Anseriformes, Galliformes, *Turnix*, *Opisthocomus*, and *Phoenicopterus* (Mindell *et al.*). Application of phylogeny and population genetics studies to broader issues include an assessment of the relevance of population-level processes to phylogeny at the species-level and above (Edwards); uses of phylogenetic hypotheses in studying evolution of behavior, morphology, and ecology (Sheldon and Whittingham); patterns of geographic variation and their potential causes (Zink); speciation (Roy *et al.*); and paleoecology and conservation biology using DNA from extinct taxa (Cooper).

All chapters have gone through a process of peer review, and I am extremely grateful to the following persons for insightful reviews of one or more chapters: Marc W. Allard, Jeremy Austin, John M. Bates, Anthony H. Bledsoe, Scott V. Edwards, Frank B. Gill, John Harshman, Peter Houde, Arnold G. Kluge, Thomas D. Kocher, Carey W. Krajewski, Mary C. McKitrick, Axel Meyer, William S. Moore, Robert B. Payne, Richard O. Prum, Thomas W. Quinn, Frederick H. Sheldon,

Michael D. Sorenson, Robert K. Wayne, and David F. Westneat. I thank Cyndy Sims Parr for valuable assistance in preparation of the index, and I am grateful to Jonathan Higgins for preparation of the cover illustration. Finally, I thank Margaret, June, and Eugene Mindell for their constant support and encouragement, which were instrumental in completion of this project.

PART I

Molecular Sequences and Evolutionary History in Birds

CHAPTER 1

Molecular Evolution of the Mitochondrial Genome

THOMAS W. QUINN
Department of Biological Sciences, University of Denver, Colorado 80208

I. INTRODUCTION

Advances in molecular techniques, particularly the development of the polymerase chain reaction (PCR; Mullis *et al.*, 1986), have made studies of vertebrate genomes increasingly practical, and have obviated the requirement for fully equipped molecular biology laboratories. Among ornithologists, this has resulted in an explosive increase in studies of systematics and population genetics. To date, many of these studies have focused on the mitochondrial genome, mainly at the level of DNA sequence determination. While several chapters in this book illustrate the enormous value of primary mitochondrial DNA (mtDNA) sequence in learning about the evolutionary history of a group of organisms, this chapter is meant to provide an overview of the evolution of the avian mitochondrial genome from a broad perspective, mainly at a level of organization above the primary sequence level. An attempt is made to illustrate similarities and differences between the mitochondrial genomes of Aves and other vertebrate classes, and to point out how some of these differences have provided unique opportunities to probe deep phylogenetic questions. It is also emphasized that the avian nuclear genome contains homologs to mitochondrial sequences in at least some species, and this may have serious implications for the interpretation of mtDNA data sets used in population genetic and systematic studies. At the same time, such occurrences present valuable new opportunities to gain a better understanding of sequence evolution in both mitochondrial and nuclear genomes.

II. MITOCHONDRIA: AN ANCIENT LEGACY

In 1922, Wallin proposed that mitochondria arose by endosymbiosis, an idea that was championed and expanded on by Margulis (1970). They proposed that mitochondria originated when a protoeukaryotic cell engulfed or was penetrated by an aerobic bacterium. The endosymbiont hypothesis was initially treated with much skepticism, but it has subsequently received strong support and is now widely accepted (Gray, 1983), with debate now centering on whether mitochondria are monophyletic or polyphyletic in origin (see review in Gray, 1989) and on the identity of the original symbionts (Yang *et al.*, 1985; Cedergren *et al.*, 1988; Andachi *et al.*, 1989; Cardon *et al.*, 1994). The acquisition of such an aerobic endosymbiont is among the most important events in the history of life, for the descendants of those associations now comprise almost all eukaryotic life, both single-celled and multicellular.

Among a number of traits that are indicative of this evolutionary heritage is the fact that mitochondria carry their own (usually circular) independently replicating genomes. The small size of this genome in animals relative to that of extant bacteria (approximately 1% as large) is presumably the result of elimination of genes that were redundant with those in the nuclear (host) genome, and of the occasional transfer of functional genes from the mitochondrial genome to the nuclear genome in our endosymbiotic ancestors [see, for example, Schuster and Brennicke (1987) and Nugent and Palmer (1991)]. The discovery of a circular genome within the mitochondria of most eukaryotes not only bolstered the endosymbiont hypothesis, but it identified a small and easily isolated source of genetic information outside of the large and complex nuclear compartment.

III. THE VERTEBRATE MITOCHONDRIAL GENOME

Since the development of recombinant DNA techniques, evolutionary biologists have made the vertebrate mitochondrial genome one of the most extensively studied on the planet. It is undeniable that a major factor in the initial selection of mtDNA for evolutionary studies was the relative ease with which it could be purified and manipulated in the laboratory, owing to its high copy number and supercoiled conformation, which allows separation from linear (nuclear) DNA. Protocols for such isolations are included in Lansman *et al.* (1981) and in Dowling *et al.* (1990). Robin and Wong (1988) estimated that there are 800 mitochondrial genomes per cell, and an average of 2.6 genomes per mitochondrial organelle, within cultured human lung fibroblast cells, while Michaels *et al.* (1982) estimated 2600 genome copies per cell in primary bovine tissue culture cells. Estimates for other tissue types vary widely.

Once purified, mtDNA sequence differences between species or individuals can be inferred indirectly using restriction endonucleases to generate discrete fragments that can then be compared via electrophoresis through agarose gels, or directly by DNA sequencing. The polymerase chain reaction (PCR) has increasingly replaced such purification methods by allowing the direct amplification and sequencing of mtDNA from unpurified sources. However, as nuclear copies of mitochondrial genes have been noted in birds, use of direct purification methods will be of continuing value (see Section VI).

As some of the first evolutionary studies using purified mtDNA were completed, it became apparent that, besides its ease of isolation and small genome size, there were many other advantages to studying the mitochondrial genome. First, it is maternally inherited (Lansman *et al.*, 1983). Second, there is no direct evidence that it can recombine with other mtDNA molecules (Clayton, 1982, 1992; Hayashi *et al.*, 1985). This means that vertebrate mtDNA is passed on through female lineages in a clonal fashion with no horizontal "mixing," and this makes it more straightforward to reconstruct an evolutionary history of this molecule than for the nuclear genome. While Gyllensten *et al.* (1991) have detected low levels of paternal "leakage" of mtDNA between two species of mouse, whether such a finding can be extrapolated to intraspecific leakage is still not clear. Kondo *et al.* (1990) showed that biparental inheritance occurred in interspecific *Drosophila* crosses, but did not occur when intraspecific crosses were done. Third, at the sequence level, mtDNA has been shown to evolve rapidly relative to DNA in the nuclear genome (Brown *et al.*, 1979). This is an asset for population studies, but eventually becomes a liability as the depth of phylogenetic comparison increases.

The first mitochondrial genomes to be sequenced in their entirety include human (Anderson *et al.*, 1981), mouse (Bibb *et al.*, 1981), cow (Anderson *et al.*, 1982), and frog (*Xenopus*; Roe *et al.*, 1985). Not only was the gene content found to be identical, but the gene order was the same, and this, in conjunction with the same finding in bony fish, led to the assumption that all vertebrate mtDNA genomes were identical in both respects (Johansen *et al.*, 1990). As greater numbers of genomes have been sequenced, it is apparent that gene order, especially of tRNA genes, is not constant. For example, among tetrapods reports of tRNA transposition have been made for marsupials (Pääbo *et al.*, 1991; Janke *et al.*, 1994), frogs (*Rana*; Yoneyama, 1987), and crocodiles (Kumazawa and Nishida, 1995; Quinn and Mindell, 1996). Lee and Kocher (1995) have shown that in an outgroup to the Osteichthyes, the sea lamprey, there have been several changes in gene order near the putative control region, some of which have included tRNAs.

All vertebrate mtDNAs contain 22 tRNAs, 13 protein-coding regions, 2 rRNAs, and 1 or 2 (lamprey) large noncoding regions. The genome is arranged in an efficient manner. Introns are absent, and intergenic spacers are small, typically less than 10 bp. In some cases genes overlap in different reading frames, and there has even been a reduction in the size of some stop codons to one or two bases (T or TA). Such codons are posttranscriptionally completed with the addition of a 3′ poly(A) sequence (Ojala *et al.*, 1981). These characteristics of vertebrate and many nonvertebrate mitochondrial genomes have led to the proposal that intermolecular selection for compactness could be the "driving force" that results in such an economical gene arrangement, assuming that smaller molecules replicate more rapidly than larger ones, all else being equal (Wallace, 1982; Rand and Harrison, 1989; Kurland, 1992). Evidence in support of this comes from observations that human mitochondrial genomes carrying deletions can, between generations, increase in frequency relative to full-sized genomes both *in vitro* (Yoneda *et al.*, 1992) and *in vivo* (Larsson *et al.*, 1990; Kobayashi *et al.*, 1992).

IV. AVIAN MITOCHONDRIAL GENOMES

A. An Altered Gene Order throughout Aves

In 1990, Desjardins and Morais published the first complete sequence of an avian mitochondrial genome. This resource, in conjunction with increasingly sophisticated methods of accessing worldwide repositories of DNA sequence data, has made it possible to use PCR to study almost any part of the genome in any avian species. It also provided our first clear view of the similarities and differences in genome structure of birds compared with other tetrapods. The most striking feature of the

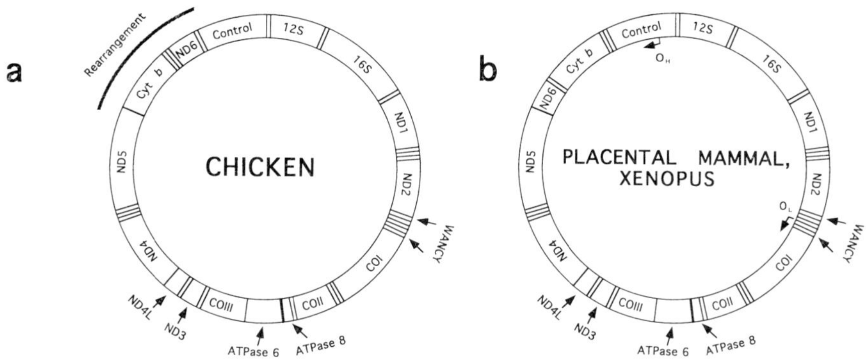

FIGURE 1.1 Schematic diagram of (a) the gene content and order of the chicken mitochondrial genome (Desjardins and Morais, 1990) and (b) that of placental mammals and *Xenopus laevis*. tRNAs are unlabeled except in the WANCY region. Abbreviations of gene names are those in common use except for the following: Cyt. *b*, cytochrome *b*; 12S, 12S rRNA; 16S, 16S rRNA; Control, control region; O_L, light-strand origin of replication; O_H, heavy-strand origin of replication. These origins are unmapped in chicken. There is a rearrangement adjacent to the control region in chicken relative to the placental mammal and *Xenopus* (highlighted with overline). A short sequence with strong secondary structure exists within the WANCY region in placental mammals and *Xenopus* (labeled O_L) but is absent in chicken [a single base is present between $tRNA^{Asn}$ (N) and $tRNA^{Cys}$ (C) in chicken]. The overlap between ATPase 6 and ATPase 8 is shown by two lines in close proximity. The proportionate sizes of genes within the chicken genome are accurate, except all tRNAs were drawn to an (average) length of 73 bp. The genes shown in (b) have been drawn to the same size as in chicken, although slight differences in gene lengths do exist.

chicken mitochondrial genome is that, while it contains the same genes as all other vertebrates, the order of those genes is unique (Fig. 1.1). Additional studies have made it apparent that this altered gene order is conserved across a wide taxonomic diversity of birds [Desjardins *et al.*, 1990 (duck); Quinn and Wilson, 1993 (goose); Wenink *et al.*, 1993 (dunlin); Wenink *et al.*, 1994 (turnstone); Moum and Johansen, 1992 (murre); Moum *et al.*, 1994 (gull, sandpiper)]. While tRNA rearrangements have been observed in other vertebrates, the avian and lamprey genomes are thus far the only ones known to have undergone major rearrangements that include protein-coding genes.

The mechanism leading to such rearrangements is unknown. Moritz *et al.* (1987) proposed that mitochondrial gene order could be changed without intermolecular recombination if tandem duplication of part of a genome was followed by deletions that include at least some of the "parental" copies. Given the documentation of large tandem duplications in reptiles (Moritz and Brown, 1986, 1987; Zevering *et al.*, 1991), amphibians (Wallis, 1987), and mammals (Poulton *et al.*, 1989), and large deletions in a similar variety of taxa including reptiles (Moritz and Brown, 1986, 1987), mammals (Shoffner *et al.*, 1989; Zeviani *et al.*, 1989; Tanaka *et al.*, 1989), and birds (Edwards, 1992; also see Avise and Zink, 1988, for another likely candidate), this seems to be a reasonable hypothesis for the avian rearrangement. In the ancestor to birds, beginning with a gene order as found in mammals and *Xenopus*, this would require a single duplication of several genes located downstream of the heavy strand origin of replication (O_H), followed by at least two independent deletion events (Desjardins and Morais, 1990; Quinn and Wilson, 1993; Fig. 1.2c). Pääbo *et al.* (1991) and Kumazawa and Nishida (1995) documented gene rearrangements among tRNA gene clusters in marsupials and crocodiles, respectively, and in each case intergenic spacers among the rearranged genes were considerably larger than in vertebrates without such rearrangements. In both cases, the authors point out that this observation is consistent with rearrangement via duplication, assuming that the large intergenic spacers are the remnants of the "extra" copies that are in the process of being (gradually) deleted. In these and in the avian rearrangement, a "replicative race" could be the force driving subsequent reduction of enlarged genomes back to the original size.

Both duplication and deletion could occur as the result of slippage during replication (Streisinger *et al.*, 1966; Levinson and Gutman, 1987). A convincing model that explains the rapid generation of length variation among relatively short tandem repeats within the control region of heteroplasmic sturgeon has been presented by Buroker *et al.* (1990). They proposed that frequent misalignment of extending nascent strands with parental strands could be enhanced by the formation of secondary structures within the tandem repeats of either strand when single stranded. Because the D-loop region is triple stranded, competitive binding of two H strands (nascent and parental) to a single parental L strand could alternately expose both H strands to a single-stranded state, thereby facilitating the formation of such single-stranded structures. In birds, similar short repeats in the control region and in association with heteroplasmy have been shown by Berg *et al.* (1995); also see Avise and Zink (1988).

For larger duplications and deletions such as envisaged here for the avian rearrangement, the development of a convincing model seems more problematic. Dispersed regions of homology might allow misalignment between parent and nascent strands of DNA during replication as shown in Fig. 1.2a and b, but here the mismatch must occur over longer distances, and the role of secondary structure in promoting mismatch is not as obvious. Perhaps competitive hybridization can occur between the nascent and parental heavy strands in binding to the light strand. This would free the nascent heavy strand for reinvasion in a new location. The fact that replication is approximately 200 times slower in mtDNA than in *Escherichia coli* (Clayton, 1982) may also facilitate such events. However, tandem duplications in humans (Poulten *et al.*, 1989) and in *Cnemidophorus* (Moritz and Brown, 1986, 1987) typically include the entire control region and hence span both sides of O_H, an observation not easily explained by such a model.

Most of the *Cnemidophorus* duplications include flanking tRNAs, perhaps implicating their involvement in gene duplication (Moritz and Brown, 1986, 1987). Desjardins and Morais (1990) suggest that the cotransposition of tRNA genes associated

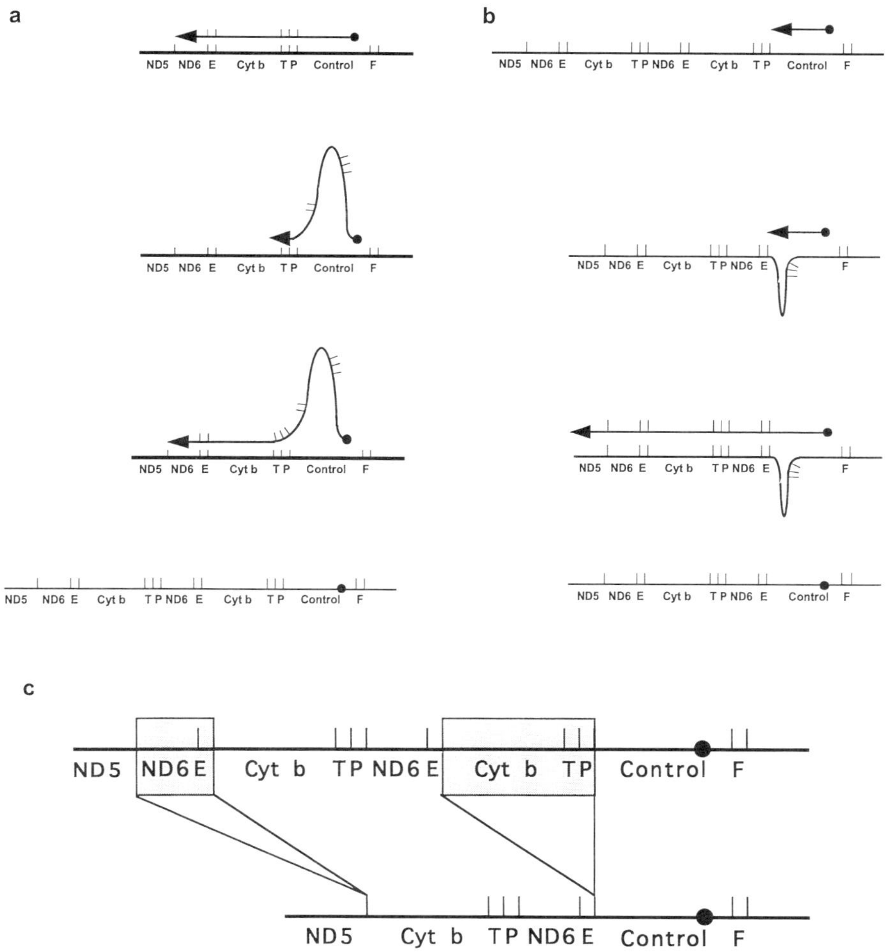

FIGURE 1.2 A model explaining how the avian rearrangement could be produced without intergenomic recombination. Cyt *b*, cytochrome *b*; E, T, P and F (respectively), tRNA^{Glu}, tRNA^{Thr}, tRNA^{Pro}, and tRNA^{Phe}. (a) DNA replication of a mitochondrial genome with the same gene order as observed in placental mammals begins from the heavy strand origin (solid circle), presumably via a primer molecule, and proceeds to the ND5–ND6 boundary (top). Melting and reannealing of the extending strand (possibly during competitive annealing of displaced parental H-strand) place the 3′ end of the nascent strand in a new position with some sequence similarity. Extension resumes to completion (only a subsection of genome is shown). The resulting molecule includes a large duplication that is resolved after one more round of replication (bottom). (Drawing is modified from Fig. 9a–c in Quinn and Wilson, 1993.) (b) Beginning with the duplicated molecule generated in part (a), a deletion occurs during replication when melting and reannealing occur such that an area at the 3′ end of the extending nascent molecule mispairs with a region of sequence similarity further downstream, perhaps by competitive invasion of a small region within the parental molecule. Extension resumes to completion to generate a large deletion that is resolved after one more round of replication (bottom). (c) Overall, the two deletions of the shaded areas would be neutral or selectively favored, and would eventually return the gene copy number to one while producing the avian gene order [the rightmost deletion was modeled in (b)]. Although only two deletions are shown, this represents the minimum number possible using this model. The deletions were not likely to have taken place simultaneously, and would probably take many generations to be completed. (Modified from Fig. 9e in Quinn and Wilson, 1993.)

with one end of cytochrome *b* (or alternatively, with ND6, depending on which one was transposed) could be relevant to the avian rearrangement. Edwards (1992) and Edwards and Wilson (1990) documented a duplication that includes the fragment $tRNA^{Pro}$-ND6-$tRNA^{Glu}$ in *Pomatostomus* (babblers). The duplicated region is situated within the 3′ end of the control region. Again, this represents a fragment flanked by tRNAs, and suggests an alternative mechanism as a possible cause of the putative avian duplication.

Once a duplication is generated, the mechanism of its reduction could also be modeled using slipped strand mispairing (Fig. 1.2b; also see Efstratiadis *et al.*, 1980). The discovery that two large independent deletions have occurred within the mitochondrial duplication of *Pomatostomus temporalus* after it diverged from other *Pomatostomus* species that carry the full complement of pseudogenes described above, and that flanking polycytosine tracts are interspersed with guanines in the undeleted genomes, led Edwards (1992) to propose the involvement of slipped strand mispairing. In humans with certain mitochondrial myopathies, mtDNA deletions are often flanked by short direct repeats (Shoffner *et al.*, 1989 and references therein). Such sites may provide areas across which mispairing of complementary regions could occur during DNA replication. To explain some unexpected observations in a series of PCR and sequencing experiments, Quinn and Mindell (1996) proposed that some mitochondria within a single tuatara (*Sphenodon punctatus*) sample carried a deletion that included cytochrome *b*, $tRNA^{Pro}$, and a small piece of the control region. This putative deletion is of particular interest since it encompasses one of the two areas shown in Fig. 1.2c as leading to the avian mitochondrial rearrangement, except that in tuatara $tRNA^{Phe}$ is missing from its "usual" location next to cytochrome *b*.

B. Other Features of the Avian Mitochondrial Genome

Desjardins and Morais (1990) have compared the avian mitochondrial genome to those of mammals and amphibians. Many similarities exist besides the common gene content, including (1) in chicken, several genes end with incomplete stop codons that are presumably completed by polyadenylation as proposed for mammals; (2) codon/anticodon rules are the same for those codons present in chicken, although the frequencies of use may differ (see Desjardins and Morais, 1990, for details); (3) guanine is relatively infrequent at third positions of codons; (4) several genes, including ATPase 6 and 8, overlap, by the same amount as in *Xenopus*, but less than in mammals, while others are butt-joined; (5) the control region includes a transcriptional promoter, which, as in amphibians, is bidirectional (L'Abbé *et al.*, 1991), and the O_H (Glaus *et al.*, 1980).

Several differences can be listed as well. In addition to an altered mitochondrial gene order, birds lack the hairpin structure that forms the light-strand origin of replication (O_L) in mammal and *Xenopus* (where it is located within a tRNA cluster called the WANCY region between $tRNA^{Asn}$ and $tRNA^{Cys}$). Thus far, which part of the avian genome serves this function remains unknown, although it is intriguing to note that Tapper and Clayton (1981) mapped two distinct 5′ end positions of the human daughter L strand, one of which is located within the complementary sequence to the $tRNA^{Cys}$ gene anticodon.

In chicken (Desjardins and Morais, 1990) and in goose (T. W. Quinn, unpublished), COI has a putative GTG initiation codon that is unusual among vertebrates and is unique for this gene. Another feature peculiar to the avian genome is the low incidence of thymine at silent positions within coding regions of cytochrome *b* and presumably other protein-coding genes coded on the same strand. In reporting this anomaly, Kocher *et al.*, (1989) point out that the extreme compositional bias created by this and the guanine deficit mentioned above may make "saturation effects" particularly severe in DNA sequence-based phylogenetic studies of birds.

V. MAJOR GENOMIC FEATURES AS CHARACTERS FOR PHYLOGENETIC ANALYSIS

A. Two Approaches to Phylogenetics Analysis Using DNA Sequence Information

Studies of systematic relationships frequently make use of direct comparison of DNA sequence information, from which estimates of branching order can be made, using an assortment of methods that include parsimony, maximum likelihood, and distance. Such approaches have provided a wealth of new information, and with the concurrent refinement of statistical or "quasistatistical" methods, allow the relative robustness of conflicting phylogenetic hypotheses to be evaluated. Despite the increasing sophistication of analytical methods, controversies over alternative phylogenetic hypotheses are not uncommon, especially in cases where several species have diverged in close temporal proximity, or in cases where divergences occurred far in the past. In the first case, resolution may not even be possible with mtDNA as the only source of information because species trees and gene trees may differ (see Avise *et al.*, 1987; Avise, 1994; and Chapter 9, by Edwards, in this volume). In the second case, resolution may be theoretically possible, but it becomes difficult in practice because of problems distinguishing "signal" from "noise." It is this second case that is addressed in the following discussion.

In analyses that attempt to distinguish between alternative branching orders of deeply divergent taxa, two problems that frequently arise are sequence alignment and sequence saturation. The problems with alignment can be minimized by careful choice of the genes/regions being compared. For instance, regions that are evolutionarily constrained, presumably by function, are easier to align than those that are not. Kumazawa and Nishida (1993) present a convincing case for their ability to align unambiguously stem regions in mitochondrial tRNA genes of vertebrates. The use of conserved protein-coding regions may also circumvent some alignment problems. Unfortunately, a balance must be reached between constraint and variability. A region that is too constrained will provide few variable or informative sites for phylogenetic analysis. In contrast, in a region that is not constrained enough, alignment of homologous bases may become difficult because of numerous insertions or deletions. In many such cases, alignments can be done reasonably well, particularly if done conservatively such that ambiguous areas are eliminated from the final analysis, preferably by a stated and objective set of rules.

Once alignment is accomplished successfully, the (perhaps) more difficult problem of sequence saturation (multiple substitution) must be considered. Multiple substitutions at any single site that occur after two taxa have diverged cannot be discerned in a pairwise comparison. In phylogenetic reconstruction, this produces noise via homoplasy that can be difficult to distinguish from the phylogenetic signal that is generated, for example, in cases where single substitutions provide informative synapomorphies. Statistical methods can be useful in clarifying the intensity of signal, but in many cases the finding is that the signal-to-noise ratio is too low to allow differentiation between competing phylogenetic hypotheses. Methods for increasing the relative weighting given to sites at which substitutions are infrequent and/or substitution types that are relatively rare (e.g., Williams and Fitch, 1990) may help to accentuate signal over noise. It is also clear from numerous studies that, for a given amount of divergence, saturation effects vary between genes and between regions within a single gene, and that careful choice of genes/regions may allow phylogenetic signal to be optimized for deeper phylogenetic comparisons (e.g., Mindell and Honeycutt, 1990; Kumazawa and Nishida, 1993). Sources of bias may also exist, such as differences in base composition between compared taxa. These also serve to increase the complexity of deeper phylogenetic comparisons, and can lead to misleading statistical support. It does not follow logically that gathering more sequence information will necessarily alleviate such problems of resolution if the basis is sequence saturation, unless a qualitatively different gene/region is chosen or a weighting scheme that corrects for such biases can be devised and justified. A poor signal-to-noise ratio is not alleviated by a larger sample size alone, especially where noise is produced in a biased fashion. Regardless of the region chosen or analyses used, it is apparent that having a clear understanding of the nature of substitutional rates and biases remains critical. Unfortunately, such an understanding is a tedious and difficult goal to attain.

The use of major genomic rearrangements or other such "genomic landmarks" as characters for phylogenetic analysis provides a complementary approach to direct sequence comparison that may be unaffected by the problems of alignment or sequence saturation. Such approaches make the seemingly reasonable but unproven assumption that major rearrangements are unlikely to occur in the same manner more than once in the time period under consideration. In theory, there are a near infinite number of "character states" that can exist (versus four or, if gaps are included, five in DNA sequence data), and hence saturation might not occur at this level. However, proof of this point must await further empirical studies, since mechanistic or selective constraints may, in fact, limit the number of possible character states. For instance, if gene rearrangement requires the involvement of flanking tRNAs as discussed earlier, the set of possible outcomes of rearrangement is drastically reduced. Similarly, rearrangements may be restricted to those regions close to origins of replication, and this would have a similar effect. Selection could also play a role. For example, insertions within the coding region of a gene may not be evolutionarily "viable."

Another difficulty in using genomic landmarks for phylogenetic analysis is that the number of informative characters that are available for comparison will be limited according to the time scale under consideration and the source and size of the genome across which distinguishing features are detected. Certainly for comparisons among vertebrates, this number remains low within the mitochondria genome, although it seems likely that more characters will become available from the (large) nuclear genome in the near future. For deeper comparisons, more characters are available, and mathematical methods to combine information from numerous characters are being developed (Sankoff *et al.*, 1992). Ultimately, given different advantages and disadvantages of using nucleotide bases versus genomic landmarks for deeper phylogenetic comparisons, one would gain more confidence that a particular branching order has been effectively solved when a variety of studies, including nonmolecular ones, either demonstrate congruence or can be combined and shown to give, collectively, strong support for a particular phylogenetic hypothesis (although this is an oversimplification in the sense that all methods are not expected to perform with uniform accuracy at all depths of comparison).

B. The Sister Taxon to Birds: Evidence from Comparisons of DNA Sequence and "Genomic Landmarks"

Literature concerning the determination of the sister taxon to birds illustrates some of the strengths and weaknesses of these two general approaches to phylogenetic reconstruction. While the classic view derived primarily from paleontological evidence places birds and crocodilians as sister taxa (Gauthier *et al.*, 1988; Donoghue *et al.*, 1989), controversial morphological analyses by Gardiner (1982) and Løvtrup (1985) placed birds and mammals as sister groups. In 1990, Hedges *et al.* published DNA sequence data from the 18S and 28S ribosomal RNA genes of a variety of tetrapods, and reviewed and reanalyzed much of the available amino acid sequence data sets pertaining to tetrapod phylogeny. Taken individually, studies of various genes supported different phylogenetic hypotheses, including a bird–crocodilian relationship (histone H2B) and a bird–mammal relationship (β-hemoglobin, myoglobin, and 18S rRNA). An additional four data sets (genes) supported differing hypotheses depending on the analysis employed.

While one interpretation of those molecular data sets supporting the controversial bird–mammal clade is that they reflect evolutionary relationships, in the case of protein-coding genes, it has been argued that selective forces have obscured phylogenetic information (Dickerson and Geis, 1983; Bishop and Friday, 1988). Hedges *et al.* (1990) originally argued that their 18S rRNA nucleotide data set provided at least some support to a bird–mammal clade, but that analysis was subsequently challenged. Marshall (1992) showed that a different conclusion is reached if weighted parsimony is used to correct for substitution bias and frequency, and Eernisse and Kluge (1993) questioned various aspects of the data set including alignments, although they arrived at a conclusion similar to that of Hedges *et al.* (1990) when just the 18S rRNA data set was used. Eernisse and Kluge (1993) went on to consider all available evidence pertaining to this phylogenetic question and reached the conclusion that a sister relationship for birds and crocodilians is best supported. The reader is referred to that paper for carefully formulated and clearly expressed suggestions on how best to combine the findings of different studies, both molecular and nonmolecular, and for insightful comments on the analysis of molecular data sets. Hedges (1994), using data from a larger number of genes (14), later decided that the original support for a mammal–bird clade provided by the 18S rRNA sequence data (Hedges *et al.*, 1990) was misleading because of a higher rate of change in the lineages leading to birds and mammals (long branches attract) and/or a common G + C substitutions bias. These studies and varying interpretations of a single data set highlight the controversies that often arise over how to correct for multiple substitutions, biases, and alignments in studying deeply divergent taxa with primary sequence information.

The first published use of major features of the mitochondrial genome for phylogenetic reconstruction involving birds, mammals, and crocodilians was made using the presence/absence of O_L as a character (Seutin *et al.*, 1994). It was already known that structures presumed to be homologous to the human O_L existed in a variety of mammals (Anderson *et al.*, 1981, 1982; Bibb *et al.*, 1981; Gadaleta *et al.*, 1989; Árnason *et al.*, 1991; Árnason and Johnsson, 1992), amphibians (Roe *et al.*, 1985; Yoneyama, 1987), and fish (Johansen *et al.*, 1990; Tzeng *et al.*, 1992). This region was also known to be absent in chicken, quail, and duck (Desjardins and Morais, 1990, 1991; Ramirez *et al.*, 1993), and hence, in the same sense as discussed below for the ND6-$tRNA^{Glu}$ region, must have been lost early within the avian clade, or perhaps before the common ancestor to Aves split from its sister taxon. Seutin *et al.* (1994) showed that O_L was also present in turtle and snake, but was absent in crocodilian and tuatara. They interpreted the shared absence of O_L in Crocodylia and Aves to be a synapomorphy linking those two lineages.

The absence of O_L in tuatara presented an enigma, as, taken in isolation, it places it in the same clade with bird and crocodile. The tuatara is usually considered to be a sister group to Squamata (snakes and lizards), forming with them, the Lepidosauria. The authors considered two hypotheses concerning this unexpected result. One was that a common ancestor to the [Lepidosauria (Crocodylia-Aves)] clade was heteroplasmic for the two character states (presence and absence of the O_L) and that the heteroplasmy was maintained for a significant length of time, and sorted out differentially among the Lepidosauria. The second hypothesis considered was that tuatara is more closely related to Crocodylia and Aves than it is to snake, although they viewed this as unlikely given numerous other synapomorphies defining the Lepidosauria. A third unmentioned hypothesis that could be forwarded is that the tuatara O_L was deleted in an independent event after its ancestor diverged from other Lepidosauria. One of the two tRNAs that flank the usual location of O_L is $tRNA^{Cys}$. In the tuatara, $tRNA^{Cys}$ is unusual, as the DHU arm has been reduced to a 7-bp loop, apparently without any secondary structure. One could speculate that this represents the remnant of a larger deletion that included O_L and a portion of the adjacent $tRNA^{Cys}$. If true, this would represent a parallel (independent) deletion. This interpretation raises the sobering possibility that even major genomic alterations may suffer from ambiguity in the correct identification of homologous versus nonhomologous evolutionary events. Seutin *et al.* (1994) also sequenced segments of ND2 and COI and found the amino acids to provide information that was also supportive of a bird–crocodilian sister relationship.

Another attempt to determine the sister taxon to birds was made by Quinn and Mindell (1996). To study the major rearrangement found adjacent to the avian control region (Fig. 1.1), they designed conserved primers to amplify and sequence gene junctions at either end of the cytochrome *b* gene in a variety of reptiles (the rearrangement was already known to be widespread among birds). While they were able to determine gene order by this method, the rearrangement occurred after the common ancestor to extant birds had split from the sampled reptilian lineages. Their study illustrates a problem with the use of such approaches for phylogenetic reconstruction relative to sequencing efforts, namely that the chances that such an event will occur along a "defining" internode within a given phylogenetic tree is very difficult to assess *a priori*. Nonetheless, it seems appropriate to investigate such major and readily interpretable genomic "landmarks" in an opportunistic fashion, particularly when they might add information to deeper phylogenetic relationships that have proven recalcitrant to other approaches.

As the divergence between bird and crocodile is believed to precede that between bird and dinosaur, this general approach could still provide an interesting synapomorphy if DNA remains intact through geological time periods, a contentious issue. Woodward *et al.* (1994) have published cytochrome *b* DNA sequence information obtained via PCR amplifications of extracts taken from two 80–85 million-year-old bone fragments presumed to be from an unidentified species of dinosaur. They were unable to show the sequence to be significantly more similar to bird or reptile than to mammal, a problem they attributed to the small length of sequence obtainable, and to sequence variability within samples, both presumably resultant from DNA damage. Subsequent analyses have raised the strong possibility that the sequence originates from accidental amplification of contaminating mammalian DNA sequence (Henikoff, 1995; Allard *et al.*, 1995), most likely of human origin (Hedges and Schweitzer, 1995; Zischler *et al.*, 1995b), and probably from human nuclear sequences that are homologous to mtDNA (Collura and Stewart, 1995; Zichler *et al.*, 1995b).

C. Other Potentially Informative Characters

With the increasing amount of mitochondrial and nuclear sequence data that is available each month, it may soon be possible to explore other types of genomic landmarks as potential sources of phylogenetic information at various depths of comparison. For instance, unusual codons, such as that proposed by Desjardins and Morais (1990) to initiate COI in birds, might be informative. Areas where genes overlap, such as ATPase 6 and 8, seem likely candidates for unusual evolutionary constraint, and there have been occasional alterations to the overlap. In birds, lamprey, and amphibians, these genes overlap by 10 bp, while in most mammals, the overlap is 40–46 bp (30 bp in fin whale). tRNAs may provide useful synapomorphies among vertebrates, as studies of reptile genomes have shown several cases of tRNA rearrangement or, assuming that "missing" tRNAs are still present elsewhere in the genome, of tRNA movement. For example, Quinn and Mindell (1996) showed that $tRNA^{Thr}$ is not adjacent to the 3′ end of cytochrome *b* in tuatara, and that a putative pseudogene of $tRNA^{Phe}$ is at the 5′ end of the control region in crocodile. Similarly, Kumazawa and Nishida (1995) found another crocodilian tRNA rearrangement, and detected the movement of a snake tRNA gene from one tRNA cluster to another. Each of these observations adds weight to the idea that mitochondrial tRNAs are relatively "mobile" through geological time. Some genes have different numbers of codons among different taxa, another potential source of characters in cases where insertion/deletion events are relatively rare. For example, both placental and marsupial mammals share what appears to be a synapomorphy defined by an extra codon near the 5′ end of cytochrome *b*. Finally, Moum *et al.* (1994) used hydropathy profile analysis to propose that an intragenic rearrangement may have occurred within the ND6 gene of mammals, raising the possibility that more characters of this type will become available as our understanding of protein structure improves.

These various types of genomic features (and more) should be considered as candidates for phylogenetic analysis, and several may prove useful at different phylogenetic depths. However, some may also prove to be useless, or even misleading for such ventures. The main lesson that emerges from considerations of DNA at its most fundamental and arguably least complex level (primary sequence) is that a thorough understanding of the forces affecting its evolution is increasingly important to have, but increasingly difficult to attain as the divergence of compared taxa deepens. While genomic landmarks are likely to enhance resolution of deeper phylogenetic questions, it also seems likely that similar and perhaps even more complex lessons will emerge from higher orders of DNA "behavior," so extreme caution is advised.

VI. INTERGENOMIC TRANSFER OF MITOCHONDRIAL SEQUENCES

A. Nuclear Homologs of Mitochondrial DNA: Not Unusual

In 1983, DNA segments within the *nu*clear genome with obvious homology to portions of the *mit*ochondrial genome were reported in yeast (Farrelly and Butow, 1983), locusts (Gellissen *et al.*, 1983), fungi (Wright and Cummings, 1983), and humans (Tsuzuki *et al.*, 1983). Lopez *et al.* (1994) used "*numts*" as a specific title for such sequences in domestic cat, and that term seems appropriate for general usage. Numts have been discovered in a variety of other eukaryotes.

Thorsness and Fox (1990) performed an elegant set of experiments in which they manipulated yeast organelles to estimate the amount of mtDNA that can potentially enter the nuclear compartment. Plasmid constructs designed to complement certain nuclear genetic defects were introduced into the mitochondria of yeast cell strains normally lacking in endogenous mtDNA. They detected approximately 2×10^{-5} transfers per cell per generation. This was a convincing demonstration that circular DNA molecules can not only move from mitochondrial organelles into the nucleus, but that they can do so with high frequency. While their study did not show, and probably rarely involved, covalent linkage of the transferred plasmids with the nuclear genome, this last step must be evolutionarily common given the large number and diversity of species in which numts have been detected. Among animals, this includes sea urchin (Jacobs *et al.*, 1983), locust (Gellissen *et al.*, 1983), rat (Hadler *et al.*, 1983; Zullo *et al.*, 1991), akodontine rodents (Smith *et al.*, 1992), "domestic" cats (Lopez *et al.*, 1994), goose (Quinn and White, 1987, Quinn, 1992), diving ducks (M. D. Sorenson and R. C. Fleischer, personal communication), tapaculo species (Arctander, 1995), humans (Tsuzuki *et al.*, 1983; Fukuda *et al.*, 1985; Nomiyama *et al.*, 1985; Kamimura *et al.*, 1989), and other primates (Van der Kuyl *et al.*, 1995; Collura and Stewart, 1995). The only vertebrate in which there has been a concerted effort to determine the frequency of mitochondrial homologs within the nuclear genome, *Homo sapiens*, has several hundred copies of mitochondrial sequence dispersed throughout the nuclear genome (Fukuda *et al.*, 1985; Kamimura *et al.*, 1989).

The examples given above include a variety of mitochondrial genes as well as the control region. In some cases the transferred fragment exists as a tandem repeat (e.g., Lopez *et al.*, 1994) while in others they are dispersed (e.g., Fukuda *et al.*, 1985). RNA intermediates were involved in some transfers (Gellissen and Michaelis, 1987; Shay and Werbin, 1992), but no such evidence exists for others. Kamimura *et al.*

(1989) reported the contiguous location of three gene segments that are normally found widely separated within the vertebrate mitochondrial genome. The central segment was inverted relative to the flanking segments. Taken together, these observations imply that a considerable variety of events may lead to the formation of numts.

B. An Avian Numt

In 1987, Quinn and White extended the documentation of numts to birds. They cloned a 5.5-kb *Hin*dIII fragment of the lesser snow goose (*Anser caerulescens caerulescens*) mitochondrial genome into a plasmid vector, and then hybridized the radioactively labeled clone to genomic Southern blots that contained *Hin*dIII-digested total DNA extract prepared from snow goose blood. While a 5.5-kb band of DNA hybridized with the probe as expected, a 3.6-kb band also showed strong signal on an autoradiographic exposure (Fig. 1.3, lane 8). When a probe of the entire mitochondrial genome was prepared and hybridized to the same blot, all expected bands of mitochondrial origin showed hybridization, but an extra 3.6-kb band was consistently seen. To verify that the 3.6-kb band was not of mitochondrial origin, the same clone was used to probe a Southern blot prepared with *Hin*dIII-digested purified mtDNA (Fig. 1.3, lane 2). As expected, a single 5.5-kb mitochondrial band showed strong hybridization, with no sign of the 3.6-kb band that hybridized when total DNA extract rather than purified mtDNA was used. To show that the 3.6-kb band was of nuclear origin, they then prepared a Southern blot that contained *Hin*dIII-digested DNA samples that were expected to have varying nuclear-to-mitochondrial ratios. Avian blood-extracted DNA is relatively rich in nuclear DNA owing to the nucleated red blood cells, and liver is relatively rich in mitochondria. A third sample highly enriched for mtDNA was prepared by homogenizing liver tissue to break open cells, and then centrifuging intact nuclei into a pellet. The supernatant, containing the smaller mitochondrial organelles and small amounts of nuclear DNA from the fraction of nuclei that invariably burst during such isolation procedures, was then extracted by standard methods. These samples were digested with *Hin*dIII and electrophoresed in order of increasing relative nuclear DNA content (enriched sample, liver sample, blood sample). This blot, when probed with the 5.5-kb clone, showed the relative intensity of the 5.5-kb mitochondrial band to decrease while the relative intensity of the 3.6-kb nuclear band increased (Fig. 1.3, lanes 6–8).

Subsequent sequencing of portions of the 5.5-kb mtDNA clone (Quinn and Wilson, 1993; T. W. Quinn, unpublished) showed it to contain, in the same order as the published chicken genome (Desjardins and Morais, 1990), part of ND5, cytochrome *b*, $tRNA^{Thr}$, $tRNA^{Pro}$, ND6, $tRNA^{Glu}$, control region, $tRNA^{Phe}$, 12S rRNA, and 16S rRNA. Presumably $tRNA^{Val}$ lies between 12S rRNA and 16S rRNA as in chicken, but this gene was not checked. Thus, the nuclear homolog must contain some subset of these sequences. Using PCR, others have also shown evi-

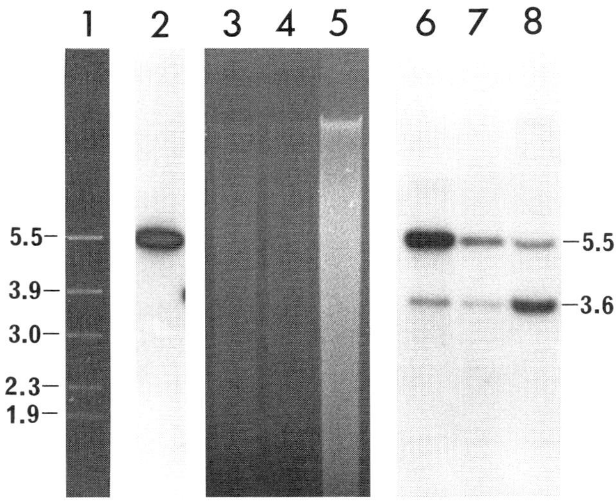

FIGURE 1.3 Demonstration of the presence of DNA sequences with homology to mtDNA sequences within the nuclear genome of the lesser snow goose (*Anser caerulescens caerulescens*). Fragment sizes are indicated in kilobases. Lane 1: 0.2 μg of purified mtDNA digested with *Hin*dIII, electrophoresed through a 1% agarose gel, and visualized with ethidium bromide. Lane 2: a Southern blot prepared from the gel shown in lane 1 was hybridized with a radioactively labeled clone of the 5.5-kb mitochondrial *Hin*dIII fragment (lane 1). Lanes 3, 4, and 5: *Hin*dIII-digested DNA samples electophoresed through a 1% agarose gel and visualized with ethidium bromide. The DNA samples originate from mitochondrially enriched liver extract (lane 3), total (unenriched) liver tissue (lane 4), and blood tissue (lane 5). Lanes 6, 7, and 8: Southern blot was prepared from the gel shown in lanes 3–5 and hybridized with a radioactivley labeled clone of the 5.5-kb mitochondrial *Hin*dIII fragment shown in lane 1 (as per lane 2) to produce this autoradiograph. Note that in addition to the 5.5-kb band of mitochondrial origin, a 3.6-kb band of presumed nuclear origin is seen. Its nuclear origin can be inferred from the fact that its intensity relative to the mitochondrial band increases with the increasing ratio of nuclear:mitochondrial DNA expected (moving from left to right across lanes 6 to 8). (Reproduction of Fig. 9 in Quinn and White, 1987.)

dence for the presence of numts in several tapaculo (*Scytalopus*) species (cytochrome *b*; Arctander, 1995) and in several diving duck (*Aythiyini*) species (control region; M. D. Sorenson and R. C. Fleischer, personal communication).

C. Numts as a Challenge for Studies of Avian Phylogenetics

The presence of numt sequences presents a special challenge for PCR-based studies of the avian mitochondrial genome. This is in part because birds have nucleated red blood cells, a frequently sampled tissue that is rich in nuclear DNA and relatively poor in mtDNA. This in turn means that PCR amplification and sequencing of mtDNA targets in blood may be misleading in cases where a "viable" target also exists in the nuclear genome. An illustration of this point is given by Quinn (1992). In an attempt to study the population genetics of snow geese across their range, PCR primers were designed to amplify the rapidly evolving 5′ end of the control region. These and other primers were used to gather sequence information from two populations using, as a template, DNA that had been extracted either from blood samples from an eastern population or from liver samples from a western population. The initial results showed a major sequence difference between these two populations, with all of the eastern samples having an identical haplotype, and the western samples having a variety of haplotypes. However, faint "shadow" bands in sequences of some samples from the eastern locales often matched western sequences at those nucleotide positions originally thought to differentiate the two populations. To investigate this further, DNA isolated from liver tissue (rather than blood) from some eastern samples was used as a template for PCR. The resultant sequences now appeared close or identical to haplotypes previously determined for western (liver) samples. Quinn (1992) concluded that predominantly nuclear sequence was being obtained from blood, and mitochondrial sequence from liver (there was some variation in the relative intensities of "shadow" bands among samples). Oligonucleotide primers that preferentially matched either the mitochondrial sequence or the nuclear sequence at the 3′ end (as determined from the blood sample sequences) produced PCR products with sequences distinctly different from each other (Fig. 1.4).

Quinn (1992) was eventually able to obtain mtDNA sequence from samples taken across the species range, but this was obviously a convoluted process. Had sequence only been obtained from a single blood sample for this species, rather than from two tissue sources from a number of different individuals, it is unlikely that its unexpected (nuclear) origin would have been detected. Hybridizing Southern blots of restriction endonuclease-digested genomic snow goose DNA with an entire mtDNA probe prepared from mouse generates extensive banding "ladders" (T. W. Quinn and B. N. White, unpublished), raising the possibility that this or another numt is arranged in a tandem repeat within the snow goose genome. Such a tandem arrangement was found in domestic cat (Lopez *et al.*, 1994) and provides, numerically, a more significant target for PCR primers.

While the development of primers that can preferentially amplify the nuclear or the mitochondrial copy from a single sample is a useful tool (Quinn, 1992; Fig. 1.4), it also raises the worrying possibility that such preferential amplification of a numt could occur by chance rather than by design. If such events occur, even a single "clean" sequence would be no guarantee that the intended mitochondrial target has been amplified. Such inadvertent preferential amplification occurred during a study of akodontine rodents (Smith *et al.*, 1992), but in this case the misamplification was detected, in part because of the unusual placement of stop codons and deletions within what was expected to be a protein-coding region. Lopez *et al.* (1994), in a study of domestic cats, noticed double bands in several positions along sequencing gels, and eventually showed them to be the result of simultaneous PCR amplification of nuclear and mitochondrial homologs of the same mitochondrial segment.

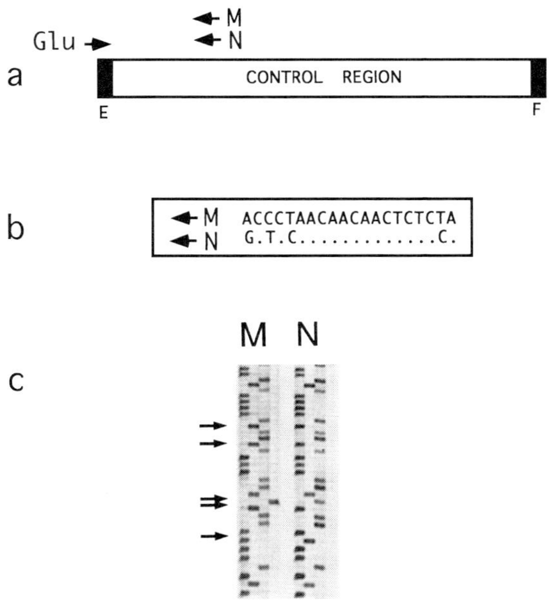

FIGURE 1.4 Use of specific PCR primers to differentially amplify mitochondrial or nuclear homologous sequences. (Adapted from Fig. 3 in Quinn and Wilson, 1993.) (a) Primer placement relative to the mitochondrial control region. "Glu" is similar in sequence to a portion of $tRNA^{Glu}$ (E), and will anneal to either the nuclear or mitochondrial homolog. "M" is identical in sequence to the mitochondrial sequence, while "N" is identical to known nuclear sequence. Further details concerning primer design are provided by Quinn and Wilson (1993). F, $tRNA^{Phe}$. (b) Comparison of the sequences of two homolog-specific primers [shown in the same orientation as in (a), written 3′ to 5′, left to right]. Dots indicate identity to the upper sequence. Note the mismatches concentrated at the 3′ end. (c) Sequences obtained from the PCR products of a single DNA sample extracted from liver tissue. PCR primer pairs used in the PCR were Glu plus M ("M" lane) or Glu plus N ("N" lane). Arrows point to obvious sequence differences. Loading order, from left to right, was GATC. (Modified from Figs. 2 and 3 in Quinn, 1992 with permission of Blackwell Science, Inc.)

Van der Kuyl *et al.* (1995) and Collura and Stewart (1995) showed that nuclear homologs are present in a variety of primates, and that those found in humans can create special difficulties in the detection of contamination in ancient PCR experiments since they are imperfectly matched with known human mitochondrial sequences. Thus, the implications of numts for other population and particularly phylogenetic studies, where species are often represented by sequence from a single sample, may be profound, depending on their ubiquity. The evidence from humans is that such transfers are common, not rare, although whether they are frequently in high copy number as in domestic cat remains unknown. If numt sequences are gathered with the assumption that they represent mitochondrial sequences, paralogy will frequently be mistaken for orthology, possibly affecting phylogenetic conclusions. Arctander (1995) provided evidence that unusual phylogenetic relationships among *Scytalopus* species, reported by Arctander and Fjeldså (1994), were likely the result of such a problem. Estimation of rates of change of mtDNA could also be altered by confusing orthology and paralogy, as it seems likely that numt sequences would evolve at the slower nuclear rate compared to mitochondrial sequences (see the next section).

One way to reduce the likelihood of such problems in modern avian samples would be to use purified mtDNA in PCR reactions. However, this would severely limit the efficiency of data collection, and furthermore, Collura and Stewart showed that in primate samples, contaminating nuclear sequences can still be amplified from such purifications. Another, more crude approach would be to use both blood and another tissue such as heart, liver, muscle, or feather in conjunction with PCR, and compare the sequencing gels from the two. In snow goose (Quinn, 1992) and in the canvasback (*Aythya valisineria*)(M. D. Sorenson and R. C. Fleischer, personal communication), a comparison between sequences obtained from blood versus liver-extracted DNA shows a consistent difference, but scenarios in which this might not be the case can also be envisioned. For example, if the 3′ end of a primer matches the nuclear sequence more precisely than the mitochondrial target, a single clean nuclear sequence may result regardless of tissue type. The use of broadly overlapping sequences generated with different sets of PCR primers might reveal such events. While a seemingly more elegant solution would be to use PCR to amplify the entire mitochondrial genome, the possibility of transposition of the entire genome has not been discounted, and there may be some danger that "jumping PCR" could occur among fragmented mtDNA genomes (Pääbo *et al.*, 1990). Nonetheless, such an approach may be worth exploring given its established feasibility (Cheng *et al.*, 1994).

D. Opportunities Provided by Numts

The presence of numts can be problematic, but they also provide a unique opportunity to better understand the differences in the mutational "spectra" of mitochondrial versus nuclear DNA sequences. While such comparisons can be made by averaging the rates of change of nonhomologous genes in the two genomes (Brown *et al.*, 1979; Vawter and Brown, 1986; also see Helm-Bychowski and Wilson, 1986), the most direct approach would be to place an identical sequence into the two environments and then to observe the changes that result. Numt sequences, in conjunction with their mitochondrial paralogs, constitute such an experiment. Arctander (1995) compared DNA sequence information of a mitochondrial and a presumptive nuclear cytochrome *b* pseudogene among eight *Scytalopus* species to estimate that the mitochondrial genome was evolving at least 13.6 times faster overall, and 39 times faster at third positions of codons than the nuclear pseudogene. Numts have also been used as a source of outgroup sequences for intraspecific comparisons of mtDNA sequence (Quinn, 1992; Zischler *et al.*, 1995a), although caution must be exercised since the possibility for gene conversion between mitochondrial and nuclear sequences has not been discounted.

VII. CONCLUSION

The impressive advances in our understanding of avian evolution that have resulted from studies based on the avian mitochondrial genome become obvious by inspecting various chapters in this book, and the published literature in general. While we still have much to learn about the genome and its evolution, our considerable knowledge is allowing increasingly sophisticated and reasoned use of its sequences for evolutionary inference at all levels. Continuing technological advances have now made publication of complete mitochondrial genomes a semiannual event, providing even more exciting prospects for evolutionary comparison.

This book is being published at a time when molecular evolutionary biologists are increasingly turning and, in some cases, returning to DNA sequence-based studies of the nuclear genome, also reflected in some of the following chapters. In birds that genome is more than 75,000 times larger than the mitochondrial genome. It is also much more complex. However, because of a different type of inheritance pattern, a different mutational environment, and distribution of its genes onto numerous unlinked chromosomes, it promises to bring with it many new insights (i.e., Kornegay *et al.*, 1993). Numt sequences may facilitate comparison and "cross-calibration" of the two genomes, as they provide an empirical view of what happens to two pieces of DNA that share a common ancestral sequence, but that have subsequently been placed within two different organelles, originating from separate kingdoms. Ultimately, information from the two genomes, regardless of the chromosome/gene/region being studied, will provide information that is synergistic in nature. Perhaps it has been stated too many times, but the technological advances of the last 15–20 years have clearly made this one of the most exciting and productive periods in which to study evolutionary biology.

ACKNOWLEDGMENTS

I thank David P. Mindell and two anonymous reviewers for their insightful comments, and M. D. Sorenson and R. C. Fleischer for providing an unpublished manuscript reporting numts in diving ducks. Support was provided by NSF through a grant to T.W.Q.

REFERENCES

Allard, M. W., Young, D., and Huyen, Y. (1995). Detecting dinosaur DNA. *Science* **268**, 1192.

Andachi, Y., Yamao, F., Muto, A., and Osawa, S. (1989). Codon recognition patterns as deduced from sequences of the complete set of transfer RNA species in *Mycoplasma capricolum*. Resemblance to mitochondria. *J. Mol. Biol.* **209**, 37–54.

Anderson, S., Bankier, A. T., Barrell, B. G., de Bruijn, M. H. L., Coulson, A. R., Drouin, J., Eperon, I. C., Nierlich, D. P., Roe, B. A., Sanger, F., Schreier, P. H., Smith, A. J. H., Staden, R., and Young, I. G. (1981). Sequence and organization of the human mitochondrial genome. *Nature (London)* **290**, 457–465.

Anderson, S., de Bruijn, M. H. L., Coulson, A. R., Eperon, I. C., Sanger, F., and Young, I. G. (1982). Complete sequence of bovine mitochondrial DNA. Conserved features of the mammalian mitochondrial genome. *J. Mol. Biol.* **156**, 683–717.

Arctander, P. (1995). Comparison of a mitochondrial gene and a corresponding nuclear pseudogene. *Proc. R. Soc. Lond. B* **262**, 13–19.

Arctander, P., and Fjeldså, J. (1994). Andean tapaculos of the genus *Scytalopus* (Aves, Rhinocryptidae): A study of speciation patterns using DNA sequence data. *In* "Conservation Genetics" (V. Loeschcke, J. Tomiuk, and S. K. Jain, Eds.). Birkhauser Verlag, Basel, Switzerland.

Árnason, Ú., and Johnsson, E. (1992). The complete mitochondrial DNA sequence of the harbor seal, *Phoca vitulina*. *J. Mol. Evol.* **34**, 493–505.

Árnason, Ú., Gullberg, A., and Widegren, B. (1991). The complete nucleotide sequence of the mitochondrial DNA of the fin whale, *Balaenoptera physalus*. *J. Mol. Evol.* **33**, 556–568.

Avise, J. C. (1994). "Molecular Markers, Natural History and Evolution." Chapman & Hall, New York.

Avise, J. C., Arnold, J., Ball, R. M., Bermingham, E., Lamb, T., Neigel, J. E., Reeb, C. A., and Saunders, N. C. (1987). Intraspecific phylogeography: The mitochondrial DNA bridge between population genetics and systematics. *Annu. Rev. Ecol. Syst.* **18**, 489–522.

Avise, J. C., and Zink, R. M. (1988). Molecular genetic divergence between avian sibling species: King and clapper rails, long-billed and short-billed dowitchers, boat-tailed and great-tailed grackles, and tufted and black-crested titmice. *Auk* **105**, 516–528.

Berg, T., Moum, T., and Johansen, S. (1995). Variable numbers of simple tandem repeats make birds of the order Ciconiiformes heteroplasmic in their mitochondrial genomes. *Curr. Genet.* **27**, 257–262.

Bibb, M. J., Van Etten, R. A., Wright, C. T., Walberg, M. W., and Clayton, D. A. (1981). Sequence and gene organization of mouse mitochondrial DNA. *Cell* **26**, 167–180.

Bishop, M. J., and Friday, A. E. (1988). Estimating the interrelationships of tetrapod groups on the basis of molecular sequence data. *In* "The Phylogeny and Classification of the Tetrapods, Vol. I," (R. S. MacIntyre, Ed.), pp. 33–58. Clarendon, Oxford.

Brown, W. M., George, M., Jr., and Wilson, A. C. (1979). Rapid evolution of animal mitochondrial DNA. *Proc. Natl. Acad. Sci. U.S.A.* **76**, 1967–1971.

Buroker, N. E., Brown, J. R., Gilbert, T. A., O'Hara, P. J., Beckenbach, A. T., Thomas, W. K., and Smith, M. J. (1990). Length heteroplasmy of sturgeon mitochondrial DNA: An illegitimate elongation model. *Genetics* **124**, 157–163.

Cardon, L. R., Burge, C., Clayton, D. A., and Karlin, S. (1994). Pervasive CpG suppression in animal mitochondrial genomes. *Proc. Natl. Acad. Sci. U.S.A.* **91**, 3799–3803.

Cedergren, R., Gray, M. W., Abel, Y., and Sankoff, D. (1988). The evolutionary relationships among known life forms. *J. Mol. Evol.* **28**, 184–186.

Cheng, S., Higuchi, R., and Stoneking, M. (1994). Complete mitochondrial genome amplification. *Nature Genet.* **7**, 350–351.

Clayton, D. A. (1982). Replication of animal mitochondrial DNA. *Cell* **28**, 693–705.

Clayton, D. A. (1992). Transcription and replication of animal mitochondrial DNAs. *Int. Rev. Cytol.* **141**, 217–232.

Collura, R. V., and Stewart, C.-B. (1995). Insertions and duplications of mtDNA in the nuclear genomes of Old World monkeys and hominoids. *Nature (London)* **378**, 485–489.

Desjardins, P., and Morais, R. (1990). Sequence and gene organization of the chicken mitochondrial genome. A novel gene order in higher vertebrates. *J. Mol. Biol.* **212**, 599–634.

Desjardins, P., and Morais, R. (1991). Nucleotide sequence and evolution of coding and noncoding regions of a quail mitochondrial genome. *J. Mol. Evol.* **32**, 153–161.

Desjardins, P., Ramirez, V., and Morais, R. (1990). Gene organization of the Peking duck mitochondrial genome. *Curr. Genet.* **17**, 515–518.

Dickerson, R. E., and Geis, I. (1983). "Hemoglobin: Structure, Function, Evolution, and Pathology." Benjamin/Cummings, Menlo Park.

Donoghue, M. J., Doyle, J. A., Gauthier, J., and Kluge, A. G. (1989). The importance of fossils in phylogeny reconstruction. *Annu. Rev. Ecol. Syst.* **20**, 431–460.

Dowling, T. E., Moritz, C., and Palmer, J. D. (1990). Nucleic acids. II. Restriction site analysis. *In*

"Molecular Systematics" (D. M. Hillis and C. Moritz, Eds.), pp. 250–317. Sinauer, Sunderland, Massachusetts.

Edwards, S. V. (1992). Gene flow and mitochondrial DNA evolution in social babblers (Aves: *Pomatostomus*). Ph.D. dissertation, University of California, Berkeley.

Edwards, S. V., and Wilson, A. C. (1990). Phylogenetically informative length polymorphism and sequence variability in mitochondrial DNA of Australian songbirds (*Pomatostomus*). *Genetics* **126**, 695–711.

Eernisse, D. J., and Kluge, A. G. (1993). Taxonomic congruence versus total evidence, and amniote phylogeny inferred from fossils, molecules, and morphology. *Mol. Biol. Evol.* **10**, 1170–1195.

Efstratiadis, A., Posakony, J. W., Maniatis, T., Lawn, R. M., O'Connell, C., Spritz, R. A., DeRiel, J. K., Forget, B. G., Weissman, S. M., Slightom, J. L., Blechl, A. E., Smithies, O., Baralle, F. E., Shoulders, C. C., and Proudfoot, N. J. (1980). The structure and evolution of the human β-globin gene family. *Cell* **21**, 653–668.

Farrelly, F., and Butow, R. A. (1983). Rearranged mitochondrial genes in the yeast nuclear genome. *Nature (London)* **301**, 296–301.

Fukuda, M., Shoji, W., Tsuzuki, T., Nomiyama, H., and Shimada, K. (1985). Mitochondrial DNA-like sequences in the human nuclear genome: Characterization and implications in the evolution of mitochondrial DNA. *J. Mol. Biol.* **186**, 257–266.

Gadaleta, G., Pepe, G., De Candia, G., Quagliariello, C., Sibisa, E., and Saccone, C. (1989). The complete nucleotide sequence of the *Rattus norvegicus* mitochondrial genome: Cryptic signals revealed by comparative analysis between vertebrates. *J. Mol. Evol.* **28**, 497–516.

Gardiner, B. G. (1982). Tetrapod classification. *Zool. J. Linnean Soc.* **74**, 207–232.

Gauthier, J., Kluge, A. G., and Rowe, T. (1988). Amniote phylogeny and the importance of fossils. *Cladistics* **4**, 105–209.

Gellissen, G., Bradfield, J. Y., White, B. N., and Wyatt, G. R. (1983). Mitochondrial DNA sequences in the nuclear genome of a locust. *Nature (London)* **301**, 631–634.

Gellissen, G., and Michaelis, G. (1987). Gene transfer: Mitochondria to nucleus. *Ann. N.Y. Acad. Sci.* **503**, 391–401.

Glaus, K. R., Zassenhaus, H. P., Fechheimer, N. S., and Perlman, P. S. (1980). Avian mtDNA: Structure, organization and evolution. *In* "The Organization and Expression of the Mitochondrial Genome" (A. M. Kroon and C. Saccone, Eds.), pp. 131–135. Elsevier/North-Holland Biochemical Press, Amsterdam.

Gray, M. W. (1983). The bacterial ancestry of plastids and mitochondria. *BioScience* **33**, 693–699.

Gray, M. W. (1989). The evolutionary origins of organelles. *Trends Genet.* **5**, 294–299.

Gyllensten, U., Wharton, D., Josefsson, A., and Wilson, A. C. (1991). Paternal inheritance of mitochondrial DNA in mice. *Nature (London)* **352**, 255–257.

Hadler, H. I., Dimitrijevic, B., and Mahalingam, R. (1983). Mitochondrial DNA and nuclear DNA from normal rat liver have a common sequence. *Proc. Natl. Acad. Sci. U.S.A.* **80**, 6495–6499.

Hayashi, J-I., Tagashira, Y., and Yoshida, M. C. (1985). Absence of extensive recombination between inter- and intraspecies mitochondrial DNA in mammalian cells. *Exp. Cell Res.* **160**, 387–395.

Hedges, S. B. (1994). Molecular evidence for the origin of birds. *Proc. Natl. Acad. Sci. U.S.A.* **91**, 2621–2624.

Hedges, S. B., and Schweitzer, M. H. (1995). Detecting dinosaur DNA. *Science* **268**, 1191.

Hedges, S. B., Moberg, K. D., and Maxson, L. R. (1990). Tetrapod phylogeny inferred from 18S and 28S ribosomal RNA sequences and a review of the evidence for amniote relationships. *Mol. Biol. Evol.* **7**, 607–633.

Henikoff, S. (1995). Detecting dinosaur DNA. *Science* **268**, 1192.

Helm-Bychowski, K. M., and Wilson, A. C. (1986). Rates of nuclear DNA evolution in pheasant-like birds: Evidence from restriction maps. *Proc. Natl. Acad. Sci. U.S.A.* **83**, 688–692.

Jacobs, H. T., Posakony, J. W., Grula, J. W., Roberts, J. W., Xin, J., Britten, R. J., and Davidson, E. H. (1983). Mitochondrial DNA sequences in the nuclear genome of *Strongylocentrotus purpuratus*. *J. Mol. Biol.* **165**, 609–662.

Janke, A., Feldmaier-Fuchs, G., Thomas, W. K., Von Haeseler, A., and Pääbo, S. (1994). The marsupial mitochondrial genome and the evolution of placental mammals. *Genetics* **137**, 243–256.

Johansen, S., Guddal, P. H., and Johansen, T. (1990). Organization of the mitochondrial genome of Atlantic cod, *Gadus morhua*. *Nucleic Acids Res.* **18**, 411–419.

Kamimura, N., Ishii, S., Liandong, M., and Shay, J. W. (1989). Three separate mitochondrial DNA sequences are contiguous in human genomic DNA. *J. Mol. Biol.* **210**, 703–707.

Kobayashi, Y., Ichihashi, K., Ohta, S., Nihei, K., Kagawa, Y., Yanagisawa, M., and Momai, M. Y. (1992). The mutant mitochondrial genes in mitochondrial myopathy, encephalomyopathy, lactic acidosis and stroke-like episodes (MELAS) were selectively amplified through generations. *J. Inher. Metab. Dis.* **15**, 803–808.

Kocher, T. D., Thomas, W. K., Meyer, A., Edwards, S. V., Pääbo, S., Villablanca, F. X., and Wilson, A. C. (1989). Dynamics of mitochondrial DNA evolution in animals: amplification and sequencing with conserved primers. *Proc. Natl. Acad. Sci. U.S.A.* **86**, 6196–6200.

Kondo, R., Satta, Y., Matsuura, E. T., Ishiwa, H., Takahata, N., and Chigusa, S. I. (1990). Incomplete maternal transmission of mitochondrial DNA in *Drosophila*. *Genetics* **126**, 657–663.

Kornegay, J. R., Kocher, T. D., Williams, L. A., and Wilson, A. C. (1993). Pathways of lysozyme evolution inferred from the sequences of cytochrome *b* in birds. *J. Mol. Evol.* **37**, 367–379.

Kumazawa, Y., and Nishida, M. (1993). Sequence evolution of mitochondrial tRNA genes and deep-branch animal phylogenetics. *J. Mol. Evol.* **37**, 380–398.

Kumazawa, Y., and Nishida, M. (1995). Variations in mitochondrial tRNA gene organization of reptiles as phylogenetic markers. *Mol. Biol. Evol.* **12**, 759–772.

Kurland, C. G. (1992). Evolution of mitochondrial genomes and the genetic code. *BioEssays* **14**, 709–714.

L'Abbé, D. L., Dunhaime, J-F., Lang, B. F., and Morais, R. (1991). The transcription of DNA in chicken mitochondria initiates from one major bidirectional promoter. *J. Biol. Chem.* **266**, 10844–10850.

Lansman, R. A., Shade, R. O., Shapira, J. F., and Avise, J. C. (1981). The use of restriction endonucleases to measure mitochondrial DNA sequence relatedness in natural populations. III: Techniques and potential applications. *J. Mol. Evol.* **17**, 214–226.

Lansman, R. A., Avise, J. C., and Huettel, M. D. (1983). Critical experimental test of the possibility of "paternal leakage" of mitochondrial DNA. *Proc. Natl. Acad. Sci. U.S.A.* **80**, 1969–1971.

Larsson, N. G., Holme, E., Kristiansson, B., Oldfors, A., and Tulinius, M. (1990). Progressive increase of the mutated mitochondrial DNA fraction in Kearns–Sayre syndrome. *Pediatr. Res.* **28**, 131–136.

Lee, W-J., and Kocher, T. D. (1995). Complete sequence of a sea lamprey (*Petromyzon marinus*) mitochondrial genome: Early establishment of the vertebrate genome organization. *Genetics* **139**, 873–887.

Levinson, G., and Gutman, G. A. (1987). Slipped-strand mispairing: a major mechanism for DNA sequence evolution. *Mol. Biol. Evol.* **4**, 203–221.

Lopez, J. V., Yuhki, N., Masuda, R., Modi, W., and O'Brien, S. J. (1994). *Numt*, a recent transfer and tandem amplification of mitochondrial DNA to the nuclear genome of the domestic cat. *J. Mol. Evol.* **39**, 174–190.

Løvtrup, S. (1985). On the classification of the taxon Tetrapoda. *Syst. Zool.* **34**, 463–470.

Margulis, L. (1970). "Origin of Eukaryotic Cells." Yale Univ. Press, New Haven, Connecticut.

Marshall, C. R. (1992). Substitution bias, weighted parsimony, and amniote phylogeny as inferred from 18S rRNA sequences. *Mol. Biol. Evol.* **9**, 370–373.

Michaels, G. S., Hauswirth, W. W., and Laipis, P. J. (1982). Mitochondrial DNA copy number in bovine oocytes and somatic cells. *Dev. Biol.* **94**, 246–251.

Mindell, D. P., and Honeycutt, R. L. (1990). Ribosomal RNA in vertebrates: Evolution and phylogenetic applications. *Annu. Rev. Ecol. Syst.* **21**, 541–566.

Moritz, C. and Brown, W. M. (1986). Tandem duplication of D-loop and ribosomal RNA sequences in lizard mitochondrial DNA. *Science* **233**, 1425–1427.

Moritz, C., and Brown, W. M. (1987). Tandem duplications in animal mitochondrial DNAs: Variation in incidence and gene content among lizards. *Proc. Natl. Acad. Sci. U.S.A.* **84**, 7183–7187.

Moritz, C., Dowling, T. E., and Brown, W. M. (1987). Evolution of animal mitochondrial DNA: Relevance for population biology and systematics. *Annu. Rev. Ecol. Syst.* **18**, 269–292.

Moum, T., and Johansen, S. (1992). The mitochondrial NADH dehydrogenase subunit 6 (ND6) gene in murres: Relevance to phylogenetic and population studies among birds. *Genome* **35**, 903–906.

Moum, T., Willassen, N. P., and Johansen, S. (1994). Intragenic rearrangements in the mitochondrial NADH dehydrogenase subunit 6 gene of vertebrates. *Curr. Genet.* **25**, 554–557.

Mullis, K., Falcoma, F., Scharf, S., Snikl, R., Horn, G., and Erlich, H. (1986). Specific amplification of DNA *in vitro*: The polymerase chain reaction. *Cold Spring Harbor Symp. Quant. Biol.* **51**, 260.

Nomiyama, H., Fukuda, M., Wakasugi, S., Tsuzuki, T., and Shimada, K. (1985). Molecular structures of mitochondrial-DNA-like sequences in human nuclear DNA. *Nucleic Acids Res.* **13**, 1649–1658.

Nugent, J. M., and Palmer, J. D. (1991). RNA-mediated transfer of the gene *coxII* from the mitochondrion to the nucleus during flowering plant evolution. *Cell* **66**, 473–481.

Ojala, D., Montoya, J., and Attardi, G. (1981). tRNA punctuation model of RNA processing in human mitochondria. *Nature (London)* **290**, 470–474.

Pääbo, S., Irwin, D. M., and Wilson, A. C. (1990). DNA damage promotes jumping between templates during enzymatic amplification. *J. Biol. Chem.* **265**, 4718–4721.

Pääbo, S., Thomas, W. K., Whitfield, K. M., Kumazawa, Y., and Wilson, A. C. (1991). Rearrangements of mitochondrial transfer RNA genes in marsupials. *J. Mol. Evol.* **33**, 426–430.

Poulton, J., Deadman, M. E., and Gardiner, R. M. (1989). Duplications of mitochondrial DNA in mitochondrial myopathy. *Lancet* **1**, 236–240.

Quinn, T. W. (1992). The genetic legacy of Mother Goose—phylogeographic patterns of lesser snow goose *Chen caerulescens caerulescens* maternal lineages. *Mol. Ecol.* **1**, 105–117.

Quinn, T. W., and Mindell, D. P. (1996). Mitochondrial gene order adjacent to the control region in crocodile, turtle and tuatara. *Mol. Phyl. Evol.* **5**, 344–351.

Quinn, T. W., and White, B. N. (1987). Analysis of DNA sequence variation. *In* "Avian Genetics: A Population and Ecological Approach" (F. Cooke, and P. A. Buckley, Eds.), pp. 163–198. Academic Press, London.

Quinn, T. W., and Wilson, A. C. (1993). Sequence evolution in and around the mitochondrial control region in birds. *J. Mol. Evol.* **37**, 417–425.

Ramirez, V., Savoie, P., and Morais, R. (1993). Molecular characterization and evolution of a duck mitochondrial genome. *J. Mol. Evol.* **37**, 296–310.

Rand, D. M., and Harrison, R. G. (1989). Molecular population genetics of mtDNA size variation in crickets. *Genetics* **121**, 551–569.

Robin, E. D., and Wong, R. (1988). Mitochondrial DNA molecules and virtual number of mitochondria per cell in mammalian cells. *J. Cell. Physiol.* **136**, 507–513.

Roe, B. A., Ma, D.-P., Wilson, R. K., and Wong, J. F.-H. (1985). The complete nucleotide sequence of the *Xenopus laevis* mitochondrial genome. *J. Biol. Chem.* **260**, 9759–9774.

Sankoff, D., Leduc, G., Antoine, N., Paquin, B., Lang, B. F., and Cedergren, R. (1992). Gene order comparisons for phylogenetic inference: Evolution of the mitochondrial genome. *Proc. Natl. Acad. Sci. U.S.A.* **89**, 6575–6579.

Schuster, W., and Brennicke, A. (1987). Plastid, nuclear and reverse transcriptase sequences in the mitochondrial genome of *Oenothera*: Is genetic information transferred between organelles via RNA? *EMBO J.* **6**, 2857–2863.

Seutin, G., Lang, B. F., Mindell, D. P., and Morais, R. (1994). Evolution of the WANCY region in amniote mitochondrial DNA. *Mol. Biol. Evol.* **11**, 329–340.

Shay, J. W., and Werbin, H. (1992). New evidence for the insertion of mitochondrial DNA into the human genome: Significance for cancer and aging. *Mutat. Res.* **275**, 227–235.

Shoffner, J. M., Lott, M. T., Voljavec, A. S., Soueidan, S. A., Constigan, D. A., and Wallace, D. C. (1989). Spontaneous Kearns-Sayre/chronic external ophthalmoplegia plus syndrome associated with a mitochondrial DNA deletion: A slip-replication model and metabolic therapy. *Proc. Natl. Acad. Sci. U.S.A.* **86**, 7952–7956.

Smith, M. F., Thomas, W. K., and Patton, J. L. (1992). Mitochondrial DNA-like sequences in the nuclear genome of an akodontine rodent. *Mol. Biol. Evol.* **9**, 204–215.

Streisinger, G., Okada, Y., Emrich, J., Newton, J., Tsugita, A., Terzaghi, E., and Inouye, M. (1966). Frameshift mutations and the genetic code. *Cold Spring Harbor Symp. Quant. Biol.* **31**, 77–84.

Tanaka, M., Sato, W., Ohno, K., Yamamoto, T., and Ozawa, T. (1989). Direct sequencing of deleted mitochondrial DNA in myopathic patients. *Biochem. Biophys. Res. Commun.* **164**, 156–163.

Tapper, D. P., and Clayton, D. A. (1981). Mechanism of replication of human mitochondrial DNA. Localization of the 5′ ends of nascent daughter strands. *J. Biol. Chem.* **256**, 5109–5115.

Thorsness, P. E., and Fox, T. D. (1990). Escape of DNA from mitochondria to the nucleus in *Saccharomyces cerevisiae. Nature (London)* **346**, 376–379.

Tsuzuki, T., Nomiyama, H., Setoyama, C., Maeda, S., and Shimada, K. (1983). Absence of mitochondrial-DNA-like sequences in the human nuclear DNA. *Gene* **25**, 223–229.

Tzeng, C.-S., Hui, C.-F., Shen, S.-C., and Huang, P. C. (1992). The complete nucleotide sequence of the *Crossostoma lacustre* mitochondrial genome: conservation and variations among vertebrates. *Nucleic Acids Res.* **20**, 4853–4858.

Van der Kuyl, A. C., Kuiken, C. L., Dekker, J. T., Preizonius, R. K., and Goudsmit, J. (1995). Nuclear counterparts of the cytoplasmic mitochondrial 12S rRNA gene: A problem of ancient DNA and molecular phylogenies. *J. Mol. Evol.* **40**, 652–657.

Vawter, L., and Brown, W. M. (1986). Nuclear and mitochondrial DNA comparisons reveal extreme rate variation in the molecular clock. *Science* **234**, 194–196.

Wallace, D. C. (1982). Structure and evolution of organelle genomes. *Microbiol. Rev.* **46**, 208–240.

Wallin, I. E. (1922). On the nature of mitochondria. *Am. J. Anat.* **30**, 203–229, 451–471.

Wallis, G. P. (1987). Mitochondrial DNA insertion polymorphism and germ line heteroplasmy in the *Triturus crustatus* complex. *Heredity* **58**, 229–238.

Wenink, P. W., Baker, A. J., and Tilanus, M. G. J. (1993). Hypervariable control-region sequences reveal global population structuring in a long-distance migrant shorebird, the dunlin (*Calidris alpina*). *Proc. Natl. Acad. Sci. U.S.A.* **90**, 94–98.

Wenink, P. W., Baker, A. J., and Tilanus, M. G. J. (1994). Mitochondrial control region sequences in two shorebird species, the turnstone and the dunlin, and their utility in population genetic studies. *Mol. Biol. Evol.* **11**, 22–31.

Williams, P. L., and Fitch, W. M. (1990). Phylogeny determination using dynamically weighted parsimony method. *Methods Enzymol.* **183**, 615–626.

Woodward, S. R., Weyand, N. J., and Bunnell, M. (1994). DNA sequence from Cretaceous period bone fragments. *Science* **266**, 1229–1232.

Wright, R. M., and Cummings, D. J. (1983). Integration of mitochondrial gene sequences within the nuclear genome during senescence in a fungus. *Nature (London)* **302**, 86–88.

Yang, D., Oyaizu, Y., Oyaizu, H., Olsen, G. J., and Woese, C. R. (1985). Mitochondrial origins. *Proc. Natl. Acad. Sci. U.S.A.* **82**, 4443–4447.

Yoneda, M., Chomyn, A., Martinuzzi, A., Hurko, O., and Attardi, G. (1992). Marked replicative advantage of human mtDNA carrying a point mutation that causes the MESAS encephalomyopathy. *Proc. Natl. Acad. Sci. U.S.A.* **89**, 11164–11168.

Yoneyama, Y. (1987). The nucleotide sequences of the heavy and light strand replication origins of the *Rana catesbeiana* mitochondrial genome. *Nippon Ika Daioaku Zasshi* **54**, 429–440.

Zevering, C. E., Moritz, C., Heideman, A., and Sturm, R. A. (1991). Parallel origins of duplications and the formation of pseudogenes in mitochondrial DNA from parthenogenetic lizards (*Heteronotia binoei*; Gekkonidae). *J. Mol. Evol.* **33**, 431–441.

Zeviani, M., Servidei, S., Gellera, C., Bertini, E., DiMauro, S., and DiDonato, S. (1989). An autosomal dominant disorder with multiple deletions of mitochondrial DNA starting at the D-loop region. *Nature (London)* **339**, 309–311.

Zischler, H., Geisert, H., von Haeseler, A., and Pääbo, S. (1995a). A nuclear "fossil" of the mitochondrial D-loop and the origin of modern humans. *Nature (London)* **378**, 489–492.

Zischler, H., Höss, M., Handt, O., von Haeseler, A., van der Kuyl, A. C., Goudsmit, J., and Pääbo, S. (1995b). Detecting dinosaur DNA. *Science* **268**, 1192–1193.

Zullo, S., Sieu, L. C., Slightom, J. L., Hadler, H. I., and Eisenstadt, J. M. (1991). Mitochondrial D-loop sequences are integrated in the rat nuclear genome. *J. Mol. Biol.* **221**, 1223–1235.

CHAPTER 2

DNA Microsatellites as Genetic Markers at Several Scales

DAVID B. MCDONALD* AND WAYNE K. POTTS†

*Department of Zoology and Physiology, University of Wyoming, Laramie, Wyoming 82071

†Department of Biology, University of Utah, Salt Lake City, Utah 84112

Avian Molecular Evolution and Systematics

I. INTRODUCTION

Understanding evolutionary processes at one level of organization often requires addressing processes at higher or lower levels of organization. For example, questions at the family (kinship) level may raise questions about genetic differences among populations. Even when a particular question does not require answers from other levels, it may raise intriguing possibilities about consequences for processes at those other levels. Given limited resources, evolutionary biologists should benefit from tools that allow one to address problems at several levels, using a single data source and a single technical tool.

DNA microsatellites are genetic markers that can be useful in addressing questions at a variety of scales, ranging from the extremely fine grained to the fairly coarse grained. More specifically, this genetic tool can help solve problems ranging from individual-specific, such as determining gender (Longmire *et al.*, 1993; Delehanty, 1995), to questions of relatedness and parentage (Amos *et al.*, 1993; McDonald and Potts, 1994; Kellogg *et al.*, 1995; Primmer *et al.*, 1995), the genetic structure of populations (Bowcock *et al.*, 1994; Taylor *et al.*, 1994; Dallas *et al.*, 1995; Estoup *et al.*, 1995; Paetkau and Strobeck, 1995; Gibbs *et al.*, 1996), and up to comparisons among species (Roy *et al.*, 1994b; Garza *et al.*, 1995). Further, it has several technical and analytical advantages that make it superior to genetic markers whose domains are far smaller. It therefore comes close to meriting the rubric "master of all trades." We present an overview of the technique, our assessment of the sorts of problems to which it is well suited, and provide examples from the literature and our own research that exemplify some of the scales at which microsatellites can provide useful answers. Because several reviews of the technique already exist (Bruford and Wayne, 1993; Schlotterer and Pemberton, 1994; Westneat and Webster, 1994), we strive to cover the essentials with a minimum of overlap. A review of molecular techniques in zoology (Fleischer, 1996) provides an overview of the role of other markers such as minisatellites and mitochondrial DNA, as well as microsatellites.

We caution against application of inappropriate models to the data. We also stress that careful attention to model assumptions is required; microsatellite data sets may not always meet model assumptions concerning, for example, the balance between drift, mutation, and migration. We provide case histories to illustrate some of the potential pitfalls. We hope to stimulate interest in the development of more sophisticated models and empirical tests of assumptions, so that the analytical techniques become maximally consistent with actual patterns observed in natural populations of birds.

II. TECHNICAL OVERVIEW

The genomes of most eukaryotes contain thousands of loci containing numerous tandem repeats of short nucleotide sequence motifs (Tautz *et al.*, 1986), such as

$(CT/GA)_n$, where n is the number of repeats (useful range from approximately 8 to 30). Tandem repeat loci are hypervariable owing to high mutation rates (10^{-5} to 10^{-3}) that either increase or decrease the number of repeat units (Wright, 1994). We discuss the mutation process below. Tandem repeat loci can be categorized arbitrarily into two groups on the basis of the size of the repeat unit; loci containing shorter repeat units (usually two to six base pairs) are called microsatellites [also referred to as simple sequence length polymorphisms (SSLPs), simple sequence repeats (SSRs), or short tandem repeats (STRs)] and loci with larger repeat units are called minisatellites (the loci used in the original form of DNA fingerprinting) (Jeffreys *et al.*, 1985). We suggest that the term "microsatellites" is preferable to "SSR" or "STR," in part because electronic literature searches return much extraneous material when using other terms (and particularly acronyms). The inclusive term for both micro- and minisatellites is VNTR (variable number tandem repeats).

VNTR alleles differ in length (number of repeats), and are therefore easier to identify than markers differing only by sequence; they can be readily discriminated on the basis of their differential electrophoretic mobility. Amplification with the polymerase chain reaction (PCR) avoids the more laborious alternative procedure of Southern blotting, while allowing precise identification of all allelic length variants on polyacrylamide gels. PCR amplification is feasible for virtually all microsatellite loci, but few minisatellites, because the size of the minisatellite tandem repeat usually exceeds PCR limitations. The advantages afforded by PCR amplification are one reason we feel microsatellites will increasingly dominate the field of genetic markers for ecological and evolutionary studies. A second reason is that the mutation rate of some minisatellite loci is so high (10^{-2} to 10^{-3}) (Jeffreys *et al.*, 1988) that even locus-specific minisatellites may be inappropriate for analyses at the scale of populations or higher. The mutation rate of microsatellite loci is estimated to range between 10^{-3} and 10^{-5} (Edwards *et al.*, 1992; Hearne *et al.*, 1992; Weissenbach *et al.*, 1992). Heterozygosity appears to be greatest for microsatellites with approximately 20 repeat units, although these larger repeats constitute only a small fraction of the total repertoire of loci with ≥ 6 repeats (Ellegren *et al.*, 1995).

The ease of generating and analyzing data for microsatellite loci is offset by the need to develop loci in new species. If one is lucky, microsatellite loci will already have been developed for the species of interest, or a related species. Such is the case for humans, and laboratory and domesticated animals and plants, where hundreds of microsatellite loci have been developed for many species, primarily for use in mapping genetic traits of scientific or economic interest. The primary reason that PCR primers developed for a microsatellite locus in one species will not work in a related species is that mutations in the flanking regions may prevent adequate annealing of the primers. The probability of this flanking region mismatch is of course related to the time since divergence of the taxa involved. In our experience, most primers will work for congeneric species and quite surprisingly a useful proportion of primers work between species in different families as documented in songbirds

(Hanotte *et al.*, 1994; McDonald and Potts, 1994; Primmer *et al.*, 1996), marine turtles (FitzSimmons *et al.*, 1995), and whales (Schlotterer *et al.*, 1991). It is likely that a series of primers for well-studied taxa such as birds will be available within the decade, such that screening the series will provide sufficient loci for most projects. Until then, most investigators will have to develop their own microsatellite loci.

Developing microsatellite loci for a new species requires that one construct a genomic library, screen the library for clones bearing one or more tandem repeats, sequence the clones, and develop PCR primers that amplify the tandem repeat. Each of these steps is relatively routine and one should be able to develop numerous microsatellite loci within a few months. Nevertheless, numerous potential pitfalls exist for the first-time user. The basic techniques have been published (Tautz, 1989; Ashley and Dow, 1994; Schlotterer and Pemberton, 1994; see the useful summary by Queller *et al.*, 1993). A publication by Strassmann *et al.* (1996) and a manual available on the Internet (see Appendix I) include complete protocols. Here we provide a brief overview of the entire procedure, including a review of relatively recent techniques for enriching the library for microsatellite loci.

A. Generating a Size-Selected Plasmid Genomic Library

A genomic library is a collection of DNA segments (clones) from the species of interest, inserted in a microbial vector. This library can be screened with a labeled probe of known sequence to select clones containing the same or similar sequences. This is the way a clone from one species can be used to clone the same gene in related species. In our case we want to clone loci bearing microsatellite repeats, so the library is probed with labeled DNA containing these repeats.

Library construction entails digesting both the plasmid vector and the genomic DNA with restriction enzymes that leave compatible ends for ligating genomic fragments into the vector. An often used combination is *Sau*IIIA for digestion of genomic DNA and *Bam*HI for digestion of the plasmid. Since sequence information immediately flanking the repeat is needed for developing PCR primers, it is useful to clone fragments that are small enough [<600 base pairs (bp)] to be sequenced in a single sequencing run. At the very least, one wants to ensure that there will be no gaps when sequencing from each end with the forward and reverse plasmid sequencing primers. To obtain fragments of the appropriate size, digested genomic DNA is run on an agarose gel and a gel slice containing fragment sizes between 300 and 600 bp is removed. Fragments smaller than 300 bp are avoided to reduce the probability that the repeat will be so close to one end of the fragment that primer development is impossible. The size-selected genomic DNA is purified and ligated into the plasmid vector. The ligation mixture is transformed into competent cells. The resulting plasmid library is ready for screening or storage.

B. Library Enrichment Techniques

Because fewer than 1 in 1000 clones will contain any given class of microsatellite, successful identification of multiple clones can be laborious and difficult. To speed the process, techniques exist for enriching the library for microsatellite-bearing clones. These techniques fall into three general categories: hybridization selection (Armour *et al.*, 1994; Kandpal *et al.*, 1994; Fleischer, 1996), primer amplification prior to cloning (Ostrander *et al.*, 1992), and triplex affinity capture (Nishikawa *et al.*, 1995). These enrichment procedures reduce the number of clones that need to be screened. Reduced screening can also be achieved by using libraries with large inserts, but the repeat may then be too far away from the cloning site to use vector primers to obtain flanking sequence. One can attack the distant flanking region problem, in turn, by using the repeat as a primer for sequencing the flanking region (Yuille *et al.*, 1991; Baron *et al.*, 1992; Koref *et al.*, 1993; Hawkins *et al.*, 1994; Rowe *et al.*, 1994).

In addition to library enrichment, a number of techniques exist for amplifying anonymous microsatellite loci, obviating the need for cloning (Charlieu *et al.*, 1992; Laurent *et al.*, 1994; Wu *et al.*, 1994; Zietkiewicz *et al.*, 1994).

C. Screening the Library for Microsatellite-Bearing Clones

Which microsatellite repeats should be used for screening? The choice is a tradeoff. Dinucleotide repeats are more numerous (Tautz *et al.*, 1986; Shriver *et al.*, 1993), and therefore easier to find, but the larger tri- and tetranucleotide repeats are easier to genotype because alleles have larger size differences. The efficiency of the screening process can be increased by probing with multiple repeats at once, which must be grouped according to similar melting temperatures. Clones are plated onto large agar plates (132 mm) to a density of about 1000 clones and replicates are made on filters. These filters are then screened with radiolabeled (or nonisotopically labeled) oligonucleotides containing the desired repeat(s). Positive clones are picked and usually subjected to a second round of screening to confirm positives. These positive clones are then sequenced to confirm the presence of a repeat and to obtain flanking sequence information for development of primers. The lower the repeat number, the less likely it is that the locus will be polymorphic. Accordingly, only clones containing eight or more repeats are usually developed. Commercial programs exist for developing primers (see also PRIMER, in Appendix I).

D. Genotyping

Alleles are resolved by acrylamide gel electrophoresis and usually visualized by radiolabeling the PCR products and exposing X-ray film to the acrylamide gel. We have described a technique for staining the acrylamide gel with ethidium bromide, allowing visualization of the DNA under ultraviolet (UV) light (Potts, 1996). By using ethidium bromide staining (as we do) a microsatellite laboratory can operate with little or no need for the extra facilities required for isotope use.

The entire PCR and gel electrophoresis process takes only 6 to 8 hr, one can run two to five 60-lane gels per day, and produce 120 to 300 genotypes per day. It is this high-efficiency genotyping that makes the somewhat laborious development of microsatellite loci worthwhile. Equipment for automating the genotyping of microsatellites has recently become commercially available (e.g., Applied Biosystems (Foster City, CA), Lycor, Pharmacia (Piscataway, NJ)], further simplifying the process.

E. Microsatellite Mutations

The mutational mechanism causing microsatellite loci to be hypervariable is not completely understood. It appears to be caused by a process referred to as "strand slippage" that occurs primarily during DNA replication (Schlotterer and Tautz, 1992). A replicating DNA strand can slip one or more repeat units within a repeat and resume perfect base pairing. The resulting bulge can then be repaired, resulting in the addition or deletion of the nonpaired bases. There is evidence that the addition or deletion usually involves a single repeat unit, with some evidence for rarer events of larger effect (Garza *et al.*, 1995). Weber and Wong (1993) described 24 microsatellite mutations occurring in the CEPH families; 20 involved changes of a single repeat unit and the remaining four changed two repeat units. This type of stepwise mutational process means that allelic variants of similar sizes are more closely related; the stepwise process can therefore provide additional information for phylogenetic reconstruction, as we develop below.

III. THE RANGE OF SCALES

In the following sections we outline the range of scales that we see as appropriate when using microsatellites as genetic markers (Table I).

A. Among Genes

Microsatellites can serve as genetic markers at the level of genes in at least two ways. One is in gender determination (Delehanty, 1995; Longmire *et al.*, 1993), and the other is in mapping genetic traits. Longmire *et al.* (1993) used microsatellite repeat sequences [e.g., $(GT)_n$] as probes, without the work of sequencing the flanking regions to develop primers. As more microsatellite flanking region sequences are

TABLE I Range of Scales at Which Microsatellites Should Prove Useful

Level of organization	Subject of inquiry	Potential problems
Among genes	Sex determination	Need locus on sex chromosome
	Gene mapping	Linkage
Among individuals	Parentage	Null alleles, intergenerational mutations
	Relatedness	Null alleles, lack of pedigrees
Among populations	Population subdivision	Variable mutation rates, variable importance of drift
	Phylogeography	Homoplasy, high mutation rate, nonlinear divergence
Among species	Fine-grained phylogeny	Homoplasy, high mutation rate, nonlinear divergence
Among higher taxa	Phylogenetics by gene arrangement	Homoplasy, nonlinearity
Temporal (0–300+ years bp)		Small, degraded DNA segments, contamination
	Temporal changes in genetic variation	
	Study of extinct/endangered populations	

published, single-locus probes from the sex-determining chromosome (W in birds) should become available for many species. Because females are the heterogametic sex in birds (WZ), a microsatellite locus on the W chromosome will serve as an unequivocal marker for females. Birds showing a band must be females, while those lacking a band could be either males or marker failures. A microsatellite marker from the Z chromosome will be informative only for heterozygous males.

Microsatellites are now the primary tool for mapping the human and mouse genomes (Dietrich *et al.*, 1996). They are used as markers linked to loci of interest, such as the "obesity gene" in mice (Zhang *et al.*, 1994). Microsatellites may eventually be important for mapping genes and gene arrangements in natural populations of birds. Current gene-mapping projects are largely restricted to studies in humans, using sibling analysis, and in laboratory mice, using inbred strains. For birds, the first applications will probably come from poultry (Cheng and Crittenden, 1994; Burt *et al.*, 1995), or possibly from natural populations for which extensive pedigrees are available. Microsatellites developed to map economically important domesticated species may then work in very closely related species for which the pattern of linkage between the microsatellite markers and the genetic traits is similar. The potential to use extensive pedigrees in birds showing genetic monogamy points to an unforeseen advantage of long-term field studies.

B. Among Individuals

Offspring are the most obvious component of fitness, and determining parentage is therefore of considerable interest to evolutionary biologists. Allozymes and multilocus minisatellite probes were an important tool in revising much of the conventional wisdom concerning avian mating systems. In many species, it became evident that superficial behavioral monogamy did not correspond to patterns of genetic parentage. In some behaviorally monogamous species, extrapair fertilizations (EPFs) and mixed paternity were found to be common [Westneat (1987), 29% of young by EPF; Stutchbury *et al.*, 1994]. In several polygynous species, such genetic markers have shown that females may rarely mate with more than a single male in a season [Hartley *et al.* (1993), 4.5% EPF; Hasselquist *et al.* (1995), 3.1% EPF], although the most successful males will have many mates.

The availability of numerous, highly polymorphic markers makes microsatellites a good choice for assessing parentage (Chakraborty *et al.*, 1988). Nevertheless, as is true of any genetic marker, microsatellites are not a panacea. As Strassmann *et al.* (1996) point out, even an error rate approaching 0.001 in scoring can produce a nonnegligible incidence of false parentage exclusions.

Relatedness among individuals is a cornerstone of much of the theory of behavioral ecology, because of the contribution of relatives to inclusive fitness. Queller *et al.* (1993) and Westneat and Webster (1994) provided useful overviews of the application of microsatellite markers to questions of relatedness among individuals. Blouin *et al.* (1996) used microsatellites to assess relatedness among mice (*Mus musculus*). In the section of case histories, we provide an example from our study of long-tailed manakins, *Chiroxiphia linearis* (McDonald and Potts, 1994).

C. Among Populations

Michod (1980) argued that the hard core of the modern synthesis was the development of the "beanbag" population genetics theory by Wright (1969; 1978), Fisher (1958), and Haldane (1966). Most of the work by these pioneers dealt with theory based on gene frequencies. The development of allozyme electrophoresis provided a laboratory method for assessing such gene frequencies in natural populations. Microsatellites share with allozymes the advantage of being locus specific, and of being inherited in Mendelian fashion. They therefore appear well suited to analysis with the full repertoire of models developed to analyze allozyme data. These include measures such as F statistics (Wright, 1978; Weir, 1996) and various genetic distance measures (Rogers, 1972; Nei, 1978; Reynolds *et al.*, 1983). Microsatellites, except those strongly linked to coding regions under selection, seem more likely to meet the assumption of neutrality that has been questioned for some allozyme analyses (Karl and Avise, 1992). The sampling necessary to examine variation among populations may entail little extra field and laboratory effort beyond that already entailed by a within-population study.

As noted above, evidence suggests that microsatellites mutate in stepwise fashion, such that similarity in repeat number indicates recent common ancestry of alleles. Slatkin (1995a,b), Goldstein *et al.* (1995a), Shriver *et al.* (1995), and Michalakis and Excoffier (1996) used this inherently phylogenetic information to generate genetic distance measures. In this respect microsatellite variation may thus resemble some phylogenetic applications of mitochondrial DNA (mtDNA). Such "phylogeographic" studies based on mtDNA combine a fine-grained phylogenetic approach with the study of geographic genetic variation (Ball and Avise, 1992; Ball *et al.*, 1988). Microsatellites in humans provide a much clearer view of such geographic pattern in human populations than does mtDNA (Bowcock *et al.*, 1994). We envision considerable theoretical attention to the special problems and prospects afforded by microsatellite data as applied to population subdivision; the application of measures of population subdivision to microsatellite data will not, however, be completely straightforward.

A few microsatellite studies of nonavian animals have addressed issues of model assumptions and their suitability of the models for microsatellite data sets. Estoup *et al.* (1995) addressed the problem of stepwise mutation versus infinite allele models of the mutation process. Their conclusion that stepwise models were not significantly better may be due, in part, to their using compound repeats comprising two or three different length motifs rather than perfect tandems. Such compound microsatellites seem unlikely to be a major element of microsatellite analyses in birds, because screening will usually be confined to perfect repeats. In a population-level study that we will explore in more detail as a case history, Allen *et al.* (1995) discussed the validity of a number of assumptions concerning the mutation process and the balance between mutation and drift.

A number of studies have used microsatellites to assess aspects of gene flow among populations. Gibbs *et al.* (1996) used microsatellite loci to examine variation among host races of the common cuckoo (*Cuculus canorus*). They showed that host specialization is not mirrored by genetic structuring in microsatellite loci or the mtDNA control region. Their results point to several avenues for further research on the way in which the parasite matches its host's egg color. Paetkau and Strobeck (1995) assessed population structure in polar bears (*Ursus maritimus*), and were able to show that despite long-distance seasonal movements, local populations maintained distinct genetic profiles and that patterns of gene flow may not be obvious from geographical proximity alone. Dallas *et al.* (1995) used microsatellites to show that subpopulations of mice (*M. musculus*) show considerable connectivity that may be due to migration or recurrent founder events from diverse source stocks.

A potentially important application of microsatellites is for conservation studies. For birds, an example would be characterizing neotropical migrants in order to link information on wintering, stopover, and breeding habitats. The study of apparent declines in neotropical migrants (Robinson *et al.*, 1995; but see James *et al.*, 1992) requires that populations be followed in all their habitats. Unfortunately, it has rarely been possible to trace migrant populations across habitats between seasons. The high degree of detail available in microsatellite data provides a potentially powerful way to assign wintering individuals to a breeding population or vice versa, as Wenink *et al.* (1993) did for dunlin (*Calidris alpina*), using mtDNA. Because PCR amplification allows analysis from minute quantities of DNA, microsatellite analyses are also feasible with noninvasive sampling such as the use of shed hair in endangered primates (Morin *et al.*, 1994). Houlden *et al.* (1996) used microsatellites to examine the effects of population bottlenecking in koalas (*Phascolarctos cinereus*) that underwent near-extinction population crashes.

D. Among Species

Microsatellite primers will often work across avian congeners or even across an entire family (Hanotte *et al.*, 1994; McDonald and Potts, 1994; Primmer *et al.*, 1996). In such cases, microsatellites may be useful in assessing relationships among species.

Microsatellite data can provide useful evidence for conservation decisions at or above the species level. Roy *et al.* (1994a) examined variation and hybridization among wolflike canids. They concluded that the red wolf (*Canis rufus*) is clearly a hybrid between the gray wolf (*Canis lupus*) and the coyote (*Canis latrans*). Exhaustive or expensive efforts to preserve this form may, therefore, be less appropriate than efforts to preserve truly distinct forms (Garcia-Moreno *et al.*, 1996) such as the Mexican wolf (*Canis lupus baileyi*).

Our case histories (Section V) include the study of Forbes *et al.* (1995) comparing levels of genetic variation in domestic sheep (*Ovis aries*) with those of Rocky Mountain bighorn sheep (*Ovis canadensis*).

E. Among Genera and Higher Taxa

Direct examination of microsatellite distance measures among genera and higher taxa will probably be less fruitful than studies at lower taxonomic levels. The problems (see Section IV) of homoplasy, constraints on repeat number, nonlinear divergence, and high mutation rate are likely to make such analyses unrewarding in comparison to those using other molecular markers more suited to these coarse-grained analyses. Microsatellites may, however, serve a key role as indirect markers for major gene rearrangements. When avian gene maps are eventually developed, avian biologists will be able to compare genomic organization at intergeneric and higher levels. Such gene rearrangements will be rare events that should characterize deep phylogenetic branching points [see T. Quinn (Chapter 1 in this volume) for an application of this method using mtDNA]. That is, the gene rearrangements would function as synapomorphies unifying groups of taxa. As noted above, microsatellites are a key marker for mapping genes.

F. Temporal Scales (Ancient DNA)

Thus far, we have addressed scale in terms of level of biological organization. Microsatellites also provide a powerful opportunity to examine problems across temporal scales. Because the primer pairs are short, and the microsatellites can be amplified by PCR, even minute quantities of ancient, degraded DNA can be analyzed (Ellegren, 1991; Roy *et al.*, 1994a; Taylor *et al.*, 1994). Genotyping of museum specimens permits assessment of variation among populations that are no longer extant, as well as among specimens from extant populations sampled one hundred or more years ago.

IV. CAUTIONS FOR DATA ANALYSIS

Because others (Bruford and Wayne, 1993; Schlotterer and Pemberton, 1994; Westneat and Webster, 1994) compared microsatellites to a variety of other genetic markers, we concentrate here on a few comparisons not covered in those sources, focusing on *caveats* particular to microsatellites.

Differences in mutation processes and rates between microsatellites and allozymes provide both opportunities and problems for analysis of population subdivision. In allozyme analyses, new alleles arise by point mutations, and allelic variation has usually been modeled under the assumption of infinite alleles. For microsatellites, the stepwise mutation process whereby repeats are usually added or subtracted one at a time suggests that similarity of length reflects allelic relatedness (Slatkin, 1995a,b). Theoretical models take advantage of the stepwise process to develop measures of genetic distance unique to microsatellites (Goldstein *et al.*, 1995a,b; Shriver *et al.*, 1995; Slatkin, 1995a,b; Michalakis and Excoffier, 1996). Because both additions and deletions are possible, however, microsatellites will exhibit length homoplasy (variants with the same overall length that are not identical by descent). Even sequencing rather than electrophoretic measurement of repeat number cannot reveal homoplasy due to addition and subsequent deletion of identical repeat units [e.g., one lineage that changes from $(CA)_{15}$ to $(CA)_{16}$, and then back to $(CA)_{15}$, while another lineage remains unchanged]. What sequencing can do, however, is reveal more complex patterns of length homoplasy resulting from nonidentical sequences with the same overall repeat size (Garza *et al.*, 1995). In some cases, such length homoplasies will derive from variation in the flanking regions, and may involve point mutations as well as, or instead of, the more frequent slippage events that presumably drive the stepwise process. Because of the high slippage mutation rate of microsatellites, a given time period will involve more divergence of the repeat units than that expected from point mutations in the flanking regions. As a result, the temporal scale over which microsatellites provide sufficient resolution may be narrower than that of markers that mutate more slowly, such as allozymes.

A further complication in analyzing microsatellite data is the presence of unamplified "null alleles" (Callen *et al.*, 1993; Paetkau and Strobeck, 1994; Pemberton *et al.*, 1995). One cause of null alleles is point mutations in the flanking region that prevent PCR amplification. As a result the single, amplified band from an individual may make it appear to be a homozygote, despite an unamplified, null allele that differs from the amplified band. This may lead to an apparent deficit of heterozygotes compared to Hardy–Weinberg expectation, and may mask divergence among populations that contain variants sufficiently different (via point mutations in the flanking region) to prevent amplification. Because of their apparent origin as point mutations in flanking regions, null alleles will tend to be particularly prevalent in applications of primers to species other than those in which they were designed. Brookfield (1996) describes a method for assessing the frequency of null alleles. Null alleles are not the only possible cause of heterozygote deficiency or excess, however. For example, Wahlund's principle can produce heterozygote deficiency when samples from divergent populations are pooled. Weir's (1996) book provides a useful compendium of several classic and newer methods of data analysis applicable to microsatellites.

A critical point in the application of any model to data is the degree to which the model's assumptions are violated in nature. Relatively little work has been done in this area (Scribner *et al.*, 1994). Different models make different assumptions concerning the relative roles of drift and mutation. Slatkin (1985) distinguished models based on the equations for differentiation in an n-island model. One such equation is

$$(f_0 - f)/(1 - f) = 1/(4Nm\alpha + 1) \tag{1}$$

where $f = [f_0 + (n - 1)f_1]/n$ is the average probability of identity of two alleles drawn at random from the population, f_0 and f_1 are the probabilities of identity of two alleles drawn randomly from the same and two different subpopulations, respectively, N is the population size, m is the migration rate, and $\alpha = [n(n - 1)]^2$. Equation (1) leads to measures such as F_{ST}. Although the mutation term, μ, is mentioned in some formulations of F_{ST} (Wright, 1969) where

$$F_{ST} = 1/[4N(m + \mu) - 1] \tag{2}$$

the μ is omitted in most or all implementations (computer programs), because the programs were developed for allozyme data, with mutation as a negligible force. An alternative formulation is

$$f_1/f_0 = 1/[1 + (n - 1)\mu/m] \tag{3}$$

which leads to measures such as Nei's genetic distance, D. Slatkin (1985) emphasized that the major distinction between the formulations is their relative emphasis on either drift [Eqs. (1) and (2); measures such as F_{ST}] or mutation [Eq. (3); measures such as D].

In most of the size-based, stepwise models developed for microsatellites, drift is considered to be negligible relative to the effect of relatively high mutation rates characteristic of microsatellites (10^{-3} to 10^{-5}). Depending on the particular history of populations, and the nature and mutation rate of the microsatellite loci analyzed, one can imagine different populations having a full spectrum of balances between the effects of drift and mutation. We provide an example of evidence for unmet assumptions in our case history of work by Allen *et al.* (1995) on gray seals (*Halichoerus grypus*).

V. CASE HISTORIES

A. Relatedness among Partners in Male Long-Tailed Manakins

Long-tailed manakins (*C. linearis*) are fruit-eating neotropical birds with a lek mating system. The essential defining characteristics of lek mating systems are that males provide no material benefits to breeding females (such as nest sites, feeding territories, or paternal care), and that females exercise some degree of choice among males, based on courtship displays or ornaments. Examples of lek-mating species include several grouse, most manakins, several birds of paradise, and two species of shorebirds [ruffs (*Philomachus pugnax*) and the buff-breasted sandpiper (*Tryngites subruficollis*)]. The five species of manakins in the genus *Chiroxiphia* have an unusual twist on the usually intense competition among males in lek-mating systems. Males cooperate for courtship display. In long-tailed manakins, two partnered males perform a unison song that resembles the word *toledo* (Trainer and McDonald, 1993; Trainer and McDonald, 1995). If the males persist in singing for several months or seasons, they may receive a female visit. Once a female arrives for a visit, the males move to a low dance perch for a dual-male backward leapfrog dance display. During the dance display, they alternate sets of as many as 100 backward leapfrog jumps with labored "butterfly" flight (McDonald, 1989a). Surprisingly, during the course of partnerships that may last as long as 8 years, only one of the two males copulates (McDonald, 1989b). It is always the alpha (dominant) male of the pair. A central problem of the senior author's long-term research on long-tailed manakins in Costa Rica has been to understand why the beta (subordinate) male cooperates in this extended courtship sequence.

A potential explanation for the cooperation between males lies in the theory of inclusive fitness (Hamilton, 1964). If partners were close relatives, a beta male that helped his partner produce more copies of their shared genes might more than offset the cost of not transmitting his own genes directly. This hypothesis predicts that partners should be close relatives. We used microsatellite DNA to assess relatedness among males. We developed four polymorphic loci with a mean heterozygosity of 0.42 and a mean of 2.8 alleles per locus. No allele had a frequency higher than 0.81. A locus is more informative (Queller and Goodnight, 1989) when it has multiple alleles of comparable frequency (e.g., four alleles, each at 0.25, rather than two alleles at 0.95 and 0.05). The four loci provided resolution sufficient to assess relatedness among partners as compared to the average, background level of relatedness among males in the local population. We found that partners were no more closely related ($r = -0.14$) than males picked at random from the local population. Indirect inclusive fitness (kin selection) cannot, therefore, be invoked as a complete or partial explanation for cooperation in this species (McDonald and Potts, 1994). The microsatellite data provided a powerful tool for rejecting an inherently plausible hypothesis for explaining cooperative behavior.

B. Gray Seals

Allen *et al.* (1995) estimated gene flow in gray seals (*H. grypus*) at two colonies separated by approximately 500 km. The estimates from different measures differed widely. Using F_{ST}, Nm was 41; with Slatkin's (1995a,b) R_{ST} it was 13.8, and with Slatkin's private allele method it was 5.6. As they pointed out, the difference clearly suggests that the microsatellite data contravene one or more assumptions of the different models. They suggested that the low F_{ST} (high Nm) might result from the possibility of back mutations returning to previous states, contravening the assumption of an infinite number of potential alleles. They also assessed the effect of including the mutation term, μ, as in Eqs. (1) and (3), which is usually omitted from implementations. For a mutation rate as high as 10^{-3} (many dinucleotide repeats will have lower mutation rates), however, inclusion of the mutation term lowered Nm only slightly (from 41 to 40.8).

C. Domestic and Bighorn Sheep

Forbes *et al.* (1995) examined microsatellite variation within and between domestic sheep (*O. aries*) and Rocky Mountain bighorn sheep (*O. canadensis*). Their major conclusion was that "classic" methods such as F_{ST} or Nei's D differed considerably from allele size-based methods such as those of Goldstein *et al.* (1995a) and Slatkin (1995a,b). The classic methods were more sensitive to population differences within species, while the size-based measures yielded results more consistent with other biogeographical and genetic analyses. Secondarily, they pointed to the problem of interpretation when markers developed in one species are applied to another. "Differences in allele size and polymorphism among taxa may be explained by bias in the cloning and characterizing [of] microsatellite loci (FitzSimmons *et al.*, 1995; Pepin *et al.*, 1995)." As pointed out by Ellegren *et al.* (1995), screening selects for repeats that are longer, and more polymorphic than the average. When the primers are applied to other taxa, the repeat length will tend to be shorter and the locus less polymorphic.

In some cases, it will be difficult to know *a priori* whether assumptions are met.

For example, one may lack the biogeographic and demographic data necessary for assessing assumptions concerning the relative importance of mutation and drift. In such cases it may be instructive to compare measures such as the stepwise models to measures such as *F* statistics at several hierarchical scales in order to assess the relative applicability of the different models (Slatkin, 1995a,b). Microsatellites are not simply glorified allozymes, and new methods will be required to deal with all the ramifications of mutation process and rate that they entail.

VI. SUMMARY AND CONCLUSIONS

The rise of PCR-based microsatellite markers provides abundant opportunities, but also calls for caution. Because of their suitability for addressing problems at a fairly wide range of levels of organization from within individuals to among populations or species, we foresee a considerable expansion of interest in the interactions across these scales. For example, how do social systems and fine-scale demography affect population subdivision? Which factor generally has the greatest impact on population subdivision in birds—sexual selection, with its potential effects on effective population size through the sex with the higher variance in reproductive success, or the sedentary habit (low dispersal distance) of many cooperative breeders? The resurgence of population-level studies that should follow in the wake of widespread development of microsatellite loci in a variety of taxa calls for development of well-documented computer software packages with which to analyze the resulting data.

The need for careful assessment of the fit between the demographic parameters of the population analyzed (migration, drift, mutation rate) and the assumptions of the models used to analyze it cannot be overemphasized. Application of an inappropriate model can lead to the wrong conclusions about the patterns and processes of genetic differentiation. We foresee a considerable need for theoretical work to explore the fit of conventional measures of population subdivision (e.g., *F* statistics) to microsatellite data sets, as well as continued exploration of the ramifications of the nature of the mutation process for genetic models and analyses.

ACKNOWLEDGMENTS

We thank M. E. Hellberg, D. P. Mindell, D. F. Westneat, and an anonymous reviewer for comments that substantially improved the manuscript.

REFERENCES

Allen, P. J., Amos, W., Pomeroy, P. P., and Twiss, S. D. (1995). Microsatellite variation in grey seals (*Halichoerus grypus*) shows evidence of genetic differentiation between two British breeding colonies. *Mol. Ecol.* **4**, 653–662.

Amos, B., Schlötterer, C., and Tautz, D. (1993). Social structure of pilot whales revealed by analytical DNA profiling. *Science* **260**, 670–672.

Armour, J. A., Neumann, R., Gobert, S., and Jeffreys, A. J. (1994). Isolation of human simple repeat loci by hybridization selection. *Hum. Mol. Genet.* **3**, 599–565.

Ashley, M. V., and Dow, B. D. (1994). The use of microsatellite analysis in population biology: Background, methods and potential applications. *In* "Molecular Ecology and Evolution: Approaches and Applications" (B. Schierwater, B. Streit, G. P. Wagner, and R. DeSalle, Eds.), pp. 185–201. Birkhauser Verlag, Basel, Switzerland.

Ball, R. M., Jr., and Avise, J. C. (1992). Mitochondrial DNA phylogeographic differentiation among avian populations and the evolutionary significance of subspecies. *Auk* **109**, 626–636.

Ball, R. M., Jr., Freeman, S., James, F. C., Bermingham, E., and Avise, J. C. (1988). Phylogeographic population structure of red-winged blackbirds assessed by mitochondrial DNA. *Proc. Natl. Acad. Sci. U.S.A.* **85**, 1558–1562.

Baron, B., Poirier, C., Simon-Chazottes, D., Barnier, C., and Guenet, J. L. (1992). A new strategy useful for rapid identification of microsatellites from DNA libraries with large size inserts. *Nucleic Acids Res.* **20**, 3665–3669.

Blouin, M. S., Parsons, M., Lacaille, V., and Lotz, S. (1996). Use of microsatellite loci to classify individuals by relatedness. *Mol. Ecol.* **5**, 393–401.

Bowcock, A. M., Ruiz-Linares, A., Tomfohrde, J., Minch, E., Kidd, J. R., and Cavalli-Sforza, L. L. (1994). High resolution of human evolutionary trees with polymorphic microsatellites. *Nature (London)* **368**, 455–457.

Brookfield, J. F. Y. (1996). A simple method for estimating null allele frequency from heterozygote deficiency. *Mol. Ecol.* **5**, 453–455.

Bruford, M. W., and Wayne, R. K. (1993). Microsatellites and their application to population genetic studies. *Curr. Opin. Gen. Dev.* **3**, 939–943.

Burt, D. W., Bumstead, N., Bitgood, J. J., de Leon, F. A. P., and Crittenden, L. B. (1995). Chicken genome mapping: A new era in avian genetics. *Trends Genet.* **11**, 190–194.

Callen, D. F., Thompson, A. D., Shen, Y., Phillips, H. A., Richards, R. I., Mulley, J. C., and Sutherland, G. R. (1993). Incidence and origin of "null" alleles in the $(AC)_n$ microsatellite markers. *Am. J. Hum. Gen.* **52**, 922–927.

Chakraborty, R., Meagher, T. R., and Smouse, P. E. (1988). Parentage analysis with genetic markers in natural populations. I. The expected proportion of offspring with unambiguous paternity. *Genetics* **118**, 527–536.

Charlieu, J. P., Laurent, A. M., Carter, D. A., Bellis, M., and Roizes, G. (1992). 3′ Alu PCR: A simple and rapid method to isolate human polymorphic markers. *Nucleic Acids Res.* **20**, 1333–1337.

Cheng, H. H., and Crittenden, L. B. (1994). Microsatellite markers for genetic mapping in the chicken. *Poul. Sci.* **73**, 539–546.

Dallas, J. F., Dod, B., Boursot, P., Prager, E. M., and Bonhomme, F. (1995). Population subdivision and gene flow in Danish house mice. *Mol. Ecol.* **4**, 311–320.

Delehanty, D. J. (1995). Genetic and morphological methods for gender identification of mountain quail. *J. Wildl. Man.* **59**, 785–789.

Dietrich, W. F., Jr., Miller, R. S., Merchant, M. A., Damronboles, D., Husain, Z., Dredge, R., Daly, M. J., Ingalls, K. A., O'Connor, T. J., Evans, C. A., DeAngelis, M. M., Levinson, D. M., Kruglyak, L., Goodman, N., Copeland, N. G., Jenkins, N. A., Hawkins, T. L., Stein, L., Page, D. C., and Lander, E. S. (1996). A comprehensive genetic map of the mouse genome. *Nature (London)* **380**, 149–152.

Edwards, A., Hammond, H. A., Jin, L., Caskey, C. T., and Chakraborty, R. (1992). Genetic variation at five trimeric and tetrameric tandem repeat loci in four human population groups. *Genomics* **12**, 241–253.

Ellegren, H. (1991). DNA typing of museum birds. *Nature (London)* **354**, 113.

Ellegren, H., Primmer, C. R., and Sheldon, B. C. (1995). Microsatellite evolution directionality or bias? *Nature Genet.* **11**, 360–362.

Estoup, A., Garnery, L., Solignac, M., and Cornuet, J. M. (1995). Microsatellite variation in honey bee (*Apis mellifera* L.) populations: Hierarchical genetic structure and test of the infinite allele and stepwise mutation models. *Genetics* **140**, 679–695.

Fisher, R. A. (1958). "The Genetical Theory of Natural Selection" (2nd ed.). Dover Press, New York.

FitzSimmons, N. N., Moritz, C., and Moore, S. S. (1995). Conservation and dynamics of microsatellite loci over 300 million years of marine turtle evolution. *Mol. Biol. Evol.* **12**, 432–440.

Fleischer, R. C. (1996). Application of molecular methods to the assessment of genetic mating systems in vertebrates. *In* "Molecular Zoology: Advances, Strategies and Protocols" (A. S. R. P. J. D. Ferraris, Ed.), pp. 133–161. John Wiley & Sons, New York.

Forbes, S. H., Hogg, J. T., Buchanan, F. C., Crawford, A. M., and Allendorf, F. W. (1995). Microsatellite evolution in congeneric mammals: Domestic and bighorn sheep. *Mol. Biol. Evol.* **12**, 1106–1113.

Garcia-Moreno, J., Matocq, M. D., Roy, M. S., Geffen, E., and Wayne, R. K. (1996). Relationships and genetic purity of the endangered Mexican wolf based on analysis of microsatellite loci. *Conserv. Biol.* **10**, 376–389.

Garza, J. C., Slatkin, M., and Freimer, N. B. (1995). Microsatellite allele frequencies in humans and chimpanzees, with implications for constraints on allele size. *Mol. Biol. Evol.* **12**, 594–603.

Gibbs, H. L., Brooke, M. D., and Davies, N. B. (1996). Analysis of genetic differentiation of host races of the common cuckoo *Cuculus canorus* using mitochondrial and microsatellite DNA variation. *Proc. R. Soc. Lond. B* **263**, 89–96.

Goldstein, D. B., Linares, A. R., Cavalli-Sforza, L. L., and Feldman, M. W. (1995a). An evaluation of genetic distances for use with microsatellite loci. *Genetics* **139**, 463–471.

Goldstein, D. B., Linares, A. R., Cavalli-Sforza, L. L., and Feldman, M. W. (1995b). Genetic absolute dating based on microsatellites and the origin of modern humans. *Proc. Natl. Acad. Sci. U.S.A.* **92**, 6723–6727.

Haldane, J. B. S. (1966). "The Causes of Evolution." Cornell University Press, Ithaca, New York.

Hamilton, W. D. (1964). The genetical evolution of social behavior. *J. Theor. Biol.* **7**, 1–52.

Hanotte, O., Zanon, C., Pugh, A., Greig, C., Dixon, A., and Burke, T. (1994). Isolation and characterization of microsatellite loci in a passerine bird: The reed bunting *Emberiza schoeniclus*. *Mol. Ecol.* **3**, 529–530.

Hartley, I. R., Shepherd, M., Robson, T., and Burke, T. (1993). Reproductive success of polygynous male corn buntings (*Milaria calandra*) as confirmed by DNA fingerprinting. *Behav. Ecol.* **4**, 310–317.

Hasselquist, D., Bensch, S., and von Schantz, T. (1995). Low-frequency of extrapair paternity in the polygynous great reed warbler, *Acrocephalus arundinaceus*. *Behav. Ecol.* **6**, 27–38.

Hawkins, G. A., Bishop, M., Kappes, S., and Beattie, C. W. (1994). Cycle sequencing of $(CA)_n$ microsatellites from cosmid inserts using degenerate primers. *Biotechniques* **16**, 418–420.

Hearne, C. M., Ghosh, S., and Todd, J. A. (1992). Microsatellites for linkage analysis of genetic traits. *Trends Genet.* **8**, 288–294.

Houlden, B. A., England, P. R., Taylor, A. C., Greville, W. D., and Sherwin, W. B. (1996). Low genetic variability of the koala *Phascolarctos cinereus* in south-eastern Australia following a severe population bottleneck. *Mol. Ecol.* **5**, 269–281.

James, F. C., Wiedenfeld, D. A., and McCulloch, C. E. (1992). Trends in breeding populations of warblers: Declines in the southern highlands and increases in the lowlands. *In* "Ecology and Conservation of Neotropical Migrants" (J. M. I. Hagan and D. W. Johnston, Eds.), pp. 43–56. Smithsonian Institution Press, Washington, D.C.

Jeffreys, A. J., Wilson, V., and Thein, S. L. (1985). Individual-specific "fingerprints" of human DNA. *Nature (London)* **316**, 76–79.

Jeffreys, A. J., Royle, N. J., Wilson, V., and Wong, Z. (1988). Spontaneous mutation rates to new length alleles at tandem-repetitive hypervariable loci in human DNA. *Nature (London)* **332**, 278–281.

Kandpal, R. P., Kandpal, G., and Weissman, S. M. (1994). Construction of libraries enriched for sequence repeats and jumping clones, and hybridization selection for region-specific markers. *Proc. Natl. Acad. Sci. U.S.A.* **91**, 88–92.

Karl, S. A., and Avise, J. C. (1992). Balancing selection at allozyme loci in oysters: Implications from nuclear RFLP's. *Science* **256**, 100–102.

Kellogg, K. A., Markert, J. A., Stauffer, J. R., and Kocher, T. D. (1995). Microsatellite variation demonstrates multiple paternity in lekking cichlid fishes from Lake Malawi, Africa. *Proc. R. Soc. Lond. B* **260**, 79–84.

Koref, M. S., Orphanos, V., and Boyle, J. M. (1993). Rapid determination of sequences flanking microsatellites using dephosphorylated cloning vectors. *Trends Genet.* **9**, 43.

Laurent, A. M., Marcais, B., Muleris, M., and Roizes, G. (1994). A rapid and simple method to isolate and characterize highly polymorphic markers from the centromeric regions of the human chromosomes. *Nucleic Acids Res.* **22**, 194–199.

Longmire, J. L., Maltbie, M., Pavelka, R. W., Smith, L. M., White, S. M., Ryder, O. A., Ellsworth, D. E., and Baker, R. J. (1993). Gender identification in birds using microsatellite DNA fingerprint analysis. *Auk* **110**, 378–381.

McDonald, D. B. (1989a). Correlates of male mating success in a lekking bird with male–male cooperation. *Anim. Behav.* **37**, 1007–1022.

McDonald, D. B. (1989b). Cooperation under sexual selection: Age-graded changes in a lekking bird. *Am. Nat.* **134**, 709–730.

McDonald, D. B., and Potts, W. K. (1994). Cooperative display and relatedness among males in a lek-mating bird. *Science* **266**, 1030–1032.

Michalakis, Y., and Excoffier, L. (1996). A generic estimation of population subdivision using distances between alleles with special reference for microsatellite loci. *Genetics* **142**, 1061–1064.

Michod, R. E. (1980). Positive heuristics in evolutionary biology. *Br. J. Phil. Sci.* **32**, 1–36.

Morin, P. A., Wallis, J., Moore, J. J., and Woodruff, D. S. (1994). Paternity exclusion in a community of wild chimpanzees using hypervariable simple sequence repeats. *Mol. Ecol.* **3**, 469–478.

Nei, M. (1978). Estimation of average heterozygosity and genetic distance from a small number of individuals. *Genetics* **76**, 379–390.

Nishikawa, N., Oishi, M., and Kiyama, R. (1995). Construction of a human genomic library of clones containing poly(dG-dA)·poly(dT-dC) tracts by Mg(2+)-dependent triplex affinity capture. DNA polymorphism associated with the tracts. *J. Biol. Chem.* **270**, 9258–9264.

Ostrander, E. A., Jong, P. M., Rine, J., and Duyk, G. (1992). Construction of small-insert genomic DNA libraries highly enriched for microsatellite repeat sequences. *Proc. Natl. Acad. Sci. U.S.A.* **89**, 3419–3423.

Paetkau, D., and Strobeck, C. (1994). The molecular basis and evolutionary history of a microsatellite null allele in bears. *Mol. Ecol.* **4**, 519–520.

Paetkau, D., and Strobeck, C. (1995). Microsatellite analysis of genetic variation in black bear populations (erratum, *Mol. Ecol.* **3**, 489–495). *Mol. Ecol.* **4**, 133.

Pemberton, J. M., Slate, J., Bancroft, D. R., and Barrett, J. A. (1995). Nonamplifying alleles at microsatellite loci: A caution for parentage and population studies. *Mol. Ecol.* **4**, 49–52.

Pepin, L., Amigues, Y., Lepingle, A., Berthier, J. L., Bensaid, A., and Vaiman, D. (1995). Sequence conservation of microsatellites between *Bos taurus* (cattle), *Capra hircus* (goat) and related species: Examples of use in parentage testing and phylogeny analysis. *Heredity* **74**, 53–61.

Potts, W. K. (1996). PCR-based cloning across large taxonomic distances and polymorphism detection: MHC as a case study. *In* "Molecular Zoology: Advances, Strategies and Protocols" (a. S. R. P. J. D. Ferraris, Ed.), pp. 181–194. John Wiley & Sons, New York.

Primmer, C. R., Møller, A. P., and Ellegren, H. (1995). Resolving genetic relationships with microsatellite markers: A parentage testing system for the swallow *Hirundo rustica*. *Mol. Ecol.* **4**, 493–498.

Primmer, C. R., Møller, A. P., and Ellegren, H. (1996). A wide-ranging survey of cross-species microsatellite amplification. *Mol. Ecol.* **5**, 365–378.

Queller, D. C., and Goodnight, K. F. (1989). Estimating relatedness using genetic markers. *Evolution* **43**, 258–275.

Queller, D. C., Strassmann, J. E., and Hughes, C. R. (1993). Microsatellites and kinship. *TREE* **8**, 285–288.

Reynolds, J., Weir, B. S., and Cockerham, C. C. (1983). Estimation of the coancestry coefficient: Basis for a short-term genetic distance. *Genetics* **105**, 767–779.

Robinson, S. K., Thompson, F. R., Donovan, T. M., Whitehead, D. R., and Faaborg, J. (1995). Regional forest fragmentation and the nesting success of migratory birds. *Science* **267**, 1987–1990.

Rogers, J. S. (1972). Measures of genetic similarity and genetic distance. *In* "Studies in Genetics" University of Texas, Austin, Texas.

Rowe, P. S., Francis, F., and Goulding, J. (1994). Rapid isolation of DNA sequences flanking microsatellite repeats. *Nucleic Acids Res.* **22**, 5135–5136.

Roy, M. S., Girman, D. J., Taylor, A. C., and Wayne, R. K. (1994a). The use of museum specimens to reconstruct the genetic variability and relationships of extinct populations. *Experientia* **50**, 551–557.

Roy, M. S., Geffen, E., Smith, D., Ostrander, E. A., and Wayne, R. K. (1994b). Patterns of differentiation and hybridization in North American wolflike canids, revealed by analysis of microsatellite loci. *Mol. Biol. Evol.* **11**, 553–570.

Schlotterer, C., and Pemberton, J. (1994). The use of microsatellites for genetic analysis of natural populations. *In* "Molecular Ecology and Evolution: Approaches and Applications" (B. Schierwater, B. Streit, G. P. Wagner, and R. DeSalle, Eds.), pp. 203–214. Birkhauser Verlag, Basel, Switzerland.

Schlotterer, C., and Tautz, D. (1992). Slippage synthesis of simple sequence DNA. *Nucleic Acids Res.* **20**, 211–215.

Schlotterer, C., Amos, B., and Tautz, D. (1991). Conservation of polymorphic simple sequence loci in cetacean species. *Nature (London)* **354**, 63–65.

Scribner, K. T., Arntzen, J. W., and Burke, T. (1994). Comparative analysis of intra- and interpopulation genetic diversity in *Bufo bufo*, using allozyme, single-locus microsatellite, minisatellite, and multilocus minisatellite data. *Mol. Biol. Evol.* **11**, 737–748.

Shriver, M. D., Jin, L., Chakraborty, R., and Boerwinkle, E. (1993). VNTR allele frequency distributions under the stepwise mutation model: A computer simulation approach. *Genetics* **134**, 983–993.

Shriver, M. D., Jin, L., Boerwinkle, E., Deka, R., Ferrell, R. E., and Chakraborty, R. (1995). A novel measure of genetic distance for highly polymorphic tandem repeat loci. *Mol. Biol. Evol.* **12**, 914–920.

Slatkin, M. (1985). Gene flow in natural populations. *Annu. Rev. Ecol. Syst.* **16**, 393–430.

Slatkin, M. (1995a). A measure of population subdivision based on microsatellite allele frequencies. *Genetics* **139**, 457–462.

Slatkin, M. (1995b). A measure of population subdivision based on microsatellite allele frequencies. *Genetics* **139**, 1463.

Strassmann, J. E., Solis, C. R., Peters, J. M., and Queller, D. C. (1996). Strategies for finding and using highly polymorphic DNA microsatellite loci for studies of genetic relatedness and pedigrees. *In* "Molecular Zoology: Advances, Strategies and Protocols" (A. S. R. P. J. D. Ferraris, Ed.), pp. 163–180. John Wiley & Sons, New York.

Stutchbury, B. J., Rhymer, J. M., and Morton, E. S. (1994). Extrapair paternity in hooded warblers. *Behav. Ecol.* **5**, 384–392.

Tautz, D. (1989). Hypervariability of simple sequences as a general source for polymorphic DNA markers. *Nucleic Acids Res.* **17**, 6463–6471.

Tautz, D., Trick, M., and Dover, G. A. (1986). Cryptic simplicity in DNA is a major source of genetic variation. *Nature (London)* **322**, 652–656.

Taylor, A. C., Sherwin, W. B., and Wayne, R. K. (1994). Genetic variation of microsatellite loci in a bottlenecked species: The northern hairy-nosed wombat *Lasiorhinus krefftii*. *Mol. Ecol.* **4**, 277–290.

Trainer, J. M., and McDonald, D. B. (1993). Vocal repertoire of the long-tailed manakin and its relation to male–male cooperation. *Condor* **95**, 769–781.

Trainer, J. T., and McDonald, D. B. (1995). Singing performance, frequency matching and courtship success of long-tailed manakins (*Chiroxiphia linearis*). *Behav. Ecol. Sociobiol.* **37**, 249–254.

Weber, J. L., and Wong, C. (1993). Mutation of human short tandem repeats. *Hum. Mol. Genet.* **2**, 1123–1128.

Weir, B. S. (1996). "Genetic Data Analysis II: Methods for Discrete Population Genetic Data" (2nd ed.). Sinauer Assoc., Sunderland, Massachusetts.

Weissenbach, J., Gyapay, G., Dib, C., Vignal, A., Morissette, J., Millasseau, P., Vaysseix, G., and Lathrop, M. (1992). A second-generation linkage map of the human genome. *Nature (London)* **359**, 794–801.

Wenink, P. W., Baker, A. J., and Tilanus, M. G. J. (1993). Hypervariable control region sequences reveal global population structuring in a long distance migrant shorebird, the dunlin (*Calidris alpina*). *Proc. Natl. Acad. Sci. U.S.A.* **90**, 94–98.

Westneat, D. F. (1987). Extra-pair fertilizations in a predominantly monogamous bird: Genetic evidence. *Anim. Behav.* **35**, 877–886.

Westneat, D. F., and Webster, M. S. (1994). Molecular analysis of kinship in birds: Interesting questions and useful techniques. *In* "Molecular Ecology and Evolution: Approaches and Applications" (B. Schierwater, B. Streit, G. P. Wagner, and R. DeSalle, Eds.), pp. 91–126. Birkhauser Verlag, Basel, Switzerland.

Wright, J. (1994). Mutation at VNTRs: Are minisatellites the evolutionary progeny of microsatellites? *Genome* **37**, 345–347.

Wright, S. (1969). "Evolution and the Genetics of Populations." Univ. of Chicago Press, Chicago.

Wright, S. (1978). "Evolution and the Genetics of Populations." Univ. of Chicago Press, Chicago.

Wu, K. S., Jones, R., Danneberger, L., and Scolnik, P. A. (1994). Detection of microsatellite polymorphisms without cloning. *Nucleic Acids Res.* **22**, 3257–3258.

Yuille, M. A., Goudie, D. R., Affara, N. A., and Ferguson-Smith, M. A. (1991). Rapid determination of sequences flanking microsatellites. *Nucleic Acids Res.* **19**, 1950.

Zhang, Y., Proenca, R., Maffei, M., Barone, M., Leopold, L., and Friedman, J. M. (1994). Positional cloning of the mouse *obese* gene and its human homologue. *Nature (London)* **372**, 425–432.

Zietkiewicz, E., Rafalski, A., and Labuda, D. (1994). Genome fingerprinting by simple sequence repeat (SSR)-anchored polymerase chain reaction amplification. *Genomics* **20**, 176–183.

Appendix I Internet and Other Resources for Microsatellite Analyses[a]

Program name	Task	Ref.	Internet access (URL)	Platform
Primer development				
GenBank	List of sequences	na	See MsatManV6	Internet
PRIMER	Primer design	na	Via e-mail to primer@genome.wi.edu	SPARC, PC, or Mac
Laboratory protocols				
MsatManV6	Detailed protocols	na	Via ftp from onyx.si.edu/protocols/MsatManV6	[Text file] MS-Word
[Book chapter]	Detailed protocols	Srassmann *et al.* (1996)	na	na
Genetic analysis (e.g., F_{ST})				
Relatedness 4.2b	*r* (coefficient of relatedness)	Queller and Goodnight (1989)	Via www from http://www.rice.edu/wasps	Macintosh
GENEPOP	Various population genetics analyses	Raymond and Rousset (1995)	ftp from ftp.cefe.cnrs-mop.fr/pub/msdos/genepop	DOS
GDA	Various population genetics analyses	Weir (1996)	Via www from http://www2.ncsu.edu/ncsu/CIL/stat_genetics[b]	Windows (PC)
MSAT	Stepwise mutation distance measures	Goldstein *et al.* (1995)	Via ftp from lotka.stanford.edu/pub/programs	Sun, DOS, Mac, DEC
WINAMOVA	Stepwise mutation distance measures	Michalakis and Excoffier (1996)	Via ftp from acasun1.unige.ch/pub/comp/win/amova http://acasun1.unige.ch/LGB/Software/Windoze/amova	Windows (PC)

[a] Internet sites and URLs may change frequently.
[b] Listing in Weir (1996) lacks /ncsu.
Abbreviation: NA, Not applicable.

CHAPTER 3

Mitochondrial Control Region Sequences as Tools for Understanding Evolution

ALLAN J. BAKER* AND H. DAWN MARSHALL†

*Centre for Biodiversity and Conservation Biology, Royal Ontario Museum, Toronto, Ontario, Canada M5S 2C6

†Department of Zoology, University of Toronto, Toronto, Ontario, Canada M5S 3G5

I. INTRODUCTION

The mitochondrial control region containing the three-strand displacement loop (D loop) characteristic of vertebrate mitochondrial DNA (mtDNA) has attracted the attention of systematists and population geneticists as a potential source of genetic markers within and among closely related species. Focus on this non-coding sequence stemmed from earlier reports suggesting that it was the most rapidly evolving region of the mtDNA molecule (Fauron and Wolstenholme, 1976; Upholt and Dawid, 1977; Walberg and Clayton, 1981; Chang and Clayton, 1985). This prediction was borne out when estimates of the rate of substitution in the human control region were found to range between 2.8 (Cann *et al.*, 1984) and five times (Aquadro and Greenberg, 1983) the rate for the rest of the mtDNA genome.

The much publicized phylogeny of humans based on hypervariable control region sequences (Vigilant *et al.*, 1991) was a major stimulus for assays of control region sequence variability in other vertebrates including birds. Sequencing and mapping studies have established that the gene order around the bird control region has been altered relative to other vertebrates (Desjardins and Morais, 1990, 1991; Ramirez *et al.*, 1993; Quinn and Wilson, 1993); the avian control region is flanked by the genes for $tRNA^{Phe}$ and $tRNA^{Glu}$. Length variation in domain III ($tRNA^{Phe}$ end) has been shown to be due to a variable number of simple tandem repeats (Wenink *et al.*, 1994; Berg *et al.*, 1995), and individuals are often heteroplasmic for repeat number (Berg *et al.*, 1995).

Applications of control region sequence data to population structure and systematics of birds are few, but have mostly been instructive in revealing the increased resolution afforded by faster mutating sequences. Using just 178 bp of the control region, Quinn (1992) was able to uncover the historical mixing of two divergent clades of the snow goose (*Anser caerulescens*), confirming results of a restriction fragment length polymorphism (RFLP) study of the whole genome (Avise *et al.*, 1992). Control-region sequences have elucidated the global phylogeography of the dunlin (*Calidris alpina*) (Wenink *et al.*, 1993; 1996), gene flow and population structure among social groups and populations of the gray-crowned babbler (*Pomatostomus temporalis*) (Edwards (1993a,b), and apparent global panmixia in knots (*Calidris canutus*) (Baker *et al.*, 1994) and possibly turnstones (*Arenaria interpres*) (Wenink *et al.*, 1994).

Although control region sequences often can be a rich source of genetic markers, they are not universally informative or highly polymorphic within species (Baker *et al.*, 1994). However, the major problems confronting avian systematists contemplating the use of control region sequences at present is the dearth of knowledge about sequence variation in more than a few species, and the appropriateness of this region as a source of genetic markers for different levels of the taxonomic hierarchy. Our objectives in writing this chapter are to summarize what is known about the control region of birds in terms of its organization, the location of markers within it, and to present exemplars mainly from our own laboratory illustrating the potential and problems of such fast-evolving sequences in elucidating population structure and molecular systematics of closely related taxa.

II. SEQUENCE ORGANIZATION AND EVOLUTION

The organization and evolution of the D loop-containing region have been described for a number of groups of animals, including vertebrates (Brown *et al.*, 1986; Mignotte *et al.*, 1987; Saccone *et al.*, 1987), mammals (Saccone *et al.*, 1991), cetaceans (Hoelzel *et al.*, 1991; Árnason *et al.*, 1993), lepidopterans (Taylor *et al.* 1993), and insects (Zhang *et al.*, 1995). Here we summarize the features of the control region of birds by aligning and comparing complete or partial sequences from a range of taxa. The reason for doing this is that knowledge of the organization of the control region is an essential precursor to using these sequences in population genetic and phylogenetic studies. For example, uncertain homology arising from hypervariability or heteroplasmy could confound measures of sequence diversity, molecular clock estimates, and phylogeny reconstruction. In addition, gene rearrangements, large tandem duplications near the control region, and duplicate copies of mtDNA genes in the nuclear genome could all complicate homology determination.

A. Control Region Sequence Data and Structural Features

Complete control region sequences are now available from 10 avian taxa comprising two Galliformes, three Anseriformes, three Charadriiformes, and two Passeriformes. Partial sequences are available for another 13 taxa from the following orders: Apterygiformes, Sphenisciformes, Columbiformes, Charadriiformes, Anseriformes, and Passeriformes (Table 1). Location of primers used to obtain these sequences are shown in Fig. 3.1, along with previously unpublished primer sequences.

1. Gene Order

The $tRNA^{Glu}$ and ND6 genes in birds are found immediately adjacent to the D-loop region of the molecule instead of being located between the ND5 and cytochrome *b* genes as in other vertebrates (Fig. 3.1). This unusual gene order has been found in the domestic chicken and other galliforms (Desjardins and Morais, 1990), the Japanese quail (Desjardins and Morais, 1991), the Peking duck (Desjardins *et al.*, 1990), and the lesser snow goose (Quinn and Wilson, 1993). In addition, we have found evidence for this rearrangement in the passerine family Fringillidae (greenfinch and common chaffinch), the Adélie penguin, and the brown kiwi, suggesting that this is a universal bird phenomenon, and must be due to an event occurring earlier in avian or reptilian history (see Quinn and Mindell, 1996). One result of this

TABLE I Control Region Sequences Included in This Study

Species	Ref.	Region[a]; size (bp)
Apterygiformes		
Apterygidae		
Apterix australis	A. J. Baker (unpublished)	I, II, and III; 772
Anseriformes		
Anatidae		
Anas acuta[b]	V. Ramirez and R. Morais (unpublished)	I: 403
Anas platyrhynchos[b]	S.T. Liu and L.Y. Lin (unpublished)	All; 1131
Anser caerulescens	Quinn and Wilson (1993)	All; 1177
Branta canadensis	A. J. Baker (unpublished)	I; 446
Cairina moschata[b]	S.T. Liu and L.Y. Lin (unpublished)	All; 1135
Charadriiformes		
Alcidae		
Alca torda	Berg *et al.* (1995)	III; 216
Cepphus grylle	M. Kidd and V. Friesen (personal communication)	All; 1121
Uria aalge	Moum and Johansen (1992)	I; 198
Uria lomvia	Moum *et al.* (1994)	I; 193
Laridae		
Larus argentatus	Berg *et al.* (1995)	III; 214
Larus fuscus	Berg *et al.* (1995)	III; 214
Scolopacidae		
Arenaria interpres	Wenink *et al.* (1994)	All; 1192
Calidris alpina	Wenink *et al.* (1994)	All; 1072
Calidris canutis	Baker *et al.* (1995)	I; 300
Columbiformes		
Columbidae		
Columba inornata[b]	M.M. Miyamoto *et al.* (unpublished)	I; 451
Galliformes		
Phasianidae		
Coturnix japonica	Desjardins and Morais (1991)	All; 1153
Gallus gallus	Desjardins and Morias (1990)	All; 1227
Passeriformes		
Fringillidae		
Carduelis chloris	H. D. Marshall and A. J. Baker (1997)	All; 1237
Fringilla coelebs	H. D. Marshall and A. J. Baker (1997)	All; 1234
Fringilla montifringilla	H. D. Marshall and A. J. Baker (1997)	II and III; 638
Fringilla teydea	H. D. Marshall and A. J. Baker (1997)	II and III; 619
Timaliidae		
Pomatostomus temporalis	Edwards (1993)	I; 399
Sphenisciformes		
Spheniscidae		
Pygoscelis adeliae	Monehan (1994)	I; 507

[a] Number refers to domain studied, as defined in text.

[b] Genbank accession numbers: *Anas acuta*: L24205; *Anas platyrhynchos*: L16770; *Cairinu Moschata*: L16769; and *Columbia inornata*: M98393.

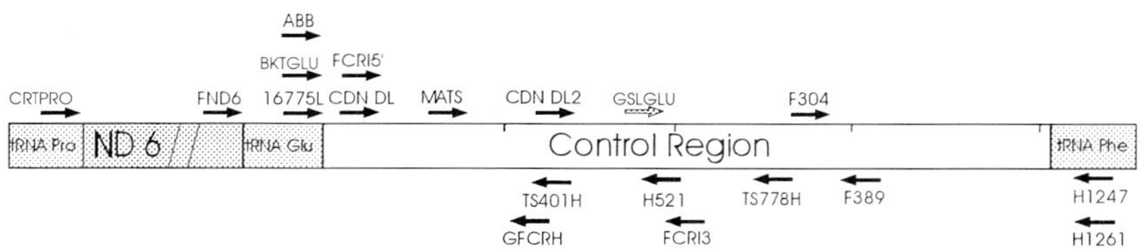

FIGURE 3.1 Schematic representation of the gene order near the control region, given for the L strand in 5′-to-3′ orientation. Forward arrows represent approximate positions of forward primers, and reverse arrows represent reverse primers. The shaded arrow indicates the suspected priming site of GSLGLU. Control region primer references or sequences are as follows (5′ to 3′): Fringillidae: CRTPRO CCA TCT CCA ACT CCC AAA GC; FND6 (P. Boag, personal communication); FCRI5′ TCA GGG TAT GTA TAA TAT GC; MATS CCA TTG TCC CCT CCA GGC GC; GFCRH CTC GTT TCC TAG GTT GGA GG; GSLGlu (P. Boag, personal communication); FCRI3′ CAC TTG CTG TGA AGA GC; F304 CTT GAC ACT GAT GCA CTT TG; F389 TAT GTC CGG CAA CCA TTA CAC TAT; H1261 AGG TAC CAT CTT GGC ATC TTC; Adélie penguin: ABB TGT TAC TTC AAC CAC AGG AAC; TS401H (Wenink *et al.*, 1994); brown kiwi: BKTGLU TAG GTC TCA ACT ACA GAA AC; TS778H (Wenink *et al.*, 1994); Canada goose: 16775L (Quinn, 1992); CDN-DL TAT GCA TAT TCG TGC ATA GA; CDN-DL2 ACG TGA AAT CAG CAA CCC G; H521 (Quinn and Wilson, 1993); H1247 (Quinn and Wilson, 1993).

rearrangement appears to be increased sequence divergence in the tRNA^{Glu} gene (17.6% different between the common chaffinch and the greenfinch, about 35% between either of these and the domestic chicken, and about 27% between the domestic chicken and the lesser snow goose) (Marshall and Baker, 1997; Quinn and Wilson, 1993). As pointed out by Quinn and Wilson (1993), this may be due to reduced functional constraint because this tRNA is no longer associated with the sense transcript of any gene.We have found that there is sufficient variation in the six bases at the 3′ end of this tRNA to prevent the primer GSLGLU (P. Boag, personal communication) from annealing at its designed priming site, exactly adjacent to the D loop. In the common chaffinch and the greenfinch, GSLGLU appears to anneal to a stretch of 10 bases about 500 bp downstream of the tRNA gene, showing 70% similarity to the 3′ 10 bases of the primer (see Fig. 3.1).

2. *Size and Base Composition*

The control region in birds is generally larger than the corresponding sequence in most mammals (e.g., 910 bp in the cow; Anderson *et al.*, 1982), but is smaller than that in *Xenopus laevis* (2134 bp; Roe *et al.*, 1985); control region length variation is thought to account for the size difference between bird and other vertebrate mt-DNAs (Desjardins and Morais, 1990). The avian control region ranges in size from 1072 bp in the dunlin to about 1240 bp in greenfinch, with the average size being 1168 bp. In pairwise comparisons of related species, this variation can be often be attributed to relatively small (1–20 bp) insertions and deletions (indels) in the 5′ and 3′ flanking regions, and varying numbers of tandem repeats in the 3′ domain

TABLE II Base Composition of Avian Control Regions and Their Respective Domains

	Nucleotide frequency			
Segment	A	C	G	T
Whole control region				
Peking duck	28.4	31.3	16.2	24.1
Muscovy duck	28.4	31.5	16.0	24.1
Lesser snow goose	28.5	30.8	14.8	26.0
Black guillemot	30.5	28.5	13.7	27.2
Chicken	26.7	26.3	13.3	33.7
Japanese quail	26.0	25.8	14.2	33.9
Turnstone	29.5	29.2	14.1	27.3
Dunlin	30.6	25.9	14.9	28.5
Common chaffinch	29.2	27.1	14.1	29.5
Greenfinch	29.1	28.6	13.0	29.3
Average:	28.69	28.50	14.43	28.36
Standard deviation:	1.393	2.085	1.004	3.252
Domain I				
Peking duck	29.8	33.0	14.6	22.6
Muscovy duck	32.2	33.6	14.5	19.8
Lesser snow goose	31.5	32.5	12.1	23.9
Chicken	26.8	29.9	13.1	30.1
Japanese quail	25.7	29.3	14.7	30.3
Turnstone	27.1	32.4	15.8	24.7
Dunlin	29.6	29.3	16.5	24.6
Common chaffinch	27.2	31.4	14.8	26.6
Greenfinch	28.2	30.9	16.1	24.8
Average:	28.68	31.37	14.69	25.27
Standard deviation:	2.102	1.525	1.325	3.168

(Continues)

(see beyond). However, between the turnstone and the dunlin there appears to be at least one large (about 65 bp) indel at the 5′ end of the control region. In addition, Quinn and Wilson (1993) described relatively large deletions in both the 5′ (61 bp) and 3′ (38 bp) flanking regions of the lesser snow goose as compared to the domestic chicken, and Desjardins and Morais (1991) discussed a 57-bp deletion in the Japanese quail relative to the domestic chicken in the 5′ portion of the control region. Ramirez *et al.* (1993) also reported large deletions in both flanking regions in the Peking duck versus the domestic chicken.

TABLE II (*Continued*)

Segment	Nucleotide frequency A	C	G	T
Domain II				
Peking duck	14.9	33.9	20.6	30.6
Muscovy duck	14.2	33.6	20.2	32.0
Lesser snow goose	17.4	32.1	19.2	31.4
Chicken	15.4	28.1	21.6	35.0
Japanese quail	14.6	30.2	20.5	34.7
Turnstone	21.0	25.9	18.8	35.1
Dunlin	22.1	24.4	18.0	34.7
Common chaffinch	23.6	26.8	19.8	29.7
Greenfinch	22.6	27.8	19.2	30.6
Average:	18.42	29.20	19.77	32.64
Standard deviation:	3.643	3.227	1.026	2.082
Domain III				
Peking duck	34.5	28.2	15.3	22.1
Muscovy duck	32.7	28.3	15.2	23.8
Lesser snow goose	32.4	28.7	14.3	24.6
Chicken	35.9	19.6	6.7	37.8
Japanese quail	36.1	17.6	8.3	38.0
Turnstone	48.7	30.0	4.1	17.2
Dunlin	49.0	24.3	4.9	21.8
Common chaffinch	37.7	26.1	3.3	32.9
Greenfinch	37.6	21.1	6.0	35.3
Average:	38.29	24.88	8.678	28.17
Standard deviation:	5.913	4.223	4.635	7.395

The base composition of control region sequences is reported in Table II. As is typical for vertebrate mtDNA, the GC asymmetry between the two strands of DNA, resulting in one being relatively "light" (low G) and the other "heavy," also occurs in the control region. Overall base composition shows a paucity of G; about 14% of the light strand of the control region is composed of G, compared to about 28% for each of the other bases. Base composition does not vary greatly among species, but it varies significantly among regions of the D loop, and also among species when regions are considered individually (see next section).

3. Species-Pair Comparisons and Domains

For descriptive purposes, the control region is often divided into three subregions: a central, more conserved domain, low in L-strand A, which is responsible for the three-strand displacement (D)-loop structure (Clayton, 1991), flanked by two variable A-rich domains showing extensive size and sequence variation (Saccone *et al.*, 1991). In mammals, the tRNA^{Phe}-adjacent domain is of varying length and greatest divergence, and contains three conserved sequence blocks [CSB-1 (not found in rats), -2, and -3], O_H, and the transcription promoters for both the heavy and light strands (HSP and LSP). The approximately 200-bp central conserved block is characterized by low L-strand A and high G content, and harbors open reading frames (ORFs) of varying lengths. The tRNA^{Pro}-adjacent domain shows the highest A and the lowest G content of the whole control region, and includes short termination-associated sequences (TASs) typified by the nucleotide motif TTACAT (Saccone *et al.*, 1987). Also found in the flanking domains and associated with the CSBs and TASs are relatively stable cloverleaf-like structures of low primary sequence similarity (Dunon-Bluteau and Brun, 1987).

Three such domains have also been demonstrated in most avian control region sequences studied to date. We define them here as the following: I, the region adjacent to the tRNA^{Glu}; II, the central conserved domain; and III, the region closest to the tRNA^{Phe} gene. To examine these regions we took a series of species-pair alignments and plotted the number of variable sites between them in nonoverlapping 50-bp windows. This enabled rough designation of the boundaries of the regions; further refinement was obtained by examination of the alignments. Species-pairs examined were: greenfinch and common chaffinch; dunlin and turnstone; turnstone and black guillemot; Muscovy duck and Peking duck; Peking duck and lesser snow goose; lesser snow goose and domestic chicken; and domestic chicken and Japanese quail. Domain structure was clearly found in all comparisons except the black guillemot versus the turnstone and the lesser snow goose versus the domestic chicken (Fig. 3.2). Domains are defined not only by their degree of variability but also by base composition (Table II). Specifically, domain I is AC rich, domain II is CT rich, and domain III is AT rich and very low in G. The increased conservation of the central domain II is shown in Table III; it has fewer indels and considerably lower among-species sequence divergence than the flanking domains.

Brown *et al.* (1986) pointed out that the central domain is well preserved evolutionarily, occurring even in *X. laevis*, which has a different mtDNA base composition from mammals, and that this region must therefore be functionally constrained.

4. Conserved Structural Features

Multiple alignment of five species of birds from three orders (Galliformes, Anseriformes, and Charadriiformes) illustrates the great primary structure variability among control regions at this taxonomic level (Fig. 3.3). Nevertheless, conserved

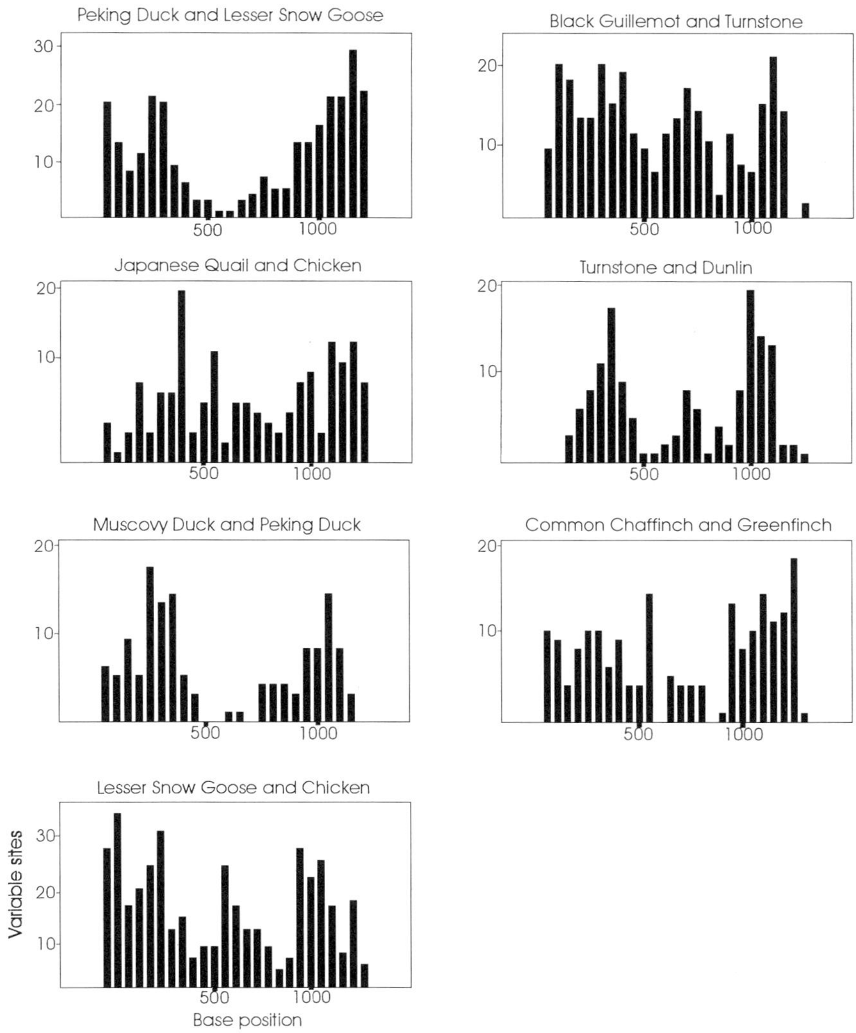

FIGURE 3.2 Plots of number of variable sites in 50 base windows along the control region for seven species-pairs. Species-pair alignments were done using Clustal V (Higgins *et al.*, 1991).

blocks [F, D, and C boxes (Southern *et al.*, 1988), and CSB-1] are clearly evident in the alignment. All three conserved boxes are located in domain II, but CSB-1 is located in domain III. These conserved structural features also occur across a broad range of birds including passerines, penguins, and kiwis. H-strand replication is thought to be initiated (and therefore O_H must occur) in the vicinity of CSB-1 in domestic chicken, human, and amphibian mtDNAs (Desjardins and Morais, 1991),

TABLE III Among-Species Variability in the Control Region of Birds

Species pair	Size compared (bp)	Percent sequence divergence	Ts/site	Tv/site	Ts/Tv	Indel/site
Whole control region						
Muscovy duck/Peking duck	1162	12.23	0.0740	0.0422	1.755	0.0430
Dunlin/turnstone	1202	14.09	0.0757	0.0458	1.655	0.1165
Chicken/Japanese quail	1233	14.21	0.0681	0.0641	1.063	0.0700
Common chaffinch/greenfinch	1251	16.25	0.0727	0.0815	0.892	0.0248
Peking duck/lesser snow goose	1196	24.73	0.1104	0.1196	0.923	0.0702
Domain I						
Muscovy duck/Peking duck	448	18.33	0.1027	0.0692	1.484	0.0625
Dunlin/turnstone	445	18.61	0.0810	0.0517	1.565	0.2292
Chicken/Japanese quail	550	14.58	0.0745	0.0527	1.414	0.1273
Common chaffinch/greenfinch	531	16.86	0.0791	0.0847	0.933	0.0282
Peking duck/lesser snow goose	396	29.92	0.1212	0.1515	0.800	0.0884
Domain II						
Muscovy duck/Peking duck	253	0.81	0.0040	0.0040	1.000	0.0198
Dunlin/turnstone	486	6.04	0.0412	0.0185	2.222	0.0123
Chicken/Japanese quail	309	8.52	0.0647	0.0324	1.600	0.0129
Common chaffinch/greenfinch	368	5.52	0.0272	0.0217	1.250	0.0000
Peking duck/lesser snow goose	288	4.24	0.0278	0.0139	2.000	0.0174
Domain III						
Muscovy duck/Peking duck	461	12.84	0.0846	0.0369	2.294	0.0369
Dunlin/turnstone	271	24.27	0.1292	0.0849	1.522	0.1181
Chicken/Japanese quail	374	18.51	0.0722	0.1070	0.675	0.0321
Common chaffinch/greenfinch	352	26.19	0.1108	0.1392	0.796	0.0455
Peking duck/lesser snow goose	512	33.12	0.1484	0.1543	0.962	0.0859

Abbrevations: Ts, Transition; Tv, transversion; Indel, insertion or deletion.

and to end in domain I. One to three copies of a direct repeat of CSB-1 have been reported in the domestic chicken, Japanese quail, Peking duck, and lesser snow goose control regions. One such repeat occurs in the black guillemot and possibly the dunlin. Also found in domain I of all birds examined are one or more putative TAS (5′ TATAT 3′ or 5′ TACAT 3′) elements located upstream from the most 5′ CSB-1-like repeat (Fig. 3.3). These TAS elements and the CSB-1-like repeats are thought to be involved in termination of D-loop synthesis (Doda *et al.*, 1981). In addition, sequences downstream and upstream of the CSB-1 and CSB-1-like repeats, respectively, are capable of forming conserved, thermodynamically stable tRNA-like cloverleaf structures thought to be associated with the start and arrest of DNA synthesis (Brown *et al.*, 1986).

5. Bidirectional Transcription Promoter

In the domestic chicken one bidirectional transcription promoter has been identified in the tRNA^{Phe}-adjacent domain, downstream from CSB-1 (Fig. 3.3). This is an AT-rich sequence containing an inverted repeat capable of forming a cruciform structure, and is flanked on either end by an octanucleotide sequence similar to the H-strand transcription start sites in mouse and *X. laevis* (L'Abbé *et al.*, 1991). A putative bidirectional transcription promoter can readily be identified in the Peking duck, the Muscovy duck, the Japanese quail, and the lesser snow goose, but is less apparent in the other sequences examined here.

6. Sequence Simplicity

The AT-rich domain III of several species is characterized by a number of repeats of a microsatellite-like motif. In particular, the sequence 5′ CAACAAA 3′ is directly repeated at least six times in four charadriiform species [lesser black-backed gull, *Larus fuscus*; herring gull, *Larus argentatus;* razorbill, *Alca torda* (Berg *et al.*, 1995), and the turnstone (Wenink *et al.*, 1994)] and 20–37 times in the cuckoo (*Cuculus canoris*; Gibbs *et al.*, 1996). Variants of this CA motif occur in the turnstone and dunlin (Wenink *et al.*, 1994), the black guillemot (V. L. Friesen, personal communication; Berg *et al.*, 1995), and an additional 11 charadriiform species (Berg *et al.*, 1995). There is also a simple sequence repeat in the third domain of the brown kiwi (A. J. Baker, unpublished data). Repeat number varies both inter- and intraspecifically, in addition to showing heteroplasmy within individuals (Berg *et al.*, 1995). These repeats occur adjacent or close to the tRNA^{Phe}, and account for some of the length variation among species in this domain. More complicated repeats also occur. For example, the motif 5′ TCATCACACATTTATCATCA 3′ is repeated twice in the razorbill (Berg *et al.*, 1995).

B. Sequence Evolution

In addition to transition and transversion substitutions and numerous small indels, length differences accumulate through variation in number of tandem repeats, and relatively large duplication or deletion events. Both inter- and intraspecific variation is more common in the two flanking domains than in the conserved central block, with tandem repeats occurring primarily in domain III and larger duplications restricted to among-species comparisons.

1. Among-Species Variation

The main features of among-species variation in the control region are summarized in Table III. In pairwise comparisons, sequence divergence ranges from about 12 to 25% overall; by domain the average divergences are approximately 20% (I), 5% (II),

```
Chicken         --------aat--tttatt--ttttaacctaactcccctactaagtgtacccccccttcccc--------aggggggtat---actatgcataatcgt 77
Peking Duck     acagctagaatagcctaataatgct--ctca---ggacc-----cccccccccttcccccccaggggttgcgggggttatttggttatgcatat-cgt 88
Snow Goose      taaccgcaaacgccccaatgattctaccccatctatgccgttatgcttaacccccccccccctcccccccgggcggggtatttggttatgcatattcgt 100
Guillemot       ctagccatcatggctaaat----ctcaaccacaggaaccaa-agacgcccccaaaatatctgagggacaccac-caccccaccc-cccacccatgtacat 93
Dunlin          tgtccaattatg--tgg-------tgcgctgcatatact------tgtcccc-----------------cc-c-catactacat-accatccatgttc-c 64
                         *              *            *            **                            *       *  ***   *

Chicken         gcatacatttatataccacatatattatggtaccggtaatatatactatatatgtactaaacccattata-tgtatacgggcattaacctatattccaca 176
Peking Duck     gcatacatttatattccccatatattaac---cta-tggtcccggtaataaacactattaaccaactatcctacatgcacggactaaacc-cat--caca 181
Snow Goose      gcatagatttatatgccccatatacatacatacta-tagtaccggtaatatacattatatacgagctatcctataagcaggtgctaaacc-cat---aca 195
Guillemot       tagtacattaacttaccccata-ac--acata----tagtgcatgt--cctataccactaatatacaaggcggcataccc-cacctcaccacatcccac- 182
Dunlin          aaatccattaatttac-----a-ac----------------cgggc--tatacacctct-------------------c-caccccc-tcac---ccgc- 114
                   *  *** *  * *     * *                           *                                              *

Chicken         tttctcccaatgtccattctatgcatgatctaggacat-actcat-ttaccctccccatagacagttccaaaccactatcaagccacctaactatgaatg 274
Peking Duck     tgtc----aacggacat--accctactatc--ggac-t-accctc-ccaacggacccagagtgaatgct---ctaatgcccaacacctcaacgcc--aca 265
Snow Goose      tgta----cacggccattaaacccttaaac--acac-t--cctac-caaaccactacaacatgaatgct---ctaggaccataccccataatacccaatg 282
Guillemot       ---ctctagagggcagttgagtcaatggacactggaatgatacacattatccacactaaaaccattataatagt-ggactgtaca--t-aataccc---- 271
Dunlin          ---ctccagagggtaaccgaagcaatgaacctaggaat-attcacacacactgtactaaacccatcaacttgttaggattataca--ttaaaactc---- 204
                         * *                 *       *            *      *     *                   *     **

Chicken         gttacaggacataaatctcactctcatgttctcccccaacaagtcacctaactatgaatggttacaggacatacatttaactaccatgttctaacccat 374
Peking Duck     -taacat-gcccccaac-cagaa-caaggccccataatgatgaat-gcttgac-agacataccc-tacca-acactccaaattcctctccacccacccat 357
Snow Goose      -taacccactcacgca-cacaa-caagacccccatattaatgaat-gcttaca-ggacataccc-taaca-a-----caa---ctctctacc-acatat 365
Guillemot       -----ctaacttacatggcagtgcttgaa-cccatatcc-tgaatgatctca--ggacaaatccataccatgt----------ctctcgctgtaccta- 350
Dunlin          ----tctaaagcgtacggcagtgctttaa-cacacgcca-tgattggtttaa--gtgcaga-cagctcgaaaa----------ctctcgaagtgcaca- 283
                                  **          * *         * *    *        *          *              *  *       *  *

Chicken         ttggttatgct-cgccgtatcagatggatttattgatcGTCCACCTCACGAGAGATCAGCAACCC-Ctgcctgtaatgtacttcatgaccagtctcaggc 472
Peking Duck     t-actcatgaagctgcgtaccagatggatttattaatcGTACACCTCACGTGAAATCAGCAATCC-Ttgcacataatgtccgacgtgactagcttcaggc 455
Snow Goose      ---ctcatgcagtt-cgtatcagatggatttattagtcGTACTCCTCACGTGAAATCAGCAACCCGTtgcacataatgtccggtatgactagcttcaggc 461
Guillemot       ---------cagctgcagactaggtc-atctattagtcGTACCTCTCACTTGAAATTAGCAACCCGAcgcatgt?agatccaacgttactagcttcagaa 439
Dunlin          ---------ccagt-cgtaccaggtt-atttattaatcGAGCTCCTCACGTGAAATCAGCAACCCGGcgtaagtaatgtcctgcgttactagcttcaggc 372
                               *  *  ** *  ** ****  ***  *  *****  ** ** ***** **   *    * *  * *    * ** **  ****
                                                                    F box

Chicken         ccattctttcccccctacacccctcgccctacttgccttccaccgtaCCTCTGGTTCCTCGGTCAGGCACATcccatgcataactcctgaactttcTCACT 572
Peking Duck     ccatacgttcccccctaaacccctcgccctcctcacatttt--tgcgCCTCTGGTTCCTCGGTCAGGGCCATcaattgggtt-cactcacct---cTCGCC 549
Snow Goose      ccatacgttcccccctaaacccctcgccctcctcacatttt--tgcgCCTCTGGTTCCTCGGTCAGGGCCATccattgggtt-cactcacctctccTTGCC 558
Guillemot       tcattcattcccccctaaacccctagcccaacttgctcttt--tgcaCCTCTGGTTCCTCGGTCAGGGCCATaacttgactagtcctctcaa---cTTGTA 534
Dunlin          ccattctttcccccctaaaccc-tagcacaacttgctcttt--tgcgCCTCTGGTTCCTATGTCAGGGCCATaaataggttagtactcataa---cTTGCT 466
                 *** * ********* **** * ** *  **  *  *     * ************  ******  ***     *  *    *
```

FIGURE 3.3 Multiple alignment of five species generated by Clustal V (Higgins *et al.*, 1991). Conserved sequence blocks (CSB-1 and F, D, and C boxes) are capitalized, possible CSB-1 repeats are underlined, possible TATAT or TACAT termination-associated sequences are in boldface, and the putative bidirectional transcription promoter sequence is double-underlined.

```
                                                                   D box

Chicken        TTTCACGAAGTCATCTG-TGGATtatct-tcccctctttagtccgtgatcgcggcatcttctc-tcttctattgctgttggttccttctcttttt---gg 666
Peking Duck    CTTCAAAGTGGCATCTG-TGGAAtactt-ccaccatctcaatgcgtaatcgcggcatcttccagcttttggcgcctctggttccttttatttttccgg 647
Snow Goose     CTTCAAAGTGGCATCTG-TG-AGTactt-tcaccttctcaatgcgtaatcgcggcatgttccagcttttggcgcctctggttcctcttatttttccgg 655
Guillemot      CTTCACCGATACATCTGGTCGGCtatttatcatcatct-cacccgtgatcgcgacat--ccgaccgtcttggcactttggttcctttt--tttttctt 629
Dunlin         CTTTACGAATACATCTGGTTGGCtatatctcaccatttcgtccgtgatcgcggcat--tccaaaattcttatactttggttcctttt--tttttggg 562
                ** *      ****** *     **  *  *  *   *     *** ****** ***    *      *  *    *   ******** *  ****
                  C box
Chicken        ggcttcttcacaggttacccttcaca-gtgcgggtgcgagtgctattcaagtgaagcctggactacacctgcgttgc-gtcc-tatcctagtcctctcgt 763
Peking Duck    ggttacctcacagctggcccttccca-gtgacttcgggggtccca--caatctaagcctggac-acacctgcgttatcgcgc-tatcctatatctcaggg 742
Snow Goose     ggttacctcacaggtggctattccca-gtgatctcgggggtccca--caatctaagcctggac-acacctgcctcacggt-c-tatcctatatctcaagg 749
Guillemot      c-t--catctcaataaacccctccagtgcaccgaggtaaatacaatcta---tagacgtgggccc?ccttg-gtattcgcgcggatcttggccctcagga 721
Dunlin         ggt--cttcacaggtggcccacccagcgcaacg-ggtaaatacaatcta----agacctgaacatctcgtgcgttgcggtctatatttggccctcagga 655
                   * ** **      *     *     *       *     * * *    *      *  * **  *     * **  *     *      **  *    ***  *

Chicken        gtccctcgatgagacgg-tttgcgtgtatggggaatcatcttgacactgatgcactttggatcgcatttggtta-tggttcttccacccccccgg--taa 859
Peking Duck    attactcaatgagacgg-ttggcgtatatggggaatcaccttgacactgatgcactttg-accacattcagttaatgctctctccacagctctatat--a 838
Snow Goose     gtccctcgatgagacgg-ttggcgtatatggggaatcatcctgacactgatgcactttg-accacattcagttaatgttacctccacc-ctccgggttaa 846
Guillemot      atacctgagtgtcatggcttaaag-gttaggggaatcatatccatactgatgcactt-gttttacacttggttatggcctcc?ccgcaagctcctactta 818
Dunlin         gttactgaatgagacggtttcaagtgtttggggaatcatatcaacactgatgcactttgttttacacttggttatggcctcccc-gcaagctgc-gtaca 753
                *  **    **  * ** **    *  *   *********      * *********** *       ** *  ****  *       *  *           *

Chicken        atggtgcTATTTAGTGAATGCTTGTCGGACATATttttatcaattttcacttcctctatttt-cttcacaaaactaggaaattcaccacaatttttctt 958
Peking Duck    atagggcTATTTAGTGAATGCTCGATGGACATACtttaa--aa---------aca-aaacca-c----cccaaccacaac--cc--cacaa--tatat-- 913
Snow Goose     atggggcTATTGGATGAATGCTCGTTGGACATAGcacaa--aa---------caacaaatca-ttaagcgcaacccctgcgctt--cacaaactaaacc- 931
Guillemot      -tgctgcTATTTAGTGAATGCTTGTTGGGCATAAtttttca-tttttcatttcctctaactttctaaacaacactagtaaatttcatctaaaaaacaa-- 914
Dunlin         -tgccatTATTTAATGAATGCTTGTGGGACATGAtttttcaattttcacttcctctaatttcttaacaacactagaagatttcaactaaact-taa-- 849
                *      ****    ******** *  ** ***                                 *            *   **              **
                          CSB-1

Chicken        tgttatttttaatttttttttatttttaaaaacatttttaaaaaactaaattacatacaaactaccgcataaaatccctcaaactatacaaacgtt 1058
Peking Duck    --atatacataca--aaacgaa-atg---cataa-----catgac----ctaaatt-tat-tagagaaactc-cagcactaaa--gacgatccaaacc-c 990
Snow Goose     -agtaaactttcactaaccacc-atgtagcatgaaccttcatcgc----ccaatcc-cag-caaactacctc-taaaactccattaatcatcccttccat 1022
Guillemot      -accatatttcatcacacattttatc-at-cg-----tcatcacacattttatcatcacattcgttacct----aca----aacggcagt--gagt-tt 995
Dunlin         -actacgtttcattacacatcttatttacaca-----tttttcatacgttatcaccgctgaaattacat--taacat---aataaccat--aaacgtt 936
                    *    *   *                                 *            *              *          *             *

Chicken        tatcgtataatatatatacattattgtttattctatcattat-tagagaaactcca-ctaccaaaaccatcattaaaacaaaaatttacatgccacttaa 1156
Peking Duck    gatgaca----atcattacttt--gaccta-acaaacattacccaat-taacc--agccacctgccccgtccacatagcttaccacaa-aagc-atggca 1078
Snow Goose     gacatcatcgaaacgacgtata--tatatatgcaaacattatacaaaataacttaaacttattagagaaactccagtactaaaaatagcaaac-acaa-a 1118
Guillemot      cattaaaaacaaaacacttacaatc-tacgttcatatattc------gaa-ctgca-taacacaaactat????g???ctcccctgaaaaaacaaacaa- 1078
Dunlin         tgggttcatctattcgaacacaaat-cccctaccaataaag------gaaactctc-ctacaaaaacaaacaaacaaacaaac--aaacaaacaaacaaa 1026
                         *                     *    *           *   *                         *          *   * *

Chicken        ctcccctcacaaacaatcgttatttatattgttaattagcaaacacaaaacccgccttctaccactataaa                              1227
Peking Duck    ctgaagctgccaaga--cggcacacgaacat---gcctgcggacaaaagacttagt--------cc-----                              1131
Snow Goose     caataatttcatata--cttcactctcactc---accactcaactatcagctaaccgc--atcccc-----                              1177
Guillemot      caa-----acaaacaa----caaacaa-ca---aacaa-caaacaaacaacaaaca--------------a                              1121
Dunlin         caa-----acaaacaa---acaaacaaaca---aacaaacaaacaaacaaaatacc--------------c                              1072
               *        *  * *      *                     **
```

FIGURE 3.3 *(Continued)*

and 23% (III). Similarly, numbers of transitions, transversions, and indels are reduced in the central domain, although transition-to-transversion ratios appear relatively constant among domains. Large deletions account for much of the length variation between similar species. For example, in a comparison of the Peking duck and the domestic chicken, a large deletion in domain I encompasses one of the CSB-1-like repeats found in the domestic chicken, while the third domain is characterized by the absence of more than 100 bases in the AT-rich region separating the CSB-1 and the transcription promoter (Ramirez *et al.*, 1993). In addition, variation in the type and number of tandem microsatellite-like repeats occurs between species. In the dunlin, for instance, the primary type of CA repeat is CAAA (18–23 copies), whereas the turnstone shows an array of repeats such as CAACAAA (12–14 copies) and CAACAAACAAA (4–8 copies) (Wenink *et al.*, 1994). More recently, Berg *et al.* (1995) reported variation in both type and number of repeats in a study of 15 charadriiform species. Clearly, the most appropriate region for among-species studies is the central conserved domain II, where alignment and homology are the least ambiguous.

2. *Within-Species Patterns of Variability and Haplotypic Diversity*

Results from six avian control region population studies are presented in Table IV. Four of these focused on domain I, a fifth on the first two domains, and a sixth looked at domains II and III. As many as 27% of surveyed sites were found to be variable (in Adélie penguins) compared to as few as 1.2% (in knots), although sample sizes varied. In all studies transitions greatly outnumbered transversions and alignment gaps, so neither homology nor site saturation appears to be a major problem at this level. Except for the knots, all species are characterized by high haplotypic diversity regardless of whether the first or third domain was studied. Thus, in most cases the two flanking domains of the control region provide a rich source of genetic markers for population studies.

III. POPULATION STRUCTURE AND INTRASPECIFIC TAXONOMY

Thorough analysis of intraspecific sequence variation leads inevitably to consideration of the population genetic processes responsible for major phylogenetic subdivisions in gene trees, and to consideration of taxonomic recognition of these discrete clades as subspecies, phylogenetic species, or biological species (Avise *et al.*, 1987; Avise, 1989; Avise and Ball, 1991). In this section we summarize studies of population structure within species of birds as recorded in their control region sequences, and examine whether or not the gene trees reflect currently held views of intraspecific taxonomy based on other characters.

TABLE IV Within-Species Variability in the Control Region of Six Species of Birds

Species	n[a]	Domain	No. of bases	No. of haplotypes	h[b]	Transition sites	Tranversion sites	Gaps
Common chaffinch	166	II and III	598	65	0.955	60	22	2
Gray-crowned babbler	163	I	400	86	0.973	88	6	3
Lesser snow goose	81	I	178	26	0.827	21	1	0
Knot	25	I	255	7	0.449	3	0	0
Dunlin	155	I and II	608	39	0.887	38	6	2
Adélie penguin	82	I	300	63	0.982	65	13	2

[a] Number of individuals sequenced.

[b] Haplotypic diversity, where $h = (1 - \Sigma x_i^2)/(n - 1)$, x_i is the frequency of a haplotype, and n is the sample size.

A. Global Phylogeography of the Dunlin

One of the clearest examples of population structure revealed by the increased resolution of control region sequences over that afforded by other mtDNA sequences is in the dunlin (*C. alpina*). Dunlins have a circumpolar breeding range in the Holarctic, and migrate along flyways to their wintering grounds in temperate and tropical regions north of the Equator (Greenwood, 1984). Both sexes display high natal philopatry to their breeding sites, thus suggesting that dunlins might be genetically structured across their breeding range.

An initial study of 73 dunlins sampled over most of this range (but lacking material from eastern Siberia) screened for sequence variation in both a short cytochrome *b* fragment (302 bp) and two control region fragments (295 and 313 bp, respectively) located in domains I and II (Wenink *et al.*, 1993). Most variability occurred in segment I in the first domain (30 of 42 variable sites in both fragments). Of the 50 haplotypes detected in the total 910 bp of mtDNA sequence, only 8 variable sites were located in the cytochrome *b* fragment whereas more than 5 times as many variable sites (42) were found in the 2 control region segments. Haplotypic diversity was also correspondingly higher for the control region; only 10 haplotypes were defined by the cytochrome *b* segment compared to 33 for the control region segments. Thus of the total of 35 haplotypes detected by the cytochrome *b* and control region sequences, only 2 unique haplotypes were added by assaying the more slowly evolving protein-coding gene. The rate of substitution (Jukes–Cantor corrected $d = 0.034 \pm 0.017$ substitutions/site) in domain I of the control region of the dunlin is similar to the rate at synonymous sites ($d_s = 0.027 \pm 0.012$) in cytochrome *b*, suggesting that the faster evolution of this control region segment emanates from lower selective constraints rather than an elevated mutation rate. Conversely, the rate of substitution ($d = 0.006 \pm 0.005$) in the much more constrained domain II of the control region approximates that at first positions in codons ($d_s = 0.007 \pm 0.005$) in cytochrome *b*.

Phylogenetic analysis of the sequences partitioned them into five major phylogeographic groups, but clearly a larger analysis was required to confirm this pattern on a global scale. When control region sequences of 155 dunlins from 15 breeding populations (including samples from eastern Siberia and a much larger representation from Europe, Iceland, and Greenland) were subsequently analyzed, the existence of the 5 major monophyletic clades was confirmed, and no additional phylogeographic groups were found (Fig. 3.4). Only six more control region haplotypes were detected in this larger sampling of dunlins. A hierarchical analysis of molecular variance (Excoffier *et al.*, 1992) grouping the total sample into five geographic regions partitioned 76.3% of the total molecular variance among regions, 2.3% among populations within regions, and 21.4% within populations (Wenink *et al.*, 1996). Despite their high potential for dispersal, dunlins have a much more subdivided population structure than do humans, for example, where 22% of the

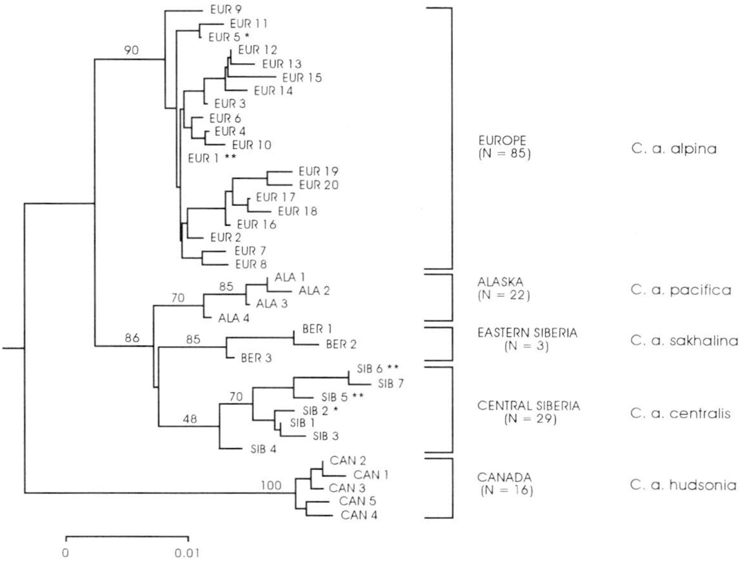

FIGURE 3.4 Neighbor-joining tree showing genealogical relationships among 39 control region haplotypes found in 155 dunlins. Phylogeographic groups and subspecies designations are shown to the right, and sample sizes are given in parentheses. Asterisks indicate haplotypes that were found in putative immigrants in other groups. The tree is rooted with a sequence from the purple sandpiper (*Calidris maritima*). (Redrawn from Wenink *et al.*, 1996. Reprinted with permission of the publisher.)

molecular variance in their control region sequences is distributed among major geographic groups (Excoffier *et al.*, 1992).

Times of divergence of the five major phylogeographic groups of dunlins were estimated on the basis of the amount of sequence divergence among them corrected for within-group variation. Estimates of divergence times of the major dunlin lineages all fall in the late Pleistocene, suggesting that they arose by fragmentation of populations in isolated tundra refugia (Wenink *et al.*, 1996).

The imprint of this historical population subdivision has most likely been maintained to the present by strong philopatry on the breeding grounds. Mixing of haplotypes indicative of gene flow was observed only in dunlins breeding in Europe and Siberia. Eight of 117 Dunlins sampled across arctic Eurasia had haplotypes that did not belong to the phylogeographic group in which they were found breeding (see Fig. 3.4), implying that they were immigrants. Cladistic measures of gene flow among Eurasian populations were derived by inferring the minimum number of historical migration events from the tree in Fig. 3.4, and translating them into values of Nm, where N is effective population size of females and m is the migration rate (Slatkin and Maddison, 1989). When values of Nm are less than about four individuals per generation, gene flow is insufficient to prevent population structure evolv-

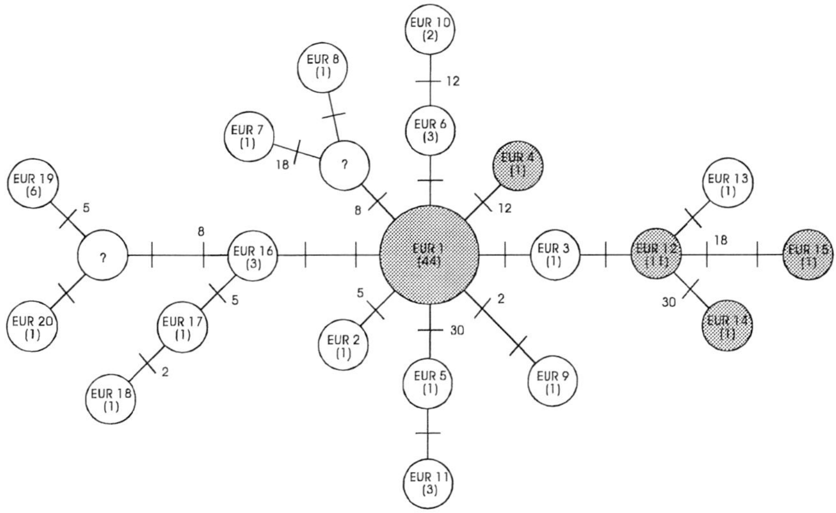

FIGURE 3.5 Parsimony network of control region haplotypes found in 85 dunlins from the European phylogeographic group. Substitutions are indicated by slashes on the network, and homoplasies are indicated with numbers as in Wenink *et al.* (1996). The frequency of each haplotype is shown in parentheses. Haplotypes found in Iceland are shaded gray.

ing by genetic drift alone (Birky *et al.*, 1983). Values of *Nm* among Eurasian locales are low for all comparisons except between northern and southern Norway, implying that gene flow is strong enough to prevent local differentiation only in Norway. Thus we might expect to see some evidence of evolving population structure in the European phylogeographic group that has been extensively sampled. Geographic localization of haplotypes supports this conjecture; for example, haplotypes EUR16–EUR20 were found only in Norway and Sweden, and haplotypes EUR12 and EUR14–15 occurred at high frequency or exclusively in Iceland (Wenink *et al.*, 1996). Corrected sequence divergence of 0.13% between the Iceland population and the rest of the European phylogeographic group suggests they diverged about 9000 years ago, and that Iceland was colonized following retreat of ice sheets about 12,000 years ago (Denton and Hugues, 1981).

The high sequence similarity among European dunlins suggests that they also have only recently been fragmented following post-Pleistocene reinvasion of new breeding grounds exposed by the melting ice sheets. The widespread distribution and abundance of haplotype EUR1 in all sampled European locales is consistent with this hypothesis. Coalescent theory predicts that the oldest haplotype in a lineage should be most widely distributed among subpopulations (Takahata, 1988), and it should also have the highest number of mutational connections to other haplotypes in a parsimony network (Crandall and Templeton, 1993) (see Fig. 3.5). EUR1 is not only central in the network but it is also basal in the neighbor-joining tree (Fig. 3.4), indicating that it is ancestral to the other European haplotypes.

The five major phylogeographic groups in the control region gene tree correspond with five subspecies defined on morphological criteria or geographic separation (see Fig. 3.4). Interestingly, two currently disputed subspecies (*Calidris alpina hudsonia* in central Canada, and *C. a. centralis* in central Siberia) are supported in the tree, whereas three others (*C. a. arctica* in Greenland, *C. a. articola* in northern Alaska, and *C. a. schinzii* in the Baltic region of Europe) are not represented by phylogenetic discontinuities. Caution is warranted in basing taxonomic conclusions solely on one gene tree, as strong selection on phenotypic characters might easily precede monophyly of haplotypes in newly isolated populations. Conversely, the previous lumping of the populations breeding in central Canada and in southern Alaska under *C. a. pacifica* on the basis of their long bills (Greenwood, 1986) is clearly invalidated by the major genetic differences between them.

B. Bottlenecking and Recent Population Expansion in Knots

A study of the control region sequences of 25 knots sampled from 10 populations representing 4 of the 5 subspecies recognized around the world (Tomkovich, 1992; Piersma and Davidson, 1992) came to the surprising conclusion that the species is effectively globally panmictic (Baker *et al.*, 1994). Furthermore, knots are depauperate genetically, as only seven haplotypes were found worldwide, all closely related and differing by one to three substitutions. Only seven variable sites were found in domain I, and none elsewhere in the control region. All substitutions were transitions. These observations most likely indicate that knots were bottlenecked down to a small population size in the late Pleistocene, and have expanded to the present broad distribution only in the last 10,000 years or so (Baker *et al.*, 1994). The alternative is that a selectively advantageous kind of mtDNA arose relatively recently and has swept through the populations, replacing older haplotypes as it did so. The control region is noncoding and sequences seem to be selectively neutral (Edwards, 1993b), but the mitochondrial genome is a single linkage group, and selection on an advantageous mutation in any gene could in theory sweep the entire haplotype toward fixation. However, the hypothesis that the low variability emanates from a severe population contraction rather than a selective sweep is also supported by an assay of 37 allozyme loci. These nuclear genes had an average heterozygosity ($\bar{H} = 0.035$) at the low end for birds (Baker and Strauch, 1988).

Under these demographic conditions, even control region sequences will not be useful in supplying genetic markers for detecting population structure and phylogenetic breaks in gene trees indicative of subspecies. Even if bottlenecking had been less severe and thus many more variable positions had been found, the average time to coalescence of haplotypes in avian mtDNA gene trees would almost certainly be much longer than the time since the population expanded (Moore, 1995). Incomplete lineage sorting would make it difficult or impossible to track the population or subspecies splits with the control region gene tree.

Another difficulty illustrated by the knot sequences is that the populations are unlikely to be in equilibrium with respect to mutation and genetic drift. Not only will this violate assumptions of methods that estimate population subdivision and gene flow, but ancestral polymorphisms distributed in the wave of population expansion will indicate that the species is globally panmictic when it actually may not be. Under such circumstances it may be possible to track population histories with even faster evolving microsatellite loci, but the number of loci required to do this could be depressingly large. For example, to track even more widely spaced speciation events in birds, Moore (1995) has estimated it may take 16 nuclear loci evolving at an appropriate rate.

C. Colonization Routes of Chaffinches in the Atlantic Islands

Chaffinches are ideal for investigating the linkage between microevolutionary processes and speciation because they provide a variety of windows through time to trace the diversification of lineages at various levels in the taxonomic hierarchy. At least two colonizations of ancestral chaffinch stock to the isolated island archipelagoes in the Atlantic from continental Europe and/or Africa seem to have occurred in the last few million years or so (Grant, 1979), an earlier one that culminated in the distinctive blue chaffinch (*Fringilla teydea*), and a later one that populated the Azores, Madeira, and the Canary Islands with well-differentiated subspecies of the common chaffinch (*Fringilla coelebs*). All island birds share a common phenotypic theme of larger body size and legs, shorter wings, and a dull orange-pink breast and blue dorsal plumage, suggesting they were derived from an expansive wave of colonization from the continents. In addition, common chaffinches appear to have expanded northward out of Africa into southern Europe within the last 100,000 years, and over the last 15,000–3000 years have colonized Europe behind the retreating ice sheets.

Studies of morphometrics (Grant, 1979; Dennison and Baker, 1991), allozymes (Baker *et al.*, 1990), and songs (Lynch and Baker, 1986, 1993, 1994) have confirmed the greater amount of population differentiation among islands in the Canaries than the Azores, consistent with reduced population sizes, lower gene flow, and enhanced genetic drift (and possibly selection) in the former archipelago. However, none of these character sets were able to resolve the origins of the island forms, nor were they able to distinguish between hypotheses of multiple invasions from several continental sources or a single invasion from Africa or Europe. Resolution of the colonization route(s) is important in testing hypotheses of convergent evolution in

TABLE V Regional Samples of Common Chaffinches Analyzed for Control Region Variation, Haplotypic Diversity (*h*), and Average Sequence Divergence (*d*)

Region	*n*	No. of haplotypes	*h*	*d*
Continent				
Southern Europe	28	16	0.86	0.46
Iberia	19	11	0.79	0.54
Northern Europe	20	11	0.82	0.46
Africa	29	13	0.83	2.3
Atlantic islands				
Azores	25	9	0.74	0.33
Madeira	10	3	0.66	1.3
Canaries	35	11	0.83	1.4

plumage coloration of island birds (assuming the Azores were colonized by European birds with olive-brown backs and brick-red breasts today represented by *Fringilla coelebs coelebs*, and the Canaries by African birds with blue backs and pale pink breasts now referred to as *Fringilla coelebs africana*). The hypothesis that the bigger blue chaffinch and the smaller common chaffinch underwent character displacement after the latter invaded the Canaries a second time also hinges on colonization routes. For example, if the source of the later invasion of the common chaffinch was from the nearest archipelago of Madeira (rather than from Africa) then character displacement is unlikely because the Madeiran birds are even smaller than their Canaries conspecifics.

We sought to reconstruct colonization routes using genetic markers obtained by sequencing 598 bp in domains II and III of the control region of 166 individuals from 19 populations of common chaffinches and an outgroup sample of blue chaffinches (Table V) (H. D. Marshall and A. J. Baker, unpublished data). The sequences are highly polymorphic; 81 variable sites were found that defined 65 haplotypes. Forty of the haplotypes were unique to individual birds, and the remaining 25 occurred in 2–21 birds. The ratio of transitions to transversions was much lower for the control region (2.73 : 1; 60 transitions, 22 transversions) than is typically found in intraspecific comparisons of coding sequences such as cytochrome *b* for passerines (20 : 1; Edwards *et al.*, 1991), indicating the faster rate of evolution in the control region.

A condensed neighbor-joining tree depicting relationships among the control region haplotypes of regional groups of chaffinches is shown in Fig. 3.6. Salient features of the tree for reconstructing the routes of colonization are as follows: (1) the continental and Atlantic island haplotypes are monophyletic and form an

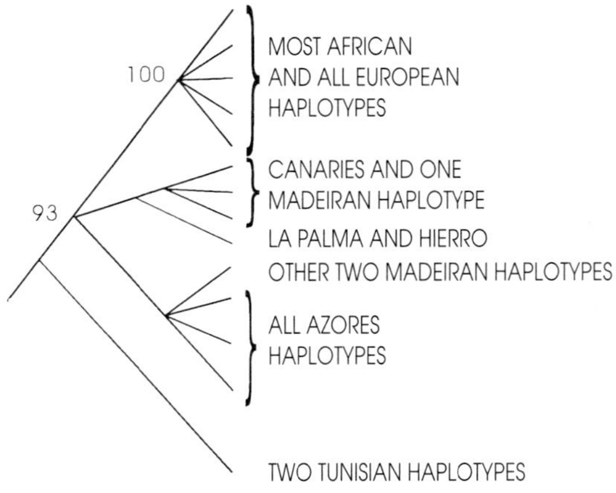

FIGURE 3.6 Neighbor-joining tree showing genealogical relationships among common chaffinch control region haplotypes from Atlantic islands, north Africa, and Europe. The tree is rooted with a sequence from the blue chaffinch.

unresolved trichotomy, consistent with only one wave of colonization; (2) the grouping of the geographically intermediate Madeiran haplotypes with both the Azores and Canaries clades suggests secondary gene flow or admixture of the island populations; (3) coalescence of the extant continental haplotypes is much more recent than colonization of the islands, judging from the much shorter branch lengths in the Africa–Europe clade, consistent with their split about 100,000 years ago; and (4) two of four haplotypes from Nefza in Tunisia are basal in the tree, suggesting an African origin for all the island and European colonists. The most parsimonious hypothesis consistent with the tree is that the Atlantic island populations are derived by a sequential wave of colonization from Africa to the Canaries, then to Madeira, and finally to the Azores. Character displacement of blue and common chaffinches following their secondary contact in the Canaries is thus plausible, but postulation of convergent phenotypic evolution of birds from different archipelagoes is unnecessary.

The control region sequences are also informative with respect to the phenotypically based intraspecific taxonomy of common chaffinches in the Atlantic islands. In the Canaries, birds from the older eastern islands of Gran Canaria, Tenerife, and Gomera are referred to as *Fringilla coelebs canariensis*. Additional taxonomic uncertainty is associated with the two populations on the younger western islands of La Palma and Hierro. Are they both referable to as *F. c. palmae* or should the Hierro population be recognized as a separate subspecies *F. c. ombiosa*? Only two kinds of sequences were found in these populations, and they cleanly distinguished the eastern and western island populations. This indicates that only two subspecies (*F. c. canariensis* and *F. c. palmae*) are warranted on the available mtDNA evidence. In addition, the recognition of separate subspecies in the Azores (*F. c. moreletti*) is well justified. Madeiran haplotypes are paraphyletic in the gene tree because of historical gene flow between the archipelagoes, but this is a classic example of when gene trees will be positively misleading about population splits and intraspecific taxonomy. The Madeiran population warrants subspecific recognition as *F. c. maderensis* on the basis of its distinctive plumage and smaller size, attributes that have presumably been maintained by selection strong enough to overcome occasional gene flow between the archipelagoes.

D. Intraspecific Variation in Canada Geese

Among birds, the Canada goose (*Branta canadensis*) shows the most extreme example of intraspecific geographic variation in body size, which along with variation in plumage characters has been the basis for splitting the species into 11 subspecies. Previous studies of relationships between subspecies using RFLP analysis of the mtDNA genome (Shields and Wilson, 1987; Van Wagner and Baker, 1990) and sequences of a 612-bp fragment of the cytochrome *b* gene (Quinn *et al.*, 1991) resolved the haplotypes into two distinct clades corresponding to large-bodied and small-bodied subspecies. However, the cytochrome *b* sequences were identical for each of the subspecies (*Branta canadensis maxima*, *B. c. moffitti*, *B. c. occidentalis*, and *B. c. fulva*) assayed by Quinn *et al.* (1991) in the big-bodied clade, despite obvious phenotypic differences between these taxa. This is consistent with reinvasion of their lower Arctic breeding grounds in the last 10,000–14,000 years or so. Nevertheless, the RFLP approach was successful in distinguishing most subspecies (Van Wagner and Baker, 1990), so we decided to sequence domain I of the control region in Canada geese to see if we could locate genetic markers for this purpose (A. J. Baker and G. F. Shields, unpublished data). Fifty-five variable sites were located in a 396-bp fragment corresponding to region I in the snow goose. As for both previous approaches, the control region sequences recovered two distinctive clades corresponding to big-bodied and small-bodied subspecies (Fig. 3.7). The increased resolution of the control region over the cytochrome *b* sequences can be seen within each of these clades where all subspecies except *B. c. moffitti* and *B. c. interior* are distinguished. However, these results are based on sample sizes of one or two birds per taxon, and we need to analyze larger samples to check the magnitude of within-species polymorphism and its possible effects on the topology of the tree (see Edwards, Chapter 9 in this volume).

Another feature of the control region tree is that the two sequences of the small-bodied *B. c. taverneri* are paraphyletic, one specimen from the state of Washington having an mtDNA sequence typical of big-bodied subspecies (possibly by hybridizing with *B. c. parvipes*) and the other possessing a haplotype typical of small-bodied subspecies. Misidentification of subspecies is also another possibility for this result. Assays of allozymes in Canada geese support the hypothesis of hybridization between subspecies in the large-bodied and small-bodied groups (Van Wagner and

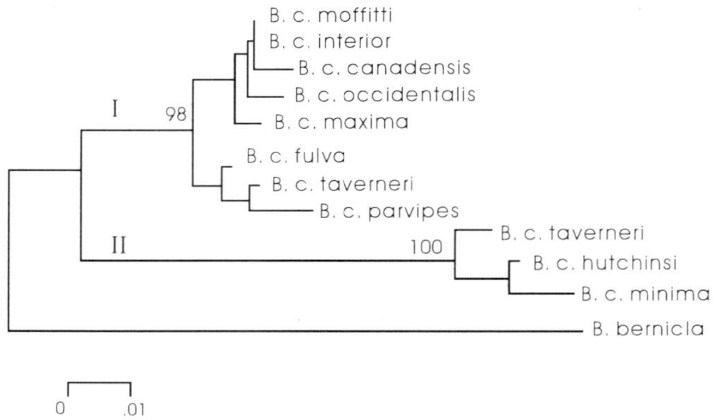

FIGURE 3.7 Neighbor-joining tree showing genealogical relationships among control region haplotypes found in 10 subspecies of Canada goose. The tree is rooted with a sequence from the Brent goose (*B. bernicla*). The big-bodied subspecies occur in clade I and the small-bodied subspecies in clade II. Bootstrap support for these two clades is indicated at their bases.

Baker, 1986, 1990), leading to male-biased gene flow as males mate with females in mixed winter flocks and return to breed at the natal area of their mates. This preserves mtDNA population structure and the inferential value of matriarchal gene trees because of haploid transmission of this genome through females, but scrambles nuclear genes and makes them unreliable for inferring taxonomic relationships.

E. Phylogeography of Gray-Crowned Babblers

Gray-crowned babblers are cooperative breeders that are widely distributed in Australia and southern Papua New Guinea. The Carpentarian biogeographic barrier in northwest Queensland marks the boundary between two subspecies, an eastern paler-breasted *Pomatostomus temporalis temporalis*, and a western *P. t. rubeculus* with a reddish-brown breast and darker upperparts (Simpson and Day, 1984). To investigate population structure in this highly social species, a 400-bp segment in domain I of the control region of 163 gray-crowned babblers sampled from 12 populations across Australia and Papua New Guinea was sequenced by Edwards (1993a). Of the 400 sites, 96 were variable across all 163 birds. Eighty-six haplotypes were detected, 44 of which occurred in the 69 *P. t. temporalis* and 42 in the 94 *P. t. rubeculus*. Haplotypes of each subspecies fall into two distinct clades in the gene tree (see Fig. 9.6 in Edwards, Chapter 9 in this volume) constructed from these sequences, corroborating intraspecific taxonomy based on phenotypic characters. Sequence divergence within subspecies was pronounced too; the maximum value was 8% between the most divergent haplotypes in *P. t. temporalis*. Because of the large amount of sequence divergence within each subspecies, it is important to correct for this source of variability when using average sequence diversity to date their divergence, as pointed out by Edwards and Kot (1995). This places the subspecies split between 275,000 and 425,000 years ago.

Cladistic analysis of gene flow based on the control region tree revealed two major surprises that were not evident from short-distance movements of adults (usually only a few territories, with an observed maximum of 25 km). First, the genealogies for each subspecies suggest that long-distance gene flow occurs frequently among populations more than 1000 km apart. Second, sequences from unrelated migrants were found in 10% of social groups. Despite this, the fraction of sequence diversity that was apportioned among populations within each subspecies was quite large (F_{ST} = 0.53 and 0.66 for *P. t. temporalis* and *P. t. rubeculus*, respectively). Hence there appears to be considerable opportunity for kin selection within each subspecies as gene flow on this scale is probably insufficient to counteract it.

F. Recent Mixing of Lineages in Adélie Penguins and in Snow Geese

Adélie penguins (*Pygoscelis adeliae*) breed colonially in suitable habitat in Antarctica, and mark–recapture studies indicate that both parents are highly philopatric to their natal areas (Ainley *et al.*, 1983). Most of the 240,000 ± 24,000 Adélies on Ross Island breed at one of three major colonies (Cape Bird, Cape Royds, and Cape Crozier) in geographically separated locales (Taylor *et al.*, 1990). The glacial history of the Ross Sea region suggests that the colonies on Ross Island have been accessible to breeding birds from about 6500 to 10,000 years ago (Young, 1981), and thus they must have colonized these breeding sites from colonies that existed elsewhere during the Pleistocene.

To investigate whether population structure has developed in Ross Island in the last 10,000 years, Monehan (1994) sequenced a hypervariable region of 300 bp in domain I of the control region of 81 Adélies from the three colonies. Seventy-six variable positions were found distributed across the sequences in 3 serial arrays and single-base substitutions. The ratio of transitions to transversions was 5 : 1 (65 transitions, 13 transversions), and there were two insertions. Haplotypic diversity was very high; 67 haplotypes were found in the 81 sampled Adélies (Table VI). The most striking feature of the sequences was that they can be divided into two distinctive types differing on average by 5.1% (Kimura two-parameter corrected distance). This basic dichotomy was recovered in a neighbor-joining tree, but haplotypes from the three colonies were scattered throughout these two clades.

A frequency or "mismatch" distribution of pairwise sequence divergences among haplotypes naturally generated a bimodal curve, one peak near the ordinate representing the close similarity among haplotypes within each clade, and the other peak reflecting the more distant comparisons among haplotypes from different clades (Fig. 3.8). One interpretation of this bimodality and high haplotypic diversity is that the Ross Island population (and its ancestral precursors) has maintained a large

TABLE VI Control Region Variation in Adélie Penguins from Three Colonies in Ross Island[a]

Region	*N*	No. of haplotypes	Diversity	Sequence divergence (%)
Cape Bird	31	28	1.024	3.88
Cape Crozier	27	23	1.014	3.11
Cape Royds	23	21	1.036	2.55
Ross Island	81	67	0.982	3.86

[a]Modified From Monehan (1994).

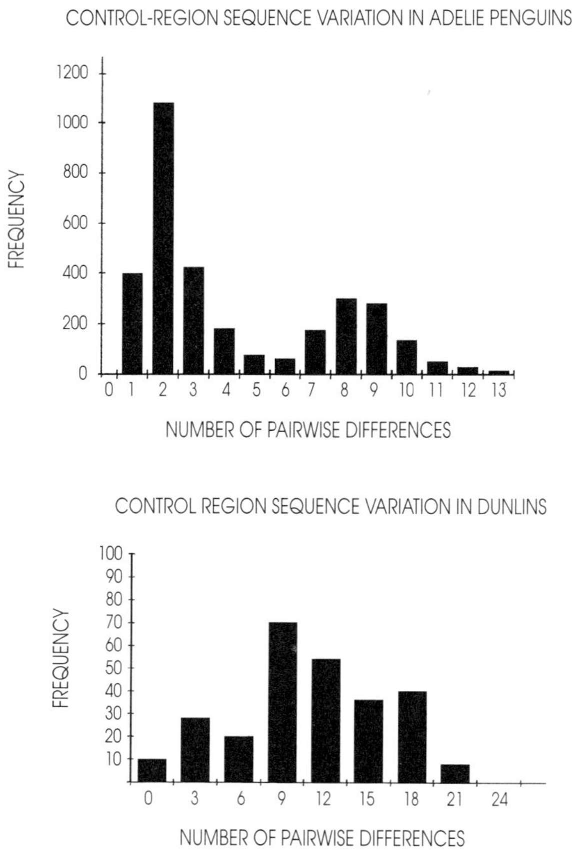

FIGURE 3.8 Mismatch distributions of pairwise comparisons among control region haplotypes found in Adélie penguins and in a sample of dunlins from Europe.

and constant size over its evolutionary history (Majoram and Donelly, 1994; Rogers and Harpending, 1992). This would require that the number of breeding females has been roughly equal to the current census number of breeding females [$N_{(c)}f$ = 240,000] for the history of the population, and this seems unlikely on purely demographic grounds. Bimodal mismatch distributions can also arise under different demographic conditions than constant population size, further weakening this line of reasoning (Slatkin and Hudson, 1991; Majoram and Donnelly, 1994). The best explanation for the widespread distribution of haplotypes from both clades across Ross Island is that of recent mixing of two populations that differentiated in allopatry during the Pleistocene. The unimodal mismatch distribution for 25 European dunlins (*C. a. alpina*) is, by contrast, indicative of an expanding population (Rogers and Harpending, 1992; Rogers, 1995).

Recent mixing of two populations that differentiated in separate Pleistocene refugia is also illustrated by an analysis of control region sequence variation in domain I of the lesser snow goose in North America (Quinn, 1992). Evidence for a vicariant origin of the populations comes from the presence of two types of sequences (clades I and II) differing by 6.7%, and which are distributed across the entire subspecies range. The sequences were not correlated with the color phase of the birds, or with sex or sampling locale. Although these divergent haplotypes could again have been maintained in a single panmictic population that has been large throughout its history, the most compelling argument against this interpretation is that the refugial populations almost certainly would have encountered bottlenecks of reduced size (Quinn, 1992).

Using a molecular clock of 20.8% divergence per million years for control region I, the two allopatric populations split about 350,000 years ago. Birds from one of these populations carrying clade II haplotypes are thought to have spread across the continent, separating into two more populations about 110,000 years ago that are today part of the eastern and western populations. Then birds with a different type of mtDNA (clade I) from the other refugium more recently spread across the continent and mixed with the clade II birds. This study is also instructive in warning against the potential pitfall of mixing nuclear and mitochondrial control region sequences; waterfowl are particularly problematic in this regard as many species we and others (Quinn and White, 1987; Quinn, 1992; K. Scribner, personal communication) have studied have translocated copies of most mitochondrial genes and the control region in the nucleus.

IV. HIGHER LEVEL SYSTEMATICS

There are no published applications of the use of control region sequences in the molecular systematics of birds at and above the species level. To investigate the utility of such fast-evolving sequences among closely related species we (Marshall and Baker, 1997) sequenced a 609-bp segment of the control region of the subspe-

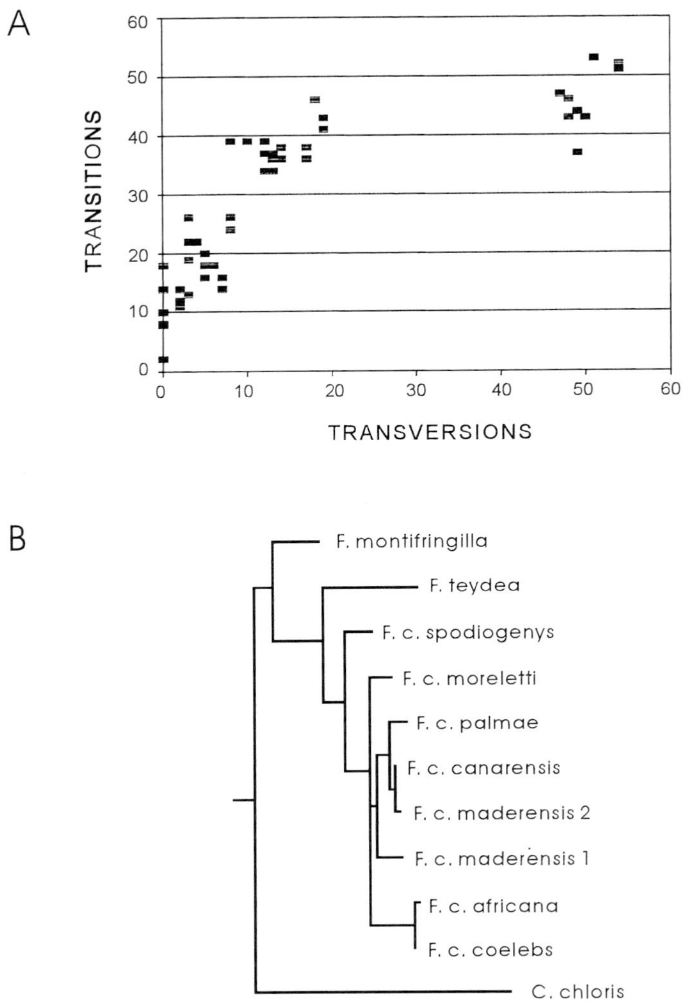

FIGURE 3.9 (A) Plot of transitions versus transversions for pairwise comparisons of control region sequences of the finches shown in the maximum likelihood tree (B). The points to the right are comparisons between *Fringilla* species and *Carduelis chloris*. (B) Maximum likelihood tree among the finch species, computed using empirical base frequencies and a transition:tranversion ratio of 1.64 computed in Phylip 3.41 (Felsenstein, 1991).

cies of common chaffinch (*F. coelebs*), and the other two species in the genus, the blue chaffinch (*F. teydea*) and the brambling (*Fringilla montifringilla*). We also sequenced the same segment from the greenfinch (*Carduelis chloris*) as an outgroup. The sequenced segment encompasses most of domains II and III of the control region.

As might be expected in comparisons between two genera, the *C. chloris* sequence differs in length from that of the species of *Fringilla*, as the former possesses three 1-base deletions, three single-base insertions, two 2-base insertions, and one 6-base insertion relative to the latter. There are only two alignment gaps within *Fringilla*. Thus in general there is no real problem associated with alignment of the segment we sequenced. The number of variable sites increases with higher levels in the taxonomic hierarchy; there are 57 variable sites within *F. coelebs* and its subspecies, 98 within *Fringilla*, and 153 when *C. chloris* is included. A plot of transitions versus transversions for the finch sequences indicates that within the genus *Fringilla* the relationship is linear, but it quickly plateaus as saturation occurs in comparisons between *Fringilla* and *Carduelis* (Fig. 3.9A). Kimura two-parameter corrected distances (×100) range from 1.4–11.0 within *Fringilla* to 15.6–19.6 between the two genera.

Alternative use of *C. chloris*, *F. montifringilla*, or *F. teydea* as an outgroup results in some instabilities of the genealogical relationships among the subspecies of *F. coelebs* in the ingroup with the neighbor-joining method, but not with maximum likelihood (Fig. 3.9B). This problem does not arise from differences in base composition of the sequences (see Hasegawa and Hashimoto, 1993; Lockhart *et al.*, 1994) because they all have similar compositions that are AT rich in the L strand. We conclude that control region sequences are likely to be most useful in higher systematics in determining phylogenetic relationships among closely related species, but even then careful consideration will need to be given to choice of outgroups and tree-building methods.

ACKNOWLEDGMENTS

We thank Oliver Haddrath and Cathy Ayley for assistance in preparing this chapter, and Paul Wenink and Scott Edwards for constructive comments. We gratefully acknowledge Monica Kidd and Vicki Friesen for generously supplying the black guillemot sequence.

REFERENCES

Ainley, D. G., Le Reseche, R. E., and Sladen, W. J. (1983). "Breeding Biology of the Adélie Penguin." Univ. California Press, Berkeley.

Anderson, S., DeBruijn, M. H. L., Coulson, A. R., Eperon, I. C., Sanger, F., and Young, I. G. (1982). Complete sequence of bovine mitochondrial DNA: Conserved features of the mammalian mitochondrial genome. *J. Mol. Biol.* **156**, 683–717.

Aquadro, C. F., and Greenberg, B. D. (1983). Human mitochondrial DNA variation and evolution: Analysis of nucleotide sequences from seven individuals. *Genetics* **103**, 287–312.

Árnason, Ú., Gullberg, A., and Widegren, B. (1993). Cetacean mitochondrial DNA control region: Sequences of all extant baleen whales and two sperm whale species. *Mol. Biol. Evol.* **10**, 960–970.

Avise, J. C. (1989). A role for molecular genetics in the recognition and conservation of an endangered species. *Trends Ecol. Evol.* **4**, 279–281.

Avise, J. C., and Ball, R. M. (1991). Mitochondrial DNA and avian microevolution. *Acta XX Congr. Int. Ornithol.* **1**, 514–524.

Avise, J. C., Arnold, J., Ball, R. M., Bermingham, E., Lamb, T., Neigel, J. E., Reeb, C. A., and Saunders, N. C. (1987). Intraspecific phylogeography: The mitochondrial DNA bridge between population genetics and systematics. *Annu. Rev. Ecol. Syst.* **18**, 489–522.

Avise, J. C., Alisauskas, R. T., Nelson, W. S., and Ankney, C. D. (1992). Matriarchal population genetic structure in an avian species with female natal philopatry. *Evolution* **46**, 1084–1096.

Baker, A. J., and Strauch, J. G. (1988). Genetic variation and differentiation in shorebirds. *Acta XX Congr. Int. Ornithol.* **2**, 1639–1645.

Baker, A. J., Dennison, M. D., Lynch, A., and Le Grand, G. (1990). Genetic divergence in peripherally isolated populations of chaffinches in the Atlantic islands. *Evolution* **44**, 981–999.

Baker, A. J., Piersma, T., and Rosenmeier, L. (1994). Unraveling the intraspecific phylogeography of knots *Calidris canutus*: A progress report on the search for genetic markers. *J. Ornithol.* **135**, 599–608.

Berg, T., Moum, T., and Johansen, S. (1995). Variable numbers of tandem repeats make birds of the order Ciconiiformes heteroplasmic in their mitochondrial genomes. *Curr. Genet.* **27**, 257–262.

Birky, C. W., Maruyama, T., and Fuerst, P. (1983). An approach to population and evolutionary genetic theory for genes in mitochondria and chloroplasts, and some results. *Genetics* **103**, 513–527.

Brown, G. G., Gadeleta, G., Pepe, G., Saccone, C., and Sbisá, E. (1986). Structural conservation and variation in the D-loop-containing region of vertebrate mitochondrial DNA. *J. Mol. Biol.* **192**, 503–511.

Cann, R. L., Brown, W. M., and Wilson, A. C. (1984). Polymorphic sites and the mechanism of evolution in human mitochondrial DNA. *Genetics* **106**, 479–499.

Chang, D., and Clayton, D. A. (1985). Priming of human mitochondrial DNA replication occurs at the light-strand promoter. *Proc. Natl. Acad. Sci. U.S.A.* **82**, 351–355.

Clayton, D. A. (1991). Replication and transcription of vertebrate mitochondrial DNA. *Annu. Rev. Cell. Biol.* **7**, 453–478.

Crandall, K. A., and Templeton, A. R. (1993). Empirical tests of some predictions from coalescent theory with applications to intraspecific phylogeny reconstruction. *Genetics* **134**, 959–969.

Dennison, M. D., and Baker, A. J. (1991). Morphometric variability in continental and Atlantic island populations of chaffinches (*Fringilla coelebs*). *Evolution* **45**, 29–39.

Denton, G. H., and Hugues, T. J. (1981). "The Last Great Ice Sheets." Wiley, Toronto.

Desjardins, P., and Morais, R. (1990). Sequence and gene organization of the chicken mitochondrial genome. A novel gene order in higher vertebrates. *J. Mol. Biol* **12**, 599–634.

Desjardins, P., and Morais, R. (1991). Nucleotide sequence and evolution of coding and noncoding regions of a quail mitochondrial genome. *J. Mol. Evol.* **32**, 153–161.

Desjardins, P., Ramirez, V., and Morais, R. (1990). Gene organization of the Peking duck mitochondrial genome. *Curr. Genet.* **17**, 515–518.

Doda, J. N., Wright, C. T., and Clayton, D. A. (1981). Elongation of displacement-loop strands in human and mouse mitochondrial DNA is arrested near specific template sequences. *Proc. Natl. Acad. Sci. U.S.A.* **78**, 6116–6120.

Dunon-Bluteau, D. C., and Brun, G. M. (1987). Mapping at the nucleotide level of *Xenopus laevis* mitochondrial D-loop H-strand: Secondary structural features of the 3′ region. *Biochem. Int.* **14**, 643–657.

Edwards, S. V. (1993a). Long-distance gene flow in a cooperative breeder suggested by genealogies of mitochondrial DNA sequences. *Proc. R. Soc. Lond. B* **252**, 177–185.

Edwards, S. V. (1993b). Mitochondrial gene genealogy and gene flow among island and mainland populations of a sedentary songbird, the grey-crowned babbler. *Evolution* **47**, 1118–1137.

Edwards, S. V., and Kot, M. (1995). Comparative methods at the species level: Geographic variation in morphology and group size in grey-crowned babblers (*Pomatostomus temporalis*). *Evolution* **49**, 1134–1146.

Edwards, S. V., Arctander, P. R., and Wilson, A. C. (1991). Mitochondrial resolution of a deep branch in the genealogical tree for perching birds. *Proc. R. Soc. Lond. B* **243**, 99–107.

Excoffier, L., Smouse, P. E., and Quattro, J. M. (1992). Analysis of molecular variance inferred from metric distances among DNA haplotypes: Application to human mitochondrial DNA restriction data. *Genetics* **131**, 479–491.

Fauron, C., M.-R., and Wolstenholme, D. R. (1976). Structural heterogeneity of mitochondrial DNA molecules within the genus *Drosophila*. *Proc. Natl. Acad. Sci. U.S.A.* **73**, 3623–3627.

Felsenstein, J. (1991). "Phylogeny Inference Package," version 3.41. Dept. Genetics, Univ. Washington, Seattle, Washington.

Gibbs, H. L., Brooks, M. L., and Davies, N. B. (1996). Analysis of genetic differentiation of host races of the common cuckoo *Cuculis canoris* using mitochodrial and microsakellike DNA variation. *Proc. Roy. Soc. London Series B. Biol. Sci.* **263**, 89–96.

Grant, P. R. (1979). Evolution of the chaffinch, *Fringilla coelebs*, on the Atlantic islands. *Biol. J. Linn. Soc.* **11**, 301–332.

Greenwood, J. G. (1984). Migration of dunlin *Calidris alpina*: A worldwide overview. *Ring. Migr.* **5**, 35–39.

Greenwood, J. G. (1986). Geographical variation and taxonomy of the dunlin *Calidris alpina. Bull. Br. Orn. Club* **106**, 43–56.

Hasegawa, M., and Hashimoto, T. (1993). Ribosomal RNA trees misleading? *Nature (London)* **361**, 23.

Higgins, D. G., Bleasby, A. J., and Fuchs, R. (1992). CLUSTAL V: Improved software for multiple sequence alignment. *CABIOS* **8**, 189–191.

Hoelzel, A. R., Hancock, J. M., Dover, G. A. (1991). Evolution of the cetacean mitochondrial D-loop region. *Mol. Biol. Evol.* **8**, 475–493.

L'Abbé, D., Duhaime, J-F., Lang, B. F., and Morais, R. (1991). The transcription of DNA in chicken mitochondria initiates from one major bidirectional promoter. *J. Biol. Chem.* **266**, 10844–10850.

Lockhart, P. J., Steel, M. A., Hendy, M. D., and Penny, D. (1994). Recovering evolutionary trees under a more realistic model of sequence evolution. *Mol. Biol. Evol.* **11**, 605–612.

Lynch, A., and Baker, A. J. (1986). Congruence of cultural and morphometric evolution in Atlantic island populations of chaffinches. *Can. J. Zool.* **64**, 1576–1580.

Lynch, A., and Baker, A. J. (1993). A population memetics approach to cultural evolution in chaffinch song: Meme diversity within populations. *Am. Nat.* **141**, 597–620.

Lynch, A., and Baker, A. J. (1994). A population memetics approach to cultural evolution in chaffinch song: Differentiation among populations. *Evolution* **48**, 351–359.

Majoram, P., and Donelly, P. (1994). Pairwise comparisons of mitochondrial DNA sequences in subdivided populations and implications for early human evolution. *Genetics* **136**, 673–683.

Marshall, H. D., and Baker, A. J. (1997). Structural variation and conservation in the mitochondrial contion region of fringilline finches (*Fringilla* spp.) and the greenfinch (*Cardnelis chloris*) *Mol. Biol. Evol.*, in press.

Mignotte, B., Dunon-Bluteau, D., Reiss, C., and Mounolou, J. C. (1987). Sequence deduced physical properties in the D-loop region common to five vertebrate mitochondrial DNAs. *J. Theor. Biol.* **124**, 57–69.

Monehan, T. M. (1994). Molecular genetic analysis of Adélie penguin populations, Ross Island, Antarctica. M.Sc. thesis, University of Auckland, Auckland, New Zealand.

Moore, W. S. (1995). Inferring phylogenies from mtDNA variation: Mitochondrial-gene trees versus nuclear-gene trees. *Evolution* **49**, 718–726.

Moum, T., and Johansen, S. (1992). The mitochondrial NADH dehydrogenase subunit 6 (ND6) gene in murres: Relevance to phylogenetic and population studies among birds. *Genome* **35**, 903–906.

Moum, T., Johansen, S., Erikstad, K. E., and Piatt, J. F. (1994). Phylogeny and evolution of the auks (subfamily Alcinae) based on mitochondrial DNA sequences. *Proc. Natl. Acad. Sci. U.S.A.* **91**, 7912–7916.

Piersma, T., and Davidson, N. C. (1992).The migrations and annual cycles of five subspecies of knots in perspective. *Wader Study Group Bull. Suppl.* **64**, 187–197.

Quinn, T. W. (1992). The genetic legacy of Mother Goose—phylogeographic patterns of lesser snow goose *Chen caerulescens caerulescens* maternal lineages. *Mol. Ecol.* **1**, 105–117.

Quinn, T. W., and Mindell, D. P. (1996). Mitochondrial gene order adjacent to the control region in crocodile, turtle, and tuatara. *Mol. Phylogenet. Evol.* **5**, 344–351.

Quinn, T. W., and White, B. N. (1987). Analysis of DNA sequence variation. *In* "Avian Genetics: A Population and Ecological Approach" (F. Cooke and P. A. Buckley, Eds.), pp. 163–198. Academic Press, London.

Quinn, T. W., and Wilson, A. C. (1993). Sequence evolution in and around the mitochondrial control region in birds. *J. Mol. Evol.* **37**, 417–425.

Quinn, T. W., Shields, G. F., and Wilson, A. C. (1991). Affinities of the Hawaiian goose based on two types of mitochondrial DNA data. *Auk* **108**, 585–593.

Ramirez, V., Savoie, P., and Morais, R. (1993). Molecular characterization and evolution of a duck mitochondrial genome. *J. Mol. Evol.* **37**, 296–310.

Roe, B. A., Ma, D. P., Wilson, R. K., and Wang, J. F. H. (1985). The complete nucleotide sequence of the *Xenopus laevis* mitochondrial genome. *J. Biol. Chem.* **260**, 9759–9774.

Rogers, A. R. (1995) Genetic evidence for a Pleistocene population explosion. *Evolution* **49**, 608–615.

Rogers, A. R., and Harpending, H. C. (1992). Population growth makes waves in the distribution of pairwise genetic differences. *Mol. Biol. Evol.* **9**, 552–569.

Saccone, C, Attimonelli, M., and Sbisá, E. (1987). Structural elements highly preserved during the evolution of the D-loop-containing region in vertebrate mitochondrial DNA. *J. Mol. Evol.* **26**, 205–211.

Saccone, C., Pesole, G., and Sbisá, E. (1991). The main regulatory region of mammalian mitochondrial DNA: Structure–function model and evolutionary pattern. *J. Mol. Evol.* **33**, 83–91.

Shields, G. F., and Wilson, A. C. (1987). Subspecies of the Canada goose (*Branta canadensis*) have distinct mitochondrial DNAs. *Evolution* **41**, 662–666.

Simpson, G., and Day, N. (1984). "The Birds of Australia." Lloyd O'Neil, Victoria, Australia.

Slatkin, M., and Hudson, R. R. (1991). Pairwise comparisons of mitochondrial DNA sequences in stable and exponentially growing populations. *Genetics* **129**, 555–562.

Slatkin, M., and Maddison, W. (1989). A cladistic measure of gene flow inferred from the phylogenies of alleles. *Genetics* **123**, 603–613.

Southern, Š. O., Southern, P. J., and Dizon, A. E. (1988). Molecular characterization of a cloned mitochondrial genome. *J. Mol. Evol.* **28**, 32–42.

Takahata, N. (1988). The coalescent in two partially isolated diffusion populations. *Genet. Res.* **52**, 213–222.

Taylor, M. F. J., McKechnie, S. W., Pierce, N., and Kreitman, M. (1993). The lepidopteran mitochondrial control region: Structure and evolution. *Mol. Biol. Evol.* **10**, 1259–1272.

Taylor, R. H., Wilson, P. R., and Thomas, B. W. (1990). Status and trends of Adélie penguin populations in the Ross Sea region. *Polar Rec.* **26**, 293–304.

Tomkovich, P. S. (1992). An analysis of the geographic variation in knots *Calidris canutus* based on museum skins. *Wader Study Group Bull. Suppl.* **64**, 17–23.

Upholt, W. B., and Dawid, I. B. (1977). Mapping of mitochondrial DNA of individual sheep and goats: Rapid evolution in the D-loop region. *Cell* **11**, 571–583.

Van Wagner, C. E., and Baker, A. J. (1986). Genetic differentiation in populations of Canada geese (*Branta canadensis*). *Can. J. Zool.* **64**, 940–947.

Van Wagner, C. E., and Baker, A. J. (1990). Association between mitochondrial DNA and morphological evolution in Canda geese. *J. Mol. Evol.* **31**, 373–382.

Vigilant, L., Stoneking, M., Harpending, H. C., Hawkes, K., and Wilson, A. C. (1991). African populations and the evolution of human mitochondrial DNA. *Science* **253**, 1503–1507.

Walberg, M. W., and Clayton, D. A. (1981). Sequence and properties of the human KB cell and mouse L cell D-loop regions of mitochondrial DNA. *Nucleic Acids Res.* **9**, 5411–5421.

Wenink, P. W., Baker, A. J., and Tilanus, M. G. J. (1993). Hypervariable control-region sequences reveal global population structuring in a long-distance migrant shorebird, the dunlin (*Calidris alpina*). *Proc. Natl. Acad. Sci. U.S.A.* **90**, 94–98.

Wenink, P. W., Baker, A. J., and Tilanus, M. G. J. (1994). Mitochondrial control region sequences in two shorebird species, the turnstone and the dunlin, and their utility in population genetic studies. *Mol. Biol. Evol.* **11**, 22–31.

Wenink, P. W., Baker, A. J., and Tilanus, M. G. J. (1996). Global mitochondrial DNA phylogeography of holarctic breeding dunlins (*Calidris alpina*). *Evolution* **50**, 318–330.

Young, E. C. (1981). The ornithology of the Ross Sea. *J. R. Soc. N.Z.* **11**, 287–315.

Zhang, D.-X., Szymura, J. M., and Hewitt, G. M. (1995). Evolution and structural conservation of the control region of insect mitochondrial DNA. *J. Mol. Evol.* **40**, 382–391.

CHAPTER 4

The Window of Taxonomic Resolution for Phylogenies Based on Mitochondrial Cytochrome b

WILLIAM S. MOORE AND VICTOR R. DEFILIPPIS

Department of Biological Sciences, Wayne State University, Detroit, Michigan 48202

Avian Molecular Evolution and Systematics

I. INTRODUCTION

A. Avian Phylogenies Inferred from Mitochondrial DNA

Mitochondrial DNA (mtDNA) seemingly has enormous value for resolving the phylogenies of recently evolved avian taxa, and numerous phylogenetic studies of avian groups have been carried out using mtDNA variation as a source of characters. Until recently these studies were based on restriction site variation (Kessler and Avise, 1984, 1985; Avise *et al.*, 1990; Zink and Avise, 1990; Zink and Dittmann, 1991, 1993a,b; Zink *et al.*, 1991a,b; Bermingham *et al.*, 1992; Tarr and Fleischer, 1993; Zink, 1993), but this technology has been supplanted by the technology of amplifying by the polymerase chain reaction (PCR) specific genes or portions of genes that are then sequenced (Saiki *et al.*, 1988; Kocher *et al.*, 1989; Kocher and White, 1989; Kocher, 1992). The mitochondrially encoded cytochrome *b* gene (cyt *b*) has been used most often in avian phylogenetic studies based on DNA sequences (Edwards and Wilson, 1990; Edwards *et al.*, 1991; Quinn *et al.*, 1991; Lanyon, 1992, 1994; Helm-Bychowski and Cracraft, 1993; Kornegay *et al.*, 1993; Kusmierski *et al.*, 1993; Avise *et al.*, 1994a,b; Krajewski and Fetzner, 1994; Lanyon and Hall, 1994; Heidrich *et al.*, 1995; Krajewski and King, 1995), although the 12S ribosome coding gene has been used also (Cooper *et al.*, 1992; Cooper, 1994; Mindell *et al.*, 1996).

Mitochondrial DNA is attractive for phylogenetic studies because of its conservative evolution with regard to gene order and, in its protein-coding genes, conservative amino acid replacement and occurrence of insertions and deletions (Brown, 1985; Desjardins and Morais, 1990, 1991), contrasted with a high rate of synonymous substitutions (Brown *et al.*, 1982; Thomas and Beckenbach, 1989; Edwards *et al.*, 1991; Irwin *et al.*, 1991). Conservation of gene order and amino acid codons makes it easy to align sequences (establish homology) among species and to design PCR primers for a diversity of species (Kocher *et al.*, 1989; Edwards *et al.*, 1991), whereas the rapid rate of silent substitution increases the probability that the molecule will contain synapomorphies that reveal recent periods of shared ancestry. The latter point is particularly important to avian systematics because birds characteristically exhibit low levels of divergence in both nuclear and mitochondrial genes. Avian taxa appear to be "shifted down" approximately one taxonomic level relative to other vertebrate taxa (Kessler and Avise, 1985); e.g., divergences between species within a mammalian genus are approximately the same as those between species in distinct avian genera (Avise and Aquadro, 1982; Shields and Helm-Bychowski, 1988).

To refine and enhance methods of phylogenetic inference on the basis of genotypic characters, it is important to distinguish the problem of resolving the gene tree from the problem of whether the gene tree, once resolved, is congruent with the species tree. Considering the gene-tree–species-tree problem first, the vertebrate mitochondrial genome comprises 13 protein-coding genes, 22 tRNA genes, 2 rRNA genes, and a control region but is inherited as a single linkage group exclusively through the female germ line. This means that only one independent gene tree can be estimated from the mitochondrial genome, no matter how many individual genes are sequenced. This is seemingly a severe limitation on the phylogenetic utility of mtDNA because lineage sorting of ancestral polymorphisms may result in a mitochondrial haplotype tree that is incongruent with the species tree (Neigel and Avise, 1986; Nei, 1987; Pamilo and Nei, 1988; Wu, 1991; Hudson, 1992; Moore, 1995). Thus, the true topology of the species tree could be further explored only by resolving trees from genes located in distinct linkage groups in the nuclear genome. For this reason, many phylogeneticists insist that a phylogeny cannot be considered resolved until multiple independent nuclear gene trees have been resolved and compared. However, Moore (1995) has shown that the mitochondrial haplotype tree has a much higher probability of congruence with the species tree than does a single nuclear gene tree and that a substantial number (16+) of independent nuclear gene trees would need to be resolved to be as confident of the species tree inferred from the mitochondrial haplotype tree. In fact, it is probable that a correctly resolved mtDNA haplotype tree will accurately reflect the species tree unless the internodes (times) between speciation events are short (Moore, 1995). The relatively high probability of congruence of the mitochondrial haplotype tree with the species tree is a consequence of the small effective size of the population with regard to the mitochondrial genome (Moore, 1995). Thus, the mitochondrial haplotype tree could be the most powerful tool for inferring species phylogenies for avian groups, provided that there is sufficient variation in mitochondrial haplotypes among avian species to allow resolution of the haplotype tree.

B. Objectives of This Study

In this chapter, we focus on problems associated with resolving mt-haplotype trees from DNA sequence data. Specifically, we will assess the value of cyt *b* as a source of characters for inferring avian phylogenies and try to determine a window in the systematic hierarchy of birds where cyt *b* sequence should result in efficient resolution of phylogenetic relationships. We focus on cyt *b* because it has been, by far, the most prevalent source of sequence data in avian studies but also because cyt *b* has developed a "bad reputation" as a source of characters for phylogenetic studies (Hillis and Huelsenbeck, 1992; Graybeal, 1993; Avise *et al.*, 1994a,b; Meyer, 1994; Honeycutt *et al.*, 1995), a reputation that seems contradicted by several avian studies. Although it is apparent from these studies that cyt *b* fails to resolve some ancient nodes but succeeds in resolving more recent relationships, the range of levels in the systematic hierarchy where it does work is not clear. Moreover, there is a danger that a general extrapolation will be made that cyt *b* is a poor choice for all phylogenetic studies—a danger of "throwing the baby out with the bath water." By determining more accurately a range of conditions where cyt *b* does produce positive results, we hope to preclude such a generalization.

We will argue that cyt *b* not only works for resolving relatively recent evolutionary history but may be the best choice for birds because of their tendency to have low rates of genic divergence at high taxonomic levels, compared to other vertebrate groups. Our analysis is based on comparisons among four cyt *b* sequence data sets: three from published phylogenetic studies (oscines [birds of paradise, and other passerines], Helm-Bychowski and Cracraft, 1993; cranes, Krajewski and Fetzner, 1994, and Krajewski and King, 1995; barbets and toucans, Lanyon and Hall, 1994) and one previously unpublished set of sequences from woodpeckers. We will use the four data sets to examine a number of parameters that are useful for characterizing patterns and extent of nucleotide substitution between diverging sequences, and we will relate these parameters to circumstances where nodes are successfully and unsuccessfully resolved. Our goals are to better understand why cyt *b* works in some circumstances but not others, and to develop guidelines that would serve in a pilot study designed to determine whether cyt *b* would likely work well for resolving the phylogeny of a particular group, or whether one would be better off deriving data from another gene.

C. Properties of DNA Sequences That Are Ideal for Recovering Phylogenies

Knowledge of what properties of DNA sequences should lead to recovery of the phylogeny would be useful for focusing attention on these properties in a pilot study and for bolstering the credibility (or incredibility) of empirical studies by identifying ranges of the parameter space where one should (or should not) be able to resolve phylogenetic relationships successfully. Unfortunately, these ideal properties are not completely known. One finds broad (but not complete) agreement that certain properties are sufficient to allow recovery of the true phylogeny by at least some of the available tree-building algorithms, but what conditions are necessary is much more contentious. Felsenstein (1988) provides an excellent review of phylogenetic inference based on DNA sequences. Springer and Krajewski (1989) reviewed phylogenetic inference based on DNA–DNA hybridization. Although their review focuses on the validity of DNA–DNA hybridization methodology, they describe an "ideal model of genomic evolution," ideal for recovering the true phylogeny by distance methods. However, much of their discussion is germane to recovering phylogenies from DNA sequence data, particularly if one opts to use a distance method. It is not our intent to review and evaluate these complex arguments. A distillation of the arguments is that recovering the true gene tree is likely if (1) nucleotide substitutions occur fast enough to "mark" periods of shared ancestry, but not so fast that these synapomorphies are obliterated by "multiple hits" after daughter species diverge from the ancestral species (Lanyon, 1988); (2) substitution rates are the same or at least similar along diverging lineages (i.e., there is a molecular clock); (3) nucleotide composition (relative frequencies of G, A, T, and C) does not change bias during the evolution of the group (Irwin *et al.*, 1991); and (4) there is ample sequence available (Saitou and Nei, 1986; Cracraft and Helm-Bychowski, 1991; Nei, 1991). The amount of sequence required to resolve the gene tree depends on the actual structure of the true tree (Saitou and Nei, 1986; Lanyon, 1988; Nei, 1991); in general, when internodes are short, more data are required.

II. MATERIALS AND METHODS

A. Producing Sequence Data

Protocols for amplifying by PCR and sequencing the mitochondrially encoded cyt *b* gene from birds are well established (Kocher *et al.*, 1989; Meyer *et al.*, 1990; Edwards *et al.*, 1991; Krajewski and Fetzner, 1994; Lanyon and Hall, 1994). Several instances have been reported of unwitting amplification of sequences from contaminant templates (see Derr *et al.*, 1992; Helm-Bychowski and Cracraft, 1993; Hackett *et al.*, 1995, for discussion). If precautions are not taken, this is easy to do because the PCR is capable of amplifying to high concentration a DNA sequence represented by a single initial copy, and even reassembling degraded templates into whole templates, which are then amplified (Pääbo *et al.*, 1990). The woodpecker sequences reported here were obtained following procedures modified from Kocher *et al.* (1989), Meyer *et al.* (1990), and Edwards *et al.* (1991). All of the modifications were additions to the protocols intended to reduce the probability of amplification from a contaminant template. These include (1) use of aerosol-resistant pipette tips (ART; Promega, Madison, WI) in all procedures involving DNA isolation and PCR; (2) exposure of DNA isolation reaction mixes to short-wave ultraviolet (UV) light (exposure is 30 min at 7.2 joules/m^2/sec, wavelength = 254 nm) before adding the tissue; and (3) a similar exposure of PCR reaction mixes to short-wave UV light before addition of the template and *Taq* polymerase (Cimino *et al.*, 1990). Finally, we sequenced two specimens for each species. This gives considerable assurance that the sequences are indeed the target species if the replicate sequences are nearly identical or more similar to each other than to other sequences. Sequencing replicate specimens is expensive, but it is common practice to sequence both the light and heavy strands from a single specimen. However, for a phylogenetic study, sequencing one strand from each of two specimens, and accepting a low level of unresolved base pairs (bp), may be a better investment than sequencing both strands from a single specimen (see below).

The woodpecker sequences reported here are for the light strand. We amplified and sequenced 1047 bp of the 1143-bp cyt *b* gene as two overlapping fragments that were sequenced using three of the four end primers from the PCR reactions and one additional internal primer. (In the following primer descriptions, the published name of the primer is italicized and the position of the terminal base at the 3′ end is given relative to the chicken mitochondrial light strand numbering convention; Desjardins and Morais, 1990). The 5′ fragment (705 bp, excluding primers) was amplified with primers *L14841* (Kocher *et al.*, 1989; 3′ = CL14990) and *H15547* (Edwards *et al.*, 1991; 3′ = CL15696) and the 3′ fragment (755 bp) with primers *CBL15311* (A. Meyer, personal communication; 5′ GCAAGCTTCTACCAT-GAGGACAAATATC 3′, 3′ = CL15311) and *H16065* (Helm-Bychowski and Cracraft, 1993; 3′ = CL16065 in tRNA^{Pro}). Sequencing primers were *L14841* for the 5′ fragment and *CBL15311*, *H16065*, and an internal primer, *L15424* (Edwards *et al.*, 1991, 3′= CL15569), for the 3′ fragment. Two specimens were sequenced for each species, except *Melanerpes carolinus*, and the authenticity of the sequences established by comparing conspecific sequences. We have amplified a second *M. carolinus* specimen, but the sequence is not complete. We have established that the partial sequences for both fragments of the second specimen are nearly identical to the first. The specimens are tabulated in Appendix I along with voucher numbers and locale data.

Sequencing was done on an Applied Biosystems (Foster City, CA) automated sequencer (Dye Deoxy Terminator Cycle sequencing kit) in the Wayne State University Center for Molecular Medicine and Genetics Core DNA Sequencing Facility (Detroit, MI).

III. RESULTS

A. Sequences

The 11 woodpecker cyt *b* sequences span 1047 bp or 349 codons (Appendix II). Sequences were not obtained for 96 bp from the 5′ end of the gene. We obtained only 946 bp for the two *Venilornis nigriceps* species (GenBank Accession Nos. U83282–U83302).

B. Phylogenetic Analysis

It is not our goal to reanalyze the three published data sets or to analyze the woodpecker sequences in exhaustive detail, but rather to examine patterns of nucleotide substitution, over the evolutionary history of these groups, that would provide insight as to when cyt *b* is useful for inferring evolutionary history. It is more important to our analyses to have trees that are comparable rather than to have the tree that is judged best by some criterion or consensus of criteria. Thus, the four phylogenies in Fig. 4.1a–d were estimated in the same "standard" way; they are neighbor-joining trees based on the Tamura–Nei estimator of nucleotide divergence, all three codon positions, and both transitions and transversions (Tamura and Nei, 1993; Kumar *et al.*, 1993). The Tamura–Nei estimator corrects for secondary

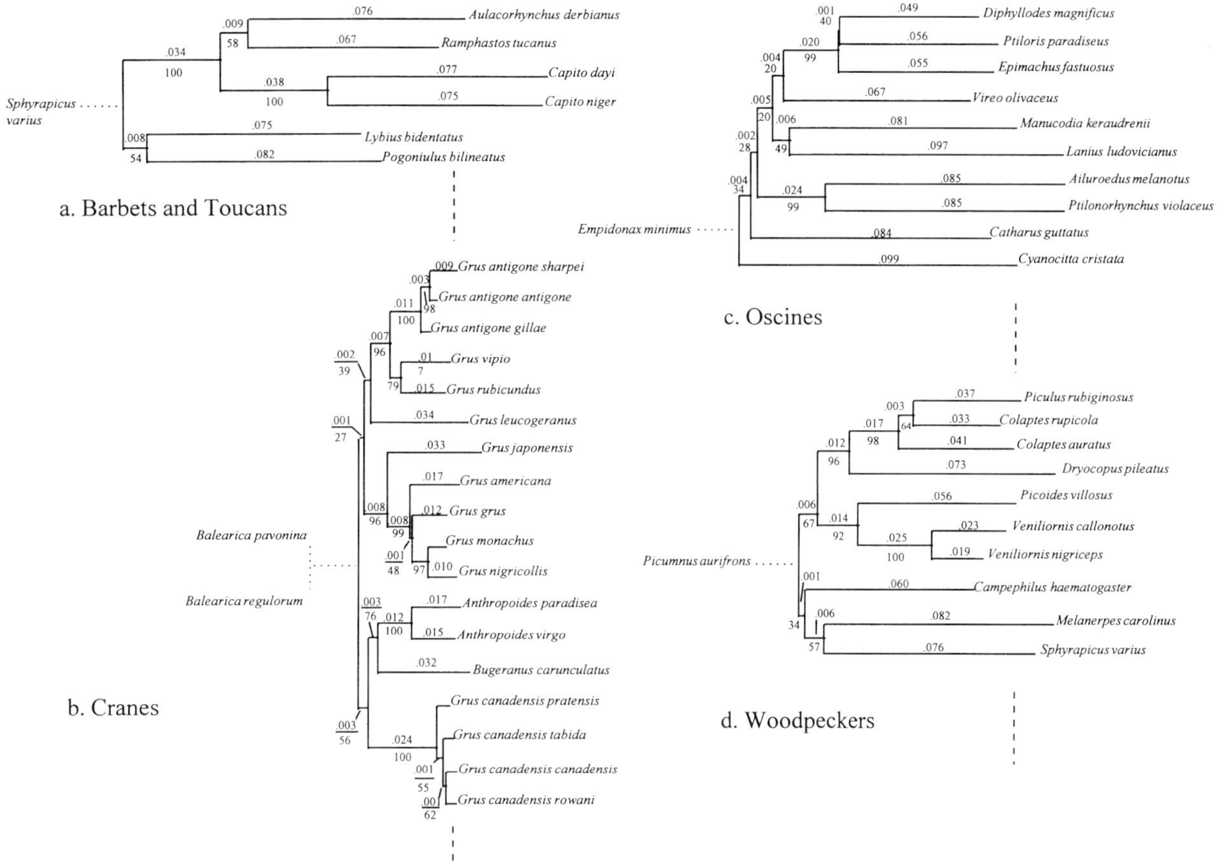

FIGURE 4.1 Four avian phylogenies based on cyt *b*. Dotted lineages to the left of each phylogeny indicate the root leading to the outgroup. Vertical dashed lines represent the present time; the average position of the branch tips for each phylogeny is set to this common time (see text). Branch lengths (substitutions/nucleotide) are above the branches; bootstrap percentages (500 replicates) supporting internodes are below the internodes.

substitutions (multiple hits); differences in substitution rates between transitions and transversions, differences in substitution rates between particular classes of transitions (A to G vs C to T) and between particular classes of transversions, and for composition bias. All of these are significant in the divergence of cyt *b* among avian lineages.

To show the relative antiquity of the taxa represented in the data sets, trees in Fig. 4.1a–d were drawn to the same scale and adjusted to a common time reference (dashed vertical line), assuming a molecular clock. This was accomplished by rooting each tree with a specified outgroup and calculating the average distance to the branch tips relative to the shortest branch for the ingroup. Each phylogeny was then positioned with this average on the common time line, which represents the present. The branch lengths are given above the branches. It is apparent that there is substantial variation in the antiquity of the clades, and overall the four phylogenies provide a good representation of bifurcations ranging from recent to ancient.

We used bootstrap proportions (*P*) in some analyses as a relative measure of the support for nodes in phylogenies (Felsenstein, 1985). Work has shown that *P* is biased such that it actually underestimates the true probability of a node in circumstances where there is strong support for the node in the data, but overestimates the true probability when the phylogenetic signal for the node is weak (Hillis and Bull, 1993; Felsenstein and Kishino, 1993). Thus, given that *P* is a biased estimator, it is biased in the direction one would hope in that it leads to conservative assertions that a node is strongly supported by the data. On the basis of simulations under a wide range of conditions, Hillis and Bull (1993) found that nodes inferred by bootstrap proportions of 70% or more actually occurred in 95% or more of the simulated phylogenies. In other words, $P = 70\%$ corresponded roughly to a 95% probability that the node was real.

1. *Oscines*

This data set (Helm-Bychowski and Cracraft, 1993) comprises 1143 bp or 381 codons, which is the entire cyt *b* gene, from 11 species. Although the phylogenetic objectives of Helm-Bychowski and Cracraft were to evaluate relationships among the birds of paradise, and their relationship to bowerbirds, the authors included several more distantly related passerines to serve as outgroup taxa. *Empidonax minimus*, a suboscine, was designated as the outgroup for purposes of rooting, whereas the ingroup comprises oscines. Overall, this is the most weakly supported of the four cyt *b* trees analyzed in this study. Helm-Bychowski and Cracraft performed parsimony analyses including both transitions and transversions (global parsimony) and transversions only. Our neighbor-joining tree (Fig. 4.1c) is based on the same data as the global parsimony tree of Helm-Bychowski and Cracraft. Their two parsimony trees differ in substantial ways and our neighbor-joining tree has various nodes in common, and in difference, with both. However, all three are identical with regard to the only two nodes that are strongly supported by bootstrap proportions: clustering of the three paradisaeinine birds of paradise (*Epimachus fastuosus*, *Ptiloris paradiseus*, and *Diphyllodes magnificus*), to the exclusion of the one manucodine bird of paradise (*Manucodia keraudrenii*), and the clustering of the two bower birds (*Ptilorhynchus violaceus* and *Ailuroedus melanotus*). The other nodes appear to be too short and/or too deep in evolutionary history to expect resolution by cyt *b* sequence (see below).

2. *Barbets and Toucans*

This data set (Lanyon and Hall, 1994) comprises 888 bp or 296 codons from seven species; a woodpecker, *Sphyrapicus varius*, was specified as the outgroup. Sequence was not obtained for 123 bp from the 5′ end of the gene and 132 bp from the 3′ end. Lanyon and Hall performed global and transversion-only parsimony analyses, each yielding the same topology. Our neighbor-joining tree (Fig. 4.1a) has a nearly identical topology to their most parsimonious tree (Fig. 3A in Lanyon and Hall, 1994), differing only in that the parsimony tree indicated that the two toucan species (*Ramphastos tucanus* and *Aulacorhynchus derbianus*) did not form a monophyletic group, whereas they do in the neighbor-joining tree. The placement of the toucans is weakly supported in both analyses. In the parsimony analysis, Lanyon and Hall (1994) noted that the toucans did join in a monophyletic group when the Old World barbets (*Pogoniulus bilineatus* and *Lybius bidentatus*) were designated as the outgroup, rather than the sapsucker (*Sphyrapicus varius*). In the neighbor-joining analysis, the branch uniting the toucans was supported by 58% of 500 bootstrap replicates. The parsimony and neighbor-joining trees are identical in supporting the monophyly of the Old World barbets, albeit weakly (54% of 500 bootstraps in the neighbor-joining tree) and strongly supporting the inference that the New World barbets (*Capito*) form a clade with the toucans and not the Old World barbets (100% of 500 bootstrap replicates support the toucan–New World barbet clade).

3. *Cranes*

This data set (Krajewski and King, 1995) comprises 1137 bp from 20 species or subspecies; 2 bp are undetermined at the 5′ end of the gene and 4 bp are undetermined at the 3′ end. (Subsequent to providing us with the sequences, Krajewski and King resolved all ambiguous bases, and increased the length of each sequence to 1143 bp for publication in their paper). Krajewski and King carried out several distance analyses involving permutations of distance estimators (Kimura two-parameter, Tamura–Nei, maximum likelihood), categories of variation (transitions plus transversions, transversions only, synonymous, nonsynonymous), and tree algorithms (least squares, neighbor-joining, maximum likelihood); they also did several parsimony analyses with a variety of weighting schemes. All of the distance analyses produced the same topology, which is presented in Fig. 4.1b. Krajewski and King's parsimony analysis, based on informative sites and equal weighting of all characters, produced a single most parsimonious tree, which differed from the distance tree topology in only one respect: joining the Americana species group (*Grus japonensis*, *G. americana*, *G. grus*, *G. monachus*, and *G. nigrocollis*) with the Anthropoides (*Anthropoides* + *Bugeranus*)–*G. canadensis* clade rather than the Antigone (*G. antigone*, *G. rubicundus*, *G. vipio*)–*G. leucogeranus* clade, as in the distance tree. However, this disparity is not statistically significant because the nodes in question are weakly supported by bootstrap replicates, 27% in the case of the Antigone–*G. leucogeranus* clade joined with the Americana species group (Fig 1b).

4. *Woodpeckers*

The neighbor-joining tree is presented in Fig. 4.1d. A branch-and-bound search found a single most parsimonious tree, which had a topology identical to that of the neighbor-joining tree. The woodpeckers comprise one of three subfamilies within the Picidae, the Picinae. The piculets (*Picumnus aurifrons*, Fig. 4.1d) comprise a second subfamily, which is thought to be the sister group of the woodpeckers (Short, 1982). We specified *P. aurifrons* as the outgroup to root the tree.

The cyt *b* tree is at odds in some respects with classic views on woodpecker relationships (Short, 1982). The cyt *b* tree strongly supports a sister group relationship between the pileated woodpecker (*Dryocopus pileatus*) and a clade comprising the genera *Colaptes* (flickers) and *Piculus*, strongly supports a sister group relationship between the genera *Picoides* and *Veniliornis*, and weakly supports a sister group relationship between sapsuckers and melanerpines. All of these relationships have been contentious historically, with at least someone advocating the relationships apparent in the cyt *b* tree at sometime (Bock and Miller, 1959; Short and Morony, 1970). These relationships are also strongly supported by a comparable analysis of 1512 bp of the mitochondrial COI gene, sequenced for the same set of specimens (except *Colaptes rupicola*). The relationship between sapsuckers and melanerpines was supported by the COI data in 96% of 500 bootstrap replicates (DeFilippis, 1995). The placement of *Campephilus* is less certain, although it does appear to have diverged early in the radiation of the woodpeckers. Finally, the moderately strong support for the pairing of *Colaptes rupicola* with *Piculus rubiginosus* rather than *Colaptes auratus* is not implausible. It is recognized that these two genera are very closely related (Short, 1982), and preliminary analyses, based on cyt *b*, involving more species of *Colaptes* and *Piculus* always result in a paraphyletic tangle of species (Moore, 1995 and unpublished).

C. Base Composition

Base composition for the three positions within codons is summarized in Table I. We point out two properties that impinge on the phylogenetic signal accumulated and retained by evolving cyt *b* sequences: (1) there is strong bias relative to a uni-

TABLE I Average Cytochrome *b* Base Composition for Avian Families[a]

Family	No. of species	A1	T1	C1	G1	A2	T2	C2	G2	A3	T3	C3	G3
Galliformes													
Phasianidae	1	26.8	22.8	29.7	20.7	20.7	39.1	27.8	12.3	34.7	10.5	51.6	3.2
Piciformes													
Picidae	11	24.13	23.53	29.38	22.95	19.23	41.17	25.67	13.91	29.83	13.07	53.10	4.00
		±1.11	±0.45	±0.73	±0.84	±0.37	±0.60	±0.67	±0.46	±2.44	±2.53	±3.74	±1.16
Lybiidae	2	24.35	24.35	29.45	21.80	19.60	39.55	28.05	12.80	36.40	12.95	49.00	1.70
		±0.78	±0.64	±0.21	±0.42	±0.00	±0.49	±0.49	±0.00	±1.27	±3.89	±0.71	±1.98
Ramphastidae	4	24.05	24.48	28.85	22.60	19.25	39.78	27.90	13.05	28.85	17.80	49.65	3.65
		±0.42	±0.71	±0.71	±0.49	±0.42	±0.07	±0.92	±0.64	±2.97	±0.78	±0.49	±1.70
Gruiformes													
Gruidae	20	26.39	22.21	30.46	21.00	20.35	39.59	27.02	13.08	38.00	12.73	46.55	2.71
		±0.40	±0.61	±0.60	±0.33	±0.14	±0.28	±0.31	±0.10	±1.01	±1.49	±1.36	±0.79
Passeriformes													
Paradisaeidae	2	26.33	21.83	29.40	22.45	20.60	41.00	25.40	13.00	42.63	14.50	40.10	2.75
		±0.57	±0.87	±1.49	±0.85	±0.12	±0.47	±0.47	±0.12	±1.09	±0.87	±0.57	±0.44
Laniidae	1	26.8	22.3	28.6	22.3	20.7	41.2	24.7	13.4	40.2	17.6	38.3	3.9
Corvidae	1	24.7	23.4	28.9	23.1	20.5	41.5	25.2	12.9	40.2	15.5	40.4	3.9
Vireonidae	1	25.5	20.2	30.7	23.6	20.5	40.4	26.0	13.1	45.6	9.8	42.7	1.8
Muscicapidae	1	25.7	21.3	29.7	23.4	21.0	40.4	26.0	12.6	37.8	12.3	46.2	3.7
Tyrannidae	1	26.3	23.9	28.7	21.1	20.0	40.5	26.6	12.9	37.4	15.3	44.2	3.2

[a] Average percentage of each nucleotide [adenine (A), thymine (T), cytosine (C), and guanine (G)], ±SD is given for each of the three positions within codons. Compositions were averaged over all species within families (number of species). The Phasianidae composition is for the chicken (Desjardins and Morais, 1990).

form distribution of 25% for each base; the bias varies markedly among positions and is most pronounced at position 3, where guanine comprises at most 4.0%; (2) variation in the pattern of bias is not great within or between families, with the exception of A and C at third positions, which vary by as much as 13% between families comprising Piciformes and those comprising Passeriformes.

D. Pairwise Comparisons among Divergent Sequences

Sequence comparison is a powerful tool for understanding processes and patterns of nucleotide substitution and how these might impinge on phylogenetic inference. For example, pairwise comparisons have provided insight on variation in substitution rates among nucleotide sites and among lineages and on conservation of gene regions. However, these comparisons are fraught with statistical problems, the most fundamental of which is stochastic dependence among observations. It is common to include data from all pairwise comparisons among the sequences that are available to a study, ignoring the fact that they have shared common ancestry to varying degrees. The problem with this is illustrated in Fig. 4.2. If, for example, a relative rates test were made comparing lineages A and C (AC) with an outgroup, O, one could test for equality of rate substitution along the lineages from n_2 to species A and C. But the distance OB is not independent of (is correlated with) OC because of the period of shared ancestry from n_2 to n_3; if by chance an improbably large number of substitutions occurred along n_2n_3, then two relative rates tests would appear significant owing to the inclusion of the same internode. The effect of not considering stochastic dependence when making all pairwise comparisons is to

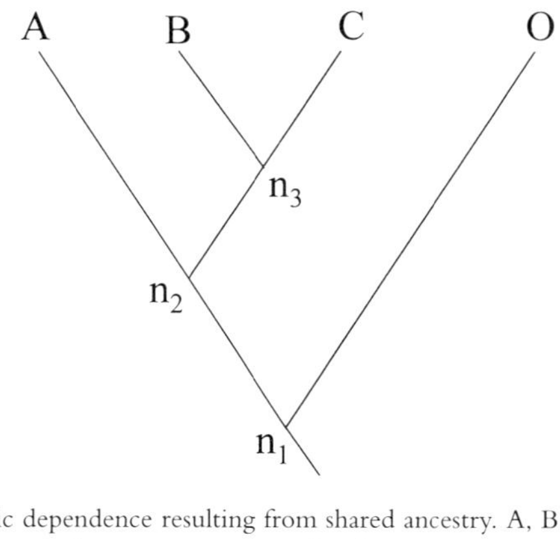

FIGURE 4.2 Stochastic dependence resulting from shared ancestry. A, B, C, and O represent contemporary sequences; n_1, n_2, and n_3 represent ancestral nodes.

make the inference, whether right or wrong, appear more certain than it really is. This problem applies to all inferences based on pairwise comparisons, be they inferences about substitution rates, transition–transversion ratios, or other parameters.

To avoid this problem we randomly selected a subset of species for pairwise comparisons from each of the four phylogenies, but with the additional constraint that the evolutionary pathway represented by each pair is independent of those of other pairs, or nearly so. This procedure avoids to a considerable extent the problem of stochastic dependence but has the disadvantage that some information is lost to the analysis. However, it is a conservative procedure in that it will tend to err on the side of not finding a significant difference when, in fact, there is one, as opposed to inferring a difference that is not really there. This procedure also presupposes that the phylogeny is correct, which is of course unknown. However, this procedure should be reasonably safe if the phylogeny is based on a data set with strong phylogenetic signal, because long internodes representing strong stochastic dependence should be relatively clearly resolved, and internodes that are not clearly resolved will tend to be short and hence contribute little to dependence among the observations.

The 24 independent pairwise comparisons are tabulated in Table II (pairs 1–24) along with corresponding estimates of some parameters that are useful in identifying the range of the taxonomic hierarchy where cyt *b* has a reasonably high probability of resolving the phylogeny. The pairs are ranked according to level of taxonomic divergence. Comparisons 23 and 24 (*Colaptes auratus* × *Belearica regulorum* and *Empidonax minimus* × *Gallus domesticus*) represent comparisons between data sets. These comparisons between remotely related taxa were made to determine the relationship of transitions to transversions at, or near, saturation. The *C. auratus* sequence was used in two "independent" comparisons, intratribal (pair 10) and infraclass (pair 23). Dependence resulting from duplicate use of this sequence should be negligible because the crane (*B. regulorum*) is so distantly related. There are six additional comparisons listed at the bottom of Table II (pairs 25–30). These are statistically dependent to varying degrees among themselves and with the 24 independent comparisons but represent pairs of taxa for which levels of divergence have been measured by DNA–DNA hybridization (Sibley and Ahlquist, 1990); the last one of these also involves a comparison between data sets. These six sequences were used in one analysis in which we compared levels of cyt *b* divergence with levels of nuclear genome divergence based on DNA–DNA hybridization.

The numbers of mismatches in the pairwise comparisons, falling in the various classes of substitutions (transitions, transversions, synonymous, and nonsynonymous), were converted to substitutions per nucleotide pair by dividing each count by the number of base pairs in the aligned sequences. It is apparent with a glance at Table II that divergence primarily entails synonymous transitions, although the relative rates at which various classes of substitutions accrue are functions of the levels of divergence. These relationships are discussed below because they are more readily seen in the graphical representations.

TABLE II Statistics for Cytochrome *b* Sequences Compared between Avian Species[a]

Pair	Taxon	bp	*d*	Ts/site	Tv/site	Ts-to-Tv ratio	Syn. sub./site	Nonsyn. subs./site	$\Delta T_{50}H$	*R*
1. *Grus antigone antigone* × *G. a. sharpei*	Species	1134	0.0116	0.005	0.006	0.86	0.008	0.004	—	5.05
2. *Capito dayi* × *C. niger*	Genus	883	0.1522	0.120	0.009	13.25	0.118	0.011	—	0.02
3. *Grus antigone gillae* × *G. leucogeranus*	Genus	1135	0.0558	0.043	0.010	4.45	0.043	0.010	—	0.03
4. *Grus vipio* × *G. rubicunda*	Genus	1135	0.0327	0.025	0.007	3.50	0.025	0.007	—	1.14
5. *Grus japonensis* × *G. canadensis pratensis*	Genus	1135	0.0763	0.064	0.006	10.43	0.061	0.010	—	1.88
6. *Grus americana* × *G. grus*	Genus	1135	0.0317	0.023	0.008	2.89	0.026	0.004	—	0.61
7. *Grus monachus* × *G. nigricollis*	Genus	1135	0.0161	0.013	0.003	5.00	0.011	0.004	—	1.14
8. *Piculus rubiginosus* × *Colaptes rupicola*	Genus	1029	0.0712	0.062	0.004	16.00	0.058	0.005	—	0.24
9. *Epimachus fastuosus* × *Ptiloris paradiseus*	Tribe	1143	0.1136	0.089	0.012	7.29	0.086	0.016	—	0.04
10. *Colaptes auratus* × *Dryocopus pileatus*	Tribe	1015	0.1368	0.101	0.020	5.15	0.104	0.011	—	1.74
11. *Veniliornis callonotus* × *Picoides villosus*	Tribe	1027	0.1034	0.083	0.010	8.50	0.079	0.010	—	0.63
12. *Anthropoides virgo* × *Bugeranus carunculatus*	Tribe	1135	0.0597	0.050	0.006	8.14	0.046	0.010	—	0.48
13. *Sphyrapicus varius* × *Melanerpes carolinus*	Subfamily	1020	0.1593	0.102	0.037	2.74	0.111	0.018	—	0.23
14. *Dphyllodes magnificus* × *Cyanocitta cristata*	Subfamily	1143	0.1741	0.101	0.050	2.02	0.122	0.028	—	1.22

15. *Ramphastos tucanus* × *Aulacorhynchus derbianus*	Subfamily	878	0.1437	0.083	0.046	1.83	0.091	0.037	4.2	0.50
16. *Campephilus haematogaster* × *Picumnus aurifrons*	Family	966	0.1886	0.105	0.059	1.77	0.101	0.046	—	—
17. *Ailuroedus melanotus* × *Ptilorhynchus violaceus*	Family	1143	0.1706	0.088	0.062	1.42	0.118	0.033	5.0	0.00
18. *Lybius bidentatus* × *Pogoniulus bilineatus*	Family	880	0.1589	0.076	0.065	1.18	0.123	0.018	—	0.27
19. *Balearica pavonina* × *Anthropoides paradisea*	Family	1135	0.1086	0.083	0.015	5.53	0.087	0.011	3.3	—
20. *Manucodia keraudrenii* × *Lanius ludovicianus*	Superfamily	1143	0.1788	0.097	0.058	1.68	0.123	0.031	9.1	1.64
21. *Vireo olivaceus* × *Catharus guttatus*	Suborder	1141	0.1723	0.074	0.079	0.93	0.120	0.033	12.8	0.25
22. *Sphyrapicus varius* × *Capito dayi*	Order	873	0.2541	0.111	0.101	1.10	0.161	0.044	16.5	—
23. *Colaptes auratus* × *Balearica regulorum*	Infraclass	1019	0.2108	0.094	0.087	1.08	0.127	0.045	26.3	—
24. *Empidonax minimus* × *Gallus domesticus*	Subclass	1140	0.2371	0.089	0.111	0.80	0.151	0.050	28.0	—
25. *Melanerpes carolinus* × *Picoides villosus*	Subfamily	1035	0.1625	0.109	0.032	3.42	0.118	0.019	5.0	—
26. *Colaptes auratus* × *Picoides villosus*	Subfamily	1020	0.1361	0.090	0.031	2.88	0.095	0.019	5.8	—
27. *Lybius bidentatus* × *Capito dayi*	Infraorder	887	0.2431	0.112	0.092	1.21	0.161	0.043	11.5	—
28. *Ramphatos tucanus* × *Capito niger*	Family	877	0.1775	0.107	0.047	2.29	0.119	0.034	6.1	—
29. *Empidonax minimus* × *Cyanocitta cristata*	Order	1139	0.2419	0.107	0.097	1.11	0.152	0.051	19.7	—
30. *Empidonax minimus* × *Balearica regulorum*	Superorder	1137	0.2208	0.094	0.094	1.00	0.142	0.046	21.6	—

[a]The first 24 comparisons are the independent comparisons: bp, base pairs; *d*, Tamura–Nei distance; Ts, transitions; Tv, transversions; Syn. sub., synonymous substitutions; Nonsyn. sub., nonsynonymous substitutions; *R*, dispersion index (see text for details). The *G. domesticus* sequence is from Desjardins and Morais (1990).

E. The Relationship between Transitions and Transversions

The ratio of transition to transversion substitutions is important for two reasons: (1) it is useful for identifying levels of divergence at which nucleotide sites are becoming saturated by multiple transitions (multiple hits), and (2) it must be known and corrected for if phylogenetic algorithms are to make accurate estimates of the branch lengths.

The ratio of transition to transversion substitutions is highest between sequences that are just beginning to diverge and decreases as time of divergence increases (Brown *et al.*, 1982). Holmquist (1983) provided a theoretical understanding of this phenomenon. Although the "instantaneous" rate of transition substitutions is always higher than that of transversions, the apparent rate at which diverging sequences accrue differences changes through time, depending also on the probabilities of the 12 possible character state changes (C→T, C→A, etc.) and the base composition of the sequences. For pairwise comparisons of diverging sequences, plots of transitions as a function of transversions have an initial phase where there is a rapid and nearly linear increase in transitions, but the curve plateaus as a significant number of sites sustain multiple transitions; at this point, sites have a much lower probability of sustaining multiple transversions and so the number of transversional differences between the sequences continues to increase while the number of transitional differences levels off (Fig. 4.3). Eventually, the transition-to-transversion ratio reaches an equilibrium value of 0.5 if G, A, T, and C are equally frequent (25% each), but when base composition is biased as in the case of the cyt *b* gene, it will equilibrate at some other value.

The empirical distribution of transitions as a function of transversions is presented in Fig. 4.3 for the 24 independent pairwise comparisons compiled in Table II. We have plotted distinctive symbols representing comparisons among taxa at various levels in the taxonomic hierarchy so that a relationship between saturation of transitions and taxonomic level might be apparent. Each taxonomic level indicated in Fig. 4.3 is the lowest level that encompasses both taxa in the comparison; e.g., a symbol indicating *tribe* involves a comparison between two species that belong to distinct genera within that tribe. Taxonomic level, of course, provides only a crude index of divergence because it varies somewhat arbitrarily among major groups. For example, the cranes *Balearica pavonina* × *Anthropoides paradisea* [Fig. 4.3, open triangle (0.015, 0.083)] are assigned to separate subfamilies (Krajewski and Fetzner, 1994), but have a level of divergence and a transition-to-transversion ratio more characteristic of distinct genera or tribes within a single subfamily.

The relationship between transitions and transversions is roughly linear through the level of tribes, at which point it plateaus rather quickly. This indicates that avian radiations retain phylogenetic information in the form of transitional substitutions at least to the tribal level (relationships among genera within tribes). Above the

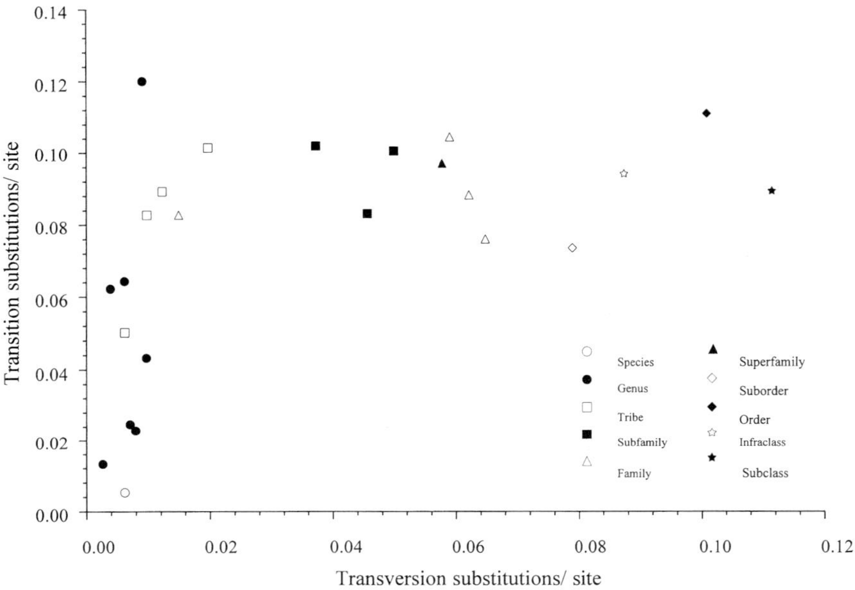

FIGURE 4.3 Transition substitutions as a function of transversions. The points represent the first 24 independent pairwise comparisons tabulated in Table II.

tribal level, transitions may still be informative, but multiple hits have significantly eroded the phylogenetic signal or, stated another way, homoplasies and autapomorphies are replacing synapomorphies. Identifying an upper limit where transitions are so saturated that they no longer retain phylogenetic information is more complex because it depends on the lengths of internal nodes in the phylogeny relative to more distal branches representing descendant lineages (see below). Transversions should continue to be informative well beyond the tribal level but, again, identifying the upper level at which synapomorphies in the form of transversions have been significantly eroded is a more complex problem that cannot be completely resolved by examining the transition-to-transversion ratio.

Accurate and precise estimation of the transition-to-transversion ratio is also fraught with statistical problems (Holmquist, 1983). Investigators commonly average the ratios for all pairwise comparisons of sequences in their data. This is incorrect. First, there is the fact that all pairwise observations are not independent. Second, and more important, distance estimates that correct for bias in the transition-to-transversion ratio require the ratio of instantaneous rates at which transition and transversion substitutions occur (Kimura, 1980), which is different from the ratio of transitions and transversions that have accumulated over a long period of time. A better way is to make pairwise comparisons among recently diverged sequences so that the problem of multiple hits is mitigated, although not eliminated.

The major problem with an estimate based on recently evolved sequences is that the numbers of observed substitutions, especially transversions, will be small and as a consequence the sampling variance of the estimate will be high (Holmquist, 1983).

We estimated the transition-to-transversion ratio for avian species by fitting a least-squares regression line, forced through the origin, to the part of the transition–transversion curve that is roughly linear. The slope of the regression line is an estimate of the ratio. We included in the estimate 13 data points where the proportion of transversion sites was less than 0.02. This includes the comparison of *Colaptes auratus* and *Dryocopus pileatus* [Fig. 4.3, open square (0.02, 0.10)] and all data points to the left. The estimated transition-to-transversion ratio is 6.25. The total numbers of transitions and transversions over these 13 comparisons are 808 and 125, respectively. We repeated the analysis with the comparison of *Capito niger* with *C. dayi* excluded. This is the solid circle in the upper left corner of Fig. 4.3, which appears to be an outlier. The ratio without the *C. niger–C. dayi* comparison is 5.77. The number of substitutions separating *C. niger* and *C. dayi* is extraordinary, but more worrisome is that its transition-to-transversion ratio lies well away from the other data, although, as mentioned above, the sampling variance in this part of the curve is high. We estimated the transition-to-transversion ratio by an alternative method. Since we had sequences for two individuals of each species of woodpecker, we summed the transitions and transversions for all intraspecific comparisons. These comparisons are independent and the level of divergence is low, so that multiple hits are minimal. The total numbers of transitions and transversions are 66 : 13 for a ratio of 5.08. As noted above, the variance of this estimate is large.

F. The Relationship between Taxonomic Divergence and Saturation of Transversions

The logic of plotting transitions as a function of transversions is that the relatively low rate of transversion substitutions provides a more or less linear scale over the time span when transitions become saturated, thus revealing the level in the taxonomic hierarchy where saturation occurs. A plausible approach to the remaining problem of determining the level at which transversions begin to saturate is to plot transversions as a function of a thermal dissociation parameter such as $\Delta T_{50}H$ from DNA–DNA hybridization studies. Thermal dissociation values are available for many pairs of avian species because of the extensive work of Sibley and Ahlquist (1990) and others (e.g., Sheldon, 1987; Bledsoe, 1987; Krajewski, 1989; Sheldon *et al.*, 1992; Bleiweiss, 1994). $\Delta T_{50}H$ and related thermal dissociation parameters measure the amount of sequence divergence in single-copy nuclear DNA, which accrues at a much slower rate than in the mitochondrial genome; thus, $\Delta T_{50}H$ could provide a relatively linear reference scale for identifying saturation of transversions in mitochondrially encoded cyt *b*. We found eight pairs of species among our set of

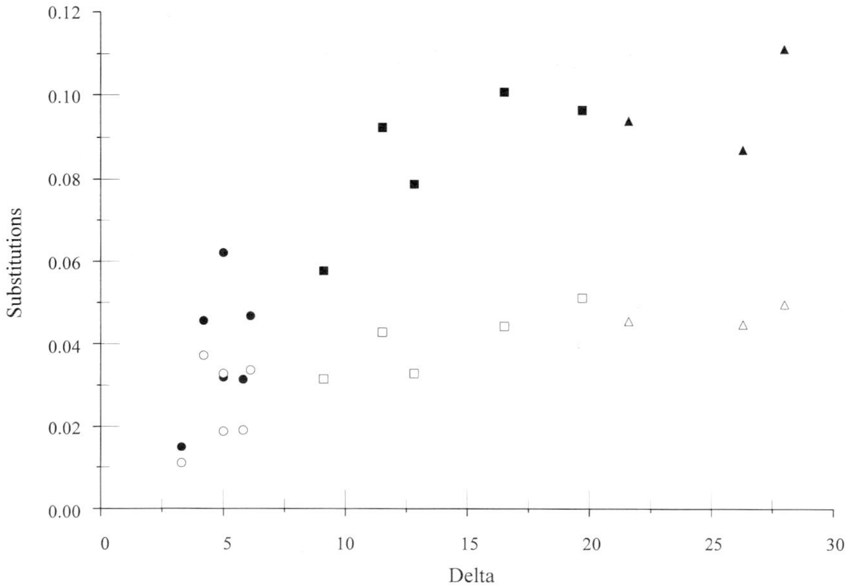

FIGURE 4.4 Transversion substitutions and amino acid substitutions in mitochondrial cyt *b* as functions of divergence in single-copy nuclear DNA as measured by $\Delta T_{50}H$ from DNA–DNA hybridization studies (Delta, horizontal axis). Solid symbols represent transversions; open symbols represent amino acids. Taxonomic levels: circles, subfamilies–families, squares, superfamilies–orders, triangles, superorders–subclasses. The vertical axis is in units of transversion substitutions/nucleotide site or amino acid substitutions/codon.

independent comparisons of cyt *b* sequences ($\Delta T_{50}H$ values listed in Table II) plus six additional pairs (Table II, pairs 25–30) among the complete sets of cyt *b* sequences for which $\Delta T_{50}H$ values could be determined from Sibley and Ahlquist (1990) for the same species or for closely related species.

Transversion substitutions in cyt *b* as a function of $\Delta T_{50}H$ are plotted in Fig. 4.4 for the pairwise comparisons. It is evident that the number of transversion substitutions between diverging sequences is fairly linear up to approximately $\Delta T_{50}H =$ 12.5–15. Again, we distinguished taxonomic levels with symbols. This indicates that the relationship is linear through most of the comparisons in the superfamily–order grouping; in fact, the first point to fall markedly below the line is the rightmost square, which compares a suboscine (*Empidonax minimus*) with an oscine (*Cyanocitta cristata*). This analysis is useful for identifying levels in the taxonomic hierarchy through which the phylogenetic information contained by transversions has not been severely eroded, but it does not identify the upper limit where so many multiple transversions have occurred that the phylogenetic signal is, for practical purposes, lost.

G. The Relationship between Taxonomic Divergence and Amino Acid Substitutions

Amino acid substitutions as a function of $\Delta T_{50}H$ are also plotted in Fig. 4.4 for the pairwise comparisons. Amino acid substitutions appear to accumulate as rapidly as transversions for a short initial period, but then the rate slows and the trajectory remains low over much of the evolution of birds.

H. Short Internodes, Long Terminal Branches, and Phylogenetic Resolution

Transitions viewed as a function of transversions, and transversions, in turn, viewed as a function of $\Delta T_{50}H$ indicate that cyt *b* retains phylogenetic signal up to at least the level of avian families and superfamilies within orders, but this does not tell the complete story. Whether the branching order of three OTUs can actually be resolved depends on the relative lengths of the internode (T_2) and the terminal branches (T_1; see Fig. 4.5a): When the internode is long and the terminal branches short, the probability of resolving the correct branching order is high because a relatively large number of synapomorphies are expected to have accrued during common ancestry and relatively little time has elapsed for subsequent substitutions to obliterate these synapomorphies.

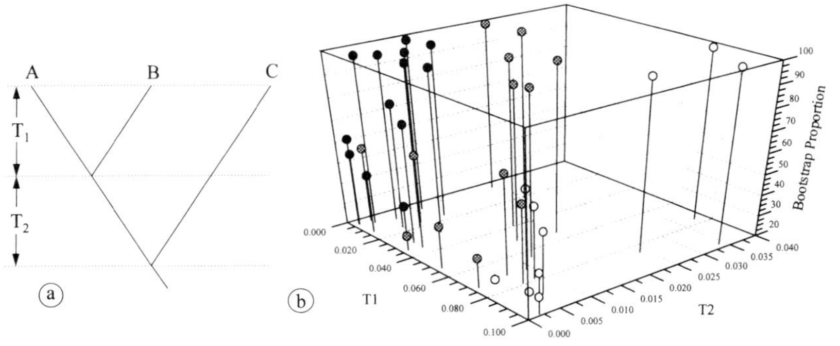

FIGURE 4.5 Resolution of phylogenetic trichotomies as a function of internode length (T_2) and terminal branch length (T_1). (a) A hypothetical trichotomy showing the relationships between taxa, T_1 and T_2. (b) Plot of bootstrap proportions supporting a trichotomy as a function of T_1 and T_2. The relative level of resolution is measured by the percentage of 500 bootstrap replicates that supported the trichotomy, based on the neighbor-joining tree (Tamura–Nei distance, including both transitions and transversions). Distinctly shaded circles represent nodes at various levels in the taxonomic hierarchy: solid circles, species–genera; cross-hatched circles, tribes–subfamilies; empty circles, families–suborders (where taxonomic level is the lowest level that encompasses all taxa in the trichotomy).

The importance of the relationship between the lengths of terminal branches and internodes was developed in a theoretical context by Lanyon (1988). Figure 4.5b is a three-dimensional plot similar to those depicted by Lanyon, in which he plotted the probabilities of resolving a trichotomy as a function of T_1 and T_2. In our application we have plotted the percentages of 500 bootstrap replicates, as a function of T_1 and T_2, that support each internode in each of the four phylogenies (Fig. 4.1a–d). For each internode, T_2 is the branch length indicated in the phylogeny, and T_1 is the average length of branches distal to the internode, averaged by clades, moving inward from the branch tips. As an example, in the woodpecker phylogeny (Fig. 4.1d) there is a group of four species that includes *Dryocopus pileatus*, two species of *Colaptes*, and one species of *Piculus*: the internode (T_2) has length 0.017, and was supported by 98% of bootstrap replicates; $T_1 = [(0.037 + 0.033)/2 + 0.003 + 0.041]/2 = 0.0395$. This data point can be located from the coordinates in Fig. 4.5b. It is among the cluster of five cross-hatched (tribe–subfamily) data points near the top–center of the space with bootstrap support near 100%; specifically, it is the leftmost point in this cluster.

Figure 4.5b is helpful in understanding circumstances where cyt *b* will likely resolve a given trichotomy and where it will not. First, following changes along the T_1 axis from back (0.000) to front (0.100) it can be seen that lower taxonomic levels (solid dots) are in the back cross-sections and the higher levels toward the front, as expected, because lower levels represent recent splits (short T_1) and higher levels represent more ancient splits (long T_1). Now, following changes along the T_2 axis, the profile for bootstrap support as a function of T_2 at a fixed value of T_1 is generally one where bootstrap values are low for low values of T_2, but increase as T_2 increases. However, bootstrap support increases more rapidly as a function of T_2 for low values of T_1 than high values of T_1; i.e., the longer the period of evolution along the lineages after they split from the internode, the more eroded is the information in the antecedent internode. Now, exploring the question of what taxonomic levels one can expect to resolve, or not resolve, it is apparent that there is no set cutoff point; at all levels, species–genera, tribes–subfamilies, families–suborders, some nodes were resolved (supported by high bootstrap values) whereas others were not. However, some generalizations can be drawn.

In making these generalizations, we use 70% bootstrap proportions (500 replicates) as a cutoff point in classifying nodes as resolved or unresolved, but acknowledge that this is somewhat arbitrary in that the parameters guiding the evolution of the avian cyt *b* gene differ from those in the simulation studies of Hillis and Bull (1993). Still, this would provide a useful ranking to identify taxonomic levels where cyt *b* works relatively well versus relatively poorly for resolving avian phylogenies. Applying the 70% criterion, the frequency of resolved nodes is highest for species–genera (9 of 13 nodes supported at 70% or more), intermediate for tribes–subfamilies (6 of 13) and lowest for families–suborders (3 of 10). Thus, cyt *b* can be expected to work well in resolving relationships among species within genera. It should continue to work well through the levels of tribes and subfamilies, but can be expected to resolve nodes at the family level and higher only if the common ancestors existed for long periods.

I. Rate Variation within Lineages

A detailed analysis of rate variation within and between the four radiations is beyond the scope of this chapter, because of the statistical complexities and uncertainties that need to be discussed. Nonetheless, it is important to know whether substitution rates within radiations vary substantially. A simple test of one prediction of the molecular clock is as follows. Neutral mutations, including silent substitutions, should occur along a lineage according to a Poisson process, and should accumulate in a stochastic, clocklike manner (Kimura, 1981, 1983; Gillespie and Langley, 1979; Gillespie, 1986, 1991). Thus, the accumulation of base substitutions along lineages that diverged from a common ancestor should have a Poisson distribution with a mean $m(t)$ that depends on the time, t, since divergence and a variance between lineages, $s^2(t)$, equal to the mean (Gillespie, 1991). Thus, $R(t) = s^2(t)/m(t)$, termed the index of dispersion, should be close to one if substitutions have occurred according to a molecular clock along lineages diverging as a "starburst" from a common ancestor. Estimation of $R(t)$, like all assessments of evolutionary rates, is fraught with statistical problems and uncertainty regarding underlying assumptions of the model. The problem of statistical independence of lineages crops up again when multiple comparisons are made, and two additional phenomena, collectively termed lineage effects by Gillespie (1989), tend to inflate estimates of $R(t)$. The two phenomena are variations in generation time and estimations based on an incorrect phylogeny.

Our pairwise comparisons are based on two lineages emerging from each of several ancestral nodes and are independent of each other (e.g., the lineages diverging from a common ancestor leading to *Piculus rubiginosus* and *Colaptes rupicola*, are independent of the pair of lineages leading to *Colaptes auratus* and *Dryocopus pileatus*). It is possible to estimate the number of substitutions along each branch and then the mean and variance for each pair. If rate variation is a pervasive property of cyt *b* evolution, R for pairwise comparisons should tend to be greater than 1. If p is the rate of nucleotide substitutions accumulated over a DNA sequence of length k along a lineage from the time it splits from a common ancestor to the present (i.e., the branch length), then the expected number of substitutions is $E(x) = E(kp) = kE(p)$, and the variance is $\text{Var}(x) = \text{Var}(kp) = k^2\text{Var}(p)$. (Note that x is corrected for multiple hits because it is based on Tamura–Nei branch distances.) An estimate of the dispersion index then is

$$R = kS_p^2/\overline{p}$$

where $\bar{p}$ is the average of the two branch lengths and S_p^2 is the variance of the two branch lengths. (Note that S_p^2 is an estimate of the variance implicit in the evolutionary model and not the sample variance.)

R values are tabulated in Table II from all of the independent pairwise comparisons (where *p* for each lineage is the sum of the branch lengths from the ancestral node to the branch tip, based on the Tamura–Nei distance and the phylogenies in Fig. 4.1). The average *R* value is 0.90. Although this analysis of rate variation is limited in scope, it is apparent that substitution rates for cyt *b* have not varied substantially within the radiations studied here. This is not surprising, given that most of the substitutions contributing to branch lengths are synonymous; indeed, it would be surprising if rate variation were substantial.

J. How Much Sequence in a Pilot Study

It would be helpful to be able to determine from a limited amount of sequence data whether cyt *b* will likely resolve relationships in a more expansive phylogenetic study. To determine this, we reduced the sequences for each of the four data sets to 282 bp that correspond to the fragment typically amplified by primers *L14841* and *H15149* (Kocher *et al.*, 1989), and then we repeated the neighbor-joining analysis. Twenty-one of 36 nodes across the four phylogenies (Fig. 4.1), which appeared in the complete data phylogenies, also appeared in the 282-bp trees, and 15 of 18 nodes in the complete data phylogenies that were supported by bootstrap proportions of 70% or more appeared in the 282-bp phylogeny. Thus, in a pilot study comprising sequences derived by amplification with two of the universal primers from Kocher *et al.* (1989), one can clearly determine whether cyt *b* is an appropriate choice.

IV. DISCUSSION

A. The Window of Taxonomic Resolution for Avian Phylogenies

In what seems to be a commentary on the sociology of molecular systematics as much as a criticism of phylogenetic studies based on cyt *b*, Meyer (1994, p. 280) noted: "For laboratories that set out to undertake molecular phylogenetic work and need to produce preliminary data for grant proposals, the cytochrome *b* gene is often the first choice. However, *the appeal of cytochrome b as an easy 'beginner's gene' for phylogenetic work is tarnished by several of its particular shortcomings.* One should not expect that this gene is going to be the right gene for all questions." More recently, in a detailed analysis of patterns of nucleotide substitutions in cyt *b* in cranes, Krajewski and King (1995) quoted the italicized part of Meyer's conclusion as the lead-in to their counterpoint ". . . that the full utility of cytochrome *b* has yet to be explored." Our study, we believe, shows that there is truth in both these conclusions. The mitochondrially encoded cyt *b* gene is good for avian systematics, but its utility is greatest in resolving the diversification of birds from the level of species (or subspecies) to subfamilies and families in some instances. At higher taxonomic levels cyt *b* might provide some resolution, but sequencing another gene would be a better investment. The "particular shortcomings" of cyt *b* manifest only when trying to resolve deeper splits in evolutionary history and, indeed, all of the disappointing phylogenetic studies based on cyt *b* involve attempts to resolve much more ancient groups (e.g., Meyer and Wilson, 1990; Kornegay *et al.*, 1993; Avise *et al.*, 1994a,b; Graybeal, 1993; Honeycutt *et al.*, 1995).

The aspects of cyt *b* divergence that impinge on the phylogenetic utility and limitations of the molecule are apparent in Tables I and II and Figs. 4.3 and 4.4 and have been discussed particularly by Irwin *et al.* (1991) and Krajewski and King (1995), and reviewed by Meyer (1994). In birds, diverging cyt *b* sequences accrue transition substitutions at a rapid, and more or less constant, rate to the level of distinct genera within tribes, and transversions continue to accrue in a similar manner to the approximate level of superfamilies. These substitutions are primarily synonymous as evidenced by comparison of the synonymous and nonsynonymous substitution columns in Table II. It is also probable that the small burst of amino acid substitutions in the early phase of divergence (Fig. 4.4) represents neutral or weakly selected substitutions. Aside from these neutral positions, which become saturated before the base of the avian tree is resolved, the remaining positions appear to be so constrained by selection that it is improbable that they would have become substituted, and hence synapomorphies, during the limited periods of shared ancestry. But, on the positive side, this period of neutral evolution spans a large and interesting portion of the systematic hierarchy of birds; a larger span than in other vertebrates because of the phenomenon of downshifting mentioned in the introduction. Stated another way, a larger fraction of avian diversity has evolved in the recent past, relative to that of other vertebrate groups.

What is the most recent level of diversification for which cyt *b* might be expected to provide resolution? The answer is probably at the level of subspecies, but at this level the problems of resolving the gene tree and congruence between the gene tree and species tree confound each other (Neigel and Avise, 1986; Pamilo and Nei, 1988; Smouse *et al.*, 1991; Moore, 1995). With regard to resolving the haplotype tree (gene tree), synonymous substitutions in the 13 mitochondrially encoded protein genes are among the fastest evolving sequences known. The mitochondrial control region has been suggested as a source of variation for studying recent evolutionary events, but to our knowledge it has not been demonstrated that the control region evolves more rapidly than the synonymous positions of the protein-coding genes. In any event, it is clear that mitochondrial haplotype trees can be resolved down to the level of individual lineages.

B. Molecular Clock

How deep in evolutionary time can cyt *b* provide useful information about avian phylogenies? Shields and Wilson (1987) calibrated the mtDNA clock for geese against the fossil record at 2.0% average divergence per million years (MY). Tarr and Fleischer (1993) reasoned to the same calibration based on divergences in a complex of Hawaiian honeycreepers and the geological history of the islands. Krajewski and King (1995) estimated a slower calibration, based on the fossil record, for cranes, 0.7–1.7% MY. Cranes have a longer generation time (average, 4–5 years to first breeding) than geese (average, 3 years), small passerines (1 year), and woodpeckers (1 year) (Table 16 in Sibley and Ahlquist, 1990). Krajewski and King (1995) noted that the apparent rate reduction in cranes could be related to longer generation time. In any case, the calibrations would be useful only over the span of time when the relationship between divergence and time is linear.

To estimate this, we plotted the Tamura–Nei distance for the independent pairwise comparisons against $\Delta T_{50}H$ and determined empirically that the relationship is linear up to $\Delta T_{50}H \cong 6$. We then determined the slope of a regression line forced through the origin (plot not shown) for values of $\Delta T_{50}H < 6$. The slope is 3.0% mtDNA sequence divergence/$\Delta T_{50}H$ unit; so, when the relationship becomes nonlinear ($\Delta T_{50}H = 6$), cyt *b* sequence divergence between pairs is approximately 18%, or 9%/lineage. For small passerines and woodpeckers this corresponds to approximately 9 million years before present (MYBP) (2% divergence between lineages or 1% along a single lineage) and 10.6–25.7 MYBP for the more slowly evolving cranes.

C. Conclusions and Advice

The mitochondrially encoded cyt *b* gene should be given consideration, beyond providing preliminary data for a grant proposal, in avian phylogenetic studies focused on taxa at the levels of approximately subfamilies or families and lower, i.e., down to species and subspecies. Indeed, at these levels it is arguably one of the best choices one can make, for several reasons. The first reason is that it is a mitochondrial gene because the mitochondrial haplotype tree has a substantially better chance of congruence with the actual species tree than a nuclear gene tree, particularly when internodes are short (Moore, 1995). For this reason and a number of others, it is prudent to begin a phylogenetic study with a mitochondrial gene sequence. Among the mitochondrially encoded genes, cyt *b* may or may not be the best choice for resolving the haplotype tree at intermediate taxonomic levels, say, families and subfamilies, but it should be as good as any gene for the lower levels because most of the variation is synonymous, and cyt *b* should accrue this type of variation as rapidly as any other gene. Protein-coding genes that are less conserved with regard to nonsynonymous substitutions may be more useful at intermediate levels because amino acid substitutions would be expected to contribute more substantially to the pool of phylogenetic information, but this has yet to be determined. The most promising candidates would include the ND and ATPase genes (Brown, 1983; Desjardins and Morais, 1990). In this same vein, if synonymous substitutions provide most of the phylogenetic information for a particular set of taxa, a mitochondrially encoded protein gene is attractive because there are 12 other protein-coding genes in the same linkage group accumulating similar variation at roughly the same rate. This provides a large, homogeneous pool of informative characters for resolving short internodes. A final reason for using cyt *b* is that it is the best known of the avian mitochondrially encoded genes, and an abundance of primers has been published.

With regard to the possibility of combining data from two or more mitochondrially encoded protein-coding genes, DeFilippis (1995) sequenced 1512 bp of the COI gene (1548 bp in total length) for the same woodpecker specimens reported here except *Colaptes rupicola*. DeFilippis did phylogenetic analyses on the cyt *b* and COI data sets separately and on the combined data sets. The neighbor-joining trees had identical topologies for the three analyses except for the placement of *Campephilus haematogaster*, which differed in all three analyses, but was weakly supported by bootstrap proportions in all three. *Campephilus haematogaster* appears to have shared common ancestry with the rest of the woodpeckers for a short time early in the radiation and may be a case in which the precise branching order cannot be resolved. The average increase in bootstrap proportions for shared nodes was 7.8% in the combined data set tree over the individual data set trees. This is not a dramatic increase, but the bootstrap support was strong in both of the individual trees.

We agree with Graybeal (1993, 1994) and Meyer (1994) in strongly encouraging pilot studies before undertaking an exhaustive molecular phylogenetic analysis. If the encompassing avian taxon is a family, even a superfamily, cyt *b* is worth further consideration. Another easily obtained, and good, indicator is Sibley and Ahlquist's (1991) "tapestry": if the taxon that encompasses the species of interest emanates in a bifurcation with a $\Delta T_{50}H > 12.5$, consider another gene. If the study group passes these tests, cyt *b* may be the best bet, but not a certain bet. Next, a pilot study should be done using the 282-bp fragment amplified with primers *L14841* and *H15149* for 5–10 carefully chosen species that likely represent the range of variation for the group. Data from the pilot study should be analyzed with regard to transition-to-transversion ratios, bootstrap support, and, importantly, the ratios of internode lengths (T_2) to branch-tip lengths (T_1). The plot of bootstrap proportions as a function of T_2 and T_1 indicates, roughly, that a trichotomy will be supported by 70%+ bootstrap proportions if the ratio of T_1 to T_2 is no more than 4:1, and the lower the ratio the better. Finally, it follows from the facts that because most of the phylogenetically informative variation in cyt *b* is drawn from a homogeneous pool of synonymous substitutions, and because the signal is readily detected in a subset of the available variation, it is unnecessary to sequence the entire gene, down to the last nucleotide. Rather than verify every nucleotide for a single specimen by sequencing both DNA strands, it is better to sequence one strand for two specimens—thus keeping the cost the same but greatly reducing the chances of reporting a contaminant sequence.

ACKNOWLEDGMENTS

We are especially thankful to Kathleen Helm-Bychowski, Joel Cracraft, Carey Krajewski, and Scott Lanyon for providing us with copies of their sequences. We thank Carey Krajewski for providing us with a copy of a paper in press. We also thank John Harshman, Axel Meyer, and an anonymous reviewer for providing valuable comments and criticisms on an early draft of this manuscript. We thank the Louisiana State University Museum of Natural Science for providing frozen tissue specimens for the woodpecker work. We thank Christina Richards for reading an early draft of the manuscript and Aamina Ahmed for laboratory assistance. This work was supported by a grant from the Systematic Biology Program, National Science Foundation, DEB-9316452, to W.S.M.

REFERENCES

Avise, J. C., and Aquadro, C. F. (1982). A comparative summary of genetic distances in vertebrates, patterns and correlations. *Evol. Biol.* **15**, 151–185.

Avise, J. C., Ankney, C. D., and Nelson, W. S. (1990). Mitochondrial gene trees and the evolutionary relationship of mallard and black ducks. *Evolution* **44**, 1109–1119.

Avise, J. C., Nelson, W. S., and Sibley, C. G. (1994a). Why one-kilobase sequences from mitochondrial DNA fail to solve the hoatzin phylogenetic enigma. *Mol. Phylogenet. Evol.* **3**, 175–184.

Avise, J. C., Nelson, W. S., and Sibley, C. G. (1994b). DNA sequence support for a close phylogenetic relationship between some storks and New World vultures. *Proc. Natl. Acad. Sci. U.S.A.* **91**, 5173–5177.

Bermingham, E., Rohwer, S., Freeman, S., and Wood, C. (1992). Vicariance biogeography in the Pleistocene and speciation in North-American wood warblers. *Proc. Natl. Acad. Sci. U.S.A.* **89**, 6624–6628.

Bledsoe, A. H. (1987). DNA evolutionary rates in nine-primaried passerine birds. *Mol. Biol. Evol.* **4**, 559–571.

Bleiweiss, R., Kirsch, J. A. W., and Matheus, J. C. (1994). DNA–DNA hybridization evidence for subfamily structure among hummingbirds. *Auk* **111**, 8–19.

Bock, W. J., and Miller, W. D. (1959). The scansorial foot of the woodpeckers, with comments on the evolution of perching and climbing feet in birds. *Am. Mus. Novit.* **1931**, 9–45.

Brown, W. M. (1983). Evolution of animal mitochondrial DNA. *In* "Evolution of Genes and Proteins" (M. Nei and R. K. Koehn, Eds.), pp. 62–88. Sinauer Associates, Sunderland, Massachusetts.

Brown, W. M. (1985). The mitochondrial genome of animals. *In* "Molecular Evolutionary Genetics" (R. J. MacIntyre, Ed.), pp. 95–130. Plenum Press, New York.

Brown, W. M., Prager, E. M., Wang, A., and Wilson, A. C. (1982). Mitochondrial DNA sequences of primates: tempo and mode of evolution. *J. Mol. Evol.* **18**, 225–239.

Cimino, G. D., Metchette, K., Isaacs, S., and Zhu., Y. S. (1990). More false-positive problems. *Nature (London)* **345**, 773–774.

Cooper, A. (1994). Ancient DNA sequences reveal unsuspected phylogenetic relationships within New Zealand wrens (Acanthisittidae). *Experientia* **50**, 558–563.

Cooper, A., Mourer-Chauvire, C., Chambers, G. K., von Haeseler, A., Wilson, A. C., and Paabo, S. (1992). Independent origins of New Zealand moas and kiwis. *Proc. Natl. Acad. Sci. U.S.A.* **89**, 8741–8744.

Cracraft, J., and Helm-Bychowski, K. (1991). Parsimony and phylogenetic inference using DNA sequences: Some methodological strategies. *In* "Phylogenetic Analysis of DNA Sequences" (M. M. Miyamoto and J. Cracraft, Ed.), pp. 184–220. Oxford Univ. Press, New York, Oxford.

DeFilippis, V. (1995). Evolution of the mitochondrial encoded cytochrome oxidase I gene versus the cytochrome *b* gene in ten species of Picidae. Masters thesis, Wayne State University.

Derr, J. N., Davis, S. K., Woolley, J. B., and Wharton, R. A. (1992). Reassessment of the 16S rRNA nucleotide sequence from members of the parasitic Hymenoptera. *Mol. Phylogenet. Evol.* **1**, 338–341.

Desjardins, P., and Morais, R. (1990). Sequence and gene organization of the chicken mitochondrial genome. *J. Mol. Biol.* **212**, 599–634.

Desjardins, P., and Morais, R. (1991). Nucleotide sequence and evolution of coding and noncoding regions of a quail mitochondrial genome. *J. Mol. Evol.* **32**, 153–161.

Edwards, S. V., and Wilson, A. C. (1990). Phylogenetically informative length polymorphism and sequence variability in mitochondrial DNA of Australian songbirds (*Pomatostomus*). *Genetics* **126**, 695–712.

Edwards, S. V., Arctander, P., and Wilson, A. C. (1991). Mitochondrial resolution of a deep branch in the genealogical tree for perching birds. *Proc. R. Soc. Lond. B* **243**, 99–107.

Felsenstein, J. (1985). Confidence limits on phylogenies: An approach using the bootstrap. *Evolution* **39**, 609–622.

Felsenstein, J. (1988). Phylogenies from molecular sequences: Inference and reliability. *Annu. Rev. Genet.* **22**, 521–565.

Felsenstein, J., and Kishino, H. (1993). Is there something wrong with the bootstrap on phylogenies? A reply to Hillis and Bull. *Syst. Biol.* **42**, 193–200.

Gillespie, J. H. (1986). Natural selection and the molecular clock. *Mol. Biol. Evol.* **3**, 138–155.

Gillespie, J. H. (1991). "The Causes of Molecular Evolution." Oxford Univ. Press, New York.

Gillespie, J. H., and Langley, C. H. (1979). Are evolutionary rates really variable? *J. Mol. Evol.* **13**, 27–34.

Graybeal, A. (1993). The phylogenetic utility of cytochrome *b*: Lessons from bufonid frogs. *Mol. Phylogenet. Evol.* **2**, 256–269.

Graybeal, A. (1994). Evaluating the phylogenetic utility of genes: A search for genes informative about deep divergences among vertebrates. *Syst. Biol.* **43**, 174–193.

Hackett, S. J., Griffiths, C. S., Bates, J. M., and Klein, N. K. (1995). A commentary on the use of sequence data for phylogeny reconstruction. *Mol. Phylogenet. Evol.* **4**, 350–353.

Heidrich, P., Konig, C., and Wink, M. (1995). Molecular phylogeny of South American screech owls of the *Otus atricapillus* complex (Aves, Strigidae) inferred from nucleotide sequences of the mitochondrial cytochrome *b* gene. *Z. Naturforsch. C* **50**, 294–302.

Helm-Bychowski, K., and Cracraft, J. (1993). Recovering phylogenetic signal from DNA sequences: Relationships within the corvine assemblage (class Aves) as inferred from complete sequences of the mitochondrial DNA cytochrome-*b* gene. *Mol. Biol. Evol.* **10**, 1196–1214.

Hillis, D. M., and Bull, J. J. (1993). An empirical test of bootstrapping as a method for assessing confidence in phylogenetic analysis. *Syst. Biol.* **42**, 182–192.

Hillis, D. M., and Huelsenbeck, J. P. (1992). Signal, noise, and reliability in molecular phylogenetic analyses. *J. Heredity* **83**, 189–195.

Holmquist, R. (1983). Transitions and transversions in evolutionary descent: An approach to understanding. *J. Mol. Evol.* **19**, 134–144.

Honeycutt, R. L., Nedbal, M. A., Adkins, R. M., and Janecek, L. L. (1995). Mammalian mitochondrial DNA evolution: A comparison of the cytochrome *b* and cytochrome *c* oxidase II genes. *J. Mol. Evol.* **40**, 260–272.

Hudson, R. R. (1992). Gene trees, species trees, and the segregation of ancestral alleles. *Genetics* **131**, 509–512.

Irwin, D. M., Kocher, T. D., and Wilson, A. C. (1991). Evolution of the cytochrome *b* gene of mammals. *J. Mol. Evol.* **32**, 128–144.

Kessler, L. G., and Avise, J. C. (1984). Systematic relationships among waterfowl (Anatidae) inferred from restriction endonuclease analysis of mitochondrial DNA. *Syst. Zool.* **33**, 370–380.

Kessler, L. G., and Avise, J. C. (1985). A comparative description of mitochondrial DNA differentiation in selected avian and other vertebrate genera. *Mol. Biol. Evol.* **2**, 109–125.

Kimura, M. (1980). A simple method for estimating evolutionary rates of base substitutions through comparative studies of nucleotide sequences. *J. Mol. Evol.* **16**, 111–120.

Kimura, M. (1981). Estimation of evolutionary distances between homologous nucleotide sequences. *Proc. Natl. Acad. Sci. U.S.A.* **78**, 454–458.

Kimura, M. (1983). "The Neutral Allele Theory of Molecular Evolution." Cambridge Univ. Press, Cambridge.

Kocher, T. D. (1992). PCR, direct sequencing, and the comparative approach. *PCR Methods Applic.* **1**, 217–221.

Kocher, T. D., and White, T. J. (1989). Evolutionary analysis via PCR. *In* "PCR Technology" (H. A. Erlich, Ed.). Stockton Press.

Kocher, T. D., Thomas, W. K., Meyer, A., Edwards, S. V., Pääbo, S., Villablanca, F. X., and Wilson, A. C. (1989). Dynamics of mitochondrial DNA evolution in animals: Amplification and sequencing with conserved primers. *Proc. Natl. Acad. Sci. U.S.A.* **86**, 6196–6200.

Kornegay, J. R., Kocher, T. D., Williams, L. A., and Wilson, A. C. (1993). Pathways of lysozyme evolution inferred from the sequences of cytochrome *b* in birds. *J. Mol. Evol.* **37**, 367–379.

Krajewski, C. (1989). Phylogenetic relationships among cranes (Gruiformes: Gruidae) based on DNA hybridization. *Auk* **106**, 603–618.

Krajewski, C., and Fetzner, J. W. (1994). Phylogeny of cranes (Gruiformes: Gruidae) based on cytochrome-*b* DNA sequences. *Auk* **111**, 351–365.

Krajewski, C., and King, D. G. (1995). Molecular divergence and phylogeny: Rates and patterns of cytochrome *b* evolution in cranes. *Mol. Biol. Evol.* (in press).

Kumar, S., Tamura, K., and Nei, M. (1993). "Mega: Molecular Evolutionary Genetics Analysis," version 1.01. University Park.

Kusmierski, R., Borgia, G., and Crozier, R. H. (1993). Molecular information on bowerbird phylogeny and the evolution of exaggerated male characteristics. *J. Evol. Biol.* **6**, 737–752.

Lanyon, S. M. (1988). The stochastic mode of molecular evolution: What consequences for systematic investigations. *Auk* **105**, 565–573.

Lanyon, S. M. (1992). Interspecific brood parasitism in blackbirds (Icteridae): A phylogenetic perspective. *Science* **255**, 77–79.

Lanyon, S. M. (1994). Polyphyly of the blackbird genus *Agelaius* and the importance of assumptions of monophyly in comparative studies. *Evolution* **48**, 679–693.

Lanyon, S. M., and Hall, J. G. (1994). Reexamination of barbet monophyly using mitochondrial DNA sequence data. *Auk* **111**, 389–397.

Meyer, A. (1994). Shortcomings of the cytochrome *b* gene as a molecular marker. *Trends Ecol. Evol.* **9**, 278–280.

Meyer, A., and Wilson, A. C. (1990). Origin of tetrapods inferred from their mitochondrial DNA affiliation to lungfish. *J. Mol. Evol.* **31**, 359–364.

Meyer, A., Kocher, T. D., Basasibwaki, P., and Wilson, A. C. (1990). Monophyletic origin of Lake Victoria cichlid fishes suggested by mitochondrial DNA sequences. *Nature (London)* **347**, 550–553.

Mindell, D. P., Knight, A., Baer, C., and Huddleston, C. J. (1996). Slow rates of molecular evolution in birds and the metabolic rate and body temperature hypotheses. *Mol. Biol. Evol.* **13** (in press).

Moore, W. S. (1995). Inferring phylogenies from mtDNA variation: Mitochondrial gene trees versus nuclear gene trees. *Evolution* **49**, 718–726.

Nei, M. (1987). "Molecular Evolutionary Genetics." Columbia Univ. Press, New York.

Nei, M. (1991). Relative efficiencies of different tree-making methods for molecular data. *In* "Phylogenetic Analysis of DNA Sequences" (M. Miyamoto and J. Cracraft, Eds.), pp. 90–128. Oxford Univ. Press, London.

Neigel, J. E., and Avise, J. C. (1986). Phylogenetic relationships of mitochondrial DNA under various demographic models of speciation. *In* "Evolutionary Processes and Theory" (S. Karlin and E. Nevo, Eds.), pp. 515–533. Academic Press, New York.

Pääbo, S., Irwin, D. M., and Wilson, A. C. (1990). DNA damage promotes jumping between templates during enzymatic amplification. *J. Biol. Chem.* **265**, 4718–4721.

Pamilo, P., and Nei, M. (1988). Relationships between gene trees and species trees. *Mol. Biol. Evol.* **5**, 568–583.

Quinn, T. W., Shields, G. F., and Wilson, A. C. (1991). Affinities of the Hawaiian goose based on two kinds of mtDNA data. *Auk* **108**, 585–593.

Saiki, R. K., Gelfand, D. H., Stoffel, S., Scharf, S. J., Higuchi, R., Horn, G. T., Mullis, K. B., and Erlich, H. A. (1988). Primer-directed enzymatic amplification of DNA with a thermostable DNA polymerase. *Science* **239**, 487–491.

Saitou, N., and Nei, M. (1986). The number of nucleotides required to determine the branching order of three species with special reference to the human–chimpanzee–gorilla divergence. *J. Mol. Evol.* **24**, 189–204.

Sheldon, F. H. (1987). Rates of single-copy DNA evolution in herons. *Mol. Biol. Evol.* **4**, 56–69.

Sheldon, F. H., Slikas, B., Kinnarney, M., Gill, F. B., Zhao, E., and Silverin, B. (1992). DNA–DNA hybridization evidence of phylogenetic relationships among major lineages of *Parus*. *Auk* **109**, 183–185.

Shields, G. F., and Helm-Bychowski, K. M. (1988). Mitochondrial DNA of birds. *In* "Current Ornithology," Vol. 5 (R. F. Johnston, Ed.), pp. 273–295. Plenum Press, New York.

Shields, G. F., and Wilson, A. C. (1987). Calibration of mitochondrial DNA evolution in geese. *J. Mol. Evol.* **24**, 212–217.

Short, L. L. (1982). "Woodpeckers of the World." Delaware Museum of Natural History, Greenville, Delaware.

Short, L. L., and Morony, J. (1970). A second hybrid Williamson's X red-naped sapsucker and an evolutionary history of sapsuckers. *Condor* **72**, 310–315.

Sibley, C. G., and Ahlquist, J. E. (1990). "Phylogeny and Classification of Birds: A Study in Molecular Evolution." Yale Univ. Press, New Haven, Connecticut.

Smouse, P. E., Dowling, T. E., Tworek, J. E., Hoeh, W. R., and Brown, W. M. (1991). Effects of intraspecific variation on phylogenetic inference: A likelihood analysis of mtDNA restriction site data in cyprinid fishes. *Syst. Zool.* **40**, 393–409.

Springer, M. S., and Krajewski, C. (1989). DNA hybridization in animal taxonomy: A critique from first principles. *Q. Rev. Biol.* **64**, 291–318.

Tamura, K., and Nei, M. (1993). Estimation of the number of nucleotide substitutions in the control region of mitochondrial DNA in humans and chimpanzees. *Mol. Biol. Evol.* **10**, 512–526.

Tarr, C. L., and Fleischer, R. C. (1993). Mitochondrial DNA variation and evolutionary relationships in the Amakihi complex. *Auk* **110**, 825–831.

Thomas, W. K., and Beckenbach, A. T. (1989). Variation in salmonid mitochondrial DNA: Evolutionary constraints and mechanisms of substitution. *J. Mol. Evol.* **29**, 233–245.

Wu, C.-I. (1991). Inferences of species phylogeny in relation to segregation of ancient polymorphisms. *Genetics* **127**, 429–435.

Zink, R. M. (1993). The geography of mitochondrial DNA variation, population structure, hybridization, and species limits in the fox sparrow (*Passerella iliaca*). *Evolution* **48**, 96–111.

Zink, R. M., and Avise, J. C. (1990). Patterns of mitochondrial DNA and allozyme evolution in the avian genus *Ammodramus*. *Syst. Zool.* **39**, 148–161.

Zink, R. M., and Dittmann, D. L. (1991). Evolution of brown towhees: Mitochondrial DNA evidence. *Condor* **93**, 98–105.

Zink, R. M., and Dittmann, D. L. (1993a). Population structure and gene flow in the chipping sparrow and a hypothesis for evolution in the genus *Spizella*. *Wilson Bull.* **105**, 399–413.

Zink, R. M., and Dittmann, D. L. (1993b). Gene flow, refugia, and evolution of geographic variation in the song sparrow (*Melospiza melodia*). *Evolution* **47**, 717–729.

Zink, R. M., Dittmann, D. L., and Rootes, W. L. (1991a). Mitochondrial DNA variation and the phylogeny of *Zonotrichia*. *Auk* **108**, 578–584.

Zink, R. M., Rootes, W. L., and Dittmann, D. L. (1991b). Mitochondrial DNA variation, population structure, and evolution of the common grackle (*Quiscalus quiscula*). *Condor* **93**, 318–329.

APPENDIX I Woodpecker and Piculet Specimens[a]

Species	Common name	Locale	Voucher number
Campephilus haematogaster	Crimson-bellied woodpecker	Esmeraldas, Ecuador	LSU 11786
		Darien, Panama	LSU 2188
Colaptes auratus	Northern flicker	Kentucky	WSU 8618
		Arizona	WSU 86101
Colaptes rupicola	Andean flicker	Pasco, Peru	LSU 8204
		Puno, Peru	LSU 3901
Dryocopus pileatus	Pileated woodpecker	Kentucky	WSU 8615
		Texas	WSU 8634
Melanerpes carolinus	Red-bellied woodpecker	Kentucky	WSU 8614
Picoides villosus	Hairy woodpecker	Arizona	WSU 86107
		California	WSU 86144
Piculus rubiginosus	Golden-olive woodpecker	Lambayeque, Peru	LSU 5162
		Lambayeque, Peru	LSU 5222
Picumnus aurifrons	Bar-breasted piculet	Bolivia	LSU 18254
		Bolivia	LSU 18479
Sphyrapicus varius	Yellow-bellied sapsucker	California	WSU 86148
		California	WSU 86149
Veniliornis callonotus	Scarlet-backed woodpecker	Lambayeque, Peru	LSU 5175
		Lambayeque, Peru	LSU 5178
Veniliornis nigriceps	Bar-bellied woodpecker	La Paz, Bolivia	LSU 1305
		Pasco, Peru	LSU 8176

[a] LSU, Louisiana State University, Museum of Natural Science; WSU, Wayne State University, Department of Biological Sciences.

APPENDIX II Mitochondrial Cytochrome *b* Sequences Aligned for 21 Specimens of Picidae, Representing 11 Species[a]

```
                                                                                                                                                  111 111 111 111
                                   111 111 111 122 222 222 223 333 333 333 444 444 444 455 555 555 556 666 666 666 777 777 777 788 888 888 889 999 999 999 000 000 000 011
                         123 456 789 012 345 678 901 234 567 890 123 456 789 012 345 678 901 234 567 890 123 456 789 012 345 678 901 234 567 890 123 456 789 012 345 678 901
Co. auratus 8618*        ??? ?TC GGA TCC CTC CTC GGT ATC TGC CTA CTA ACT CAA ?TC ATC ACA GGC CTC CTA CTT G?C ACA CAC TAC ACC GCC GAC ACA ACC CTA GCC TTC TCA TCC GTC GCC CAC
Co. auratus 86101        AAT TTC GGA TCC CTC CTC GGC ATC TGC CTA CTA ACT CAA ATC ATC ACA GGC CTC CTA CTT ?CC ACA CAC TAC ACC GCC GAC ACA ACC CTA GCC TTC TCA TCC GTC GCC CAC
Ca. haematogaster 11786* ??? ?TC GGA TCT CTC CTC GG? ATC TGC CTA ATA ACA CAA ATC ATT ACA GGC CTT CTA TTG GCC ACT CAC TAT ACT GCC GAC ACA ACC CTA GCC TTC TCA TCC GTC GCC CAC
Ca. haematogaster 2188   AA? TTC GGA TCT CTC CTC GGC ATC TGC CTA ATA ACA CAA ATC ATT ACA GGC CTT CTA TTG GCC ACT CAC TAT ACT GCC GAC ACA ACC CTA GCC TTC TCA TCC GTC GCC CAC
D. pileatus 8634*        AAT TTT GGA TCC CTC CTC GGC ATC TGC CTG ATA GTA CAA ATT ATC ACA GGC CTC CTC CTG GCC ACT CAC TAC ACC GCC GAC ACA ACC TTA GCC TTC TCA TCC GTC GCC CAC
D. pileatus 8615         ??? ??? GGA TCC CTC CTC GG? ATC TGC CTG ATA G?? CAA ATT ATC ACA GGC CTC CTC CTG GCC ACT CAC TAC ACC GCC GAC ACA ACC TTA GCC TTC TCA TCC GTC GCC CAC
M. carolinus 8614*       ??C TTC GGA TCT CTC CTA GGC ATC TGC CTA GTT ACA CAA ATC ATT ACA GGC CTA CTA CTC GCC ACC CAC TAC ACT GCT GAC ACA ACC CTA GCC TTC TCA TCC GTT GCC CAC
Picum. aurifrons 18254*  ??? TT? GGC TCC CTT CTA GGC ATT TGC CTA ATA ACA CAA ATC ATC ACA GGC CTA CTT ATG GCC ACC CAT TAT ACA G?C GAC ACG AAC CTG GCC TTC TCA TCC GTC GCC CAC
Picum. aurifrons 18479   ??G TTT GG? TCT C?? CT? GGG ATT ?G? CTA ATA ACA CAA ?TC A?C ACA GGT CTA CTT ATG GCC ACC CAT A?T ?CA GAC GAC ACG A?C CTG GTC TTC TCA TCC G?C GAC ?AC
Picul. rubiginosus 5162* AAA TTC GGA TCC CTC CTC GGC ATC TGC CTA CTA ACA CAA ATC ATT ACA GGC CTT TTA CTC GCT ACA CAC TAC ACC GCC GAC ACA ACC CTG GCC TTC TCA TCC GTC GCC CAC
Picul. rubiginosus 5222  ??? TTC GGA TCC CTC CTC GGC ATC TGC CTA CTA ACA CAA ATC ATT ACA GG? CTT TTA CTC GCT ACA CAC TAC ACC GCC GAC ACA ACC CTG GCC TTC TCA TCC GTC GCC CAC
Pico. villosus 86144*    AA? TTT GGA TCC CTC TTA GGC ATC TGC CTA GTA ACA CAG ATT GTC ACA GGC CTC CTA CTT GCT ACC CAC TAC ACC GCC GAC ACA ACC CTA GCC TTC TCA TCC GTC GCC CAC
Pico. villosus 86107     AA? TTT GGA TCC CTC TTA GGC ATC TGC CTA GTA GCA CAG ATT GTC ACA GGC CTC CTA CTT GCC ACC CAC TAC ACC GCC GAC ACA ACC CTA GCC TTC TCA TCC GTC GCC CAC
S. varius 86148*         ??? TTC GGA TCC CTT CTC GGA ATC TGC TTG TTA ACA CAA ATC ATT ACA GGC CTA CTA CTT GCC ACC CAC TAC ACT G?A GAC ACC ACC CTA GCC T?C TCG TCT GTC GCC CAC
S. varius 86149          ?CC TTC GGA TCC CTT CTC GGA ATC TGC TTG TTA ACA CAA ATC ATT ACA GGC CTT CTA CTT GCC ACC CAC TAC ACT G?A GAC ACC ACC CTA GCC T?C TCG TCT GTC GCC CAC
V. callonotus 5178*      AAC TTT GGA TCC CTC CTA GGC ATT TGC CTA GTA ACA CAA ATT GTC ACA GGC CTC CTG CTT GCT ACC CAT TAC ACT GCC GAC ACA ACC CTA GCC TTT TCT TCC GTC GCC CAC
V. callonotus 5175       AAC TTT GGA TCC CTC CTA GGC ATT TGC CTA GTA ACA CAA ATT GTC ACA GGC CTC CTG CTT GCT ACC CAT TAC ACT GCC GAC ACA ACC CTA GCC TTT TCT TCC GTC GCC CAC
V. nigriceps 8176*       AAA TTT GGA TCT CTT CTA GGC ATT TGC CTA ATA ACA CAA ATT GTC ACA GGC CTC CTG CTT GCT ACC CAC TAC ACT GCC GAC ACA ACC CTA GCC TTC TCT TCC GTC GCC CAT
V. nigriceps 1305        AA? TTT GGA TCT CTC CTA GGC ATT TGC CTA ATA ACA CAA ATT GTC ACA GGC CTC CTG CTT GCC ACC CAC TAC ACT GCC GAC ACA ACC CTA GCC TTT TCT TCC GTC GCC CAT
Co. rupicola 8204*       ?AT TTC GGA TCC CTC CTC GGC ATC TGC TTA CTA ACA CAG ATC ATC ACA GGC CTC TTA CTT GCC ACA CAC TAC ACC GCC GAC ACA ACC CTG GCC TTC TCA TCC GTC GCC CAC
Co. rupicola 3901        ?AT TTC GGA TCC CTC CTC GGC ATC TGC TT? CTA ACA CAG ATC ATC ACA GGC CTC TTA CTT GCC ACA CAC TAC ACC GCC GAC ACA ACC CTG GCC TTC TCA TCC GTC G?C CAC

                         111 111 111 111 111 111 111 111 111 111 111 111 111 111 111 111 111 111 111 111 111 111 111 111 111 111 111 111 111 122 222 222 222 222 222 222 222
                         111 111 112 222 222 222 333 333 333 344 444 444 445 555 555 555 666 666 666 677 777 777 778 888 888 888 999 999 999 900 000 000 001 111 111 111 222
                         234 567 890 123 456 789 012 345 678 901 234 567 890 123 456 789 012 345 678 901 234 567 890 123 456 789 012 345 678 901 234 567 890 123 456 789 012
Co. auratus 8618*        ACC TG? CGA AAT GTC CAA TAC GG? TGA ?TA ATC CGA AAC CTC CAC G?C AAT GGT GCC TCA TTC TTT TT? AT? T?T ATC TAC C?C CAC ATT GG? CG? GGA TTT TAC TAC GGA
Co. auratus 86101        ACC TGC CGA AAT GTC CAA TAC GGC TGA CTA ATC CG? AAC CTC CAC GCC AAT GGT GCC TCA TTC TTT TTC ATC TGC ATC TAC CTC CAC ATT GGA CGC GGA TTT TAC TAC GGA
Ca. haematogaster 11786* ACA TGC CG? A?? GTC CAA TAC GGT TGA ?TA ATC CG? AAT CTC CAC G?T AAC GGA GCC TCA TTC TTC TTC ATC TGA ATC TAC CTT CAC ATC GGA CGT GG? TTC TAC TAC GGA
Ca. haematogaster 2188   ACA TGC CGC AAC GTC CAA TAC GG? TGA CTA ATC CGC AAT CTC CAC GCT AAC GGA GCC TCA TTC TTC TTC ATC TGC ATC TAT CTT CAC ATC GGA CGT GGA TTC TAC TAC GGA
D. pileatus 8634*        ACC TGC CGA AAC GTT CAA TAC GGC TGA CTA ATT CGA AAC CTT CAT GCC AAC GGA GCC TCA TTT TTC TTC ATT TGC ATC TAC CTC CAC ATT GGA CGA GGA TTC TAC TAC GGG
D. pileatus 8615         ACC TG? CGA AAC GTT CA? TAC GG? TGA CTA ATT CGA AAC CTT CAT GCC AAC GGA GCC TCA TTT TTC TTC ATT TG? ATC TAC CTC CAC ATT GGA CGA GGA TTC TAC TAC GGG
M. carolinus 8614*       ACA TGC CGC AAC GTC CAA CAT GGC TGA CTT ATC CGC AAC CTT CAC GCT AAC GGA GCC TCA TTC TTC TTC ATC TGC ATC TAT CTG CAC ATC GGA CGG GGA CTC TAC TAT GGG
Picum. aurifrons 18254*  ACA TGC CGA AA? GTC CA? TAC GGT TGA ?TT ATC CG? AAC CTA CAC GCC AAC GG? GCC TCA TTC TTC TTT ATC TGA ATC TAC TTC CAC ATC GGA CGA GGA TTT TAC TAC GGA
Picum. aurifrons 18479   ACA TGC CGA A?? GTC CAG T?? GGT TG? CTT ATC CGC AA? C?A CAC G?C AGC GGC GCC TCA ?TC TTC TTT ATC T?A ATC TAC TAC C?C ATC GGG CGA GGA TTT T?? TAC GGA
Picul. rubiginosus 5162* ACC TGC CGA AAC GTT CAA TAC GGG TGA CTA ATC CGA AAC CTC CAC GCC AAC GGC GCC TCA TTC TTT TTC ATC TGT ATC TAC CTC CAC ATT GGA CGT GGA TTC TAC TAC GGA
Picul. rubiginosus 5222  ACC TGC CGA AAC GTT CAA TAC GGG TGA CTA ATC CGA AAC CTC CAC GCC AAC GG? GCC TCA TTC TTT TTC ATC TGT ATC TAC CTC CAC ATT GGA CGT GGA TTC TAC TAC GGA
Pico. villosus 86144*    ACA TGC CGC AAC GTT CAA TAT GGC TGA CTA ATC CGC AAC CTC CAT GCC AAC GGG GCC TCA TTC TTC TTC ATC TGC ATT TAC TTA CAC ATC GGA CGA GGA TTC TAC TAC GGA
Pico. villosus 86107     ACA TGC CGC AAC GTT CAA TAT GGC TGA CTA ATC CGC AAC CTC CAT GCC AAC GGG GCC TCA TTC TTC TTC ATC TGC ATT TAC TTA CAC ATC GGA CGA GGA TTC TAC TAC GGA
S. varius 86148*         ACA TGC CG? A?C GTC CAA TAC GGA TGA CTC ATC CGA AAC CTC CAC GCC AAC GGA GCC TCA TTC TTT TTC ATC TGT ?TC TAC CTA CAC ATC GGC CGG GGA TTC TAC TAC GGA
S. varius 86149          ACA TGC CGG AAC GTC CAA TAC GGA TGA CTC ATC CGA AAC CTC CAC GCC AAC GGA GCC TCA TTC TTT TTC ATC TGT ATC TAC CTA CAC ATC GG? CG? GG? TTC TAC TAC GGA
V. callonotus 5178*      ACA TGC CGC AAT GTC CAA TAC GGC TGA CTA ATC CGC AAT CTC CAT GCC AAC GGG GCC TCA TTC TTC TTC ATC TGT ATC TAC TTA CAC ATC GGG CGC GGG TTC TAC TAC GGA
V. callonotus 5175       ACA TGC CGC AAT GTC CAA TAC GGC TGA CTA ATC CGC AAT CTC CAT GCC AAC GGG GCC TCA TTC TTC TTC ATC TGT ATC TAC TTA CAC ATC GGG CGC GGA TTC TAC TAC GGA
V. nigriceps 8176*       ACA TGC CGC AAC GTC CAA TAC GGC TGA CTA ATC CGC AAC CTC CAT GCC AAC GGG GCC TCA TTC TTC TTC ATC TGT ATC TAC TTA CAC ATC GGA CGC GGA TTC TAC TAC GGA
V. nigriceps 1305        ACA TGC CGC AAC GTT CAA TAC GGC TGA CTA ATC CGT AAC CTC CAT GCC AAC GGG GCC TCA TTC TTT TTC ATC TGC ATC TAC TTA CAC ATC GGA CGT GGA TTC TAC TAC GGA
Co. rupicola 8204*       ACC TGC CGA AAC GTC CAA TAC GG? TGA CTA ATC CGA AAC CTT CAC GCC AAC GGC GCC TCA TTC TTT TTC ATC TGT ATC TAC CTC CAC ATC GGA CGA GGA TTC TAC TAC GGA
Co. rupicola 3901        ACC TGC CGA AAC GTT CAA TAC GG? TGA CTA ATC CGA AAC CTC CAC GCC AAC GG? GCC TCA TTC TTT TTC ATC TGT ATC TAC CTC CAC ATT GGA CGA GGA TTC TAC TAC GGA
```

(*Continues*)

APPENDIX II (*Continued*)

	222	222	222	222	222	222	222	222	222	222	222	222	222	222	222	222	222	222	222	222	222	222	222	222	222	223	333	333	333	333	333	333	333	333	333	333	333
	222	222	233	333	333	334	444	444	444	555	555	555	566	666	666	667	777	777	777	888	888	888	899	999	999	990	000	000	000	111	111	111	122	222	222	223	333
	345	678	901	234	567	890	123	456	789	012	345	678	901	234	567	890	123	456	789	012	345	678	901	234	567	890	123	456	789	012	345	678	901	234	567	890	123
Co. auratus 8618*	TCC	TAC	CTG	TTT	AAA	GAA	ACC	TGA	AAC	ACC	GGA	GTC	ATT	CTC	CTC	CTC	ACC	CTC	ATA	G?C	ACT	GCC	TTC	GT?	GGC	TAC	GTT	CTG	CCC	TGA	GGA	CAA	ATA	TCA	TTC	TGA	GGT
Co. auratus 86101	TCC	TAC	CTG	TTT	AAA	GAA	ACC	TGA	AAC	ACC	GGA	GTC	ATT	CTC	CTC	CTC	ACC	CTC	ATA	G?C	ACT	GCC	TTC	GTA	GGC	TAC	GTT	CTG	CCC	TGA	GGA	CAA	ATA	TCA	TTC	TGA	GGT
Ca. haematogaster 11786*	TCC	TAT	CTA	TTC	AAA	GAG	ACC	TGA	AAC	ACC	GGA	GTC	ATT	CTC	CTC	CTC	ACC	CTT	ATA	GCA	ACT	GCC	TTC	G?C	GGT	TAC	GTC	CTC	CCT	TGA	GGA	CAA	ATG	TCG	TTC	TGA	GGG
Ca. haematogaster 2188	TCC	TAT	CTA	TTC	AAA	GAG	ACC	TGA	AAC	ACC	GGA	GTC	ATT	CTC	CTC	CTC	ACC	CTT	ATA	GCA	ACT	G?C	TTC	GTC	GGT	TAC	GTC	CTC	CCT	TGA	GGA	CAA	???	???	???	???	???
D. pileatus 8634*	TCC	TAC	CTA	TTT	AAA	GAA	ACC	TGA	AAC	ACT	GGG	GTC	ATC	CTC	CTT	CTC	ACC	CTT	ATA	GCC	ACT	GCT	TTC	GTC	GGC	TAT	GTC	CTC	CCT	TGA	GGG	?AA	ATA	TCA	TTC	TGA	GGC
D. pileatus 8615	TCC	TAC	CTA	TTT	AAA	GAA	ACC	TGA	AAC	ACT	GGG	GTC	ATC	CTC	CTT	CTC	ACC	CTT	ATA	G?C	ACT	G?T	TTC	GTC	GGC	TAT	GTC	CTC	CCT	TGA	GGG	CAA	ATA	TCA	TTC	TGA	GGC
M. carolinus 8614*	TCA	TAC	CTA	TTT	AAA	GAA	ACG	TGA	AAC	ACC	GGA	GTC	ATC	CTT	CTT	CTC	ACC	CTC	ATA	GCC	ACT	GCC	TTC	GTC	GGC	TAC	GTT	CTC	CCC	TGA	GGA	CAG	ATA	TCA	TTC	TGA	GGG
Picum. aurifrons 18254*	TCT	TAC	CTC	T?C	AAA	GAG	A?C	TGA	AAC	ACA	GG?	GTC	ATC	CTC	CTA	CTC	ACC	CTA	ATA	GCC	ACC	GCC	TTC	GTC	GGT	TAC	GTC	CTC	CCC	TGA	GGA	CAA	ATG	TCA	TTC	TGA	GGG
Picum. aurifrons 18479	TCT	?AC	C?C	T?C	AAA	GAG	ACC	TGA	A?C	ACA	GGT	GTC	ATC	CTC	C?A	CTC	ACC	CTA	?TA	G?C	ACC	G??	T?C	GTC	GGT	TAC	GTC	CTC	CCC	TGA	GG?	CAA	AT?	T?A	TTC	TGA	GGG
Picul. rubiginosus 5162*	TCC	TAC	CTA	TTC	AAA	GAA	ACC	TGA	AAC	ACC	GGA	GTC	ATT	CTC	CTC	CTC	ACC	CTT	ATA	GGC	ACT	GCC	TTC	GTA	GGC	TAT	GTC	CTG	GCC	TGA	GGA	CAA	ATA	TCG	TTC	TGA	GGC
Picul. rubiginosus 5222	TCC	TAC	CTA	TTC	AAA	GAA	ACC	TGA	AAC	ACC	GGA	GTC	ATT	CTC	CTC	CTC	ACC	CTT	ATA	GGC	ACT	GCC	TTC	GTA	GGC	TAT	GTC	CTG	?CC	TGA	GGA	?AA	ATA	TCG	TTC	TGA	GGC
Pico. villosus 86144*	TCC	TAC	CTA	TTC	AAG	GAG	ACC	TGA	AAT	ACA	GGA	GTT	ATT	CTC	CTT	CTT	ACC	CTT	ATA	GCC	ACC	GCC	TTC	GTC	GGC	TAC	GTC	CTC	CCC	TGA	GGA	CAA	ATA	TCG	TTC	TGA	GGG
Pico. villosus 86107	TCC	TAC	CTA	TTC	AAG	GAG	ACC	TGA	AAT	ACA	GGA	GTT	ATT	CTC	CTT	CTT	ACC	CTT	ATA	GCC	ACC	GCC	TTC	GTC	GGC	TAC	GTC	CTC	CCC	TGA	GGA	CAA	ATA	TCG	TTC	TGA	GGG
S. varius 86148*	TCC	TAC	CTA	TTT	AAA	GAG	ACT	TGA	AAT	ACT	GGA	GTT	ATT	CTT	CTT	CTA	ACC	CTT	ATA	GCA	ACT	GCC	TTC	G?C	GGC	TAC	GTT	CTC	CCA	TGA	GGA	CAA	ATA	TCC	TTC	TGA	GG?
S. varius 86149	TCC	TAC	CTA	TTT	AAA	GAG	A?T	TGA	AAT	ACT	GGA	GTT	ATT	CT?	CTT	CTA	ACC	CTT	ATA	G?A	ACT	GCC	TTC	G?C	GG?	TAC	GT?	CTC	CCA	TGA	GGA	?AA	AT?	TCC	TTC	TGA	GGC
V. callonotus 5178*	TCC	TAC	CTA	TTT	AAA	GAG	ACC	TGA	AAC	ACA	GGA	GTC	ATC	CTT	CTC	CTC	ACC	CTC	ATA	GCC	ACC	GCC	TTC	GTC	GGC	TAC	GTC	CTC	CCT	TGA	GGA	CAA	ATA	TCA	TTC	TGA	GGG
V. callonotus 5175	TCC	TAC	CTA	TTT	AAA	GAG	ACC	TGA	AAC	ACA	GGA	GTC	ATC	CTT	CTC	CTC	ACC	CTC	ATA	GCC	ACC	GCC	TTC	GTC	GGC	TAC	GTC	CTC	CCT	TGA	GGA	CAA	ATA	TCA	TTC	TGA	GGG
V. nigriceps 8176*	TCC	TAC	TTA	TTT	AAA	GAA	ACC	TGA	AAC	ACA	GGA	GTC	ATC	CTT	CTC	CTC	ACC	CTC	ATA	GCC	ACC	GCC	TTC	GTC	GGC	TAC	GTC	CT?	CCC	TGA	GGA	CAA	ATA	TC?	??C	TGA	GGG
V. nigriceps 1305	TCC	TAC	TTA	TTT	AAA	GAA	ACT	TGA	AAC	ACG	GGA	GTC	ATC	CTT	CTC	CTC	ACC	CTC	ATA	GCC	ACC	GCC	TTC	GTC	GGC	TAC	GTC	CTC	CCC	TGA	GGA	CAA	ATA	TCA	TCC	TGA	GGA
Co. rupicola 8204*	TCC	TAC	CTG	TTC	AAA	GAA	ACC	TGA	AAC	ACT	GGA	GTC	ATT	CTC	CTC	CTC	ACC	CTC	ATA	GCC	ACT	GCT	TTC	GTA	GGC	TAT	GTC	CTG	CCC	TGA	GGG	CAA	ATA	TCG	TTC	TGA	GG?
Co. rupicola 3901	TCT	TAC	CTA	TTC	AAA	GAA	ACC	TGA	AAC	ACT	GGA	GTC	ATT	CTC	CTC	CTC	ACC	CTC	ATA	G?C	ACT	GCT	TTC	GTA	GGC	TAT	GTC	CTG	CCC	TGA	GGG	CAA	ATA	TCG	TTC	TGA	GGT

	333	333	333	333	333	333	333	333	333	333	333	333	333	333	333	333	333	333	333	333	333	333	444	444	444	444	444	444	444	444	444	444	444	444	444	444	444
	333	333	444	444	444	455	555	555	556	666	666	666	777	777	777	788	888	888	889	999	999	999	000	000	000	011	111	111	112	222	222	222	333	333	333	344	444
	456	789	012	345	678	901	234	567	890	123	456	789	012	345	678	901	234	567	890	123	456	789	012	345	678	901	234	567	890	123	456	789	012	345	678	901	234
Co. auratus 8618*	GCA	ACC	GTC	ATT	ACA	AAC	CTC	TTC	TCG	G?C	CTC	CCC	TAC	ATC	GGC	CAA	ACC	ATT	GTC	GAA	TGA	GCC	TGA	GGT	GGT	TTC	TCC	GTA	GAC	AAC	CCC	ACC	CTC	ACC	CGA	TTC	TTC
Co. auratus 86101	GCA	ACC	GTC	ATT	ACA	AAC	CTC	TTC	TCG	G?C	CTC	CCC	TAC	ATC	GGC	CAA	ACC	ATT	GTC	GAA	TGA	?CC	TGA	GGT	GGT	TTC	TCC	GTA	GAC	AAC	CCC	ACC	CTC	A?C	CGA	TTC	TTC
Ca. haematogaster 11786*	GGA	ACC	TTT	ATC	ACA	AAC	CTA	TTC	TCA	GCC	CTC	CCC	TA?	ATC	GGC	CAA	ACC	ATT	GTC	GAA	TGG	GCC	TGA	GG?	GGA	TT?	TC?	GTG	GA?	AAT	CCT	ACC	CTC	ACT	CGA	TTC	TTC
Ca. haematogaster 2188	???	???	GTT	ATC	ACA	AAC	CTA	TTC	TCA	GCC	CTC	CCC	TAC	ATC	GGC	CAA	ACC	ATT	GTC	GAA	TGG	GCC	TGA	GGC	GGA	TT?	TCC	GTG	GA?	AAT	CCT	ACC	CTC	ACT	CGA	TTC	TTC
D. pileatus 8634*	GCT	ACC	GTT	ATT	ACT	AAT	CTT	TTC	TCA	GCC	CTC	CCC	TAC	ATC	GGC	CAA	ACC	ATC	GTT	GAA	TGG	GCC	TGA	GGC	GGA	TTC	TCT	GTA	GAC	AAC	CCC	ACC	CTC	ACC	CGA	TTT	TTC
D. pileatus 8615	GCT	ACC	GTT	ATT	ACT	AAT	CT?	T?C	TCA	G?C	CTC	CCC	TA?	ATC	GGC	CAA	ACC	ATC	GTT	GAA	TGG	G?C	TGA	GG?	GGA	TTC	TCT	GTA	GAC	AAC	CCC	ACC	CTC	ACC	CGA	TT?	TTC
M. carolinus 8614*	GCT	ACT	GTT	ATC	ACA	AAC	CTA	TTT	TCA	GCT	TTC	CCT	TAC	ATC	GGC	CAA	ACC	ATT	GTC	GAA	TGG	GCC	TGA	GGA	GGA	TTC	TCT	GTA	GAC	AAC	CCA	ACA	CTC	ACC	CGA	TTC	TTC
Picum. aurifrons 18254*	GCG	ACC	GTA	ATC	ACA	AAC	CTA	TTC	TCA	GCA	TTC	CCC	TAC	ATC	GGC	CAA	ACC	ATC	GTC	GAA	TGA	GCT	TGA	GGG	GGA	TTC	TCA	GTA	GAC	AAC	CCA	ACC	CTT	ACC	CGA	TTC	TTC
Picum. aurifrons 18479	GCG	ACC	GTA	ATC	ACA	AAC	CTA	TTC	TCA	GCA	TTC	CCC	TAC	ATC	GGC	CAA	ACC	ATC	GTC	GAA	TGA	GCT	TGA	GGG	GGA	TTC	TCA	GTA	GAC	AAC	CCA	ACC	CTT	ACC	CGA	TTC	TTC
Picul. rubiginosus 5162*	GCA	A?C	GTT	ATT	ACA	AAC	CTC	TTC	TCA	GCC	CTC	CCT	TAC	ATT	GGC	CAG	ACT	ATC	GTT	GAA	TGA	GCT	TGA	GGT	GGG	TTC	TCC	GTA	GAC	AAC	CCC	ACC	CTT	ACC	CGA	TTC	TTC
Picul. rubiginosus 5222	GCA	A?C	GTT	ATT	ACA	AAC	CTC	TTC	TCA	GCC	CTC	CCT	TAC	ATT	GGC	CAG	ACT	ATC	GTT	GAA	TGA	GCT	TGA	GGT	GGG	TTC	TCC	GTA	GAC	AAC	CCC	ACC	CTT	ACC	CGA	TTC	TTC
Pico. villosus 86144*	GCA	??C	GTT	ATT	ACA	AAC	TTA	TTT	TCA	GCC	CTC	CCT	TAC	GTA	GGA	CAA	ACT	ATC	GTA	GAA	TGA	GCC	TGA	GGG	GGA	TTC	TCC	GTA	GAC	AAC	CCC	ACC	CTC	ACC	CGA	TTC	TTT
Pico. villosus 86107	G??	??C	GTT	ATT	ACA	AAC	TTA	TTT	TCA	GCC	CTC	CCT	TAC	GTA	GGA	CAA	ACT	ATC	GTA	GAA	TGA	GCC	TGA	GGG	GGA	TTC	TCC	GTA	GAC	AAC	CCC	ACC	CTC	ACC	CGA	TTC	TTT
S. varius 86148*	G??	A?C	GTC	ATC	ACA	AAC	CTA	TTC	TCA	GCC	TTC	CCT	TAC	ATT	GGC	CAA	ACC	ATT	GTT	GAA	TGA	GCC	TGA	GGG	GGA	TTT	TCC	GTA	GAC	AAC	CCT	ACA	CTC	ACC	CGA	TTT	TTT
S. varius 86149	GCA	ACC	GTC	ATC	ACA	AAC	CTA	TTC	TCA	GCC	TTC	CCT	TAC	ATT	GGC	CAA	ACC	ATT	GTT	GAA	TGA	GCC	TGA	GGG	GGA	TTT	TCC	GTA	GAC	AAC	CCT	ACA	CTC	ACC	CGA	TTT	TTT
V. callonotus 5178*	GCA	A?C	GTT	ATT	ACA	AAC	CTA	TTC	TCA	GCC	CTC	CCC	TAC	GTA	GGA	CAA	ACT	ATC	GTC	GAA	TGA	GCC	TGA	GGG	GGA	TTC	TCT	GTA	GAC	AAC	CCC	ACT	CTC	ACC	CGA	TTC	TTC
V. callonotus 5175	GCA	A?C	GTT	ATT	ACA	AAC	CTA	TTC	TCA	GCC	CTC	CCC	TAC	GTA	GGA	CAA	ACT	ATC	GTC	GAA	TGA	GCC	TGA	GGG	GGA	TTC	TCT	GTA	GAC	AAC	CCC	ACT	CTC	ACC	CGA	TTC	TTC
V. nigriceps 8176*	GCG	ACC	GTT	ATT	ACA	AAC	CTA	TTC	TCA	GCC	CTT	CCC	TAC	GTA	GGA	CAA	ACT	ATC	GTC	GAA	TGA	GCC	TGA	GGA	GGA	TTC	TCT	GTA	GAC	AAC	CCT	ACT	CTC	ACC	CGA	TTC	TTC
V. nigriceps 1305	GCA	ACC	GTT	ATT	ACA	AAT	TTA	TTC	TCA	GCC	CTC	CCC	TAC	GTA	GGA	CAA	ACT	ATC	GTC	GAA	TGA	GCC	TGA	GGA	GGA	TTC	TCT	GTA	GAC	AAC	CCC	ACT	CTC	ACC	CGA	TTC	TTC
Co. rupicola 8204*	GCA	ACC	GTC	ATC	ACA	AAC	CTC	TTC	TCA	GCC	CTC	CCC	TAC	ATT	GGC	CAG	ACC	ATT	GTT	GAA	TGA	GCC	TGA	GGC	GGA	TTC	TCC	GTA	GAT	AAC	CCC	ACC	CTC	ACT	CGA	TTC	TTC
Co. rupicola 3901	GCA	ACC	GTC	ATC	ACA	AAC	CTC	TTC	TCA	GCC	CTC	CCC	TAC	ATT	GGC	CAG	ACC	ATC	GTT	GAA	TGA	GCC	TGA	GGT	GGA	TTC	TCC	GTA	GAT	AAC	CCC	ACC	CTC	ACT	CGA	TTC	TTC

(*Continues*)

	444 444 444 444 444 444 444 444 444 444 444 444 444 444 444 444 444 444 455 555 555 555 555 555 555 555 555 555 555 555 555 555 555 555 555 555 555
	444 445 555 555 555 666 666 666 677 777 777 778 888 888 888 999 999 999 900 000 000 001 111 111 111 222 222 222 233 333 333 334 444 444 444 555 555
	567 890 123 456 789 012 345 678 901 234 567 890 123 456 789 012 345 678 901 234 567 890 123 456 789 012 345 678 901 234 567 890 123 456 789 012 345
Co. auratus 8618*	GCT CTC CAC TTC CTC CTC CCA TTC TTA ATC GCA GGC CTC ACC CTA ATT CAC TTC ACC TTC CTC CAC GAA TCC GGC TCA AAC AAC CCC CTC GGG ATC GTA TCT GAC TGT GAC
Co. auratus 86101	G?C CTC CAC TTC CTC CTC CCA TTC TTA ATC G?A GGC CTC ACC CTA ATT CAC TTC ACC TTC CTC CAC GAA TCC GGC TCA AAC AAC CCC CTC GGG ATC GTA TCT GAC TGT GAC
Ca. haematogaster 11786*	GCC TTA CAC TTC CTC CTC CCA TTC ATA ATC GCA GGC CTC ACC CTA GTC CAT TTC ACC TTT CTC CAC GAA TCA GGC TCA AAC AAC CCT CTA GGA ATT GTA TCC GAC TGC GAC
Ca. haematogaster 2188	GCC TTA CAC TTC CTC CTC CCA TTC ATA ATC GCA GGC CTC ACC CTA GTC CAT TTC ACC TTC CTC CAC GAA TCA GGC TCA AAC AAC CCT CTA GGA ATT GTA TCC GAC TGC GAC
D. pileatus 8634*	GCA CTT CAC TTC CTT CTT CCA TTC CTA ATC GCA GGT CTT ACC CTA ATC CAC TTT ACC TTC CTT CAC GAA TCA GGC TCG AAC AAC CCC CTT GGA ATC GTG TCC GAT TGT GAT
D. pileatus 8615	G?A CTT CAC TTC CTT CT? CCA T?C ATA ATC GCA GGT CTC ACC CTA ATC CAC TT? ACC TTC CTC CAC GAA TCA GGC TCG AAC AAC CCC CTC GGA ATC GTG TCC GAT TGT GA?
M. carolinus 8614*	GCC CTG CAC TTT CTC CTT CCA TTC ATA ATC GCA GGA CTT ACT CTA ATC CAT TTT ACC TTC CTC CAC GAA TCA GGC TCC AAC AAC CCC CTA GGA ATC GTA TCT GAT TGT GAC
Picum. aurifrons 18254*	GCC CTA CAC TTC CTC CTC CCC TTC ATA ?TC GCC GGG CTA ACC CTA ATC CAC TTC ACC TTC CTC CAC GA? TCG GG? TCC AAC AAC CCC CTC GGA ATC A?A TCC GAC TG? GA?
Picum. aurifrons 18479	GCC CTA CAC TTC CTC CTC CCC TTC ATA ATC GCC GGC CTA ACC CTA ATC CAC TTC ACC TTC CTC CAC GAA TCG GGC TCC AAC AAC CCC CTC GGA ATC ACA TCC GAC TG? GAC
Picul. rubiginosus 5162*	GCC CTC CAC TTC CTT CTC CCA TTC CTA ATC GCA GGT CTC ACC CTA ATC CAC TTC ACC TTC CTC CAC GAA TCT GGT TCA AAC AAC CCC CTC GGA ATC GTA TCC GAC TGT GAT
Picul. rubiginosus 5222	GCC CTC CAC TTC CTT CTC CCA TTC ?TA ATC GCA GGT CTC ACC CTA ATC CAC TTC ACC TTC CTC CAC GAA TCT GGT TCA AAC AAC CCC CTC GGA ATC GTA TCC GAT TGT GAT
Pico. villosus 86144*	G?C CTA CAC TTC CTC CTC CCA TTC CTA ATC GCA GGA CTC ACC CTA ATC CAT TTC ACT TTC CTC CAC GAA TCT GGC TCA AAT AAC CCC CTC GGA ATT GTA TCC GAC TGT GAC
Pico. villosus 86107	G?C CTA CAC TTC CTC CTC CCA TTC CTA ATC GCA GGA CTC ACC CTA ATC CAT TTC ACT TTC CTC CAC GAA TCT GGC TCA AAC AAC CCC CTC GGA ATT GTA TCC GAC TGT GAC
S. varius 86148*	G?T CTA CAC TTC CTT CTC CCA TTC ATA ATT GCT GGA CTC ACC CTA ATT CAC TTC ACC TTC CTC CAC GAA TCA GGC TCT AAC AAC CCC CTT GGA ATC GTA TCC GAT TGC GAC
S. varius 86149	G?T CTA CAC TTC CTT CTC CCA TTC ATA ATT GCT GGA CTC ACC CTA ATT CAC TTC ACC TTC CTC CAC GAA TCA GGC TCT AAC AAC CCC CTT GGA ATC GTA TCC GAT TG? GAC
V. callonotus 5178*	GCC CTA CAC TTT CTC CTC CCA TTC CTA ATT GCA GGA CTC ACC CTA ATC CAC TTC ACT TTC CTC CAC GAA TCC GGC TCG AAC AAC CCC CTA GGA ATC GTA TCC GAC TGC GAT
V. callonotus 5175	G?C CTA CAC TTT CTC CTC CCA TTC CTA ATT GCA GGA CTC ACC CTA ATC CAC TTC ACT TTC CTC CAC GAA TCC GG? TCG AAC AAC CCC CTA GGA ATC GTA TCC GAC TGC GAT
V. nigriceps 8176*	GCC CTA CAC TTT CTC CTC CCA TTC TTA ATT GCA GGG CTC ACC CTA ATC CAC TTC ACC TTC CTC CAC GA? TCT GGC TCG AAC AAC CCC CTC GGA GTC GTA TCC GAC TGC GAT
V. nigriceps 1305	G?C CTA CAC TTT CTC CTC CCA TTC TTA ATT G?A GGA CTC ACC CTA ATT CAC TTC ACT TTC CTC CAC GAA TCC GG? TCG AAC AAT CCC CTC GGA ATC GTA TCC GAC AG? GAT
Co. rupicola 8204*	GCC CTC CAC TTC CTC CTC CCA TTC CTA ATC GCA GG? CTT ACT CTA ATC CAT TTC ACC TTC CTC CAC GAA TCT GG? TCA AAC AAC CCT CTT GGA ATC GTA TCC GAC TGT GAC
Co. rupicola 3901	G?C CTC CAC TTC CTC CTC CCA TTC CTA ATC GCA GG? CTT ACT CTA ATC CAC TTC ACC TTC CTC CAC GAA TCT GGC TCA AAC AAC CCT CTT GGA ATC GTA TCC GAC TGT GAC

	555 555 555 555 555 555 555 555 555 555 555 555 555 556 666 666 666 666 666 666 666 666 666 666 666 666 666 666 666 666 666 666 666 666 666 666 666
	555 566 666 666 667 777 777 777 888 888 888 899 999 999 990 000 000 000 111 111 111 122 222 222 223 333 333 333 444 444 444 455 555 555 556 666 666
	678 901 234 567 890 123 456 789 012 345 678 901 234 567 890 123 456 789 012 345 678 901 234 567 890 123 456 789 012 345 678 901 234 567 890 123 456
Co. auratus 8618*	AAA ATC CCC TTC CAC CCC TAC TTC TCC GTA AAA GAC ATC CTA GGC TTC ATA TTC ATA CTC CTC CCC CTC ACG G?C CTA GCA CTA TTC TCA CCC AAC CTC CTA GGA GAC CCA
Co. auratus 86101	AAA ATC CCC TTC CAC CCC TAC TTC TCC GTA AAA GAC ATC CTA GGC TTC ATA TTC ATA CTC CTT CCC CT? A?G GCC CTA G?A CTA TTC TCA ?CC AAC CTC CTA GGA GAC CCA
Ca. haematogaster 11786*	AAA ATT CCC TTC CAC CCA TAC TTC TCC ACA AAA GAC ATC CTA GGA TTC ATA TTC ATA CTC CTC CCC CTC GTA GCC CTC GCA CTA TTC TCA CCT AAC CTG TTA GGT GAC CCC
Ca. haematogaster 2188	AAA ATT CCC TTC CAC CCA TAC TT? TCC ACA AAA GAC ATC CTA GGA TTC ATA TTC ATA CTC CTC CCC CTC GTA G?C CTC GCA CTA TTC TCA ?CT AAC CTG TTA GGG GA? ???
D. pileatus 8634*	AAA ATC CCC TTC CAC CCC TAC TTC TCC GTA AAA GAC ATC CTG GGC TTC ATA TTC ATG CTC CTC CCC CTT GTA ACC CTA GCA CTA TTC TCA CCC AAT CTC CTG GGG GAC CCA
D. pileatus 8615	AA? ATC CCC TTC CAC CCC TAC TTC TCC GTA AAA GAC ATC CTG GG? TTC ATA TTC ATG CTC CTC CCC CTT GTA ?CC CTA GCA CTA TTC TCA CCC AAT CTC ?TG GGG GAC CCA
M. carolinus 8614*	AAA ATC CCC TTC CAC CCT TAC TTC ACC GTA AAG GAC ATT CTT GGA TTC ATA CTT ATA TTC CTC CCC CTC ACA ACC CTA GCA CTA TTC TCC ?CC AAC CTT TTA GGT GAC CCA
Picum. aurifrons 18254*	AAA ATC CCA TTT CAT CCG TAC ??? ??C CTA AAA GAC ATC CTA GGA TTG GTA TTC GTG CTT CTC CCT TTA ACA ACC CTT TCC CTA TTC TTA CCC AAC CTC CTA GGT GAC CCA
Picum. aurifrons 18479	AAA ATC CCA T?? ??? ??? ??? ??? ??C CTA AAA GAC ATC CTA GGA TTG GTA TTC GTG CTT CTC CCT TTA ACA ACC CTT TCC CTA TTC TTA CCC AAC CTC CTA GGT GAC CCA
Picul. rubiginosus 5162*	AAA ATC CCC TTC CAC CCC TAC TTC TCT GTA AAA GAC ATC CTA GGC TTC ATA TTC ATA CTC CTC CCC CTC ACG G?C TTA GCA CTA TTC TCA CCC AAC CTC TTA GGA GAC CCA
Picul. rubiginosus 5222	AAA ATC CCC TTC CAC CCC TAC TTC TCT GTA AAA GAC ATC CTA GGC TTC ATA TTC ATA CTC CTC CCC CTC ACG GGC TTA GCA CTA TTC TCA CCC AAC CTC TTA GGA GAC CCA
Pico. villosus 86144*	AAA ATT CCC TTC CAC CCC TAC TTC TCC GTA AAA GAC ATC CTA GGA TTC ATA TTC ATG CTC CTC CCC CTC ACA GGC CTA GCA CTA TTT TCA CCC AAC CTC TTA GGC GAC CCA
Pico. villosus 86107	AAA ATT CCC TTC CAC CCC TAC TTC TCC GTA AAA GAC ATC CTA GGA TTC ATA TTC ATA CTC CTC CCC CTC ACA G?C CTA GCA CTA TTT TCA CCC AAC CTC TTA GGC GAC CCA
S. varius 86148*	AAA ATC CCA TTC CAC CCA TAC TTC TCT GTA AAA GAC ATC CTA GGA TTT ATA CTT ATA CTA CTC CCC CTT ACA GCC CTA GCA CTA TTC TCG CCC AAC CTC TTA GGT GAC CCA
S. varius 86149	AAA ATC CCA TTC CAC CCA TAC TTC TCT GTA AAA GAC ATC CTA GGA TTT ATA CTT ATA CTA CTC CCC CTA ?CA G?C CTA G?A CTA TTC TCG CCC AAC CTC TTA GGT GAC CCA
V. callonotus 5178*	AAA ATC CCC TTC CAC CCT TAC TTC TCC GTA AAA GAT ATC CTA GGA TTC ATA TTC ATA CTC TTC CCC CTC GCA ACC CTA GCA CTA TTC TCA CCC AAC CTC CTA GGA GAC CCA
V. callonotus 5175	AAA ATC CCC TTC CAC CCT TAC TTC TCC GTA AAA GAT ATC CTA GGA TTC ATA TTC ATA CTC TTC CCC CTC GCA ACC CTA GCA CTA TTC TCA CCC AAC CTC CTA GGA GAC CCA
V. nigriceps 8176*	AAA ATC CCT TTC CAC CCC TAC TTC TCC GTA AAA GAT ATC CTA GGA TTC ATA TTC ATA CTC CTC CCC CTC GTA ACC CTA GCA CTA TTC TCA CCC AAC CTC CTA GGA GAC CCA
V. nigriceps 1305	AAA ATC CCC TTT CA? CCC TAC TTC TCC ?TA AAA GAT ATC CTA GGA TTC ATA TTC ATA CTC CTC CCC CTC GT? ?CC CTA GC? CTA TTC TCA CCT AAC CTC CTA GGA GAC C?G
Co. rupicola 8204*	AAA ATC CCC TTC CAC CCC TAC TTC TCT GTA AAA GAC ATC CTA GGC TTC ATA TTT ATA CTC CTC CCC CTC ACA ACC CTA GCA CTA TTC TCA CCT AAC CTC CTA GGA GAC CCA
Co. rupicola 3901	AAA ATC CCC TTC CAC CCC TAC TTC TCT GTA AAA GAC ATC CTA GGC TTC ATA TTC ATA CTC CTC CCC CTC ACA ACC CTA GCA CTA TTC TCA CCC AAC CTC TTA GGG GAC CCA

(*Continues*)

APPENDIX II *(Continued)*

```
                              666 666 666 666 666 666 666 666 666 666 666 777 777 777 777 777 777 777 777 777 777 777 777 777 777 777 777 777 777 777 777 777 777 777 777 777 777
                              666 777 777 777 788 888 888 889 999 999 999 000 000 000 011 111 111 112 222 222 222 333 333 333 344 444 444 445 555 555 555 666 666 666 677 777 777
                              789 012 345 678 901 234 567 890 123 456 789 012 345 678 901 234 567 890 123 456 789 012 345 678 901 234 567 890 123 456 789 012 345 678 901 234 567
Co. auratus 8618*             GAA AAC TTC AC? CCC GCA AAC CCC CTA GTC ACT CCC CCC CAC ATC AAA CCC GAA TGG TAC TTC CTA TTC GCA TA? GCC ATT CTA CGC TCT ATT CCT AAC AAA CTA GGA GGC
Co. auratus 86101             GAA AAC TTC ACG CCC GCA AAC CCC CTA GTC ACT CCC CCC CAC ATC AAA CCC GAA TGG TAC TTC CTA TTC G?A TAT G?C ATT CTA CGC TCT ATT CCT AAC AAA CTA GGA GGC
Ca. haematogaster 11786*      GAA AAC TTC ACA CC? GC? AAC CCC CTG G?T ACA CC? CCC CAT AT? AAA CC? GAA TGA TAC TTC CTA TTT GGA TA? GTC ATC CTA CGC TCA ATC CCC AAC AAA ?T? GGA GGT
Ca. haematogaster 2188        ??? ??? ??? ??? ??? ??? ??? ??? ??? ??? ACA CCC CCC CA? ATT AAA CCC GAA TGG TAC TTC CTA TTT G?A TA? GCC ATC CTA CGC TCA ATC CCC AAC AAA CT? GGA GGT
D. pileatus 8634*             GAG AAC TTC ACA CCC GCA AAC CCA CTA GTC ACA CCC CCC CAC ATT AAA CCC GAA TGA TAC TTC CTA TTC G?A TAC GAC ATC CTA CGC TCC ATC CCC AAC AAG CTA GGA GGT
D. pileatus 8615              GAG AAC TTC ACA CCC GCA ??? CC? CTA G?C ACA CCC CCC CAC ATT AAA CCC G?A TGA TAC TTC CTA TTC G?A ?AC G?C ATC CTG CGC TCC ATC CCC AAC AAG CTA GGA GGC
M. carolinus 8614*            GAA AAC TTC ACA CCA GCA AAC CCC CTA GTA ACG CCC CCC CAC ATC AAA CCC GAA TGA TAC TTC TTA TTT GCT TAT GCT ATT CTC CGC TCT ATC CCA AAT AAA CTT GGA GGA
Picum. aurifrons 18254*       GAA AAC TTC ACA CCA GCA AAC CCC CTG ATC ACA CCA CCT CAC ATC AAA CCT GAA TGA TGC TCC CTA TTC CCA TAT GCC ATT TTA CAC TCA ATC CTT AAC AAA CTA GGA GG?
Picum. aurifrons 18479        GAA AAC TTC ACA CCA GCA AAC CCC CTG ATC ACA CCA CCT CAC ATC AAA CCT GAA TGA TGC TCC CTA TTC CCA TAT GCC ATT TTA CAC TCA ATC CTT AAC AAA CTA GGA GGG
Picul. rubiginosus 5162*      GAA AAC TTC ACA CCC GCA AAC CCC TTG GTC ACT CCC CCC CAC ATC AAG CCC GAA TGA TAC TTC CTG TTC GCA TAT GCC ATC CTA CGC TCC ATC CCC AAC AAA CTG GGA GGC
Picul. rubiginosus 5222       GAA AAC TTC ACA CCC GCA AAC CCC TTG GTC ACT CCC CCC CAC ATC AAG CCC GAA TGA TAC TTC CTG TTC GCA TAT GCC ATC CTA CGC TCC ATC CCC AAC AAA CTG GGA GGC
Pico. villosus 86144*         GAA AAC TTC ACA CCC GCA AAC CCC CTA GTT ACA CCC CCC CAT ATT AAA CCC GAA TGG TAC TTC CTA TTT GCA TAC GCC ATT CTA CGC TCC ATC CCT AAC AAA CTA GGA GGA
Pico. villosus 86107          GAA AAC TTC ACA CCC GCA AAC CCC CTA GTT ACA CCC CCT CAT ATT AAA CCC GAA TGG TAC TTC CTA TTT GCA TAC GCC ATT CTA CGC TCC ATC CCT AAC AAA CTA GGA GGA
S. varius 86148*              GAA AAC TTC ACA CCT GCA AAT CCC CGT GTC ACA CCC CCA ?AT ATC A?G CCT GAA TGA TAC TTT CTA T?C GTA TAC GCT ATC CTA CGC TCT ATC CCC AAT AAA ?TG GGA GGA
S. varius 86149               GAA AAC T?C ACA CCT GCA AAT CCC CGT G?C ACA CCC CCA ?AT ATC AAG CCT GA? TGA TAC TTT CTA T?C GTA TAC GCT ATC CTA CGC TCT ATC CCC AAT AAA CTG GG? GGA
V. callonotus 5178*           GAA AAC TTC ACA CCC GCA AAC CCC CTA GTC ACA CCC CCT CAC ATC AAA CCA GAA TGG TAC TTC CTA TTT GCA TAT GCT ATC CTA CGC TCA ATC CCC AAT AAA CTA GGA GGA
V. callonotus 5175            GAA AAC TTC ACA CCC GCA AAC CCC CTA GTC ACA CCC CCT CAC ATC AAA CCA GAA TGG TAC TTC CTA TTT GCA TAT GCT ATC CT? CGC TCA ATC CCC AAT AAA CTA GGA GGA
V. nigriceps 8176*            GAA AAT TTC ACA CCC GCA AAC CCC CTA GTC ACA CCC CCC CAC ATC AAA CCA GAG TGG TAC TTC CTA TTT G?A TAT GCC ATC CTA CGC TCA ATC CCC AAT AAA CTA GGA GGA
V. nigriceps 1305             GAA AAT TT? ACG CCC GCA AAC CCC CTA GT? ACA CCC CCC CAC ATC AAA CCA GAA TGG TAC TTC CTA TTT GCA TAT GCT ATC CTA CGC TCA ATC CCC AAT AAA CTA GGA GGA
Co. rupicola 8204*            GAA AAC TTC ACA CCC GCA AAC CCC CTA GTC ACT CCC CCC CAC ATC AAG CCC GAA TGA TAC TTC CTA TTT G?A TAT GCC ATC CTA CGC TCC ATC CCC AAC AAA CTA GGG GGT
Co. rupicola 3901             GAA AAC TTC ACA CCC GCA AAC CCC CTA GTC ACT CCC CCC CAC ATC AAA CCC GAA TGA TAC TTC CTA TTT G?A TAT G?C ATC CTA CGC TCC ATC CCC AAC AAA CTA GGG GGT

                              777 777 777 777 777 777 777 788 888 888 888 888 888 888 888 888 888 888 888 888 888 888 888 888 888 888 888 888 888 888 888 888 888 888 888 888 888
                              778 888 888 888 999 999 999 900 000 000 001 111 111 111 222 222 222 233 333 333 334 444 444 444 555 555 555 566 666 666 667 777 777 777 888 888 888
                              890 123 456 789 012 345 678 901 234 567 890 123 456 789 012 345 678 901 234 567 890 123 456 789 012 345 678 901 234 567 890 123 456 789 012 345 678
Co. auratus 8618*             GTC CTA GCC CTA GCC GCC TCT GTC CTA ATC CTA TTT CTA GCC CCT CTC CTC CAC ATA TCC AAG CAA CGC ACA ATG GCC TTC CGA CCA TTC TCC CAA TTC CTG TTT TGA ATA
Co. auratus 86101             GTC CTA GCC CTA GCC GCC TCC GTC CTA ATC CTA TTT CTA GCC CCT CTC CTC CAC ATA TCC AAG CAA CGC ACA ATG GCC TTC CGA CCA TTC TCC CAA TTC CTG TTT TG? ATA
Ca. haematogaster 11786*      GTC CTC GCT CT? GCT GCC TCC GT? CTA ATT CTA TTC ?TA GC? CCC CTC CTC CAC GTA TCC AAA CA? CGT ACT ATA GCC TTC ?GG CCC CTC TCC CAA CTC ?TA TTC TGA ATA
Ca. haematogaster 2188        GTC CTC GCT CT? GCT GCC TCC GT? CTA ATT CTA TT? CTA GCC CCC CTC CTC CAC ?TA TCC AAA CAA CGT ACC ATA GCC TTC CGG CCC CTC TCC CAA CTC CTA TTC TGA ?TA
D. pileatus 8634*             GTC CTG GCC CTA GCC GCC TCC GTC CTA GTC TTA TTC CTA A?A CCC CTC CTC CAT ATG TCC AAA CAG CGC ACA ATA GCC TTC CGA CCC CTC TCC CAA CTC CTA TTC TGA ACG
D. pileatus 8615              GTC CTG GCC CTA GCC GCC TCC GTC CTA GTC TTA TTC CTA AAA CCC CTC CTC CAT GTG TCC AAA CAG CGC ACA ATA GCC TTC CGA CCC CTC TCC CAA CTC CTA TTC TGA ACG
M. carolinus 8614*            GTC CTG GCC TTA GCC GCC TCC GTC CTA ATC CTA TTC CTA ATT CCC CTC CTC CAC ACA TCC AAA CAA CGT ACA ATA GCC TTC CGA CCC CTA TCC CAA CTC CTA TTC TGA ?CC
Picum. aurifrons 18254*       ATG CTA GCC TTA GCC GCC TCC ATC CTA GCT CTC TTC CTA CCT CAA TTT CTA CGC ATG TCC AAA CAG GAC ATA ATG GGC TTC TGG GCC CTC TCC CAA CTT CTT TTC TGA ACC
Picum. aurifrons 18479        ATG CTA GCC TTA GCC GCC TCC ATC CTA GCT CTC TTC CTA CCT CAA TTT CTA C?C ATG TCC AAA CAG GAC ATA ATG GGC TTC TGG GCC CTC TCC CAA CTT CT? TTC TGA ACC
Picul. rubiginosus 5162*      GTC CTA GCC CTA GCT GCC TCC GTC CTA ATC CTA TTC CTA GCC CCC CTC CTC CAC ACA TCC AAA CAA CGT ACA ATA GCC TTC CGA CCA CTC TCC CAA TTC CTA TTC TGA ATG
Picul. rubiginosus 5222       GTC CTA GCC CTA GCT GCC TCC GTC CTA ATC CTA TTC CTA GCC CCC CTC CTC CAC ACA TCC AAA CAA CGT ACA ATA GCC TTC CGA CCA CTC TCC CAA TTC CTA TTC TGA ATG
Pico. villosus 86144*         GTC CTA GCC CTA GCC GCC TCA GTC CTA ATC CTA TTC CTA GCC CCC CTC CTC CAC ACA TCT AAA CAG CGC ACA ATA GCC TTC CGA CCC CTC TCC CAA ATC CTG TTC TGA GTA
Pico. villosus 86107          GTC CTA GCC CTA GCC GCC TCA GTC CTA ATC CTA TTC CTA GCC CCC CTC CTC CAC ACA TCT AAA CAG CGC ACA ATA GCC TTC CGA CCC CTC TCC CAA ATC CTG TTC TGA GTA
S. varius 86148*              GTC CTA GCT CTA GCC GCC TCC GTC CTA ATC CTA TTC CTA GCC CCC CTC CTC CAC GTA TCT AAA CAA CGC ACA ATA GCC TTC CGG CCC CTC TCA CAA CTC CTG TTC TGA GCC
S. varius 86149               GTC CTA GCT CTA GCC GCC TCC GTC CTA ATC CTA TTC CTA GCC CCC CTC CTC CAC GTA TCT AAA CAA CGC ACA ATA ?CC TTC ?GG CCC CTC TCA CAA CTC CTG TTC TGA GCC
V. callonotus 5178*           GTC CTA GCC TTA GCT GCC TCA GTC CTA ATT CTA TTC CTA GCC CCC CTC CTT CAC ACA TCC AAA CAG CGC ACA ATA GCT TTC CGA CCC CTT TCC CAA CTT CTA TTC TGA ACA
V. callonotus 5175            GTC CTA GCC TTA GCT GCC TCA GTC CTA ATT CTA TTC CTA GCC CCC CTC CTT CAC ACA TCC AAA CAG CGC ACA ATA GCT TTC CGA CCC CTT TCC CAA CTT CTA TTC TGA ACA
V. nigriceps 8176*            GTC CTA GCC CTA GCT GCC TCA GTC CTA ATT CTA TTC CTA GCC CCC CTC CTT CAC ACA TCC AAA CAA CGC ACA ATA GCC TTC CGA CCC TTT TCC CAG CTT CTA TTC TGA ACA
V. nigriceps 1305             GTC CTA GCC CTA GCT GCC TCA GTC CTA ATT CTA TTC CTA GCC CCC CTC CTT CAT ACA TCC AAA CAA CGC ACG ATA GCC TTC CGA CCC TTT TCC CAA CTC CTA TTC TGA ATA
Co. rupicola 8204*            GTC CTA GCC CTA GCC GCC TCC GTT CTA GTC CTA TTC TTA GCC CCC CTC CTC CAC ACA TCC AAA CAG ?G? ACG ATA GCC TTC CG? CCA CTC TCC CAA TTC CTA TTT TG? ATG
Co. rupicola 3901             GTC CTA GCC CTA GCC GCC TCC GTT CTA GTC CTA TTT TTA GCC CCC TTT CTC CAC ACA TCC AAA CAG CGC ACG ATA GCC TTC CGA CCA CTC TCC CAA TTC CTA TTC TGA ATG
```

(Continues)

APPENDIX II *(Continued)*

	888	888	888	889	999	999	999	999	999	999	999	999	999	999	999	999	999	999	999	999	999	999	999	999	999	999	999	999	999	999	999	999	999	999	999	999	999
	899	999	999	990	000	000	000	111	111	111	122	222	222	223	333	333	333	444	444	444	455	555	555	556	666	666	666	777	777	777	788	888	888	889	999	999	999
	901	234	567	890	123	456	789	012	345	678	901	234	567	890	123	456	789	012	345	678	901	234	567	890	123	456	789	012	345	678	901	234	567	890	123	456	789
Co. auratus 8618*	CTC	GTC	GCG	AAC	CTC	CTC	ATC	CTT	ACC	TGA	ATC	GGC	AGC	CAG	CCA	GTA	GGA	CAC	CCC	TTC	ATC	ATC	ATC	GGC	CAG	TTA	GCC	TCT	ATC	ACC	TAC	TTC	TCC	ACA	ATC	CTC	ATC
Co. auratus 86101	CTC	GTC	GCG	AAC	CTC	CTC	ATC	CTT	ACC	TGA	ATC	GGC	AGC	CA?	CCA	GTA	GGA	CAC	CCC	TTC	ATC	ATC	ATC	GGC	CAG	TTA	?CC	TCT	ATC	ACC	TAC	TTC	TCC	ACA	ATC	CTC	ATC
Ca. haematogaster 11786*	CTA	GTC	G?C	AAC	CTC	TTC	ATT	CTC	AC?	TG?	AT?	GGC	AGC	CAA	CCA	GTA	GGA	CAC	CCC	TT?	ATC	ATT	A?T	GGC	CA?	?TA	?CC	TCC	?TT	ACC	?AC	TTC	??C	A?A	ATT	?TC	A?C
Ca. haematogaster 2188	CTA	GTC	GCC	AAC	CTC	CTC	AT?	CT?	AC?	TG?	ATT	GGC	AGC	CAA	CCA	GTA	GGA	CAC	CCC	TTC	ATC	ATT	A?T	GGC	CAA	?TA	?CC	TCC	ATT	ACC	TAC	TTC	TCC	ACA	ATT	?T?	A?C
D. pileatus 8634*	CTC	GTT	GCC	AAC	CTC	CTC	ATT	CTC	ACC	TGA	ATT	GGG	AGC	CAA	CCA	GTA	GAA	CAC	CCC	TTC	ATC	ATC	ATC	GGC	CA?	CTA	GCC	TC?	ATC	ACC	TAC	TTC	TCC	ACA	ATC	CTC	ATT
D. pileatus 8615	CT?	GTT	GCC	AAC	CTC	CTC	ATT	CTC	ACC	TGA	ATT	GG?	AGC	CAA	CCG	GTA	GA?	CAC	CCC	TTC	ATC	ATC	ATC	GGC	CAA	CTA	GCC	TCC	ATC	ACC	TAC	TTC	TCC	ACA	ATC	CTC	ATT
M. carolinus 8614*	CTA	?CC	GCC	AAC	CTT	CTC	ATC	CTT	ACC	TGA	GTC	GGC	AGC	CAA	CCA	GTA	GGA	CAC	CCC	TTT	ATC	ATT	ATT	GGC	CAA	CTA	?CC	TCC	ATC	GCC	TAC	TTC	TTC	ACC	ATC	CTG	ATT
Picum. aurifrons 18254*	CTA	GTT	?CC	AAC	CTT	CTC	ATC	CTA	ACC	TGG	GTA	GGC	AGC	CAA	CCA	G?A	GGA	CAT	CCC	TTT	ATT	ATC	ATT	GGC	CAA	CTA	?CC	TCC	ATC	ACA	?AC	TTC	ACC	ACC	ATT	CTA	GTC
Picum. aurifrons 18479	CTA	GTT	?CC	AAC	CTT	CTC	ATC	CTA	ACC	TGG	GTA	GGC	AGC	C?A	CCA	GTA	GGA	CAT	CCC	TTT	ATT	ATC	ATT	GGC	CAA	CTA	GCC	TCC	ATC	ACA	TAC	TTC	ACC	ACC	ATT	CTA	?TC
Picul. rubiginosus 5162*	CTT	GTC	GCA	AAC	CTC	CTC	ATC	CTT	ACC	TGA	ATC	GGC	AGC	CAA	CCA	GTA	GAA	CAT	CCC	TTC	ATC	ATT	ATC	GGC	CAG	TTA	?CC	TCC	ATC	ACC	TAC	TTC	TCC	ACA	ATC	CTC	ATC
Picul. rubiginosus 5222	CTT	GTC	GCA	AAC	CTC	CTC	ATC	CTT	ACC	TGA	ATC	GGC	AGC	CAA	CCA	GTA	G?A	CAT	CC?	TTC	ATC	ATT	ATC	GGC	CAG	?TA	?CC	TCC	ATC	ACC	TAC	TTC	TCC	ACA	ATC	CTC	ATC
Pico. villosus 86144*	CTA	GTC	GCC	AAC	CTC	CTC	GTC	CTT	ACC	TGA	ATC	GGC	AGC	CAA	CCA	GTA	GAA	CAT	CCC	TTC	ATC	ATT	ATC	GGC	CAA	TTA	GCC	TCC	ATC	GCC	TAT	TTT	TTC	ACA	ATC	CTA	ATC
Pico. villosus 86107	CTA	GTC	GCC	AAC	CTC	CTC	ATC	CTT	ACC	TGA	ATC	GGC	AGC	CAA	CCA	GTA	GAA	CAT	CCC	TTC	ATC	ATT	ATC	GGC	CAA	TTA	?CC	TCC	ATC	GCC	TAT	TTT	TTC	ACA	ATC	CT?	ATC
S. varius 86148*	TTA	GTT	?CC	AAC	CTC	CTC	ATC	CTT	ACC	TGA	G?T	GGC	AGC	CAA	CCA	G?A	GGA	CAC	CCC	TTC	ATC	ATT	ATT	GGT	CAA	CTA	GCC	TCC	ATT	ACC	TAC	TTC	CTC	ACA	ATC	CTA	TTC
S. varius 86149	TTA	GTT	?CC	AAC	CTC	CTC	ATC	CTT	ACC	TGA	G?T	GGC	AGC	CAA	CCA	G?A	GGA	CAC	CCC	TTC	ATC	ATT	ATT	GGT	CAA	CTA	GCC	TCC	ATT	ACC	TAC	TTC	CTC	ACA	ATC	CTA	TTC
V. callonotus 5178*	TTA	GTC	GCT	AAC	CTC	CTC	ATC	CTC	ACC	TGA	ATC	GGC	AGC	CAA	CCA	GTA	G?A	CAT	CCC	TTC	ATC	ATT	ATT	GGC	CA?	?TA	GCC	TC?	??T	?CC	TAC	TTC	TC?	?CG	ATT	CTA	ATC
V. callonotus 5175	TTA	GTC	GCT	AAC	CTC	CTC	ATC	CTC	ACC	TGA	ATC	GGC	AGC	CAA	CCA	GTA	G?A	CAT	CCC	TTC	ATC	ATT	ATT	GGC	C??	?TA	GCC	TCC	ATT	?CC	TAC	TTC	TCT	?CG	ATT	CTA	ATC
V. nigriceps 8176*	TTA	GTC	GCC	AAC	CTC	CTC	ATC	CTC	ACC	TGA	ATC	GG?	AG?	CAA	CCA	GTA	GAA	CAT	CCC	T??	???	???	???	???	???	???	???	???	???	???	???	???	???	???	???	???	???
V. nigriceps 1305	CTA	GTC	GCC	AAC	CTC	CTC	ATC	CTC	ACC	TGA	ATC	GG?	AG?	CAA	CCA	G?A	GAA	CAT	CCC	T??	???	???	???	???	???	???	???	???	???	???	???	???	???	???	???	???	???
Co. rupicola 8204*	CTC	GTT	G?A	AAC	CTC	CTC	ATC	CTC	ACC	TGA	ATC	GGC	AGC	CAA	CCA	GTA	GGA	CAC	CCC	TTC	ATC	ATC	ATC	GGC	CAG	TTA	GCC	TCT	ATC	ACC	TAC	TTC	TCC	ACA	ATC	CTT	ATC
Co. rupicola 3901	CTC	GTT	GGA	AAC	CTC	CTC	ATC	CTC	ACC	TGA	ATC	GGG	AGC	CAA	CCA	GTA	GAA	CAC	CCC	TCC	ATC	ATC	ATC	GGC	CAG	TTA	GCC	TCT	ATC	ACC	TAC	TTC	TCC	ACA	ATC	CTT	ATC

	111	111	111	111	111	111	111	111	111	111	111	111	111	111	111	111
	000	000	000	000	000	000	000	000	000	000	000	000	000	000	000	000
	000	000	000	011	111	111	112	222	222	222	333	333	333	344	444	444
	012	345	678	901	234	567	890	123	456	789	012	345	678	901	234	567
Co. auratus 8618*	CTC	TTC	CCA	CTC	ATC	AGC	ACC	TTA	GGA	AAC	?AA	CT?	CTG	GAC	CTC	TAA
Co. auratus 86101	CTC	TTC	CCA	CTC	ATC	AGC	ACC	TTA	GGA	AAC	?AA	CT?	CTG	GAC	CTC	TAA
Ca. haematogaster 11786*	CTT	T?C	CCC	CT?	A?T	AGC	ACT	??A	G?A	?AC	A?A	CT?	?TA	AAC	TTC	TAA
Ca. haematogaster 2188	CTT	TTC	CCC	CTC	ATT	AGC	ACT	?TA	G?A	A?C	?AA	CTG	?T?	AAC	?TC	TAA
D. pileatus 8634*	CTC	TTC	CCT	CTT	?TC	AGC	ACC	CTA	GGA	AAC	AA?	CT?	CTG	?AC	CTC	TAA
D. pileatus 8615	CTC	TTC	CCT	CTT	GGC	AGC	ACC	CTA	GGA	AAC	AAG	CTG	CTG	A?C	CTC	TAA
M. carolinus 8614*	CTC	TTT	CCC	CTA	ATC	AGC	ACC	CTA	GAG	?AC	?AA	CTA	TTA	AAT	TTC	TAA
Picum. aurifrons 18254*	CTA	TTC	CCG	CTT	ATT	AGT	GTA	ATA	GGA	AAT	AAG	CTA	CTA	AAC	TAC	TAA
Picum. aurifrons 18479	CTA	TTC	CCG	CTT	ATT	AGT	?T?	ATA	GGA	AAT	AAG	CTA	CTA	AAC	??C	TAA
Picul. rubiginosus 5162*	CTC	TTT	CCC	CTC	ATC	AGC	ACC	TTA	GGA	GAT	AAA	CTG	CTG	??C	CTC	TAA
Picul. rubiginosus 5222	CTC	TTT	CCC	CTC	ATC	AGC	ACC	TTA	GGA	GAT	AAA	CTG	CTG	AGC	CTC	TAA
Pico. villosus 86144*	CTT	CTC	CCT	CTT	ATC	AGC	TCC	CTA	GAG	GAC	?AA	TTA	CTA	AAC	TTC	TAA
Pico. villosus 86107	CTT	CTC	CCT	CTT	ATC	AGC	TCC	CTA	GAG	?AC	?AA	TTA	CTA	?AC	TTC	TAA
S. varius 86148*	CTC	TTC	CCC	CTT	ACT	AGC	GCT	CTA	GGA	AAC	AAA	CTA	CTA	AAT	TTC	TAA
S. varius 86149	CTC	TTC	CCC	CTT	ACT	AGC	GCT	CTA	GGA	AAC	AAA	CTA	CTA	AAT	TTC	TAA
V. callonotus 5178*	CT?	?TC	CCC	CTT	ATC	AGC	CTC	CTA	GAG	?AC	??A	CT?	CTT	?AC	TTC	TAA
V. callonotus 5175	CT?	CTC	CCC	CTT	ATC	AGC	?TC	CTA	GA?	??C	?AA	CTA	CT?	?AC	TTC	TAA
V. nigriceps 8176*	???	???	???	???	???	???	???	???	???	???	???	???	???	???	???	???
V. nigriceps 1305	???	???	???	???	???	???	???	???	???	???	???	???	???	???	???	???
Co. rupicola 8204*	CTC	TTC	CCC	CTC	ATC	AGC	ACC	TTA	GGA	AAC	?AA	TT?	CTG	?AC	CTC	TAA
Co. rupicola 3901	CTC	TTC	CCC	CTC	ACC	AGC	ACC	TTA	GGA	?AC	?AA	TT?	CTG	GAC	CTC	TAA

[a] Positions are numbered vertically; position 1 corresponds to position 14989 in the chicken sequence (Desjardins and Morais, 1990). The asterisks (*) indicate sequences used in the phylogenetic analysis.

CHAPTER 5

Phylogeny and Evolution of 12S rDNA in Gruiformes (Aves)

PETER HOUDE,* ALAN COOPER,†,1 ELIZABETH LESLIE,*,2 ALLAN E. STRAND,* AND GABRIEL A. MONTAÑO*

* *Department of Biology, New Mexico State University, Las Cruces, New Mexico 88003*
† *Molecular Genetics Laboratory, National Zoological Park, Washington D.C. 20008*

I. INTRODUCTION

The mid- to late nineteenth century was the heyday of comparative vertebrate taxonomy (e.g., Garrod, 1874; Gadow, 1892). Many subfamilial relationships were self-evident even before the Darwinian revolution. At higher taxonomic ranks,

[1] Present address: Institute of Molecular Medicine, Oxford University, OX3 9DU United Kingdom.
[2] Present address: Neuroscience Laboratory, University of Michigan, Ann Arbor, Michigan 48109.

however, there is a discontinuity in morphological similarity that still obfuscates the relationships of many families and orders of birds (e.g., compare Cracraft, 1981; Olson, 1985; Sibley and Ahlquist, 1990).

Gruiformes present many unsolved mysteries of systematics and biogeography. They include many highly diverged, depauperate or monotypic families scattered patchily the world over. Most previous attempts to resolve their phylogeny have yielded conflicting results.

Molecular data hold great promise for resolving problem phylogenies. The 12S rDNA is used here to address gruiform phylogeny because it includes both evolutionarily labile and conserved regions, hence it is believed to have a broad window of resolution for addressing recent and very ancient divergences (Kocher *et al.*, 1989; Mindell and Honeycutt, 1990; Simon *et al.*, 1994). We achieve some resolution of gruiform phylogeny and contribute some basic description of the evolution of the 12S rDNA gene. Complementary analysis of morphological characters is treated elsewhere (Houde, in preparation).

A. The Birds

Gruiformes traditionally include rails, coots, and gallinules (Rallidae, 120 species worldwide; e.g., *Gallinula*, *Laterallus*, *Rallus*), roatelos (Mesitornithidae, 3 species, Madagascar, e.g., *Mesitornis*), hemipodes (Turnicidae, 15 species, from Africa through southern Eurasia to Australia, e.g., *Turnix*), the Australian plains-wanderer[3] (Pedionomidae, 1 species, Australia, not used in this study), finfoots and sungrebe (Heliornithidae, 3 species, pantropical, e.g., *Heliornis*, *Podica*), the kagu (Rhynochetidae, 1 species, New Caledonia, i.e., *Rhynochetos*), the sunbittern (Eurypygidae, 1 species, South America, i.e., *Eurypyga*), trumpeters (Psophiidae, 3 species, South America, i.e., *Psophia*), seriemas (Cariamidae, 2 species, South America, e.g., *Cariama*), the limpkin (Aramidae, 1 species, Neotropics, i.e., *Aramus*), cranes [Gruidae, 14 species, cosmopolitan except South America, i.e., *Anthropoides* (= *Grus*), *Balearica*, *Grus*], and bustards [Otididae, 23 species, from Africa through southern Eurasia, to Australia, e.g., *Ardeotis* (= *Choriotis*)] (Sibley and Monroe, 1990). We use the familial nomenclature of Wetmore (1960).

At the morphological extremes of this assemblage are the small quail like hemipodes and the large long-legged cranes. Seriemas, trumpeters, kagu, and limpkin are lanky and superficially cranelike, with stubby tails (except seriemas). Seriemas, trumpeters, sunbittern, finfoots, kagu, and some rails are forest dwellers. Hemipodes, bustards, and cranes inhabit prairies or steppes, although cranes also inhabit wetlands. Limpkin, finfoots, and rails prefer watery habitats.

Phylogenetic Issues

Most of what is commonly believed about the relationships of Gruiformes can be traced to descriptive comparative anatomy and phyletic inferences made more than

[3] Lowercase lettering for proper names of birds has been used throughout this chapter to conform with the editorial policy of this book.

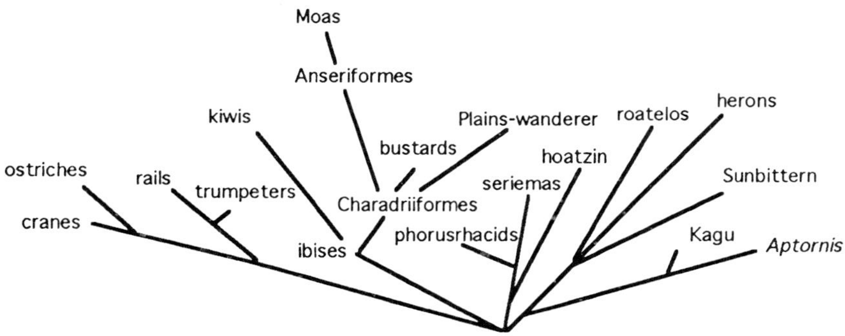

FIGURE 5.1 Relationships of Gruiformes and others according to Olson's (1985) phenetic assessment of morphology and paleontology (tree interpreted from text only). Branch lengths are not proportional to distance.

a century ago (e.g., Fürbringer, 1888; Gadow, 1892). Within the past three decades, however, new issues in gruiform phylogeny and systematics have been raised and old ones rekindled.

Olson worked mainly with rails (Olson, 1973, 1975, 1977, 1985), maintaining that trumpeters are sister to rails and resurrecting an old idea that the extinct flightless *Aptornis* (= *Apterornis*) from New Zealand is closer to kagu than to rails (Beddard, 1898 *vs* Brodkorb, 1967). As strong proponents of gruiform polyphyly (see Fig. 5.1), Olson and Steadman (1981) asserted that Australian plains-wanderer and bustards are Charadriiformes, not Gruiformes (Olson, 1985). Olson further suggested that sunbittern and/or roatelos are relicts of an ancient assemblage that includes herons (Olson, 1979, 1985). Olson considered ibises to be intermediate between Charadriiformes and Gruiformes. Olson and Steadman marshaled an eclectic assortment of (mostly) osteological traits in support of their hypotheses.

Cracraft erected a group, "Psophii," to which limpkin, then cranes are sisters, respectively (Cracraft, 1981, 1982; see Fig. 5.2). In it, trumpeters and seriemas form a sister clade to sunbittern and kagu plus *Aptornis*. The effort by Cracraft was the first attempt at a cladistic analysis of gruiform osteology, but no formal analysis was available for his 26 characters. Cracraft, like Olson, advocated a closer relationship between *Aptornis* and kagu than between kagu and sunbittern.

Sibley and colleagues came the closest to treating all traditionally recognized Gruiformes in a single analysis (Sibley and Ahlquist, 1985, 1990; see Fig. 5.3). They corroborated the treatment by Steadman and Olson of plains-wanderer. However, their more contemporary, noncommittal reconstructions were to replace the nearly fully resolved dichotomous trees of their earlier works (Sibley *et al.*, 1993). In their treatise, *Phylogeny and Classification of Birds*, hemipodes were placed as sister to all neognathous birds except fowl and waterfowl. They separated rails from all other Gruiformes at the subordinal rank "Ralli" (Sibley and Ahlquist, 1990). Roatelos were unstudied but also placed in their own suborder. Last, the suborder "Grues" consisted of a ladderized tree beginning apically with cranes, then limpkin plus sun

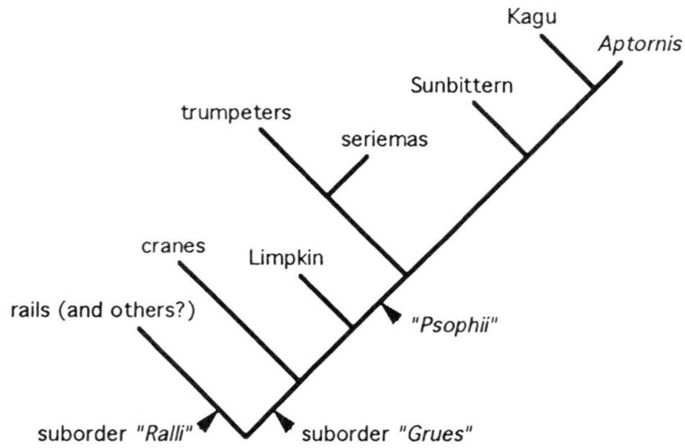

FIGURE 5.2 Relationships of Gruiformes and others according to Cracraft's (1982) cladistic analysis of morphological characters. Note the fundamental dichotomy of Gruiformes into suborders Ralli and Grues. Branch lengths are not proportional to distance.

grebe, trumpeters, seriemas or kagu or both, bustards, and then finally sunbittern at the base. It will be of interest below that in one figure they illustrated a sister relation of bustards and seriemas, even though in others they did not (1990: Fig. 335 *vs* Fig. 363). They stridently advocated a close sister relationship for limpkin and sun grebe, but not for kagu and sunbittern. However, both sungrebe and kagu were removed from their 1993 publication. In it, the only interfamilial relationship they advocated was between trumpeters and cranes. They found Gruiformes and Charadriiformes to be broadly indivisible (i.e., possibly not mutually monophyletic).

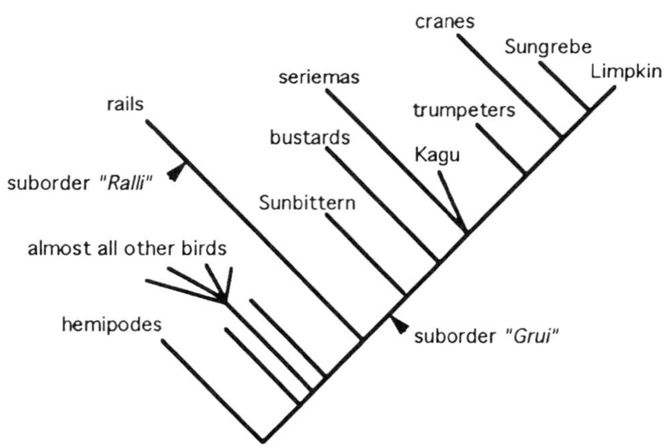

FIGURE 5.3 Relationships of Gruiformes and others according to Sibley and Ahlquist's (1985; 1990) reconstruction from DNA hybridization. Note the exclusion of hemipodes from Gruiformes and the fundamental dichotomy of Gruiformes into suborders Ralli and Grui. Branch lengths are not proportional to distance.

Houde argued that the treatment by Sibley and colleagues of the Neotropical sungrebe and limpkin had profound implications for interpretation of character polarity and biogeography (Houde, 1994; Houde *et al.*, 1995). Sibley and Ahlquist (1990) intimated that the two might be more closely related to one another than to the other two species of finfoots (are to sungrebe), which are endemic to Africa and Asia. Morphological character transformations constrained to this or similar topologies are one-third to one-half as parsimonious as on unconstrained morphology trees. Houde eventually dismissed the previous DNA hybridization results as irreproducible, rejected the monophyly of the limpkin–sungrebe clade, supported the monophyly of finfoots, and reaffirmed their longheld sister relationship to rails (Gadow, 1892; Sibley and Ahlquist, 1972; Olson, 1973; Cracraft, 1982; Houde, 1994).

Several questions are within the scope of the present analysis. (1) Are traditionally recognized Gruiformes monophyletic? In particular, should bustards and hemipodes be included in a monophyletic Gruiformes or Charadriiformes? Are seriemas related to secretary-bird? Are sunbittern or roatelos related to herons? (2) Apart from roatelos, does the first branch in Gruiformes separate all raillike birds from all cranelike birds, as in the subordinal classifications of Cracraft and Sibley and Ahlquist? (3) Are the Psophii of Cracraft monophyletic? If not, then are trumpeters sister to seriemas (as by Cracraft), cranes (as by Sibley and Ahlquist), or rails (as by Olson)? (4) Are finfoots monophyletic, and are they most closely related to the limpkin or to rails? (5) Are sunbittern and kagu sister taxa? (6) Is the fossil *Aptornis* more closely related to the kagu or to rails?

B. The Gene

12S rDNA is the smaller (about 1 kilobase) of two mitochondrial ribosomal DNAs. rDNA "gene" products are nonprotein-coding rRNAs that complex with proteins to form a ribosome. rRNAs fold onto themselves, like peptides, with evolutionarily conserved secondary structure (Fig. 5.4). The 12S gene can be subdivided into four principal domains, each of which includes self-complementing "stem" and single-stranded "loop" regions. Substitutions within stem regions may be selected against or precipitate compensatory (nonindependent) substitutions to maintain functional stem structure. Single-stranded regions may be involved in temporary base pairing with tRNAs and DNA templates during translation (Watson *et al.*, 1987), and in binding and cross-linking the many proteins that collectively make up the larger ribosomal particles (Noller *et al.*, 1990).

The popular wisdom that loops are evolutionarily more labile than stems is not entirely accurate (Vawter and Brown, 1993; Simon *et al.*, 1994). Some loops contain regions of variable length, but so do some stems. There are motifs within loops that are invariant, from microbes to vertebrates; while some positions within stems are among the most variable of sites in the gene.

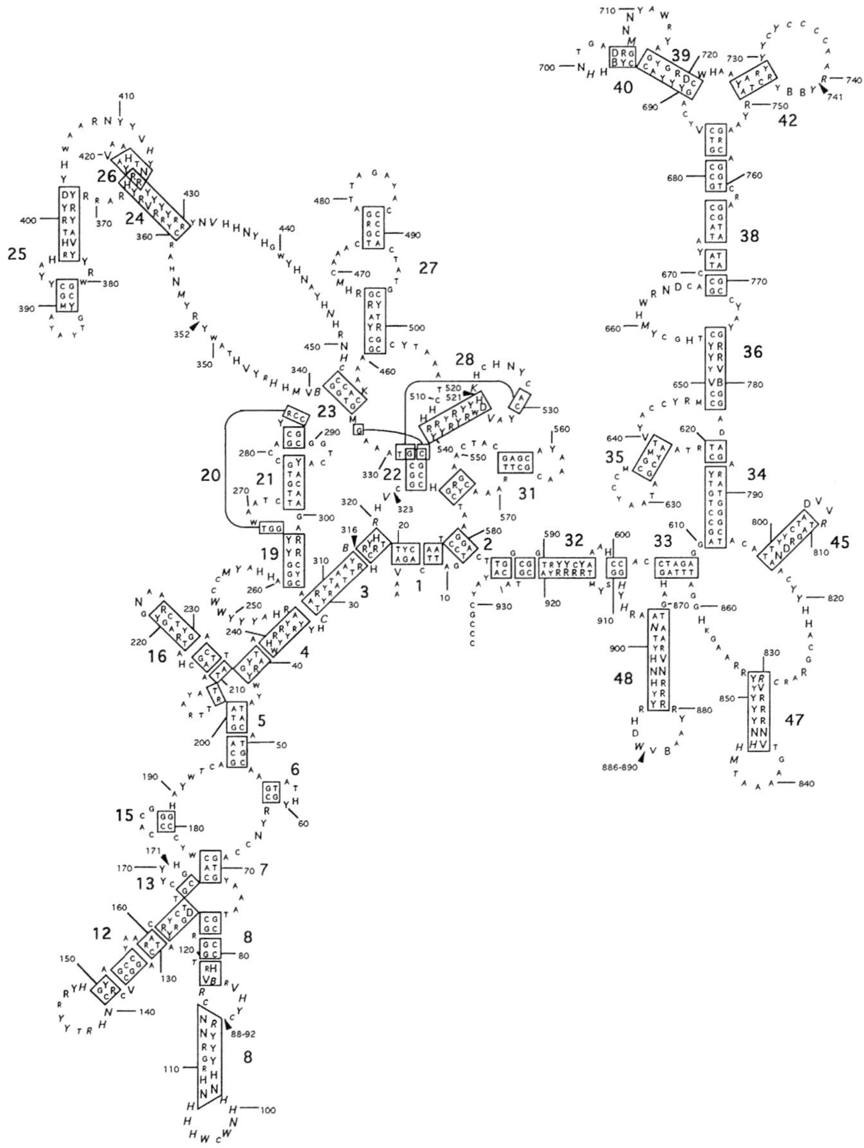

FIGURE 5.4 Mitochondrial 12S rDNA, hypothesized structure for domains I–III modified from Sullivan *et al.* (1995), with stems boxed and numbered according to Van de Peer *et al.* (1994). The sequence shown is a strict consensus for gruiform birds. Arrows indicate gaps in sequence numbering for alignment with outgroups. Font size of sequence is proportional to site diversity index (Shannon and Weaver, 1949; see Section II,A,6). Insertion/deletion sites are indicated by italics.

The rate of evolution of the 12S rDNA gene is believed to be appropriate to the level of phyletic divergence we aim to address, i.e., within late Cretaceous and Tertiary times (Mindell and Honeycutt, 1990). rDNA has been used for phylogenetic inference at greater and lesser taxonomic ranks (e.g., Kocher *et al.*, 1989; Simon *et al.*, 1994; Cummings *et al.*, 1995).

II. METHODS

A. DNA Sequence Data

1. Sources

Most DNA samples were obtained from ultrafrozen or chemically preserved soft tissues [Gruiformes: *Aramus guarauna*, *Ardeotis* (= *Choriotis*) sp., *Heliornis fulica*, *Mesitornis unicolor*, *Laterallus melanophaius*, *Rallus longirostris*, *Turnix* sp.; Charadriiformes: *Larus heermani*; Ciconiiformes: *Phimosus infuscatus*] or whole blood [Gruiformes: *Anthropoides* (= *Grus*) *virgo*, *Balearica pavonina*, *Cariama cristata*, *Eurypyga helias*, *Gallinula chloropus*, *Grus canadensis*, *Psophia crepitans*], but a few were obtained from museum skins and skeletons [Gruiformes: *Podica senegalensis*, *Rhynochetos jubatus*, *Aptornis* (= *Apterornis*) sp.]. Calcium salts were completely removed from bone by chelation with EDTA. DNA was isolated from tissues by proteinase K digestion, phenol–chloroform extraction, and ethanol precipitation and, when necessary, purification by glass milk or anion-exchange column (Sambrook *et al.*, 1989). The 12S rDNA gene was molecularly cloned by polymerase chain reaction (PCR) (Mullis and Faloona, 1987; Saiki *et al.*, 1988). The gene was amplified intact by priming on conserved flanking tRNAs that readily permit amplification by PCR in novel organisms (e.g., Kocher *et al.*, 1989), except when DNA derived from museum specimens was used (Houde and Braun, 1988; Cooper *et al.*, 1992). Sequencing templates were constructed by λ exonuclease digestion of asymmetrically phosphorylated PCR products (Higuchi and Ochman, 1989), which were sometimes gel purified (Kretz *et al.*, 1989). The method of sequencing was direct PCR sequencing using dideoxy chain termination (Sanger *et al.*, 1977), as modified (Engelke *et al.*, 1988; Sheen and Seed, 1988). Sequencing primers were spaced at internal sites across both strands, specific to chicken sequence, in conserved regions chosen by alignment of sequences of diverged lineages (e.g., Anderson *et al.*, 1981; Clary and Wolstenholme, 1985; Desjardins and Morais, 1990). Sequences were verified from both strands, except in the most 5′ region (varies between taxa). Negative controls lacking template DNA were run to check against contaminating DNA in reaction mixtures and pipettors. Negative-control DNA extracts of museum specimens were carried through every step from extraction to sequencing. DNA extraction and PCR setup were performed in a remote, dedicated PCR-free laboratory.

Approximately 870 bases representing domains I–III were obtained for all ingroup taxa except *Aptornis*, *Podica*, and *Rhynochetos* (GenBank accession numbers U76011–76027). The latter are represented by 673, 336, and 388 bases, respectively.

Unpublished complete 12S rDNA sequences of outgroups stone curlew (Charadriiformes: Burhinidae: *Burhinus oedinemus*), night heron (Ciconiiformes: Ardeidae: *Nyctannasa violacea*), and secretary bird (Falconiformes: Sagittariidae: *Sagittarius serpentarius*) were kindly provided by D. P. Mindell, and the complete sequence of chicken (Galliformes: Phasianidae: *Gallus gallus*; Desjardins and Morais, 1990), partial sequence of sandpiper, gull, and murre [Charadriiformes: Scolopacidae: *Calidris maritima* (X76362), Laridae: *Larus canus* (X76361), and Alcidae: *Uria aagle* (X76435), respectively; Moum *et al.*, 1994] and stork [Ciconiiformes: Ciconiidae: *Ciconia nigra* (L33370); Hedges and Sibley, 1994] were obtained from GenBank.

2. *Sampling Considerations*

We used 17 gruiform species to represent an order that includes 196. Thus, autapomorphies of the species sampled may be mistaken for synapomorphies of families (e.g., Patterson *et al.*, 1993). This problem is diminished because 8 of the 12 gruiform families include 3 or fewer species. Rallidae are the only family with more than 25. Our sampling in no way represents the nominal diversity within Rallidae but should address its interfamilial relations since rallid monophyly is supported by both molecular and morphological studies (Olson, 1973; Sibley and Ahlquist, 1990). We sampled 32% of all genera and 74% of the nonrallid genera in the order. All traditionally recognized gruiform families except the plains-wanderer are represented, making this the most comprehensive investigation of gruiform molecular systematics to date. There is agreement from molecular and morphological phenetic studies that the plains-wanderer is not gruiform (Olson and Steadman, 1981; Sibley and Ahlquist, 1985, 1990).

We sampled one or two individuals (rarely three) per species, with about equal frequency across taxa. Moore (1995) showed that internode lengths in four recently evolved woodpecker species were almost always longer than coalescence time for mitochondrial DNA (mtDNA). Thus, lineage sorting (Avise *et al.*, 1984) should rarely if ever corrupt phylogeny reconstruction for groups with equal or greater internode lengths (i.e., supraspecific levels; age of divergence is inconsequential). Gene phylogeny should adhere to organismal phylogeny, and between-species variation should exceed within-species variation. Accordingly, the two most closely related species in this study [*Grus canadensis* and *Anthropoides* (= *Grus*) *virgo*] display numerous transition substitutions. We never detected sequence differences in any two samples of one species, except one apparently heteroplasmic individual with a single transition substitution.

Small samples increase the risk that species misidentifications will go unnoticed. We detected one mislabeled specimen (*Balearica* mislabeled as *Cariama*) only because we sequenced other samples of both species. Voucher specimens (a live bird in this case) are no assurance against sample mislabeling.

3. Sequence Alignment

Sequence alignment was initiated with a pairwise similarity measure (MacVector 4.14; Needleman and Wunsch, 1970) and was improved by individual discretion (see below). Sequences were fitted to a map of secondary structure to identify complementary positions (e.g., Kjer, 1995). In so doing, discrepancies between opposite strands resulting from "compressions" (i.e., bases missing on one strand but not the other) were discovered and resolved (Fig. 5.5). The mapping of sequences onto structural models also served to monitor the possible existence of nuclear pseudogenes of mtDNA sequences (Fukuda *et al.*, 1985). The hypothesis of an endosymbiont origin of mitochondria predicts the existence of nuclear copies of mitochondrial genes because the mitochondrial genomes themselves are depauperate in housekeeping genes (Gray *et al.*, 1984). Inasmuch as nuclear pseudogenes are released from selective constraints, loss of conserved binding motifs and stem complementarity would be conspicuously absent in nuclear copies of mitochondrial rDNA. We detected no such instances.

Further improvement of alignment was made according to the principle of interactive phylogenetic weighting (Feng and Doolittle, 1987; Hein, 1990; Konings *et al.*, 1987; Lake, 1991; Mindell, 1991; Thorne and Kishino, 1992). Regions of

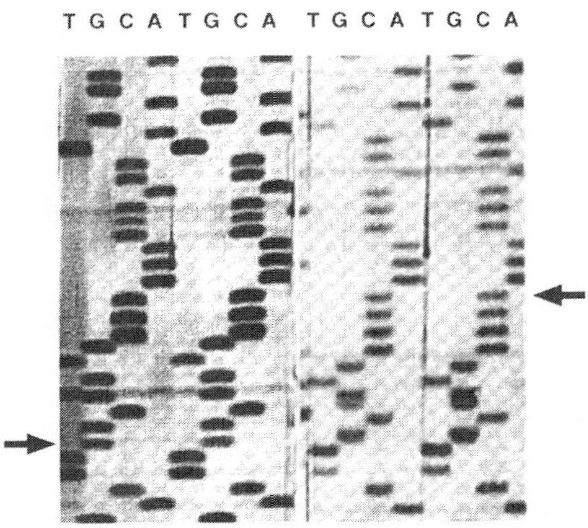

FIGURE 5.5 Sequencing artifact. Autoradiograph of sequencing gel showing a common sequencing artifact in mitochondrial 12S rDNA, domain III, stem 32 (*Eurypyga helias* shown). *Left*: Double loading of L strand. *Right*: Double loading of H strand, reverse complemented; arrows indicate G and C bases not evident on opposite strands.

variable length were subjected to successive bouts of phylogeny reconstruction separately from the complete data set using maximum parsimony following alterations of alignment (Section II,A,6). We wanted to determine how "badly" (i.e., counter to available phylogenetic information) alignments could be contrived before traditionally recognized monophyletic families no longer associated with themselves in phylogeny reconstruction (i.e., Gruidae, Rallidae, and Heliornithidae). Some effect of variable alignment was observed, but most often alignment had little or no consequence on phylogeny reconstruction in this study. In this data set synapomorphies of close taxa usually provided sufficient phylogenetic signal to reconstruct sister relationships, whether the synapomorphies are aligned to gaps or to a background of sequence "noise" of questionable homology (i.e., randomized sequence).

When no phylogenetic information was available, we strived to minimize any impact of alignment on phylogenetic inference. When phylogenetic information was available, we made alignments according to a parsimony principle of invoking the fewest number of changes between sequences from well-supported sister taxa. This could just as likely involve the insertion of gaps in nonhomologous positions to maintain sequence alignment as insertion of gaps at homologous positions. Sequence alignments were finalized according to a distance optimality criterion based on majority segregation of purines vs pyrimidines (i.e., minimizing inferred transversions across all taxa without reference to a hypothesis of phylogeny).

4. Transformation Weighting

As transformations saturate, the observed ratio of transversions to transitions deviates significantly from the instantaneous ratio (Brown *et al.*, 1982; Mindell and Honeycutt, 1990; Knight and Mindell, 1993). In other words, the ratio of transversions to transitions appears much closer to 1 : 1 for deep divergences than for shallow divergences. In spite of this, approximately the same intrinsic difference in rates of transversions to transitions probably occurred throughout the evolutionary history of a group, i.e., all levels of divergence. Thus, weighting schemes for phylogeny reconstruction should attempt to employ the instantaneous ratio rather than one averaged across all levels of divergence, including those saturated. Transversion weighting makes the difference between the recovery or lack of recovery of the traditionally recognized monophyletic clades Gruidae, Rallidae, and Heliornithidae in many of our phylogeny reconstructions. The monophyly of these families is supported in whole or part by a variety of morphological and DNA studies, employing both phenetic and cladistic methodologies (Olson, 1973, 1985; Sibley and Ahlquist, 1990; Krajewski, 1989; Houde, 1994; Krajewski and Fetzner, 1994).

We estimated a ratio of instantaneous rate of transversions to transitions from two most parsimonious phylogenies, one large and not known to be correct (including all 17 gruiform taxa herein), and the other small but believed to be correct (a ladderized tree of *Grus*, *Anthropoides*, *Balearica*, *Aramus*, *Psophia*, and *Gallus*). Both trees produced identical results on relative transformation rates. The ratio of observed transversions to transitions was expressed as a function of total substitutions. In several cases where no transversions were observed, they were assigned a value of 1 to preclude the biologically meaningless ratio of infinity. A second-order regression was fitted to the plots and the intercept at one substitution was calculated. The instantaneous ratio is 6:1, even though there was substantial scatter of observed values near the origin (from 2:1 to 19:0).

We also estimated the transformation ratio by fitting a two-parameter model of transversion and transition rates to both the large and small phylogenies (Kimura, 1980). We used TREECALC (Milligan, 1994) to find the transformation ratio with the maximum likelihood (Felsenstein, 1981) for the sequence matrix given a user-specified topology with specified branch lengths. Like the first approach, this estimates the instantaneous rate of transformations rather than an average rate of change over the entire phylogeny. The resulting ratio of 7.3:1 from both large and small phylogenies is in fairly good agreement with the previous estimate. We observed no difference in topology of optimal trees obtained by changing the weighting of transversions from 6 to 7.3 and we used the larger value for the phylogeny reconstructions presented here.

In reality, A/Y transversions appear to outnumber G/Y transversions in our data by about an order of magnitude. In light of the high frequency of transitions, the few G/Y transversions may primarily represent A/Y transversion followed by A/G transition. A weighting scheme of 6:1 for A/Y transversions and 60:1 for G/Y transversions produced similar bootstrap support for the same groups as the 7.3:1 weighting for all transversions, but performed worse in recovering Gruidae monophyly.

Gaps were treated in two ways in parsimony analyses: as missing data and with an intermediate weight of 4 (to satisfy the triangle inequality for weights of all transformation types) with 8:1 transversion-to-transition weights.

5. Position Weighting

The number of substitutions per site was estimated from the most parsimonious tree of 17 ingroup taxa obtained with 7.3-to-1 transversion weighting (Table I). Small

TABLE I Frequencies of Substitutions per Site

			Number of substitutions per site							
	Total	Variable	1	2	3	4	5	6	7	8
Number of sites	934	395	104	90	76	62	36	20	6	1
Percentage of sites	100	42.3	11.1	9.6	8.1	6.6	3.9	2.1	0.6	0.1
Percentage of substitutions	100	100	9.4	16.2	20.5	22.3	16.2	10.8	3.8	0.7

differences in tree topology (i.e., maximum likelihood and neighbor-joining trees) have almost no effect whatsoever on the number of substitutions per site. We nevertheless acknowledge that observed values both overestimate and underestimate actual values.

Four weighting schemes were applied in subsequent bouts of phylogeny reconstruction: equal (weight = 1), reverse (weight = x), inverse (weight = $1/x$), and quadratic weights (weight = $1/x^2$, where x = number of substitutions per site). Each was used with and without transformation weighting, on complete and partitioned data sets, and in jackknife and bootstrap analyses. Reverse and inverse weighting produced identical trees. Equal weights performed best overall at recovering traditionally recognized families. Quadratic weighting performed worst.

Stems were not weighted differently from loops to account for compensatory substitutions because stems, loops, and bulges evolve at about the same rate (Vawter and Brown, 1993).

Since among-site evolutionary rate variation is known to complicate phylogeny reconstruction (e.g., Milkman and Crawford, 1983; Huelsenbeck and Hillis, 1993), intuition dictates that position weighting would improve phylogenetic estimates. It did not appear to work well in this study.

6. Data Partitioning

Nucleotide data were analyzed in total and in subsets. Variable length and flanking regions that were subject to alternative alignment consisted of sites 82–116, 140–149, 245–259, 320–323, 339–347, 352–358, 409–424, 433–438, 520–528, 707–711, 808, 809, 817–822, and 902–908. Each of the aforementioned position and transformation weighting schemes was used in phylogeny reconstruction with and without these sites removed from the data set. Inclusion of variable-length regions tended to improve tree resolution and bootstrap support of nodes. Many synapomorphies of the traditionally recognized families occur within variable length regions.

We analyzed bases 500–920 (the "12Sa-b" region of Kocher *et al.*, 1989) separately to see whether this region was representative of the entire gene. In short, the answer is no. Although bootstrap values for some clades increased compared with those obtained from the entire data set, the monophyly of cranelike birds was lost.

We partitioned data according to number of changes per site and analyzed each class individually and in groups. This approach addresses among-site evolutionary rate variation and saturation of most-variable sites. Popular wisdom holds that sites that change most are most homoplasious (e.g., Sullivan *et al.*, 1995). Thus, one might rationalize their removal. We were surprised to discover, however, that phylogeny reconstruction using sites that change least yield thousands of equally parsimonious trees. Least-variable sites appear to be the only ones lacking phylogenetic information.

The inconsequence of saturation in our data is suggested by a regression of site consistency index, CI, to site diversity index, H' (Shannon and Weaver, 1949). H' is the sum of all nucleotide frequencies at a given position times the natural log of that nucleotide frequency. It is a measure of the amount of variation observed at a position. The maximum value is obtained by equal frequency of each of the four bases at a position; the minimal value is obtained by invariant sites. Site diversity is not a measure of substitution rate. A single substitution can result in either high or low H', depending on where in the phylogeny it occurred. The slope of the regression of H' and CI is -0.09 ($P = 0.0006$; variable sites only), indicating that CI does not vary as a function of site diversity in these data. In other words, most-diverse positions are no more or less consistent than least-diverse positions. Substitution rate may be correlated with consistency, but substitution rate and CI cannot be legitimately regressed because their calculations are not independent.

B. Phylogeny Reconstruction

Trees are constructed and evaluated using parsimony [MP, in MacClade 3.03, (Maddison and Maddison, 1992), and PAUP 3.1.1 (Swofford, 1993); all searches performed with heuristic algorithm and optimized by accelerated transformation], dynamically weighted parsimony (DWP; Williams and Fitch, 1990), maximum likelihood (ML, in PHYLIP 3.5; Felsenstein, 1989), and neighbor-joining [NJ, in MEGA 1.01, (Saitou and Nei, 1987; Kumar *et al.*, 1993) and PHYLIP 3.5 (Felsenstein, 1981, 1989)]. Figures of trees were created with TreeView (Page, 1996). Empirical base frequencies used in ML and NJ reconstructions are as follows: A, 0.32033; C, 0.28244; G, 0.19667; T, 0.20056.

The recovery of traditionally recognized groups—cranes and limpkin (Gruidae plus Aramidae), rails (Rallidae), and finfoots (Heliornithidae)—within larger phylogeny reconstructions is our standard for the reliability of reconstruction and weighting methods. The monophyly of each of these has traditionally been accepted (e.g., Wetmore, 1960), and is supported at least in part by DNA hybridization and sequences and morphology (Olson, 1973, 1985; Krajewski, 1989; Sibley and Ahlquist, 1990; Krajewski and Fetzner, 1994; Houde, 1994; Houde *et al.*, 1995; and reviews therein). We acknowledge the potential circularity of seeking answers to phylogenetic questions in trees that use the recovery of accepted groups as a standard for evaluating trees.

Dynamically weighted parsimony and NJ using gamma distances are methods that are specifically designed to cope with among-site evolutionary rate variation; yet, these performed worst at recovering expected clades. DWP approximates trees that are otherwise obtained from inverse and quadratic position weighting schemes in MP. We were particularly frustrated by the profound effect that seed trees have on final trees in the WTSUBS program. Had this not been a factor, then we would not have felt constrained by the limits on taxa (i.e., nine) that are allowable in an exhaustive search using the ALLTOPS program. We abandoned the NJ routine in MEGA after discovering that it could not accept IUPAC symbols. Two-parameter distances (Kimura, 1980) come closer than gamma distances (Tajima and Nei, 1984) to producing trees that recover traditionally recognized family groups.

All trees are rooted using chicken (*Gallus*). Although outgroups are ideally close sisters to the ingroup and chicken is not close to Gruiformes, certainty of outgroup status is paramount. Charadriiformes would be the obvious choice for outgroups if traditional classifications truly embody evolutionary history. However, all of the other "outgroup" taxa we examined in this study (Charadriiformes, Ciconiiformes, and Falconiformes) were chosen specifically to test for their potential relations as unrecognized ingroups. Chicken is the only taxon available that is known with certainty to be an outgroup.

III. PHYLOGENETIC INFERENCES

A. Results

Certain groupings of taxa were partly or entirely consistent in spite of reconstruction methods, weighting and data partitioning, and low bootstrap support (Felsenstein, 1985). The only clades with MP bootstrap values in excess of 95% in the complete data set with all weighting and data partitioning regimes were *Anthropoides–Grus*, *Rallus–Laterallus–Gallinula*, and *Eurypyga–Rhynochetos*. All NJ bootstrap values for the same clades were lower, suggesting poorer resolution. We believe the MP tree using 7.3:1 transversion weighting and no position weighting is the best estimate of phylogeny (Fig. 5.6). We restrict all further discussion to that tree unless stated otherwise. In it, *Heliornis–Podica* receives 92% bootstrap support (if gaps are also assigned weight, then support for finfoots is $> 95\%$). Figure 5.6 summarizes the results of jackknife (Lanyon, 1985) and bootstrap support for each node in the parsimony analysis under six different weighting schemes on the complete ingroup matrix. Most values are low, especially for deep nodes, and indicative of high levels of homoplasy. Higher bootstrap values for trees constructed with subsets of ingroup taxa are discussed further in the context of particular issues.

Maximum likelihood and NJ trees differ from the MP tree in movement of several branches by one node. Maximum likelihood also groups *Ardeotis* with finfoots (Fig. 5.7). Neighbor-joining groups *Psophia* and *Turnix* as sisters to cranes, and groups *Balearica* with *Aramus*, rather than with gruines (Fig. 5.8). Neighbor joining with variable-length regions of sequence omitted and 9:1 transversion weighting restores gruid monophyly, retains the position of *Psophia*, and places *Turnix* as sister to cranes plus rails. Maximum likelihood and NJ trees are both about 0.5% longer than the MP tree for character data. All branches on the ML tree are significantly positive ($p < 0.01$). All branches on the NJ tree are positive, but many internodes are extremely short (e.g., *Ardeotis–Cariama*).

Cranelike versus raillike birds do not comprise a fundamental dichotomy (i.e.,

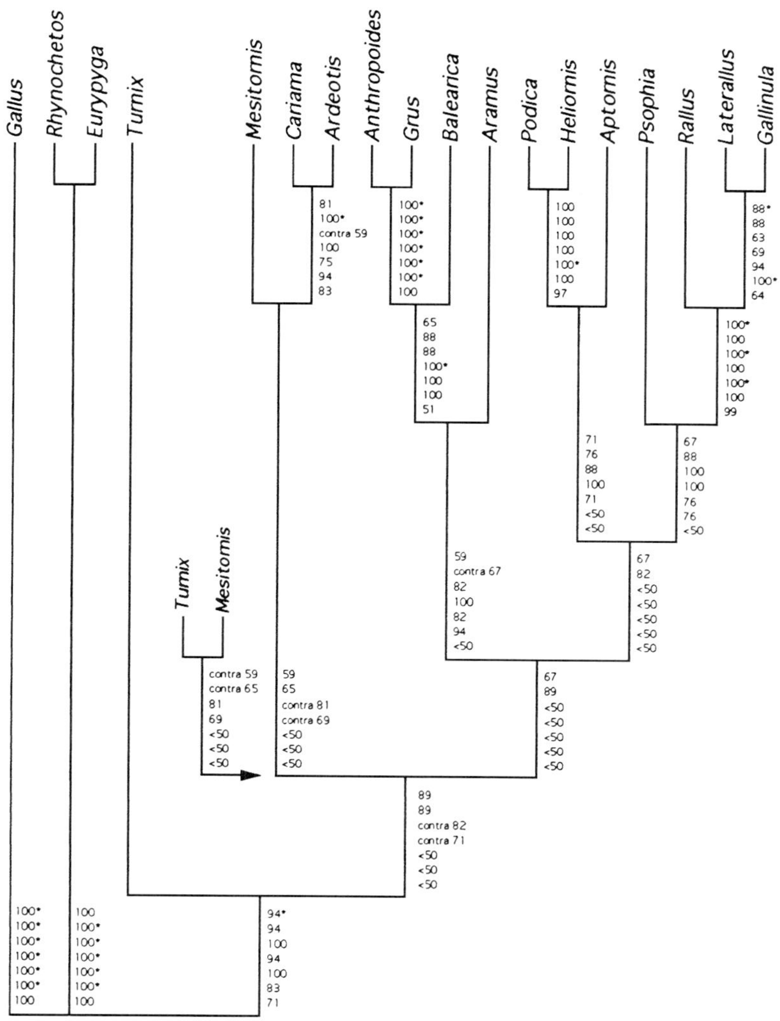

FIGURE 5.6 Parsimony jackknife and bootstrap tree of Gruiformes mitochondrial 12S rDNA, domains I–III, obtained with 7.3:1 transversion weighting using PAUP 3.1.1 (Swofford, 1991). First six numbers shown on each branch are jackknife (Lanyon, 1985) consensus values: (1) 7.3:1 transversion weighting, no position weighting, all characters; (2) 7.3:1 transversion weighting, inverse position weighting, all characters; (3) 7.3:1 transversion weighting, no position weighting, variable length regions excluded; (4) 7.3:1 transversion weighting, inverse position weighting, variable length regions excluded; (5) 8:1 transversion weighting, gaps weighted 4, no position weighting; and (6) 8:1 transversion weighting, gaps weighted 4, inverse position weighting, respectively. Seventh number is the highest bootstrap support obtained under any of the six weighting schemes (100 replicates). Asterisks indicate all weighting schemes that provided bootstrap values of 95% or higher or the single weighting scheme that yielded the highest bootstrap value if less than 95%. Jackknife values preceded by "contra" indicate majority consensus levels that contradict the topology shown. Alternate branching topology and values are shown for *Mesitornis* and *Turnix*. Variable length regions include sites 82–116, 140–149, 245–259, 320–323, 339–347, 352–358, 409–424, 433–438, 520–528, 707–711, 808, 809, 817–822, and 902–908.

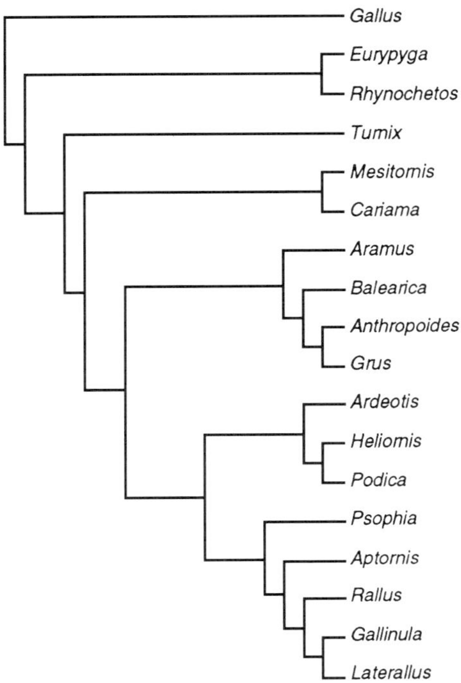

FIGURE 5.7 Maximum likelihood tree of Gruiformes mitochondrial 12S rDNA, domains I–III, obtained with 7.3 : 1 transversion weighting using PHYLIP 3.5 (Felsenstein, 1989); ln(likelihood) = −6669.4. All branches significantly positive ($p < 0.01$). Empirical base frequencies used are as follows: A, 0.32033; C, 0.28244; G, 0.19667; T, 0.20056.

most basal split) in Gruiformes in any of our reconstructions. Nine-taxon trees (*Anthropoides*, *Grus*, *Gallinula*, *Laterallus*, *Eurypyga*, *Rhynochetos*, *Gallus*, and any two of *Balearica* and/or *Aramus* and/or *Psophia*; 7.3 : 1 transversion parsimony; 1000 replicates) yielded bootstrap values >98% in support of a separation of cranes, trumpeter, and rails from sunbittern and kagu (Fig. 5.9). In almost all reconstructions from the complete ingroup data set, seriema, bustard, roatelo, and hemipode are outside of the cranes–rails clade.

Bootstrap support for the trumpeter–rails clade is lacking in the complete data set but rises to 90% in the "12Sa-b" subset with 4 : 1 transversion weighting. The ML and MP trees agree, but NJ always groups trumpeter with cranes. Parsimony tree lengths are trumpeter–cranes (1215, 30815, unordered and 7.3 : 1 transversion weights, respectively), trumpeters–rails plus finfoots (1210, 30629), and trumpeters–rails (1214, 30437) (*Aptornis* excluded for simplicity).

Our data support the monophyly of finfoots and their position in the clade of raillike birds, rather than with limpkin among the cranes as suggested by Sibley and Ahlquist (1985, 1990). But trumpeter is important for their placement near rails.

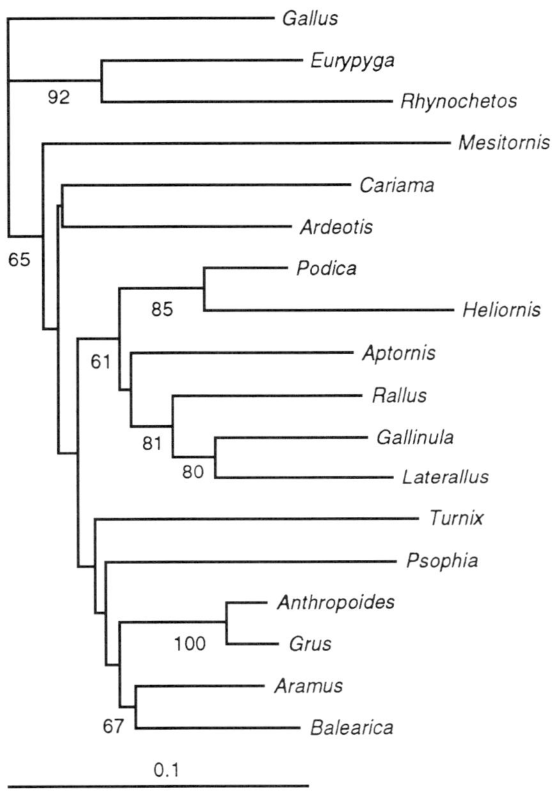

FIGURE 5.8 Neighbor-joining bootstrap tree of Gruiformes mitochondrial 12S rDNA, domains I–III, obtained with 7.3:1 transversion weighting and two-parameter distances (Kimura, 1980) using PHYLIP 3.5 (Felsenstein, 1989). All bootstrap values >50% are labeled (100 replicates). Branch lengths are calculated by least-squares fitting of distances to bootstrap tree. Empirical base frequencies used are as follows: A, 0.32033; C, 0.28244; G, 0.19667; T, 0.20056.

Finfoots group with raillike birds, except (1) when both trumpeter is removed and *Aptornis* is included, with or without transversion weighting (in which case they group with limpkin, sister to cranes), or (2) when gaps are weighted (in which case they group with roatelo, sister to cranes plus limpkin).

Unweighted trees placing *Aptornis* with kagu are 27 unordered steps longer (kagu positioned as in Fig. 5.6) or 102 steps longer (kagu positioned as in Cracraft's Psophii) than a grouping of *Aptornis* with rails. Bootstrap support for a sunbittern–kagu clade exclusive of *Aptornis* is 95–100% in virtually every tree examined. In a 6-taxon tree (*Aptornis*, *Eurypyga*, *Gallinula*, *Gallus*, *Laterallus*, *Rhynochetos*; 7.3:1 transversion parsimony; 1000 replicates) of only bases 200–920 for which the *Aptornis* sequence is complete, bootstrap support is 99% for rails, 99% for rails plus *Aptornis*, and 98% for sunbittern–kagu (Fig. 5.10). The decay index for the *Aptor-*

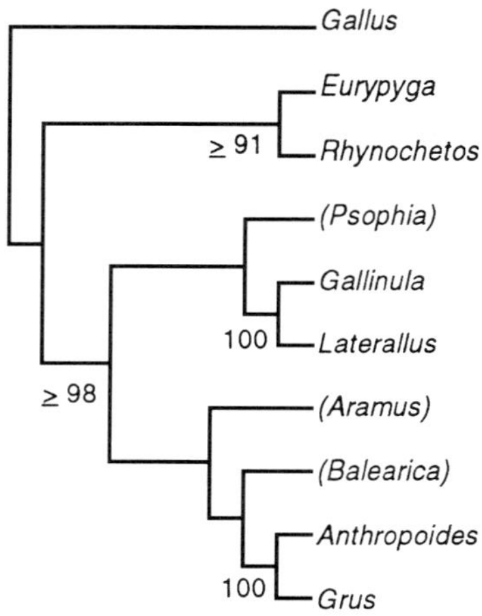

FIGURE 5.9 Consensus 9-taxon parsimony bookstrap tree of *Anthropoides*, *Grus*, *Gallinula*, *Laterallus*, *Eurypyga*, *Rhynochetos*, *Gallus*, and any 2 of *Balearica* and/or *Aramus and/or Psophia* (1000 replicates, heuristic parsimony with 7.3:1 transversion weighting of mitochondrial 12S rDNA, domains I–III, using PAUP 3.1.1; Swofford, 1991). Bootstrap values in excess of >90% in support of a cranes–rails clade, exclusive of sunbittern–kagu, are labeled. The *Psophii* of Cracraft are polyphyletic. Cranes and rails do not form a fundamental dichotomy in Gruiformes, as is widely presumed (compare to Figs. 5.2 and 5.3).

nis–rails node is 6% in this tree (expressed as a percentage to account for 7.3:1 transversion weighting). Although optimal transversion parsimony trees indicate a sister relationship between *Aptornis* and finfoots in the full data set, bootstrap support for this grouping is always ≤ 82%.

Bustard, seriema, roatelo, and hemipode tend to group in consistent positions, but none of the positions are well supported and some weighting regimes produce

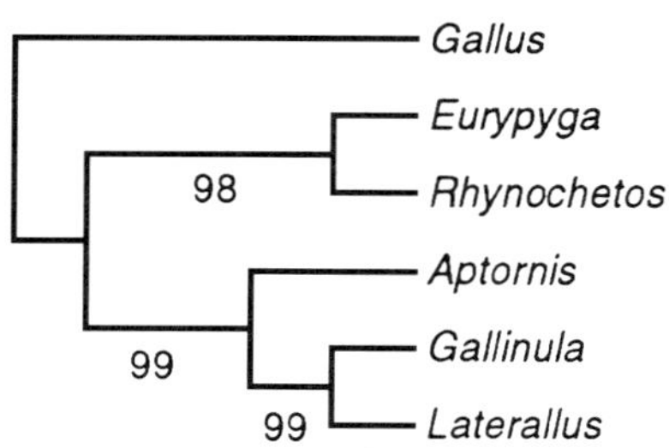

FIGURE 5.10 Parsimony bootstrap tree showing the relationship of *Aptornis* to rails, not kagu (1000 replicates heuristic parsimony with 7.3:1 transversion weighting, sites 200–920 of mitochondrial 12S rDNA, domains I–III, for which *Aptornis* sequence is complete, using PAUP 3.1.1; Swofford, 1991; compare to Fig. 5.2). Bootstrap values are labeled.

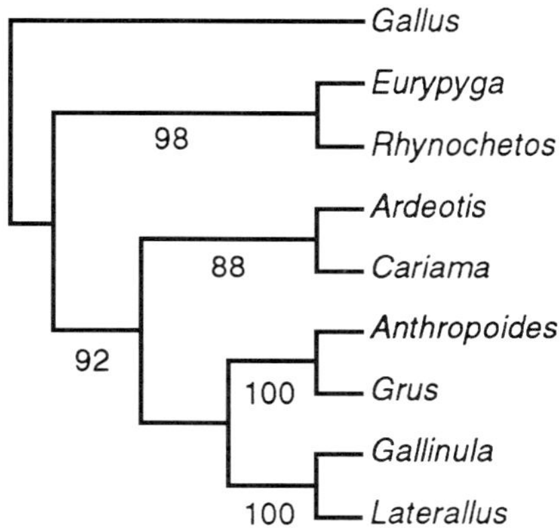

FIGURE 5.11 Parsimony bootstrap tree suggesting a relationship of bustards and seriemas (1000 replicates heuristic parsimony with 7.3:1 transversion weighting of mitochondrial 12S rDNA, domains I–III, using PAUP 3.1.1; Swofford, 1991; compare to Fig. 5.2). Bootstrap values are labeled.

different results. Most consistent is a tendency for seriema and bustard to form a clade that is sister to the cranes plus raillike birds. In a 9-taxon tree (*Anthropoides, Grus, Laterallus, Gallinula, Ardeotis, Cariama, Eurypyga, Rhynochetos, Gallus*; 7.3:1 transversion parsimony; 1000 replicates) bootstrap support for the bustard–seriema clade is 88% and its sister relation to cranes–rails is 92% (Fig. 5.11).

Seriema, bustard, roatelo, all Charadriiformes, heron, and ibis form a clade when all outgroups are included in transversion parsimony. Relaxation of weighting removes half of the Charadriiformes and both Ciconiiformes from this clade. None of these nodes receive strong bootstrap support.

Hemipode is sister to all Gruiformes except sunbittern–kagu. In an 8-taxon tree (*Anthropoides, Grus, Laterallus, Gallinula, Turnix, Eurypyga, Rhynochetos, Gallus*; 7.3:1 transversion parsimony; 1000 replicates), bootstrap support for a sister relationship of hemipode to cranes plus rails is 95% (Fig. 5.12).

The shortest MP tree without transversion weighting includes roatelo as a member of the raillike clade, sister to trumpeter. This, however, is only 2–10 steps shorter than 4 dissimilar trees, grouping roatelo with seriema and/or bustard or hemipode. Roatelo has a strong tendency to cluster with Charadriiformes (*Larus, Uria, Calidris*) and ibis (*Phimosus*), and always groups with gull–murre when they are included, regardless of weighting regime. The highest bootstrap values we obtained in support of roatelo–Charadriiformes is 87% (all taxa; 8:1 transversion parsimony, gaps weighted 4, inverse position weighting; 100 replicates).

B. Discussion

Several groups are consistently supported by our analyses, in spite of what may first appear to be conflicting results from different methods of reconstruction and low

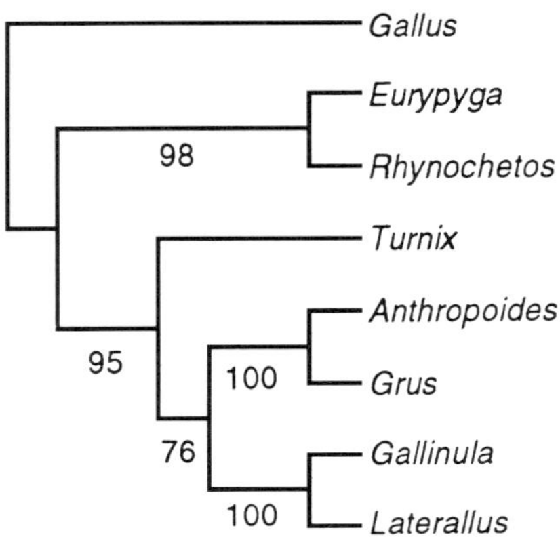

FIGURE 5.12 Parsimony bootstrap tree showing the relationship of hemipodes among Gruiformes (1000 replicates heuristic parsimony with 7.3 : 1 transversion weighting of mitochondrial 12S rDNA, domains I–III, using PAUP 3.1.1; Swofford, 1991; compare to Fig. 5.3). Bootstrap values are labeled.

bootstrap support. Many of the nodes with low bootstrap support in the complete data set receive high support with fewer taxa. While the complete phylogeny (Fig. 5.6) is not robust, it agrees with well-supported trimmed trees (Figs. 5.9–12). Questions posed in Section I,A are addressed in order here.

1. Are traditionally recognized Gruiformes monophyletic? Taken at face value, our data suggest that Gruiformes are not monophyletic. Kagu and sunbittern always assume a position more basal than putative outgroups, except chicken. Bustards, seriemas, and roatelos form a clade with Charadriiformes, so Gruiformes may be paraphyletic. This accords with the description by Olson (1985) of the osteology of bustards as resembling glareolid Charadriiformes and the inability of Sibley *et al.* (1993) to consistently separate Gruiformes from Charadriiformes. We did not address the monophyly of Charadriiformes in our study.

Hemipodes are not distant outgroups to Gruiformes as proposed by Sibley and Ahlquist (1990), but stone curlew (*Burhinus*) appears to be closely related to them in the absence of transformation weighting.

Seriema never groups with secretary bird, as suggested by Verheyen (1957). Instead, they appear to be ecologically, behaviorally, and morphologically convergent.

Neither sunbittern nor roatelos group with herons, as suggested by Olson (1979, 1985). However, roatelos are the most problematic taxon studied. Their position is unstable and poorly supported in our reconstructions. This is unfortunate since morphology has thus far provided little guidance on their relationships and we provide their first molecular data here. Preliminary observations without formal character analysis reveal no compelling morphological synapomorphies to unite roatelos with either Charadriiformes or Gruiformes (P. Houde, personal observation). Their alliance with Charadriiformes among Gruiformes in our gene phylogeny may be spurious.

Roatelos have probably survived a long period of isolation in Madagascar. Their pairwise genetic distances to ingroups are larger than for any other pair of taxa, even exceeding all other chicken–Gruiformes distances. Thus, roatelos either represent an ancient lineage, unrelated to Gruiformes, or an unusually rapid rate of evolution has erased any evidence of that relationship in their 12S rDNA. If the latter, then this is quite the opposite of the evolutionary slow-down of some other insular Malagasy endemics (Bonner *et al.*, 1981).

2. Apart from roatelos, does the first branch in Gruiformes separate all rail-like birds from all cranelike birds, as in the subordinal classifications of Cracraft (1982) and Sibley and Ahlquist (1990)? No, all methods of reconstruction strongly support a clade that includes cranes and raillike birds to the exclusion of seriemas, hemipodes, kagu, and sunbittern, and probably roatelos and bustards. The cranes and raillike birds appear to be a monophyletic group.

3. Are the Psophii of Cracraft monophyletic? Clearly not, as trumpeters and *Aptornis* are among the cranes–rails clade, seriemas appear to be sisters to bustards, and kagu and sunbittern are far removed from all others. Reconstructions showing monophyly of the Psophii of Cracraft are 60 unordered steps longer than optimal trees. While *Aptornis* is closely related to rails, the same is only weakly supported for trumpeters. Our result explains the difficulty people have had in distinguishing the position of trumpeters relative to rails and cranes. Trumpeters truly are intermediate between the other two. Olson's (1973) analogy of trumpeters to primitive rails agrees well with our result (paraphyly of trumpeters is not implied).

4. Are finfoots monophyletic, and are they more closely related to the limpkin or to rails? Finfoots are monophyletic and group among raillike birds. This is corroborated by our own DNA hybridization experiments and cladistic analysis of morphology (Houde, 1994; Houde *et al.*, 1995). Although only weakly upheld by the data set at hand, this consensus of our findings should dispel any lingering acceptance of the limpkin–sungrebe clade proposed by Sibley and Ahlquist (1985, 1990). The fact that they are as close as they are is a surprise, indeed, and underscores the previously unrecognized phyletic proximity of cranelike and raillike birds.

5. Are sunbittern and kagu sister taxa? Yes, this relationship is supported strongly by all analyses. Kagu-like fossils from the early Eocene of Wyoming and middle Eocene of Germany (Hesse, 1988, 1992) place peculiar Amazonian and New Caledonian distributions of these monotypic families into perspective. The distributions of fossils and neotaxa suggest a fairly cosmopolitan pantropical–temperate distribution of "eurypygoids" in the warm forests of the early Tertiary. The modern kagu and sunbittern appear to be relicts of this radiation surviving in the most isolated forest refugia.

6. Is the fossil *Aptornis* more closely related to the kagu or to rails? There is no question that *Aptornis* is much closer to rails than to kagu. The original description by Parker (1866) of the skull of *Aptornis* in comparison to trumpeters was remarkably insightful (considering trumpeters to be like primitive rails, as by Olson, 1973). The phylogenetic informativeness of postcranial morphological characters is all but obliterated by gigantism and the shift of locomotory dependence from the wings to the legs in this flightless bird.

Many of our results conflict markedly with the DNA hybridization studies of Sibley and Ahlquist (1985, 1990). The 12S rDNA sequences support a cranes–rails clade and a sunbittern–kagu clade, and a sister relationship of trumpeters and finfoots to rails. Sibley and Ahlquist proposed that all other Gruiformes except hemipodes are closer to cranes than rails are to cranes, that kagu is closer to everything except rails and hemipodes than it is to sunbittern, and that trumpeter is sister to cranes (Fig. 5.3). We also disagree with their placement of hemipodes far from Gruiformes. They presented a dendrogram that supports the sister relationship of bustards and seriemas that we found, but they did not speculate on its veracity (Sibley and Ahlquist, 1990: Fig. 335).

IV. MOLECULAR EVOLUTION

A. Sequence Divergence

Pairwise Kimura distances (Table II) provide a rough guide to relative amounts of sequence divergence in the 12S rDNA of Gruiformes, although they cannot be considered uniformly proportional to divergence times. Temporal calibrations of sequence divergence apply neither across taxa nor across genes, and perhaps not through time itself (Ayala, 1986; Britten, 1986; Sheldon, 1987; Mindell *et al.*, 1996). Saturation of sequence divergence by multiple hits on individual sites is a convincing mechanism for compression of genetic distance relative to time (Mindell and Honeycutt, 1990). Opposite effects could hypothetically result from other as yet poorly characterized factors, including pervasive environmental mutagens, aspects of the natural history and population structure of a species, phyletic radiation, lateral gene transfer by viruses, or genome transfer by hybridization.

All sites are not equally available to substitution, and the choices of which are used and how they are weighted significantly alter divergence estimates (Pesole *et al.*, 1992). For this reason, we present the highest (variable sites only, 7.3 : 1 transversion weighting) and lowest (all sites, no weighting) pairwise distances (Table II), rather than intermediate distances (i.e., all sites with weighting and variable sites without weighting). Although the number of sites unavailable to vary may be overestimated by considering only those that are observed to vary, distances based on all sites definitely underestimate invariant sites. Moreover, the divergence of chicken from Gruiformes is ancient (e.g., probably Cretaceous), so most sites available for variation should show it.

TABLE II Kimura Two-Parameter Distance Matrix[a]

Gallus	—	0.464	0.493	0.452	0.451	0.658	0.720	0.714	0.841	0.571	0.764	0.529	0.835	0.604	0.586	0.529	0.742	0.677
Anthropoides	0.127	—	0.058	0.222	0.212	0.364	0.465	0.396	0.535	0.411	0.514	0.435	0.703	0.331	0.453	0.477	0.678	0.386
Grus	0.136	0.023	—	0.233	0.204	0.360	0.435	0.356	0.464	0.377	0.505	0.401	0.730	0.346	0.458	0.484	0.739	0.421
Balearica	0.123	0.075	0.080	—	0.231	0.441	0.440	0.442	0.507	0.394	0.514	0.362	0.780	0.444	0.438	0.527	0.936	0.537
Aramus	0.124	0.073	0.071	0.077	—	0.396	0.416	0.352	0.465	0.357	0.420	0.359	0.655	0.330	0.433	0.496	0.788	0.428
Psophia	0.159	0.112	0.113	0.127	0.119	—	0.528	0.502	0.678	0.566	0.675	0.566	0.783	0.553	0.782	0.615	0.893	0.635
Rallus	0.166	0.131	0.126	0.124	0.120	0.145	—	0.355	0.328	0.468	0.650	0.418	0.825	0.591	0.634	0.596	0.969	0.545
Gallinula	0.165	0.116	0.109	0.126	0.106	0.140	0.112	—	0.294	0.364	0.594	0.351	0.914	0.518	0.566	0.564	1.025	0.682
Laterallus	0.172	0.141	0.129	0.134	0.127	0.166	0.101	0.093	—	0.472	0.624	0.327	0.933	0.595	0.639	0.691	1.221	0.881
Aptornis[b]	0.157	0.131	0.121	0.125	0.116	0.162	0.140	0.116	0.137	—	0.547	0.319	0.788	0.439	0.513	0.554	0.982	0.614
Heliornis	0.170	0.136	0.137	0.134	0.118	0.162	0.153	0.146	0.146	0.153	—	0.287	1.114	0.549	0.649	0.781	1.168	0.819
Podica[b]	0.127	0.118	0.111	0.100	0.103	0.140	0.114	0.097	0.090	0.093	0.089	—	0.774	0.466	0.644	0.780	1.381	0.949
Mesitornis	0.181	0.171	0.180	0.179	0.163	0.180	0.177	0.190	0.185	0.194	0.201	0.166	—	0.750	0.849	1.011	0.884	0.852
Ardeotis	0.153	0.103	0.108	0.127	0.102	0.149	0.149	0.137	0.148	0.133	0.141	0.122	0.176	—	0.424	0.502	0.709	0.581
Cariama	0.149	0.130	0.133	0.125	0.124	0.184	0.157	0.145	0.155	0.149	0.155	0.149	0.184	0.124	—	0.664	0.690	0.687
Eurypyga	0.138	0.131	0.136	0.140	0.136	0.157	0.150	0.144	0.162	0.151	0.172	0.163	0.198	0.134	0.160	—	0.452	0.653
Rhynochetos[b]	0.178	0.171	0.181	0.201	0.190	0.199	0.209	0.200	0.228	0.225	0.217	0.208	0.201	0.181	0.177	0.140	—	0.907
Turnix	0.163	0.117	0.127	0.148	0.126	0.164	0.144	0.167	0.189	0.175	0.178	0.193	0.193	0.152	0.169	0.161	0.217	—

[a] Above diagonal: 7.3:1 transversion weighting, variable sites only; below diagonal: unweighted, all sites included.

[b] Missing data.

Statistically significant differences in evolutionary rates among Gruiformes are identified by ranking unweighted two-parameter distances within monophyletic groups, summed across all outgroups ("multiple comparisons for ranked data in randomized block" of Zar, 1974; "SNK" of Houde, 1987). Although not significant, among the cranes genetic distances involving *Grus* appear low while those of crowned crane appear high, in accordance with the observations of Krajewski (1989). Limpkin exhibits significantly lower distances than crowned crane ($\alpha =$ 0.05; 14 outgroups). Accordingly, in a 6-taxon tree rooted to chicken, trumpeter undergoes 57 substitutions, limpkin only 32, crowned crane 55, and the two gruine cranes have 53 and 54. The vastly different branch lengths within the cranes–limpkin clade may complicate the recovery of the expected topology of limpkin sister to cranes *vs* sister to crowned crane or outgroups (Nei, 1991; Huelsenbeck and Hillis, 1993).

Rail and sungrebe distances appear high, although not so much as trumpeter. Among rails, *Gallinula* distances are significantly lower than *Laterallus* ($\alpha = 0.001$) and *Rallus* ($\alpha = 0.05$; 15 outgroups). No other robust relative rate tests are justifiable because of missing data in the only other clades for which independent evidence of monophyly is available. However, roatelo has the highest distance to chicken of all Gruiformes.

Neighbor-joining places trumpeter with cranes but MP and ML include it with the rails, so we tested its distances using both phylogenies. When included in the cranes clade, trumpeter distances are observed to be significantly higher than all the others ($\alpha = 0.001$; 13 outgroups). Trumpeter distances are not different than any rails when included in that clade ($\alpha = 0.05$; 10 outgroups).

B. Evolutionary Dynamics of 12S rDNA

One cannot help but be struck by the conservation of both sequence and secondary structure between rDNAs of such disparate groups as bacteria and vertebrates (Van de Peer *et al.*, 1994). Yet, on this broad evolutionary scale one also appreciates the elongation, shortening, and complete loss of some stems. The small-scale events that lead to such large-scale patterns require comparison of the genes in both relatively closely and distantly related organisms (Kjer, 1995; Hickson *et al.*, 1996). Here, we describe some of the small-scale variation in the 12S rDNA of Gruiformes. The variation we note has negative implications for the general practice of matching sequences to structural maps of rDNA constructed from unrelated organisms. Our stem nomenclature follows Van de Peer *et al.* (1994).

1. *Stem Migration*

We noted movements of complementary bases upstream and downstream within stems. This "stem migration" seems to result more from substitutions that affect complementarity than from insertions and deletions. It may result in extended or reduced base complementarity, but usually involves the migration of one or both sides of the stem region. These discrepancies between taxa are reflected in the irregularly boxed stems 8 and 26 of Fig. 5.4.

Stem 8 is the most variable (Figs. 5.4 and 5.13). Its distal segment is flanked on both sides by single-stranded loops of variable length. Similarities in sequence and stem position are apparent among closely related taxa, but identification of homologous sites is difficult across all Gruiformes. The stem consists of five pairs of complementary bases in all taxa except the hemipode (one mismatch) and trumpeter (which seems to have six). The upstream side of the stem in rails has a pattern of YYCCT, seriema and bustard have CCTTA, limpkin and crowned crane have CCTAR, and gruine cranes have CCTAT. The pattern in trumpeter (GCCTAC) is the same as in gruines (i.e., CCTAY), except an additional purine is added to the 5′ end. A 5′ purine is otherwise a uniquely shared character of the stems of sunbittern and kagu (RCCTT).

Alignment of the upstream CCT of stem 8 in all taxa invokes considerable stem migration. Phylogentic weighting produces an alternate alignment that is both more parsimonious and invokes less stem migration. On the downstream side of stem 8 most taxa have a homologous sequence of AGG (i.e., aligned), complementing the upstream CCT (minor differences in hemipodes and roatelos). The stem of sunbittern and seriemas (kagu data missing) migrates 1 base downstream, while that of rails migrates 2 bases downstream relative to the AGG sequence and the stem of other Gruiformes.

In roatelo, stem 24 migrates upstream 1 base on the downstream side of the stem by transition substitutions. In crowned crane, roatelo, and sunbittern, stem 26 appears to migrate upstream 1 base on the upstream side of the stem, and upstream 2 bases on the downstream side of the stem. It is difficult to infer the location of stem 26, however, because it consists of only two pairs of bases. This is reflected in the ambiguity of Fig. 5.4, in which the stem is shown to overlap stem 24. Hypothetically, such overlap could accurately reflect temporally alternate or tertiary structural interactions as in stem 22, but there is no independent evidence for such phenomena here.

All the birds examined here differ from mammals in stems 24, 27, and 47. The entirety of the avian stems 24 and 27 migrate one position proximal compared with mammals (i.e., upstream on upstream side, downstream on downstream side). Stem 47 is inferred to have elongated distally relative to mammals (i.e., downstream on upstream side, upstream on downstream side).

2. Compensatory Substitutions

We made anecdotal notes on frequency of compensatory substitutions while counting putative synapomorphies for clades of interest. Unlike insertions and deletions within stems (Section IV,B,3), most substitutions within stems precipitate compen-

```
                   80        90        100       110       120            730       740       750          830       840
                    .         .         .         .         .              .         .         .            .         .
Gallus      GCCCCCaaacct-----TTCTT--------cccAAGCA-aAA-G   CtcaaTAGCc-----------cctcGCTAa   cgaaa-AAGGATGtgaa
Anthropoides GCCCTCggac-------CCTAT----ctcttATAGGca-aGA-G   CacaaTAGCcccc------g--cccGCTAa   cggaa-GGGGGtGtgaa
Grus        GCCCTCggac-------CCTAT----ctcctATAGGca-aGA-G   CacaaCAGCccccc-----a--cccGCTAa   cggaa-GGGGGtGtgaa
Balearica   GCCCTCaatc-------CCTAG-----tcccCTAGGca-gGA-G   CacaaTAGCccctcccc--a--cgcGCTAa   cgaaa-AGGGGACtgaa
Aramus      GCCCTCaacc-------CCTAA----caatcTTAGGca--AA-G   CacaaCAGCcc-------aa--cccGCTAg   cggaa-GGGGGtGtgaa
Psophia     GCCCtCagccc-----GCCTACttat-----GTAGGCacaGAtG   CttaaTAGCcc------aag--cccGCTAg   cgaaa-GGGGGCatgaa
Rallus      GCCCTGag---------TTCCTttgtc----atAGGAA-gCA-G   CacaaCAAC----------a-tcccGCTAa   cgaaa-AGAGGCCtgaa
Gallinula   GCCCTCac---------CCCCTac-tcaaaaacAGGGA-aGA-G   CacaaTAGC----------a-tcccGCTAa   cgaaa-AGGGGCCtgaa
Laterallus  GCCCACac---------CTCCTccatcaa--acAGGAG-aGA-G   CacaaCAGC----------a-tgccGCTAa   cgaaa-AGAGATCtgaa
Aptornis                                                  CacaaCAGCc---------a--cccGCTAa   cgaaa-GAGGAAAtgaa
Heliornis   GCCCaCgact-------TCTAAa--------TTAGAtt-gGa-G   CtcaaCAGCc---------a--cccGCTAg   cggaa-AGGGAAAtgaa
Podica                                                    CacaaTAGCc---------a--cccGCTAa   cgaaa-AGGGAGAtgaa
Mesistornis GCCCTCagc--------CCTTA--------tTAAGAca-cGA-G   CataaCAGCc--------ag--tttGCTAa   cggaa-AAGGGTAtgaa
Ardeotis    GCCCa-agccc------CCTtAcata-----TtAGGcc-aAA-G   CataaTAGCccc---------tcccGCTAg   cgaaa-GGGAGTAtgaa
Cariama     GCCCaCagcc------GCCTTa--------ccAAGGCa-aAa-G   CacaaTAGCccc----------cgcGCTAa   cgaaa-GCGGGTGtgaa
Eurypyga    GCCCCT-gtc------ACCTTt-------atcAAGATg-aAG-G   CaaaaTAGCct--------a-cctcGCTAa   cggaacAGGGGCAtgaa
Rhynochetos                                               CaaaaTAGCtc-----------ttcGCTAa   cggagc-GGGGTGtgaa
Turnix      GCCCTctc---------CCCtCttcacttatGcGGGaa--AA-G   CacaaTAGTtt-------aa--cccACTAg   cggaa-GAAAGTGtgaa
consensus   GCCCHBrvhyc.....RYYYHNhhnwcwhhhNHRGRNNcrVRtG   CwhaaYARYyycyccccaar.ybbyRCTAr   cgrarcRVRRRNVtgaa
stem         8               8               8    8  8   39   42                 42                   47
```

FIGURE 5.13 Partial sequence alignment of Gruiformes mitochondrial 12S rDNA. The consensus sequence is for Gruiformes only, but gaps in consensus sequence are shown to permit alignment with outgroups. Stems are underlined and labeled. Upper-case lettering, canonical pair in stem; upper-case italics, noncanonical pair in stem; lower-case, nonpairing base. Positions 77–120 illustrate migration of stem 8 and an uncompensated insertion within the stem at position 119 in *Posphia*. Positions 721–750 illustrate replication slippage in the distal loop of stem 42 as a mechanism for sequence length variation. Positions 824–840 illustrate shortening of stem 47 in *Rhynochetos* by an uncompensated deletion at position 830.

satory substitutions on the complementary side. Of those that we tracked, no fewer than 64% of compensations that occur do so within the period of one internode in the phylogeny reconstruction. Some that are located in the inferred distal extension of stem 47 remain uncomplemented over several internode periods, but such delays are not unique to this stem. Noncanonical pairing or nonpairing may be favored in these instances, or they may reflect our inability to detect all substitutions in the depths of the tree.

We may have observed an early event in the process of compensatory substitution. This involves an apparent example of heteroplasmy in *Rallus*. Both C and T bands occur at site 372 in an otherwise clear and unambiguous sequence autoradiograph. This site is in a complementary position to site 402 in stem 24, where *Rallus* is unique in possessing a transversion substitution from A to T. The C represents a transition substitution from the T state at site 372. Neither C nor T complement with site 402. We speculate that the heteroplasmy represents temporary relaxation of selection for sequence conservation related to the process of compensation. Stated differently, all noncomplementary bases may be tolerated equally, transitions would likely be the first form of variants to appear, and complementary bases would be selectively favored when they arise, eventually stabilizing the sequence.

3. Insertions and Deletions

We noted uncompensated insertions and deletions in stems that create or eliminate bulges (Figs. 5.4 and 5.13). Crowned crane and limpkin share a deletion of a 1-base bulge on downstream side of stem 3 (position 315). Trumpeter possesses a 1-base uncomplemented addition in stem 8 (position 119). Kagu exhibits a unique 1-base uncompensated deletion on the first base of the upstream side of stem 47 (position 830). All three rails exhibit a synapomorphic 1-base uncompensated deletion on the downstream side of stem 48 (position 902).

Replication slippage may accelerate length variation in polynucleotides, and di- and trinucleotide repeats in a variety of systems (Tautz *et al.*, 1986; Hancock and Dover, 1990; Weston-Hafer and Berg, 1991; Degoul *et al.*, 1991; Wolfson *et al.*, 1991). The terminal loop of stem 42 varies from 5 to 12 bases in Gruiformes (Figs. 5.4 and 5.13). Cranes synapomorphically possess the most bases here. The loop begins with a variable number of pyrimidines (up to eight; almost all Cs) that immediately follow a C in the stem. The stem of hemipode ends with a T, and its loop begins with poly(T) instead of poly(C). The next segment of the loop includes up to three purines (mostly As), followed by a variable number of bases of which most are pyrimidines, especially Cs. This appears to be an example of length variation by slippage. Slippage seems to occur more readily than transitions, which in turn exceed transversions. Accordingly, crowned crane is the only member of the cranelike birds to exhibit a substitution among the 3- to 9-base poly(C) of the first segment, and it is a transition. Replication slippage may also be involved in the length variation at positions 83–107, 140–145, 659–660, and 817–820.

After a long history of substitutions has overwritten the earmarks of replication slippage, it may be impossible to distinguish chance similarities from homology. In variable-length regions, it may be prudent not to attempt to force sequences into alignment in the interest of phylogenetic inference.

4. Among-Site Rate Heterogeneity

Site variability, or substitution rate per site, differs profoundly between higher taxonomic groups. This is somewhat surprising in light of the sequence conservation between bacteria and vertebrates at some positions. Sullivan *et al.* (1995) identified 29 most-variable sites in their study of 12S rDNA in sigmodontine rodents, a group that is considerably less diverged among themselves than are Gruiformes. While the maximum number of changes per site they observed was three, we infer sites to have changed as many as eight times in our phylogeny of Gruiformes. Of 21 most-variable sigmodontine sites of which we could determine the homology in Gruiformes, 8 are invariant in Gruiformes, 3 changed only once, 3 twice, 3 three times, 3 four times, and 1 changed six times. This is clearly a different distribution of among-site rate variability between the two groups. Thus, the evolutionary dynamics of 12S rDNA may differ substantially between birds and mammals, and possibly contribute to the higher levels or resolution in some nonavian 12S rDNA reconstructions of even deeper divergences than those in Gruiformes (e.g., Cummings *et al.*, 1995).

Phylogenetic informativeness is often thought of as inversely related to the rate at which changes occur, and weighting schemes are designed accordingly. But they need not be. The lengths of the monotonous polynucleotides in loop 42 may have evolved rapidly but they are obviously highly correlated with phylogeny (Section IV,B,3). For example, in the first segment rails have one C residue, finfoots have two, trumpeters and limpkin have three, and cranes have five to eight. Further similarities are exhibited by sunbittern–kagu–hemipode and seriema–bustard and finfoot–*Aptornis*–roatelo.

C. Character Evolution of 12S rDNA

Many regard nucleotide substitutions as stochastic or effectively neutral within the recognized constraints of positional variation in evolutionary rates and differences in rates between evolutionary lineages. By limiting consideration of frequency of nucleotide substitution merely to positional effects of evolutionary rate, one may overlook the ways in which nucleotide subsitutions may be functionally correlated or adaptively specialized (Margoliash *et al.*, 1976; Hancock and Dover, 1990; Irwin *et al.*, 1991; Gillespie, 1991; Ma *et al.*, 1993).

Positions of synapomorphously and homoplasiously shared derived characters of individual clades may be identified on a structural map of the gene, and tested for fit to models of expected distributions of nucleotide substitutions. Derived characters that are concentrated in functional or structural regions of a gene and that differ significantly from expected distributions of nucleotide substitutions might represent adaptive specializations.

1. Expected Probability of Substitution

The expected probability of substitution for any site is estimated by dividing the number of times that site changed by the sum of all nucleotide substitutions at all sites in a most-parsimonious phylogeny of many taxa. Expected substitution frequencies for a functional or structural region is simply the sum of expected substitution probabilities of all sites included in a region that is otherwise defined by independent criteria (e.g., protein-binding domain or secondary structure).

We partitioned the 12S rDNA molecule into large regions on the basis of protein binding, and small regions corresponding to stem/loop structure. The large regions include sites 339–540 in domain II, which broadly encompass the binding domain of ribosomal proteins S6+18, and sites 610–670 and 770–930 in domain III, which collectively include the binding domains of proteins S7 and S19 (Noller *et al.*, 1990; Ehresmann *et al.*, 1990). The small regions include positions 339–452 and 796–861 within the larger sets.

We summed all nucleotide substitutions (from 0 to 8 per site; Section II,A,5; Table I) in each these regions in the most parsimonious phylogeny and divided those sums by the total number of nucleotide substitutions in the phylogeny ($n = 1110$) to calculate the percent of substitutions detected in each region, our estimate of expected substitution probabilities for each region. Across all Gruiformes, only 21.5% of all sites but 34.3% of all substitutions occur in large region domain II, 27.8% of sites but 26.5% of substitutions in large region domain III, 12.1% of sites but 23.0% of substitutions in small region domain II, and 7% of sites but 10.0% of substitutions in small region domain III. The expected probability of substitution is greater in domain II than in domain III.

2. Distribution of Synapomorphies

Casual observation suggested to us that substitution frequencies were different in some clades than the whole of Gruiformes. We used chi-square (Zar, 1974) to test whether the relative frequencies of substitutions (i.e., synapomorphies) on individual branches differed significantly from the whole phylogeny in each of these regions (Section IV,C,1; substitutions defining the clade were subtracted from the whole to ensure independence of sets being compared). We performed this test on both the branches uniting trumpeter with raillike birds and uniting trumpeter with cranelike birds, since our data support the former sister relationship only weakly. Parsimony bootstrap values support inclusion of trumpeters in the cranes–rails group, but does not significantly favor a grouping with either cranes or rails (Fig. 5.9).

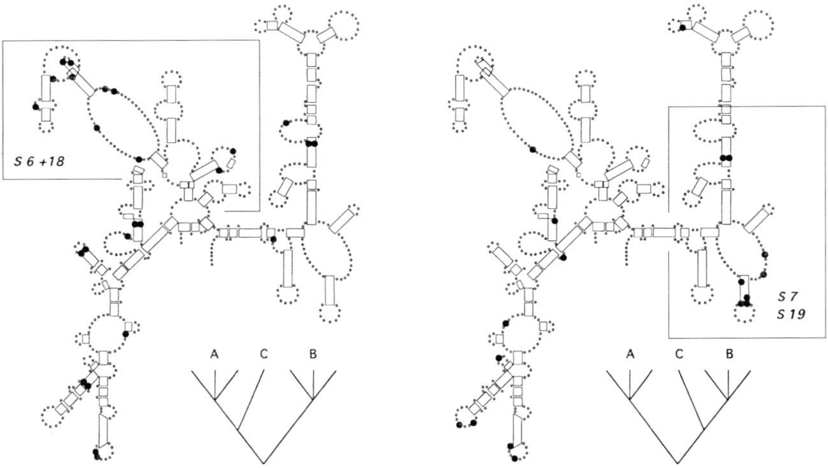

FIGURE 5.14 Putative synapomorphies (large dots) in mitochondrial 12S rDNA, domains I–III, of trumpeter (*Psophia*) with mutually exclusive hypothetical sister taxa cranelike birds (left) and raillike birds (right). (A) Cranes and limpkin; (B) rails, finfoots, and *Aptornis*; (C) trumpeter. Protein-binding domains of S6+18, and S7 and S19, are broadly circumscribed by boxes. The distribution of substitutions is significantly different from expected frequencies in boxed regions and between the two hypotheses of phylogeny. See text for details.

Raillike birds ("rails" below) consisted of a star phylogeny for *Aptornis*, *Gallinula*, *Heliornis*, *Laterallus*, *Podica*, *Rallus*. Cranelike birds ("cranes" below) consisted of a star phylogeny for *Anthropoides*, *Balearica*, *Aramus*, and these were rooted by a star clade of *Ardeotis*, *Cariama*, *Mesitornis*, *Turnix*, sisters sunbittern–kagu and finally chicken. Twenty-five putative synapomorphies unite trumpeter with "cranes" and 18 are shared by trumpeter and "rails" (Fig. 5.14). We did not consider a phylogeny with trumpeters as sister to both "cranes" and "rails" because this parsimony reconstruction was 2.2% longer than the others.

Significant differences from expected frequencies of substitutions were apparent in the large region of domain II in the branch uniting trumpeters–"rails," and in all of the small regions for both trumpeters–"rails" and trumpeters–"cranes" ($p < 0.025$). Eliminating compensatory substitutions from consideration does not change the results. Thus the evolutionary rate in some functional and structural regions differs significantly between some gruiform clades, as both phylogenies gave significant results.

3. *Distribution of Synapomorphy versus Homoplasy*

Patterns of molecular synapomorphy and homoplasy can be elucidated by comparing mutually exclusive sets of putative synapomorphies indicated by alternate hypotheses of phylogeny for a single taxon (C) relative to its candidate sisters (A and

B; Fig. 5.14). The putative synapomorphies of C to each of clades A and B are mutually exclusive as long as A and B are sisters to one another (exclusive of C).

We tested for differences in frequencies of putative synapomorphies in functional and structural regions between the mutually exclusive hypothetical clades, trumpeters–"rails" and trumpeters–"cranes" (Fig. 5.14). This asks whether trumpeters–"cranes" and trumpeters–"rails" tree topologies affect the positions of putative synapomorphies differently. Using a chi-square contingency table with Yates correction (Yates, 1934; Zar, 1974), we found that the two phylogenies exhibit significant differences from one another in the frequencies of synapomorphies in the regions examined except for large region, domain II (large region, domain III, $p = 0.001$; small region, domain II, $p = 0.025$; small region, domain III, $p = 0.05$).

What value could there be in making a comparison with at least one tree that must be erroneous? Minimally, we have shown that synapomorphies and homoplasies are distributed differently; but we believe we have done more. Because (1) sets of synapomorphies defining two mutually exclusive hypothetical clades must in reality include at least one set of homoplasies, (2) both sets represent derived states (i.e., not plesiomorphies, as determined from outgroups), (3) distributions of those sets differ significantly between alternative phylogenies, and (4) those differences correspond to functional and structural regions of the gene, both genic divergence and convergence are observed to localize in functional and structural regions. This demonstrates variation in among-site evolutionary rates between sister taxa that could possibly represent (different) adaptive specializations at the molecular level.

Convergence may be implicated when homoplasiously shared characters coincide with discrete structural or functional parts of an organism or a gene, as they are in this example. We emphasize a distinction between suites of convergent characters and homoplasious noise. The former may be under the influence of a unifying selective agent, possibly affecting secondary structure and molecular interactions. We cannot distinguish whether our observation results from selection or is merely a byproduct of other factors.

There may be explanations for the different groupings of shared characters in the trumpeters–"rails" and trumpeters–"cranes" trees that do not invoke adaptive specialization. Because the rate of evolution in domain II exceeds that of domain III overall (Section IV,C,1), then the higher than expected number of shared characters in domain III of trumpeters–"rails" might represent phylogenetic signal while the higher than expected number of shared characters in domain II of trumpeters–"cranes" might be an attraction of long branches (Felsenstein, 1978). This explanation may not account for the less than expected numbers of shared characters in domain II of trumpeters–"rails" and domain III of trumpeters–"cranes."

We used Mann–Whitney U (Zar, 1974) to test whether synapomorphies of trumpeters–"cranes" (domain II) differ from synapomorphies of trumpeters–"rails" (domain III) in evolutionary rate (i.e., number of substitutions per site; Table I; both large and small regions were tested). This differs from the observation that domain II has a higher expected substitution probability overall than domain III because it examines the evolutionary rate on a site-by-site basis. We also tested whether all synapomorphies of trumpeters–"cranes" differed from all synapomorphies of trumpeters–"rails" in evolutionary rate. No test detected significant differences in substitution rates between sites defining the trumpeters–"cranes" and trumpeters–"rails" clades ($p > 0.2$). This suggests that the difference in distributions of synapomorphies and homoplasies we observed cannot be attributed simply to rate differences among sites.

V. IMPLICATIONS OF 12S EVOLUTION FOR PHYLOGENETIC INFERENCE

The lack of a clear resolution of many aspects of gruiform phylogeny from 12S rDNA is disappointing. 12S rDNA strongly supports relationships of some recently diverged taxa, but not of more distant taxa. Even some traditionally accepted family groups do not receive robust support from our data. What is the cause for the lack of phylogenetic resolution in these data?

12S rDNA is not entirely saturated by homoplasious substitutions at the levels of gruiform divergence because it performs well at resolving much older divergences. Yet it exhibits sufficient noise to hinder resolution of a phylogeny that may be characterized by relatively short internodal branches (Fig. 5.8). Low jackknife values are symptomatic of such noise (Fig. 5.6). Our phylogenetic hypotheses must be interpreted cautiously. A gene phylogeny may not always accurately reflect organismal phylogeny (Avise *et al.*, 1984; Wu, 1991; but see Moore, 1995). Discrepancies in phylogenetic reconstructions derived from different genes demonstrate that all cannot accurately reflect organismal phylogeny (Felsenstein, 1988; Bremer, 1988; Pamilo and Nei, 1988; Wheeler and Honeycutt, 1988; Hendy and Penny, 1989; Doyle, 1992; Sanderson and Doyle, 1992).

We chose 12S rDNA partly because it includes both evolutionary labile and conserved regions, and therefore should have a broad window of resolution for addressing recent and ancient divergences. However, among-site variation in evolutionary rate, among other factors, proves to impede rather than enhance the successful recovery of gene phylogenies. Phylogeny reconstruction is sensitive to substitution bias (Brown *et al.*, 1982; Knight and Mindell, 1993), differences in evolutionary rates at the level of the organism (Britten, 1986; Sheldon, 1987), gene (Ayala, 1986) or nucleotide position (Milkman and Crawford, 1983), homoplasious evolution (Wilkinson, 1991), composition bias (Collins *et al.*, 1994), and differences in branch lengths, tree symmetry, and number of taxa (Nei, 1991; Huelsenbeck and Hillis, 1993). The factors that influence the reliability of tree-building methods is well understood for only simple conditions (Hendy and Penny, 1989; Rohlf *et al.*, 1990; Nei, 1991; Navidi and Beckett-Lemus, 1992; Huelsenbeck and Hillis, 1993; Kim *et al.*, 1993; Zharkikh and Li, 1993).

Whatever lack of resolution is symptomatic of 12S rDNA data, it probably does not accrue from positional rate variation as traditionally perceived because neither position weighting nor data partitioning improves tree resolution. Instead, it may result from differences in evolutionary rates between taxa and differences in positional rates between taxa. The conserved sequences and secondary structures of small subunit rDNA shared by prokaryotes and vertebrates belie dramatic differences in evolutionary lability of homologous regions between different lineages. This is evident at a large scale by the observation that a quarter to a third of the most variable sites in rodents are invariant in Gruiformes. It is evident at the small scale by significant differences in regional substitution rates between sister clades. High rates of substitution, thus, are not confined to particular sites across taxa; they are found in different locations in different lineages.

VI. SUMMARY

We performed phylogenetic reconstructions using 12S rDNA sequences from representatives of all the currently recognized families of Gruiformes. We found rails closer to cranes than many other Gruiformes widely believed to be close to cranes. We suggest that trumpeters are closer to rails than to cranes, but suggest that they are intermediate between the two. Among a clade of rail relatives are the sungrebe and finfoots and the fossil *Aptornis*. Kagu and sunbittern are each the only close relative of the other, and are the most distant of all Gruiformes examined. We make several observations and inferences on the evolutionary dynamics and character evolution of the 12S rDNA molecule, including (1) variation in secondary structure resulting from stem migration and uncompensated insertions and deletions within stems, (2) replication slippage as a mechanism of sequence length variation in loops, (3) differences in per-site substitution rates between birds and mammals, (4) the process of compensatory substitution in stems, and (5) differences in distributions of synapomorphies and homoplasies that are spatially correlated with functional and structural regions of 12S rDNA. A robust but simplified estimate of the instantaneous ratio of rates between transversions to transitions is calculated for the 12S rDNA of Gruiformes.

ACKNOWLEDGMENTS

We thank Carey Krajewski, Fred Sheldon, and David P. Mindell for detailed and thoughtful reviews that greatly improved this chapter. We thank William J. Boecklen, David Daniel, Brook G. Milligan, and David Swofford for insightful and stimulating discussions, statistical advice, and computer programming. For specimens we are indebted to George M. Gee, Steven Goodman, Gary R. Graves, Steven A. Nisbett, Storrs L. Olson, Enrique Ortiz, Charles Pickett, Marco Romero, Frederick H. Sheldon, and Richard L. Zusi. David P. Mindell graciously provided unpublished sequences of relevant outgroups. We also thank all those who have contributed sequences to GenBank for the benefit of the scientific community at large. This work was supported by New Mexico State University. G. A. Montano was supported by a MARC traneeship, NIGMS grant T34GM06667.

REFERENCES

Anderson, S., Bankier, A. T., Barrell, B. G., deBruijn, M. H. L., Coulson, A. R., Drouin, J., Eperon, I. C., Nierlich, D. P., Rose, B. A., Sanger, F., Schreier, P. H., Smith, A. J. H., Staden, R., and I. G. Young (1981). Sequence and organization of the human mitochondrial genome. *Nature (London)* **290**, 457–465.

Avise, J. C., Neigel, J. E., and Arnold, J. (1984). Demographic influences of mitochondrial DNA lineage survivorship in animal populations. *J. Mol. Evol.* **20**, 99–105.

Ayala, F. J. (1986). On the virtues and pitfalls of the molecular evolutionary clock. *J. Heredity* **77**, 226–235.

Beddard, F. E. (1898). "The Structure and Classification of Birds." Longmans, Green, and Co., London.

Bonner, T. I., Heinemann, R., and Todaro, G. J. (1981). A geographic factor involved in the single copy DNA sequence of primates. *In* "Evolution Today" (G. G. E. Scudder and J. L. Reveall, Eds.), pp. 293–300. Carnegie Mellon University Press, Pittsburgh, Pennsylvania.

Bremer, K. (1988). The limits of amino acid sequence data in angiosperm phylogenetic reconstruction. *Evolution* **42**, 795–803.

Britten, R. J. (1986). Rates of DNA sequence evolution differ between taxonomic groups. *Science* **231**, 1393–1398.

Brodkorb, P. (1967). Catalogue of fossil birds. 3. Ralliformes, Ichthyornithiformes, Charadriiformes. *Bull. Florida State Mus.* **11**, 99–220.

Brown, W. M., Prager, E. M., Wang, A., and Wilson, A. C. (1982). Mitochondrial DNA sequences of primates, tempo and mode of evolution. *J. Mol. Evol.* **18**, 225–233.

Clary, D. O., and Wolstenholme, D. R. (1985). The mitochondrial DNA molecule of *Drosophila yakuba*: Nucleotide sequence, gene organization, and genetic code. *J. Mol. Evol.* **22**, 252–271.

Collins, T. M., Wimberger, P. H., and Naylor, G. J. P. (1994). Composition bias, character-state bias, and character-state reconstruction using parsimony. *Syst. Biol.* **43**, 482–496.

Cooper, A., Mourer-Chauviré, C., Chambers, G. K., von Haesler, A., Wilson, A. C., and Pääbo, S. (1992). Independent origins of New Zealand moas and kiwis. *Proc. Natl. Acad. Sci. U.S.A.* **89**, 8741–8744.

Cracraft, J. (1981). Toward a phylogenetic classification of the recent birds of the world (class Aves). *Auk* **98**, 681–714.

Cracraft, J. (1982). Phylogenetic relationships and trans-Atlantic biogeography of some gruiform birds. *Geobios. Mem. Spec.* **6**, 393–402.

Cummings, M. P., Otto, S. P., and Wakeley, J. (1995). Sampling properties of DNA sequence data in phylogenetic analysis. *Mol. Biol. Evol.* **12**, 814–882.

Degoul, F. F., Nelson, I., Amselem, S., Romero, N., Obermaier-Kusser, B., Ponsot, G., Marsac, C., and Lestienne, P. (1991). Different mechanisms inferred from sequences of human mitochondrial DNA deletions in ocular myopathies. *Nucleic Acids Res.* **19**, 493–496.

Desjardins, P., and Morais, R. (1990). Sequence and gene organization of the chicken mitochondrial genome: A novel gene order in higher vertebrates. *J. Mol. Biol.* **212**, 599–634.

Doyle, J. J. (1992). Gene trees and species trees, molecular systematics as one-character taxonomy. *Syst. Bot.* **17**, 144–163.

Ehresmann, B., Ehresmann, C., Romby, P., Mougel, M., Baudin, F., Westhof, E., and Ebel, J.-P. (1990). Detailed structures of rRNAs: New approaches. *In* "The Ribosome, Structure, Function, and Evolution" (W. E. Hill, A. Dahlberg, R. A. Garrett, P. B. Moore, D. Schlessinger, and J. R. Warner, Eds.), pp. 148–159. American Society for Microbiology, Washington, D.C.

Engelke, D. R., Hoener, P. A., and Collins, F. S. (1988). Direct sequencing of enzymatically amplified human genomic DNA. *Proc. Natl. Acad. Sci. U.S.A.* **85**, 544–548.

Felsenstein, J. (1978). Cases in which parsimony and compatibility methods will be positively misleading. *Syst. Zool.* **27**, 401–410.

Felsenstein, J. (1981). Evolutionary trees from DNA sequences: A maximum likelihood approach. *J. Mol. Evol.* **17**, 368–376.

Felsenstein, J. (1985). Confidence limits on phylogenies: An approach using the bootstrap. *Evolution* **39**, 783–791.

Felsenstein, J. (1988). Phylogenies from molecular sequences, inference and reliability. *Annu. Rev. Genet.* **22**, 521–565.

Felsenstein, J. (1989). PHYLIP—phylogeny inference package (version 3.2). *Cladistics* **5**, 164–166.

Feng, D., and Doolittle, R. F. (1987). Progressive sequence alignment as a prerequisite to correct phylogenetic trees. *J. Mol. Evol.* **25**, 351–360.

Fukuda, M., Wakasugi, S., Tsuzuki, T., Nomiyama, H., Shimada, K., and Miyata, T. (1985). Mitochondrial DNA-like sequences in the human nuclear genome. Characterization and implications in the evolution of animal mitochondrial DNA. *J. Mol. Biol.* **186**, 257–266.

Fürbringer, M. (1888)."Untersuchungen zur Morphologie und Systematik der Vögel, zugleichein Beitrag zur Anatomie der Stutz-un Bewegungsorgane." T. van Holkema, Amsterdam.

Gadow, H. (1892). On the classification of birds. *Proc. Zool. Soc. Lond.* **1892**, 229–256.

Garrod, A. (1874). On certain muscles of birds and their value in classification. 2. *Proc. Zool. Soc. Lond.* **8**, 111–123.

Gillespie, J. H. (1991). "The Causes of Molecular Evolution." Oxford University Press, New York.

Gray, M. W., Sankhoff, D., and Cedergren, R. J. (1984). On the evolutionary descent of organisms and organelles, a global phylogeny based on a highly conserved structural core in small subunit ribosomal RNA. *Nucleic Acids Res.* **12**, 5837–5852.

Hancock, J. M., and Dover, G. A. (1990). "Compensatory slippage" in the evolution of ribosomal RNA genes. *Nucleic Acids Res.* **18**, 5949–5954.

Hedges, S. B., and Sibley, C. G. (1994). Molecules vs. morphology in avian evolution: The case of the "pelecaniform" birds. *Proc. Natl. Acad. Sci. U.S.A.* **91**, 9861–9865.

Hein, J. (1990). Unified approach to alignment and phylogenies. *Methods Enzymol.* **183**, 626–644.

Hendy, M. D., and Penny, D. (1989). A framework for the study of evolutionary trees. *Syst. Zool.* **38**, 297–309.

Hesse, A. (1988). Die Messelornithidae—Eine neue Familie der Kranichartigen (Aves, Gruiformes, Rhynocheti) aus dem Tertiar Europas und Nordamerikas. *J. Ornithol.* **129**, 83–95.

Hesse, A. (1992). A new species of *Messelornis* (Aves, Gruiformes, Messelornithidae) from the Middle Eocene Green River Formation. *Nat. Hist. Mus. Los Angeles Co. Sci. Ser.* **36**, 171–178.

Hickson, R. E., Simon, C., Cooper, A., Spicer, G. S., Sullivan, J., and Penny, D. (1996). Conserved sequence motifs, alignment, and secondary structure for the third domain of animal 12S rRNA. *Mol. Biol. Evol.* **13**, 150–169.

Higuchi, R. G., and Ochman, H. (1989). Production of single-stranded DNA templates by exonuclease digestion following the polymerase chain reaction. *Nucleic Acids Res.* **17**, 5865.

Houde, P. (1987). Critical evaluation of DNA hybridization in avian systematics. *Auk* **104**, 17–34.

Houde, P. (1994). Evolution of the Heliornithidae: Reciprocal illumination by morphology, biogeography and DNA hybridization (Aves, Gruiformes). *Cladistics* **10**, 1–19.

Houde, P., and Braun, M. J. (1988). Museum collections as a source of DNA for studies of avian phylogeny. *Auk* **105**, 773–776.

Houde, P., Sheldon, F. H., and Kreitman, M. (1995). A comparison of solution and membrane-bound DNA × DNA hybridization, as used to infer phylogeny. *J. Mol. Evol.* **40**, 678–688.

Huelsenbeck, J. P., and Hillis, D. M. (1993). Success of phylogenetic methods in the four-taxon case. *Syst. Biol.* **42**, 247–264.

Irwin, D. M., Kocher, T. D., and Wilson, A. C. (1991). Evolution of the cytochrome *b* gene of mammals. *J. Mol. Evol.* **32**, 128–144.

Kim, J., Rohlf, F. J., and Sokal, R. R. (1993). The accuracy of phylogenetic estimation using the neighbor-joining method. *Evolution* **47**, 471–486.

Kimura, M. (1980). A simple method for estimating rate of base substitutions through comparative studies of nucleotide sequences. *J. Mol. Evol.* **16**, 111–120.

Kjer, K. M. (1995). Use of rRNA secondary structure in phylogenetic studies to identify homologous positions: An example of alignment and data presentation from the frogs. *Mol. Phylogenet. Evol.* **4**, 314–330.

Knight, A., and Mindell, D. P. (1993). Substitution bias, weighting of DNA sequence evolution, and the phylogenetic position of Fea's viper. *Syst. Biol.* **42**, 18–31.

Kocher, T. D., Thomas, W. K., Meyer, A., Edwards, S. V., Pääbo, S., Villablanca, F. X., and Wilson, A. C. (1989). Dynamics of mitochondrial DNA evolution in animals, amplification and sequencing with conserved primers. *Proc. Natl. Acad. Sci. U.S.A.* **86**, 6196–6200.

Konings, D. A. M., Hogeweg, P., and Hesper, B. (1987). Evolution of the primary and secondary structures of the E1a mRNAs of the adenovirus. *Mol. Biol. Evol.* **4**, 300–314.

Krajewski, C. (1989). Phylogenetic relationships among cranes (Gruiformes, Gruidae) based on DNA hybridization. *Auk* **106**, 603–618.

Krajewski, C., and Fetzner, J. W., Jr. (1994). Phylogeny of cranes (Gruiformes, Gruidae) based on cytochrome-*b* DNA sequences. *Auk* **111**, 351–365.

Kretz, K. A., Carson, G. S., and O'Brien, J. S. (1989). Direct sequencing from low-melt agarose with Sequenase. *Nucleic Acids Res.* **17**, 5864.

Kumar, S., Tamura, K., and Nei, M. (1993). "MEGA, Molecular Evolutionary Genetics Analysis," version 1.01. Pennsylvania State University, University Park, Pennsylvania.

Lake, J. A. (1991). The order of sequence alignment can bias the selection of tree topology. *Mol. Biol. Evol.* **8**, 378–385.

Lanyon, S. M. (1985). Detecting internal inconsistencies in distance data. *Syst. Zool.* **34**, 397–403.

Ma, D.-P., Zharkikh, A., Graur, D., VandeBerg, J. L., and Li, W.-H. (1993). Structure and evolution of opossum, guinea pig, and porcupine cytochrome *b* genes. *J. Mol. Evol.* **36**, 327–334.

Maddison, W. P., and Maddison, D. R. (1992). "MacClade: Analysis of Phylogeny and Character Evolution," version 3. Sinauer Associates, Sunderland, Massachusetts.

Margoliash, E., Ferguson-Miller, S., Kang, C. H., and Brautigan, D. L. (1976). Do evolutionary changes in cytochrome *c* structure reflect functional adaptations? *Fed. Proc.* **35**, 2124–2130.

Milkman, R., and Crawford, I. P. (1983). Clustered third-base substitutions among wild strains of *Escherichia coli*. *Science* **221**, 378–380.

Milligan, B. G. (1994). Estimating evolutionary rates for discrete characters. *In* "Models and Phylogeny Reconstruction" (R. W. Scotland, D. J. Sibert, and D. M. Williams, Eds.), pp. 299–311. Clarendon Press, Oxford.

Mindell, D. P. (1991). Aligning DNA sequences, homology and phylogenetic weighting. *In* "Phylogenetic Analysis of DNA Sequences" (M. M. Miyamoto and J. Cracraft, Eds.), pp. 73–89. Oxford University Press, New York.

Mindell, D. P., and Honeycutt, R. L. (1990). Ribosomal RNA in vertebrates: Evolution and phylogenetic applications. *Annu. Rev. Ecol. Syst.* **21**, 541–566.

Mindell, D. P., Knight, A., Baer, C., and Huddleston, C. J. (1996). Slow rates of molecular evolution in birds and the metabolic rate and body temperature hypothesis. *Mol. Biol. Evol.* **13**, 422–426.

Moore, W. S. (1995). Inferring phylogenies from mtDNA variation, mitochondrial-gene trees versus nuclear-gene trees. *Evolution* **49**, 718–726.

Moum, T., Johansen, S., Erikstad, K. E., and Piatt, J. F. (1994). Phylogeny and evolution of the auks (subfamily Alcinae) based on mitochondrial DNA sequences. *Proc. Natl. Acad. Sci. U.S.A.* **91**, 7912–7916.

Mullis, K. B., and Faloona, F. A. (1987). Specific synthesis of DNA *in vitro* via a polymerase-catalyzed chain reaction. *Methods Enzymol.* **155**, 335–350.

Navidi, W. C., and Beckett-Lemus, L. (1992). The effect of unequal transversion rates on the accuracy of evolutionary parsimony. *Mol. Biol. Evol.* **9**, 1163–1175.

Needleman, S. B., and Wunsch, C. D. (1970). A general method applicable to the search for similarities in the amino acid sequence of two proteins. *J. Mol Biol.* **48**, 443–453.

Nei, M. (1991). Relative efficiencies of different tree-making methods for molecular data. *In* "Phylogenetic Analysis of DNA Sequences" (M. M. Miyamoto and J. Cracraft, Eds.), pp. 90–128. Oxford University Press, New York.

Noller, H. F., Moazed, D., Stern, S., Powers, T., Allen, P. N., Robertson, J. M., Weiser, B., and Triman, K. (1990). *In* "The Ribosome, Structure, Function, and Evolution" (W. E. Hill, A. Dahlberg, R. A. Garrett, P. B. Moore, D. Schlessinger, and J. R. Warner, Eds.), pp. 73–92. American Society for Microbiology, Washington, D.C.

Olson, S. L. (1973). A classification of the Rallidae. *Wilson Bull.* **85**, 381–416.

Olson, S. L. (1975). A review of the extinct rails of the New Zealand region. *Nat. Mus. N.Z. Rec.* **1**, 63–79.

Olson, S. L. (1977). A synopsis of the fossil Rallidae. *In* "Rails of the World" (by S. D. Ripley), pp. 339–373. D. R. Godine, Boston.

Olson, S. L. (1979). Multiple origins of the Ciconiiformes. *Proc. Colonial Waterbird Group* **1978**, 165–170.

Olson, S. L. (1985). The fossil record of birds. *In* "Avian Biology" (D. S. Farner, J. R. King, and K. C. Parkes, Eds.), Vol. 8, pp. 79–252. Academic Press, New York.

Olson, S. L., and Steadman, D. W. (1981). The relationships of the Pedionomidae (Aves, Charadriiformes). *Smithsonian Contrib. Zool.* **337**, 1–25.

Page, R. D. M. (1996). TreeView: An application to display phylogenetic trees on personal computers. *Comput. Appl. Biosci.* **12**, 357–358.

Pamilo, P., and Nei, M. (1988). Relationships between gene trees and species trees. *Mol. Biol. Evol.* **5**, 568–583.

Parker, W. P. (1866). On the skull of *Aptornis*. *Phil. Trans. R. Soc.* (not seen).

Patterson, C., Williams, D. M., and Humphries, C. J. (1993). Congruence between molecular and morphological phylogenies. *Annu. Rev. Ecol. Syst.* **24**, 153–188.

Pesole, G., Sbisa, E., Preparata, G., and Saccone, C. (1992). The evolution of the mitochondrial d-loop region and the origin of modern man. *Mol. Biol. Evol.* **9**, 587–598.

Rohlf, F. J., Chang, W. S., Sokal, R. R., and Kim, J. (1990). Accuracy of estimated phylogenies: Effects of tree topology and evolutionary model. *Evolution* **44**, 1671–1684.

Saiki, R. K., Gelfand, D. H., Stoffel, S., Scharf, S. J., Higuchi, R., Horn, G. T., Mullis, K. B., and Erlich, H. A. (1988). Primer-directed enzymatic amplification of DNA with a thermostable DNA polymerase. *Science* **239**, 487–491.

Saitou, N., and Nei, M. (1987). The neighbor-joining method, a new method for reconstructing phylogenetic trees. *Mol Biol. Evol.* **4**, 406–425.

Sambrook, J., Fritsch, E. F., and Maniatis, T. (1989). "Molecular Cloning: A Laboratory Manual," 2nd ed. Cold Spring Harbor Laboratory Press, Cold Spring Harbor, New York.

Sanderson, M. J., and Doyle, J. J. (1992). Reconstruction of organismal and gene phylogenies from data on multigene families: Concerted evolution, homoplasy, and confidence. *Syst. Biol.* **41**, 4–17.

Sanger, F., Nicklen, S., and Coulson, A. R. (1977). DNA sequencing with chain-terminating inhibitors. *Proc. Natl. Acad. Sci. U.S.A.* **74**, 5463–5467.

Shannon, C. E., and Weaver, W. (1949). "The Mathematical Theory of Communication." University of Illinois Press, Urbana, Illinois; as cited by Pielou, E. C. (1977). *In* "Mathematical Ecology," p. 298. John Wiley & Sons, New York.

Sheen, J.-Y., and Seed, B. (1988). Electrolyte gradient gels for DNA sequencing. *Biotechniques* **6**, 942–944.

Sheldon, F. H. (1987). Rates of single-copy DNA evolution in herons. *Mol Biol. Evol.* **4**, 56–69.

Sibley, C. G., and Ahlquist, J. E. (1972). A comparative study of the egg white proteins of non-passerine birds. *Bull. Peabody Mus. Nat. Hist.* **39**, 1–276.

Sibley, C. G., and Ahlquist, J. E. (1985). The relationships of some African birds, based on the genetic material, DNA. *In* "Proceedings of the International Symposium on African Vertebrates, Systematics, Phylogeny, Evolution, and Ecology" (K. O. Schuchmann, Ed.) pp. 115–161. Zool. Forsch. Mus. Alexander Koenig, Bonn.

Sibley, C. G., and Ahlquist, J. E. (1990). "Phylogeny and Classification of Birds." Yale University Press, New Haven, Connecticut.

Sibley, C. G., and Monroe, B. L., Jr. (1990). "Distribution and Taxonomy of the Birds of the World." Yale University Press, New Haven, Connecticut.

Sibley, C. G., Ahlquist, J. E., and DeBenedictis, P. (1993). The phylogenetic relationships of the rails, based on DNA comparisons. *J. Yamashina Inst. Ornithol.* **25**, 1–11.

Simon, C., Frati, F., Beckenbach, A., Crespi, B., Liu, H., and Flook, P. (1994). Evolution, weighting, and phylogenetic utility of mitochondrial gene sequences and compilation of conserved polymerase chain reaction primers. *Ann. Entomol. Soc. Am.* **87**, 651–701.

Sullivan, J., Holsinger, K. E., and Simon, C. (1995). Among-site variation and phylogenetic analysis of 12S rRNA in sigmodontine rodents. *Mol. Biol. Evol.* **12**, 988–1001.

Swofford, D. L. (1991). When are phylogeny estimates from molecular and morphological data incongruent? *In* "Phylogenetic Analysis of DNA Sequences" (M. M. Miyamoto and J. Cracraft, Eds.), pp. 295–333. Oxford University Press, New York.

Swofford, D. L. (1993). "PAUP, Phylogenetic Analysis Using Parsimony," version 3.1. Illinois Natural History Survey, Champaign, Illinois.

Tajima, E., and Nei, M. (1984). Estimation of evolutionary distance between nucleotide sequences. *Mol. Biol. Evol.* **1**, 269–285.

Tautz, D., Trick, M., and Dover, G. A. (1986). Cryptic simplicity is a major source of genetic variation. *Nature (London)* **322**, 652–656.

Thorne, J. L., and Kishino, H. (1992). Freeing phylogenies from artifacts of alignment. *Mol. Biol. Evol.* **9**, 1148–1162.

Van de Peer, Y., Van den Broeck, I., De Rijk, P., and De Wachter, R. (1994). Database on the structure of small ribosomal subunit RNA. *Nucleic Acids Res.* **22**, 3488–3494.

Vawter, L., and Brown, W. M. (1993). Rates and patterns of base change in the small subunit ribosomal RNA gene. *Genetics* **134**, 597–608.

Verheyen, R. (1957). Contribution au démembrement de l'ordo artificiel des Gruiformes (Peters 1934). *Bull. Inst. Sci. Nat. Belgique* **33**, 1–19.

Watson, J. D., Hopkins, N. H., Roberts, J. W., Argetsinger-Steitz, J., and Weiner, A. M. (1987). "Molecular Biology of the Gene," 4th ed. Benjamin-Cummings, Menlo Park, California.

Weston-Hafer, K., and Berg, D. E. (1991). Limits to the role of palindromy in deletion formation. *J. Bacteriol.* **173**, 315–318.

Wetmore, A. (1960). A classification of birds of the world. *Smithsonian Misc. Coll.,* **139**, 1–37.

Wheeler, W. C., and Honeycutt, R. L. (1988). Paired sequence difference in ribosomal RNAs: Evolutionary and phylogenetic implications. *Mol. Biol. Evol.* **5**, 90–96.

Wilkinson, M. (1991). Homoplasy and parsimony analysis. *Syst. Zool.* **40**, 105–109.

Williams, P. L. and Fitch, W. M. (1990). Phylogeny determination using dynamically weighted parsimony method. *Methods Enzymol.* **183**, 615–626.

Wolfson, R., Higgins, K. G., and Sears, B. B. (1991). Evidence for replication slippage in the evolution of *Oenothera* chloroplast DNA. *Mol. Biol. Evol.* **8**, 709–720.

Wu, C.-I, (1991). Inferences of species phylogeny in relation to segregation of ancient polymorphisms. *Genetics* **127**, 429–435.

Yates, F. (1934). Contingency tables involving small numbers and the chi-square test. *J. R. Statist. Soc. Suppl.* **1**, 217–235; as cited in Zar, J. H. (1974). "Biostatistical Analysis." Prentice-Hall, Englewood Cliffs, New Jersey.

Zar, J. H. (1974). "Biostatistical Analysis." Prentice-Hall, Englewood Cliffs, New Jersey.

Zharkikh, A., and Li, W.-H. (1993). Inconsistency of the maximum-parsimony method: The case of five taxa with a molecular clock. *Syst. Biol.* **42**, 113–125.

CHAPTER 6

Phylogeny of the Pelecaniformes: Molecular Systematics of a Privative Group

DOUGLAS SIEGEL-CAUSEY

School of Biological Sciences, University of Nebraska, Lincoln, Nebraska 68588

I. INTRODUCTION

There is only one evolutionary history. A quick glance, however, through the current systematics literature will indicate that some troublesome taxa are notable for the degree of divergent and incongruent phylogenies associated with them. Incongruency in phylogenetic reconstruction is troublesome, because it is the outward manifestation of intrinsic contradictions in the methodologies or philosophies employed. Perhaps the most troublesome taxon in birds at present is the Pelecaniformes, a traditional order of pelican-like birds commonly hailed as an exemplar of a natural group.

Pelecaniformes and the other large waterbird orders (i.e., Sphenisciformes, Gaviiformes, Podicipediformes, Procellariiformes, Ciconiiformes) are considered by most systematists to comprise the early branching clades of nonratite birds. Previous studies of the biology and higher level relationships among these orders have helped influence our concepts about macroevolutionary paterns among birds. Exceedingly detailed anatomical and morphological studies from the beginnings of avian systematics research (e.g., Garrod, 1873; Mivart, 1878; Fürbringer, 1888; Gadow, 1889; Beddard, 1899) established the relationships and methods of character analysis and interpretation that prevail today.

Critical assessment of the higher order relationships among nonpasserines is one of the most critical issues in avian systematics (Sheldon and Bledsoe, 1993) and a crucial problem at present is whether the Pelecaniformes is monophyletic. This is a relatively small group (about 70 species in 6 families) of pelican-like waterbirds having a cosmopolitan distribution—they are absent only from deserts and polar regions—and one that has generated controversy since the beginnings of avian systematic research. Molecular studies (Sibley and Ahlquist, 1990; Hedges and Sibley, 1994; fig. 6.1) suggest that Pelecaniformes is not monophyletic and the former members are instead grouped with other early branching groups. If these molecular-based studies are validated and Pelecaniformes is dismantled, then the traditional morphological framework that forms the basis for most phylogenetic assumptions in avian systematics is clearly in need of reexamination as it pertains to modern systematics research. Most avian phylogenies at present are based on morphology or other traditional characters, but while the hypotheses may be explicit, their empirical and philosophical foundations are not.

Phylogenetic hypotheses are the critical framework for understanding macroevolutionary patterns and interpreting comparative evolutionary data. Quantitative comparative analysis of the patterns in molecular, behavioral, and morphological evolution requires detailed phylogenies, particularly when character states and transformation series are incompletely known. For these reasons, a better understanding of higher order phylogenies is crucial when group monophyly is problematic and major evolutionary change has occurred.

A. The Problem of the Putatively Privative Pelecaniformes

Understanding the phylogenetic relationships of the pelecaniform birds, therefore, is central to the larger question of understanding the higher level relationships among nonpasserine birds. Traditionally, this order comprises six families, e.g., Phaethontidae (tropicbirds), Fregatidae (frigatebirds), Pelecanidae (pelicans), Sulidae (gannets), Phalacrocoracidae (cormorants), and Anhingidae (darters). Pelecaniform affinities have been established using conventionally distinctive features including totipalmate feet or steganopody (four toes joined by a web), gular pouch under the mandible, a prelanding vocalization, and several others. Procellariiform birds are considered to be the sister-group to the traditional Pelecaniformes, but

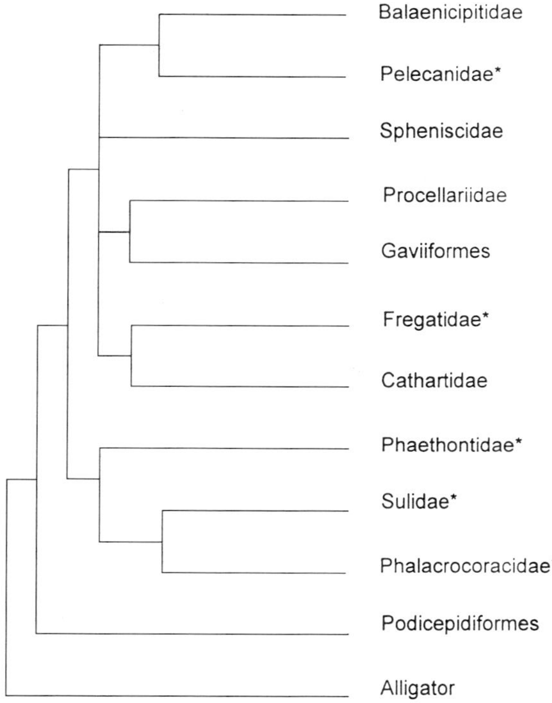

FIGURE 6.1 Distance tree for pelecaniforms (taxa marked by asterisk) and relatives inferred from DNA sequences of mitochondrial 12S and 16S rRNA genes (1.7 kb total). The tree was contructed by neighbor joining (Saitou and Nei, 1987); bootstrap values were determined by 2000 replications. (Reproduced with permission from Hedges and Sibley, 1994.)

there is no agreement among studies on this or on most other aspects of higher order relationships among the early branching clades of waterbirds (Cracraft, 1985; Olson, 1979; Sibley *et al.*, 1988; Sibley and Monroe, 1990; Hedges and Sibley, 1994).

Pelecaniformes, in turn, have been considered by many authorities to be the sister-group to herons and storks (Ciconiiformes *sensu latu*), and since the earliest studies systematists have utilized pattern analyses of pelecaniform morphology and behavior as insights into the historical and developmental trajectories of avian evolution. For example, most higher order avian taxa are defined by traditional characters such as an acarinal sternum in ratites, tubular nostrils in procellariiforms, and holorhinal bills in charadriiforms. If these and other traditional characters used for grouping prove to be convergent, homoplasious, or otherwise unuseful, then current notions concerning use and stability of traditional characters in systematic research are problematic. Furthermore, the issue of pelecaniform monophyly is pivotal to the larger issue of higher level relationships among basal waterbirds. If hypotheses concerning the paraphyly of pelecaniform birds are correct, then traditional ideas about early patterns of morphological and behavioral evolution in birds need to be completely reassessed. This will have great impact on current studies underway in later branching clades of birds. Equally important, rearrangement of the basal orders will impact future systematic studies of other avian taxa by change in the status of putative sister-groups, the patterns of character state evolution, and the polarity of transformation series.

Group membership in the Pelecaniformes has remained contentious from the beginning. Many of the dissenting early authorities believed that tropicbirds and/or frigatebirds were inappropriately conjoined with the others; however, most agreed that the Sulidae, Phalacrocoracidae, and Anhingidae (the "core Pelecaniformes") were closely related. Unfortunately, in the more than 30 systematic studies done since Huxley's (1867) treatment of the order, only three have employed reproducible methodology—and these results are together ambiguous. Cracraft (1981, 1985) pioneered numerical cladistics for birds and undertook one of the first quantitative analyses in his study on the monophyly of the Pelecaniformes. Cracraft employed both morphology and behavior and used cladistic methods as a means to build trees and to test for group monophyly. Cracraft concluded that the order was monophyletic, that the traditional relationship among families (e.g., Wetmore, 1960) was correct, that tropicbirds and frigatebirds were members of the order, and that *Balaeniceps rex* was not (see below). Soon after, Sibley and colleagues (Sibley *et al.*, 1988; Sibley and Ahlquist, 1990), employing DNA–DNA hybridization techniques and UPGMA clustering methods, presented novel results for pelecaniform relationships. They concluded that Pelecaniformes were paraphyletic. The original six families comprising the order were now distributed among three large taxon groups and some with startling sister-group relationships (e.g., pelicans and herons; frigatebirds and a clade comprising penguins, loons, and albatrosses). Most recently, Hedges and Sibley (1994) recovered the 12S and 16S mitochondrial DNA nucleotide sequences of representative pelecaniform taxa and outgroups, and generally affirmed the results obtained earlier by Sibley and colleagues (Fig. 6.1).

This sharp divergence between morphological and molecular-based research is troubling, because we expect that phenotypic and genotypic characters generally should correlate given the same evolutionary history. While most morphological and molecular studies on the same groups do differ in some details, few deviate as much as those done on Pelecaniformes (Bledsoe and Raikow, 1990). One possible cause for the disparity between the morphological results compared to the molecular data may be that Pelecaniformes as traditionally constituted is a privative group. *Privative groups* are those founded on the absence of traits or on traits so ill-defined that no inclusive group exclusively possesses them. While steganopody, precourtship "wing flipping," or lack of an incubation patch may appear to be well-characterized traits, they may in fact be descriptive only of homoplasious similarities, or worse, of a conflation of similar but distinct traits, each with independent histories (see Siegel-Causey and Kharitonov, 1997).

Aristotle objected to privative groups because they could not be logically subdivided and because no group could be characterized by the absence of nothing. Further problems obtain with privative groups, notably instability in higher order relationships and assignment of group membership (Lynch and Renjifo, 1990). If privative, at best the assemblage currently considered as constituting Pelecaniformes presents many problems in interpreting the nature of morphological character states; at worst, it is an artifical taxon confounding our understanding of morphological evolution in birds.

B. The Enigma of *Balaeniceps*

Another long-standing controversy that relates directly to the issue of pelecaniform monophyly and higher level relationships is the status of the shoebill, *Balaeniceps rex*. The controversy began one year after its description in 1850 by Gould (1852). He considered this "new and most remarkable form" to be a long-legged type of pelecanid, but Jardine (1851) considered the differences as outweighing the similarities to pelicans and instead declared it a member of the Ciconiiformes (herons, storks). Subsequent authorities agreed and differed only in whether it was a heron, a stork, or a monotypic family. Cottam (1957) reasserted Gould's original impression of pelecanid affinities by examining similarities in cranial osteology. Cottam identified several osteological features shared by *Balaeniceps* and pelecanids, and not shared with other putative sister-groups. For example, in shoebills and pelicans the external nares are posterior to the internal nares, the prevomer is weakly developed, the palatines are ankylosed, the hypocleideum is fused to the sternal carina (also found in frigatebirds), the stomach has a pyloric chamber, and the syrinx lacks intrinsic muscles. Cottam's findings were ignored by most except for Olson (1979), and Cracraft (1985) concluded that the pelecaniform characters adduced by Cottam were either convergent or primitive. Cottam's work preceded the advent of modern phylogenetic systematics, so the analysis lacks currently accepted rudiments such as character assessment and polarization by comparison to outgroups. Subsequent molecular evidence (e.g., Hedges and Sibley 1994), however, support Cottam's conjecture and pair the shoebill and pelicans as sister taxa (Fig. 6.1).

Several points can be made about pelecaniform relationships as they are known today. First, more recent results suggest that Pelecaniformes is not monophyletic, particularly in that the group relationship of tropicbirds and frigatebirds is controversial. Second, the phylogenetic relationship of the shoebill to Pelecaniformes is unknown. Third, knowledge of pelecaniform relationships is central to understanding evolutionary patterns in the early evolutionary history of birds, with respect not only to past and current treatment of morphological data, but in what claims can be made about early evolutionary branching patterns in birds.

This chapter describes preliminary results of the molecular systematics of the traditional Pelecaniformes. Results are compared with previous molecular studies and with reanalysis of existing data as a means to understand some of the disparity in findings. Some potential problems in the characterization of morphological and behavioral traits traditionally used in phylogenetic analyses are discussed, and some possible new approaches in character analysis are suggested.

II. METHODS

A. Data Sets and Analysis

Morphological data used here are primarily those published in morphological, osteological, and behavioral studies (Cottam, 1957; van Tets, 1965; Cracraft, 1985; Saiff, 1978; Siegel-Causey, 1986a,b, 1988, 1991, 1992). Numerous characters have been proposed, but many were judged to be homoplasious, ill-defined, or traits with indeterminable polarities (Siegel-Causey, 1996). Two analyses were performed, one with all 20 characters, and another with two classic characters removed (e.g., 19, steganopody; 20, gular pouch; Table I).

Molecular data sets are from molecular studies by the author and Hedges and Sibley (1994). Partial DNA sequences of three mitochondrial genes (12S rRNA, 16S rRNA, and cytochrome B mtDNA), totaling about 1.5 kb of aligned sequence, were obtained from 2 to 5 individuals of the following 12 species of birds: Magellanic penguin (Speniscidae: *Spheniscus magellanicus*), Laysan albatross (Diomedeidae: *Diomedea immutabilis*), short-tailed shearwater (Procellariidae: *Puffinus tenuirostris*), Leach's storm petrel (Hydrobatidae: *Oceanodroma leucorhoa*), white-billed tropicbird (Phaethontidae: *Phaethon lepturus*), magnificent frigatebird (Fregatidae: *Fregata magnificens*), white pelican (Pelecanidae: *Pelecanus erythrorhynchus*), gannet (Sulidae: *Morus bassana*), double-crested cormorant (Phalacrocoracidae: *Phalacrocorax auritus*), anhinga (Anhingidae: *Anhinga anhinga*), shoebill (Balaenicepididae: *Balaeniceps rex*), and black stork (Ciconiidae: *Ciconia nigra*). The sequenced regions correspond to sites 1765–2040 (12S rRNA), 2800–3750 (16S rRNA), and 14,905–15,275 (cytochrome B) in the published sequence of chicken *Gallus gallus*; Desjardins and Morais, 1990).

DNA was recovered from frozen tissue obtained by the author, by collection or loan, using standard protocols. Tissue was ground to a powder in liquid nitrogen; approximately 100 mg of tissue was digested in a buffer containing proteinase K and nonionic detergents or containing guanidinium isothiocyanate. Following incubation, the solution was extracted twice with equal volumes of phenol:chloroform:isoamyl alcohol (24:24:1). The DNA was precipitated with ethanol and suspended in 50 μl of Tris-EDTA buffer.

Gene regions were amplified by polymerase chain reaction (PCR) using conserved primer regions among vertebrates (Kocher *et al.*, 1989; Hedges and Sibley, 1994). Single-stranded DNA was made by asymmetric amplification with one of the two primers limiting, and purification of the template and sequencing were by methods described elsewhere (Siegel-Causey, 1997). The resulting PCR product

TABLE I Data Set of Traditional Characters Used in Analyses[a]

Taxon	Character[b]																			
	1	2	3	4	5	6	7	8	9	10	11	12	13	14	15	16	17	18	19	20
Outgroup	0	0	0	0	0	0	0	0	0	0	0	0	0	0	0	0	0	0	0	0
Diomedeidae	1	1	1	1	0	0	0	0	0	0	0	0	0	0	0	0	0	0	0	0
Ciconiidae 1	1	1	1	0	0	0	0	0	0	0	0	0	0	0	0	0	0	0	0	0
Phaethontidae	1	1	1	1	0	0	0	0	0	0	0	0	0	0	0	0	0	0	1	1
Fregatidae	1	1	0	1	1	1	0	1	0	0	0	0	0	0	0	0	0	0	1	1
Pelecanidae	1	1	0	1	1	1	1	1	0	0	0	0	0	0	0	0	0	0	1	1
Balaeniceps	1	1	1	1	1	1	1	1	0	0	0	0	0	0	0	0	0	0	0	0
Sulidae	1	1	0	1	1	1	1	0	1	1	1	1	1	1	0	0	0	0	1	1
Phalacrocoracidae	1	1	0	1	1	1	1	0	1	1	1	1	1	1	1	1	1	1	1	1
Anhingidae	1	1	0	1	1	1	1	0	1	1	1	1	1	1	1	1	1	1	1	1

[a]Characters were based on traditional analyses of morphology and behavior.

[b]Character descriptions are as follows: (1) vascular notches in cranial metotic process (0, absent; 1, present) (Saiff, 1988); (2) cranial nerve IX foramen (0, indistinct; 1, separate) (Saiff, 1988); (3) cranial nerve V foramen (0, posterior or subequal; 1, anterior) (Saiff, 1988); (4) bony nostril (0, present; 1, absent, reduced, or indistinct) (Cracraft, 1985); (5) incubation (0, never by feet; 1, usually by feet) (Cracraft, 1985); (6) stapedial canal (0, present; 1, absent) (Saiff, 1988); (7) eustachian canal (0, open; 1, partially or completely closed) (Cracraft, 1985); (8) hypocleideum (0, always free; 1, usually fused to sternal carina) (Cottam, 1957); (9) maxillopalatines (0, large; 1, reduced) (Cracraft, 1985); (10) braincase width to depth (0, subequal; 1, wider than deep) (Cracraft, 1985); (11) orbital process of quadrate (0, large; 1, reduced) (Cracraft, 1985); (12) lateral wall of presphenoid sinus (0, present; 1, absent, reduced, or indistinct) (Cracraft, 1985); (13) hop display (0, present; 1, absent) (van Tets, 1965); (14) sky-pointing display (0, absent; 1, present) (van Tets, 1965); (15) postorbital processes (0, present; 1, absent) (Cracraft, 1985); (16) lateral wall of presphenoid sinus (0, without bony ring; 1, with bony ring) (Cracraft, 1985); (17) interorbital septum (0, present; 1, absent) (Cracraft, 1985); (18) occipital style (0, absent; 1, present) (Cracraft, 1985); (19) totipalmate foot (0, absent; 1, present) (Aristotle); (20) gular pouch (0, absent; 1, present) (Aristotle).

was visualized with ethidium bromide staining to verify product band size and purified from primers (Wizard PCR preps; Promega, Madison, WI). Cycle sequencing was carried out using nonradioactive labels (Genius System; Boehringer Mannheim, Indianapolis, IN). The cycle sequencing product was run out on a 6% denaturing polyacrylamide gel. Sequences discussed in this chapter have been deposited in the GenBank database (accession Nos. L33368–L33397, U83149–U83160, U83203, U83204).

Alignments of putative homologous sequences were done using GCG (Genetics Computing Group, Madison, WI) with multiple alignment parameters of fixed and floating gap penalty equal to 10 and pairwise parameters of gap penalty equal to 3 and *k*-tuple equal to 1. Insertion and deletion gaps were coded as missing (Swofford, 1993). Phylogenetic signal within the data was assessed using the g_1 statistic of the random tree distribution (Hillis and Huelsenbeck, 1992). Character states were polarized by a conservative reference to outgroup taxa; i.e., characters with ambiguous or polymorphic states among outgroup taxa were discarded (see Wiley *et al.*, 1991). Parsimony analyses were performed using the heuristic search option in PAUP with random addition of taxa (Swofford, 1993). The neighbor-joining method (Saitou and Nei, 1987) using the Jukes and Cantor (1969) and Kimura (1980) models of nucleotide substitution and maximum-likelihood method was implemented using PHYLIP (Felsenstein, 1991).

III. RESULTS

A. Molecular Evidence

Using the sequences and alignment procedures described above, a random tree distribution was generated using PAUP and having a $g_1 = -0.615$, which indicates that these data are significantly skewed at $P < 0.01$ (Hillis and Huelsenbeck, 1992).

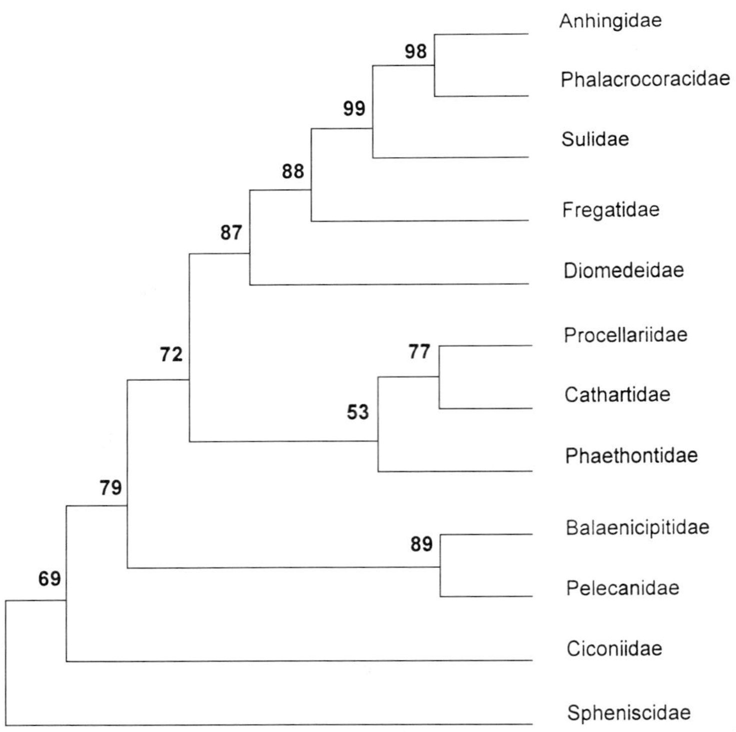

FIGURE 6.2 The single most-parsimonious tree (length = 1149) based on mitochondrial 12S–16S rRNA and cytochrome B nucleotide sequences for pelecaniforms and relatives resulting from maximum parsimony analysis using a heuristic search by PAUP (CI = 0.519, RC = 0.192). Majority-rule bootstrap percentile values generated by 2000 replications are shown on branches.

A heuristic search with random sequence addition and tree-bisection-reconnection branch swapping resulted in a single most-parsimonious tree (Fig. 6.2), with a length of 1149 steps, a CI excluding uninformative characters of 0.519, and a rescaled consistency index (RC) of 0.192. The confidence in the parsimony tree was assessed by 2000 bootstrap replications; all but two bootstrap percentile (BP) values were greater than 50. The nodes joining the putative clade comprising Hydrobatidae and Cathartidae, and this with the Phaethontidae, were found in only 41 and 45% of the replications, respectively. The "core" pelecaniform taxa (Anhingidae, Phalacrocoracidae, Sulidae) were robustly supported by bootstrap replications with BP values of 98 and 100. Other groupings were less well supported by the data (e.g., the clade comprising Phaethontidae + Diomedeidae + Fregatidae + "core" pelecaniforms was found in 51% of replications), but most bootstrap percentile values exceeded 80. One of the most robust groupings (BP = 95) was the clade Pelecanidae + Balaenicipitidae, which in this analysis was revealed as the sister-group to the remaining traditional pelecaniform taxa.

B. Morphological and Behavioral Evidence

Analyses using all characters and with two characters removed produced similar results; the larger data set yielded a proportionally longer and less robust tree (length = 23; CI = 0.899; RC = 0.891; $g_1 = -1.433$, $P < 0.01$). The two characters in question, characters 19 (steganopody) and 20 (gular pouch), are traditionally invoked for pelecaniform monophyly. To examine more closely the nature of the other characters used in the analysis, the following discussion is limited to the data set with characters 19 and 20 removed.

Similar reconstruction methodology was used for the reduced morphological data set as with the molecular data; a single tree was found (length = 19 steps; CI = 0.944; RC = 0.914; $g_1 = -1.811$, $P < 0.001$). The "core" pelecaniform taxa were similarly well supported by the traditional data, with 100% of the bootstrap replications recovering the clade comprising Sulidae + Phalacrocoracidae + Anhingidae (Fig. 6.3). The sister-group to this clade (Fregatidae + Balaeniciptidae + Pelecanidae) was less supported in terms of bootstrap replications (BP = 68) in a polytomy; however, members uniquely shared a single morphological synapomorphy (character 8: hypocleideum fused with sternal carina). The Phaethontidae, the remaining family of traditional pelecaniform taxa, was not found to be a member of the preceding clade. All branchings except for that including *Balaeniceps* were supported by three to seven synapomorphies.

IV. DISCUSSION

The preliminary analyses of the molecular data and the traditional data reported here agree in several important aspects. First, there is joint support for a "core" clade

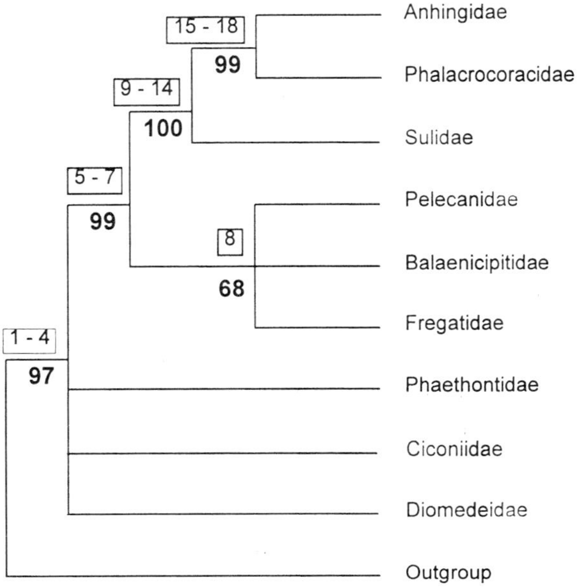

FIGURE 6.3 The single most-parsimonious tree (length = 19) based on morphological and behavioral characters for pelecaniforms and relatives resulting from maximum parsimony reconstruction using a heuristic search by PAUP (CI = 0.944, RC = 0.914). Majority-rule bootstrap percentile values generated by 2000 replications are shown on branches.

of pelecaniform taxa, i.e., Anhingida + Phalacrocoracidae + Sulidae. This finding is reassuring in that it follows that of nearly every previous analysis done for Pelecaniformes. While consensus does not necessarily equate to reliability, it strongly suggests that morphology, behavior, and DNA sequence data are indicative of a common evolutionary pattern. Second, a close relationship between shoebills and pelicans is supported by both molecular and morphological data. It thus appears that Cottam's (1957) conjecture may be correct in that *Balaeniceps rex* is not the shoebill stork, but the shoebill pelican!

Except for the "core" pelecaniform taxa, and the close relationship between shoebills and pelicans, nearly every modern analysis of pelecaniform relationships has produced different results. Nonetheless, what is quite apparent is that none of the results discussed here supports monophyly of the Pelecaniformes. If the traditional Pelecaniformes are thus paraphyletic, this brings into question how traditional pelecaniform characters are characterized.

Closer examination of the data supporting the basal branches obtained in the tree based on traditional data (Fig. 6.3) illustrates some of the problems associated with the conflict among previous studies of Pelecaniformes. The four characters supporting the branch including pelecaniform taxa except for tropicbirds appear problematic. For example, character 5 [incubating by feet rather than by incubation patch (character 41 of Cracraft 1985)] is by definition acceptable, but the behavior clearly is associated with the loss of an incubation patch. Furthermore, incubation behavior in Pelecaniformes is not uniform and there is variation among species and even individuals in whether the eggs are covered by the feet during incubation or whether they are held between the feet and body (Siegel-Causey, 1987, 1988; van Tets, 1965). Similarly, character 6 [lack of a stapedial canal in the opisthotic cranium (Saiff, 1978)] represents the loss or absence of a character that is otherwise present in outgroups.

Six of the eight characters supporting basal branches in the morphological tree shown in Fig. 6.3 (i.e., characters 3–8) can be questioned on the grounds that the characters represent loss of features, that they possess a derived state synapomorphy that is homoplasious in other related taxa, or that they comprise a potentially confounded set of independent traits. The two characters most commonly invoked as indicative of pelecaniform monophyly are those that have been least studied of all (i.e., characters 19 and 20). Surprisingly, a concerted search of the considerable literature on pelecaniform anatomy did not reveal a single study on these two features, except for repeated mentions that they exist. The aberrant forms of the gular and the nonconforming appearance of the webbing in Fregatidae and Phaethontidae argue for a closer examination of this morphology and for consideration of the possibility that these two morphological traits have a more complex evolutionary pattern than thought previously. As suggested at the beginning, the traditional avian order of Pelecaniformes may be privative, and therefore only an arrangement of taxa associated by artificial characteristics.

Analagous findings were reported by Hedges and Sibley (1994) using a different set of taxa and molecular data (i.e., mitochondrial 12S and 16S genes). Tree topologies differed substantially from those reported here, although they also found strong support for the "core" pelecaniform taxa, and association between the shoebill and pelicans. In addition, they found no support for monophyly of the Pelecaniformes, and none for association of Fregatidae with other traditional pelecaniform taxa.

Data set incongruity as commonly discussed with respect to morphological vs molecular data (e.g., Patterson *et al.*, 1993; Miyamoto and Fitch, 1995) may lay more in the unsuitability of particular characters used in traditional analyses, and in the conceptual and methodological gulf separating distance and cladistic analyses. Although molecular studies of Pelecaniformes are still in an elementary stage, the results to date are heuristic and indicate the need for reexamination of the morphological characters traditionally utilized in systematics research, and further investigation into the status of Pelecaniformes as a natural taxon.

ACKNOWLEDGMENTS

I benefited from discussion with R. C. Bromley, B. C. Livezey, J. D. Lynch, and E. O. Wiley concerning the nature of characters and methodology, and two anonymous reviewers. I am grateful to F. C.

Sheldon (Louisiana State University Museum of Zoology) for selected pelecaniform tissue. Research was supported in part by the University of Nebraska Center for Biotechnology, and by NSF Grants BSR 8722562 and DEB 9420041.

REFERENCES

Beddard, F. E. (1899). "The Structure and Classification of Birds." Longmans & Green, London.

Bledsoe, A. H., and Raikow, R. J. (1990). A quantitative assessment of congruence between molecular and nonmolecular estimates of phylogeny. *J. Mol. Evol.* **30**, 425–433.

Cottam, P. A. (1957). The pelecaniform characters of the skeleton of the shoe-bill stork, *Balaeniceps rex*. *Bull. Br. Mus. Nat. Hist. (Zool.)* **5**, 51–71.

Cracraft, J. (1981). Toward a phylogenetic classification of recent birds of the world (class Aves). *Auk* **98**, 681–714.

Cracraft, J. (1985). Monophyly and phylogenetic relationships of the Pelecaniformes. *Auk* **98**, 681–714.

Desjardins, P., and Morais, R. (1990). Sequence of mitochondrial DNA in chicken. *J. Mol. Evol.* **32**, 153–161.

Felsenstein, J. (1991). "PHYLIP: Phylogenetic Inference Package," version 3.4. Department of Genetics, Univ. Washington, Seattle, Washington.

Fürbringer, M. (1888). "Üntersuchungen zur Morphologie und Systematik der Vögel." von Hokema, Amsterdam.

Gadow, H. (1889). On the taxonomic value of the intestinal convolutions in birds. *Proc. Zool. Soc. Lond.* **1889**, 303–316.

Garrod, A. H. (1873). On certain muscles of the thigh of birds, and on their value in classification. Part I. *Proc. Zool. Soc. Lond.* **1873**, 626–644.

Gould, J. (1852). On a new and most remarkable form in ornithology. *Proc. Zool. Soc. Lond.* **1852**, 1–2.

Hedges, S. B., and Sibley, C. G. (1994). Molecules vs. morphology in avian evolution: The case of the "pelecaniform" birds. *Proc. Natl. Acad. Sci. U.S.A.* **901**, 9861–9865.

Hillis, D. M. (1991). Discriminating between phylogenetic signal and random noise in DNA sequences. *In* Phylogenetic Analysis of DNA Sequences" (M. M. Miyamoto and J. Cracraft, Eds.), pp. 278–294. Oxford Univ. Press, Oxford.

Hillis, D. M., and Huelsenbeck, J. P. (1992). Signal, noise, and reliability in molecular phylogenetic analyses. *J. Hered.* **83**, 189–195.

Huxley, T. E. (1867). On the classification of birds; and on the taxonomic value of the modifications of certain of the cranial bones observable in that class. *Proc. Zool. Soc. Lond.* **1867**, 415–472.

Jardine, W. (1851). Ornithology in 1850; *Balaeniceps rex*. *Contr. Ornithol.* **1851**, 11–14.

Jukes, T. H., and Cantor, C. R. (1969). Evolution of protein molecules. *In* Mammalian Protein Metabolism" (H. M. Munro, Ed.), pp. 21–132. Academic Press, New York.

Kimura, M. (1980). A simple method for estimating evolutionary rate of base substitutions through comparative studies of nucleotide sequences. *J. Mol. Evol.* **16**, 111–120.

Kocher, T. D., Thomas, W. K., Meyer, A., Edwars, S. V., Pääbo, S., Villablanca, F. X., and Wilson, A. C. (1989). Dynamics of mitochondrial DNA evolution in animals: Amplification and sequencing with conserved primers. *Proc. Natl. Acad. Sci. U.S.A.* **86**, 6196–6200.

Lynch, J. D., and Renjifo, J. M. (1990). Two new toads (Bufonidae: *Rhamphophryne*) from the northern Andes of Colombia. *J. Herpetol.* **24**, 364–371.

Mivart, St. G. (1878). On the axial skeleton of the Pelecanidae. *Trans. Zool. Soc. Lond.* **1878**, 173–275.

Miyamoto, M. M., and Fitch, W. M. (1995). Testing species phylogenies and phylogenetic methods with congruence. *Syst. Biol.* **44**, 64–76.

Olson, S. L. (1979). Multiple origins of the Ciconiiformes. *Proc. Colonial Waterbird Group* **1978**, 165–170.

Patterson, C., Williams, D. M., and Humphries, C. J. (1993). Congruence between molecular and morphological phylogenies. *Annu. Rev. Ecol. Syst.* **24**, 153–188.

Saiff, E. (1978). The middle ear of the skull of birds: The Pelecaniformes and Ciconiiformes. *Zool. J. Linn. Soc.* **63**, 315–370.

Saitou, N., and Nei, M. (1987). The neighbor-joining method: A new method for reconstructing phylogenetic trees. *Mol. Biol. Evol.* **4**, 406–425.

Sheldon, F. H., and Bledsoe, A. H. (1993). Avian molecular systematics, 1970s and 1990s. *Annu. Rev. Ecol. Syst.* **24**, 243–278.

Sibley, C. G., and Ahlquist, J. E. (1990). "Phylogeny and Classification of the Birds of the World." Yale Univ. Press, New Haven, Connecticut.

Sibley, C. G., Monroe, B. L., Jr. (1990). "Distribution and Taxonomy of Birds of the World." Yale Univ. Press, New Haven, Connecticut.

Sibley, C. G., Ahlquist, J. E., and Monroe, B. L., Jr. (1988). A classification of the living birds of the world based on DNA–DNA hybridization studies. *Auk* **105**, 409–423.

Siegel-Causey, D. (1986a). The courtship behavior and mixed-species pairing of king and imperial blue-eyed shags (*Phalacrocorax albiventer* and *P. atriceps*). *Wilson Bull.* **98**, 571–580.

Siegel-Causey, D. (1986b). The behaviour and phylogenetic status of the Magellanic cormorant *Phalacrocorax magellanicus*. *Notornis* **33**, 249–257.

Siegel-Causey, D. (1987). The behaviour of the red-footed cormorant *Phalacrocorax gaimardi*. *Notornis* **34**, 1–9.

Siegel-Causey, D. (1988). Phylogeny of the Phalacrocoracidae. *Condor* **90**, 885–905.

Siegel-Causey, D. (1991). Systematics and biogeography of North Pacific shags, with a description of a new species. *Occ. Pap. Univ. Kansas Mus. Nat. Hist.* **140**, 1–17.

Siegel-Causey, D. (1992). Systematics and Classification. *Auk* **109**, 939–944.

Siegel-Causey, D. (1997). Phylogeny of the Anhingidae. *Mem. Carnegie Mus. Nat. Hist.* (in press).

Siegel-Causey, D., and Kharitonov, S. P. (1997). Evolution and ecological history of coloniality in waterbirds. *In* "Evolution and Ecology of Coloniality in Waterbirds" (F. Cézilly, H. Hafner, and D. N. N. Nettleship, Eds.). Oxford University Press, Oxford (in press).

Swofford, D. (1993). "PAUP: Phylogenetic Analysis Using Parsimony," version 3.1. Illinois Natural History Survey, Illinois.

van Tets, G. F. (1965). A comparative study of some social communication patterns in the Pelecaniformes. *AOU Ornithological Monographs No. 2.*

Wetmore, A. (1960). A Classification for the birds of the world. *Smithsonian Misc. Coll.* **139**, 1–37.

Wiley, E. O., Siegel-Causey, D., Brooks, D. R., and Funk, V. A. (1991). The compleat cladist, a handbook for phylogenetic techniques. *Spec. Publ. Univ. Kansas Mus. Nat. Hist.* **19**, 1–158.

CHAPTER 7

The Phylogeny of Ratite Birds: Resolving Conflicts between Molecular and Morphological Data Sets

KRISTA LEE,* JULIE FEINSTEIN,† AND JOEL CRACRAFT†[1]

*Museum of Natural Science, Louisiana State University, Baton Rouge, Louisiana 70803

†Department of Ornithology, American Museum of Natural History, New York, New York 10024

I. INTRODUCTION

Perhaps no group of birds has engendered more controversy over their relationships and evolutionary history than have the paleognaths, including both the flightless

[1] To whom correspondence should be addressed.

ratites and the tinamous. The ratite birds, in particular, have interested and perplexed avian systematists for well over 100 years. Early workers, noting the absence of a keel on the sternum, placed the ratites together (Merrem, 1813; Lesson, 1831), and this view was later reinforced by Huxley (1867), who noted that they share a so-called "dromaeognathous palate."

Other avian systematists, on the other hand, denied a close relationship, believing instead that the similarities among ratites were due to convergence (Fürbringer, 1888, 1902; Parker, 1895). Certainly a primary reason for discounting a close relationship among the ratites was that their widely disjunct distributions—interpreted within the context of a stable-continent paradigm—could seemingly be explained only by assuming independent origins from separate, volant ancestors.

Most subsequent investigators during the first half of this century did not accept paleognath monophyly, thinking instead that all the major groups of paleognaths arose independently from various carinate ancestors (see Sibley and Ahlquist, 1990, for a review). Using behavioral similarities, Meise (1963) was among the first to present an explicit hypothesis of paleognath monophyly as well as for the relationships among the families. A decade later, Cracraft (1974) presented a cladistic hypothesis of paleognath monophyly and ratite interrelationships using an osteological data set.

Although most subsequent authors have supported or accepted the strict monophyly of the paleognaths (Ho *et al.*, 1976; Prager *et al.*, 1976; Rich, 1979; de Boer, 1980; Sibley and Ahlquist, 1981, 1990; Stapel *et al.*, 1984, Cracraft, 1981, 1986, 1988; Bledsoe, 1988; Cracraft and Mindell, 1989; Cooper *et al.*, 1992), a minority of opinion has questioned that hypothesis (Feduccia, 1980, 1985; Houde and Olson, 1981; Olson, 1985; Houde, 1986, 1988).

Within the framework of paleognath monophyly, the debate has shifted to controversies over the interrelationships among the ratites. At the heart of this debate are apparent conflicts over, first, what morphological characters appear to be telling us about those interrelationships, and second, the seemingly disparate relationships implied by several different molecular data sets, on the one hand, and morphology, on the other.

Figure 7.1 presents six of the most recent hypotheses of paleognath interrelationships. A cladistic analysis of osteological data (Cracraft, 1974; Fig. 7.1A) confirmed earlier suppositions that the tinamous are the sister-group to the ratites and that emus and cassowaries are closely allied. The morphological data also linked rheas and ostriches on the basis of a large suite of shared derived characters and were consistent with the hypothesis that the kiwis of New Zealand were at the base of the ratite tree and apparently related to New Zealand moas (the latter group not shown on Fig. 7.1A). This morphological hypothesis was challenged by Sibley and Ahlquist (1981) on the basis of DNA hybridization data and by the use of a "personal communication" from H. H. Bledsoe as well as from J. J. Baker and C. McGowan that a reinterpretation of the morphological data is not congruent with the Cracraft (1974) hypothesis but is consistent with the DNA hybridization

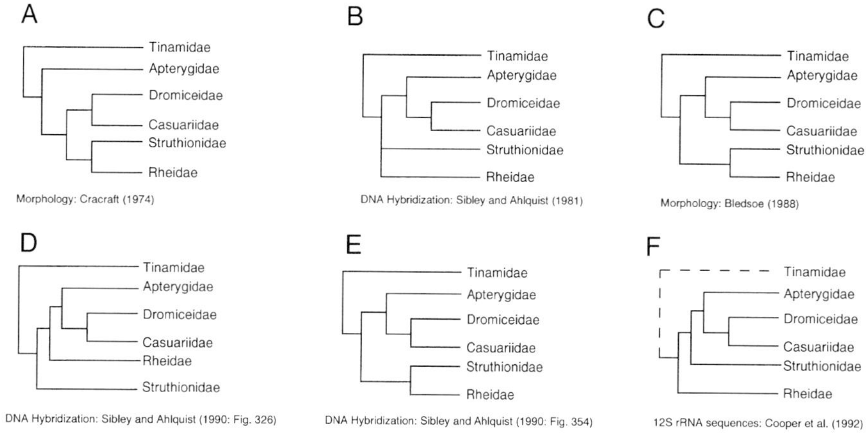

FIGURE 7.1 Six hypotheses for the relationships among ratite birds. See text.

tree (Fig. 7.1B). The reinterpretation of Bledsoe was published 7 years later (Bledsoe, 1988; Fig. 7.1C), and Bledsoe's tree is identical to that of Cracraft (1974) with the exception of the placement of the kiwis close to emus and cassowaries rather than at the base of the ratites. Baker and McGowen's analysis has not yet been published; nevertheless, Sibley and Ahlquist (1990, p. 283) cited once again a personal communication to the effect that those authors were unable to duplicate Cracraft's results.

The DNA hybridization data themselves have remained consistently ambiguous. The hypothesis of Sibley and Ahlquist (1981; Fig. 7.1B) was not duplicated by a subsequent study (Sibley and Ahlquist, 1990), and in the latter at least two different hypotheses were inferred from the DNA hybridization data depending on different assumptions of the clustering algorithms (Fig. 7.1D and E). One of these hypotheses placed rheas and ostriches together, just as inferred from morphology. The primary difference, however, between these results and those of morphology was the placement of the kiwis as the sister-group to the emus and cassowaries on the molecular tree. A similar result was also found by Cooper *et al.* (1992; Fig. 7.1F), using sequences from a small fragment of mitochondrial 12S rRNA.

The conflicts among these hypotheses exemplify the tone of the debates that often arise when morphological and molecular data give different results. Within ornithology, as elsewhere within systematics, it has often been the case that the morphological data are dismissed out of hand with the claim of "convergence" (e.g., Sibley and Ahlquist, 1990, p. 283), even when the molecular data themselves have either been poorly or incorrectly analyzed or, when correctly analyzed, are ambiguous. Nor does it help if the molecular data are dismissed by those convinced that morphology is a more persuasive indicator of relationships. In either case, questions still remain: Are rheas and ostriches related, or is one or the other linked to the other ratites by a short internode? Are kiwis close to emus and cassowaries or are they basal to other ratites? If the former, then how does one explain the pervasive reversal of many morphological characters to a more primitive condition? If kiwis are basal, what factors within the molecular data are leading them to cluster kiwis with emus and cassowaries? Answers to these questions have ramifications for data analysis well beyond the paleognaths.

In an effort to investigate paleognath interrelationships further we have augmented both morphological and molecular data sets. The morphological characters described by Cracraft (1974) and Bledsoe (1988) were reevaluated to verify codings and diagnoses and to attempt a resolution of the discrepancies between their analyses. Several new postcranial characters are described and, in addition, new characters based on an examination of cranial morphology are added to the osteological database. To examine the results provided by molecules, a new data set has been assembled on the basis of nucleotide sequences from the complete 16S rRNA and cytochrome *b* genes, complete $tRNA^{Lys}$, as well as large portions of the 12S rRNA, cytochrome oxidase I (COI), and cytochrome oxidase II (COII) genes of the mitochondrial genome. In addition, a 361-base pair (bp) fragment of the 12S rRNA gene published by Cooper *et al.* (1992) is also included in the analysis. In all, 5444 bp of sequence are used in this study, more than in any previous comparative analysis of avian sequences.

II. MATERIALS AND METHODS

A. Taxa

1. Osteological Analysis

Skeletons were examined from the following institutions: the Field Museum of Natural History (FMNH; Department of Geology and the Division of Ornithology within the Department of Zoology, Chicago, IL); the American Museum of Natural History (AMNH; Department of Ornithology and the Department of Vertebrate Paleontology, New York, NY). These include Tinamidae: *Tinamus tao kleei* (FMNH 315145), *Tinamus major* (FMNH 104192, AMNH 3675), *Crypturellus cinnamomeus* (FMNH 104259), *Crypturellus undulatus* (FMNH 290488, AMNH 6481, 6479), *Crypturellus noctivagus* (AMNH 10444, 10443), *Nothoprocta cinerescens* (AMNH 6505), *Eudromia elegans* (AMNH 8678), *Rhynchotus rufescens* (AMNH 6605); Dinornithidae: *Megalapteryx didinus* (FMNH PA177), *Emeus crassus* (FMNH PA34), *Dinornis maximus* (FMNH PA35 AMNH VP7303), *Dinornis robustus* (AMNH VP80, VP81), *Pachyornis elephantopus* (AMNH VP7307, VP7313), *Emeus* sp. (AMNH VP69), *Euryapteryx* sp. (AMNH VP7309); Struthionidae: *Struthio camelus* (FMNH 106776, 313619, 104586, AMNH 1503, 1294, 4474, 3199); Rheidae: *Rhea americana* (FMNH 105749, 104061, 105636, AMNH 2875, 3783, 6470),

Rhea sp. (AMNH 2300); Casuariidae: *Casuarius casuarius* (FMNH 314889, 93274, AMNH 1106), *Casuarius unappendiculatus* (FMNH 104271), *Casuarius* sp. (AMNH 1517, 1554); Dromaiidae: *Dromaius novaehollandiae* (FMNH 104536, 313620, AMNH 1513, 3742, 18458); Apterygidae: *Apteryx australis* (FMNH 85775–a juvenile, 314890, AMNH 3738, 4437, 5372, 18456), *Apteryx* sp. (AMNH 3739); Galliformes/Anseriformes: Megapodiidae: *Megapodius freycinet* (FMNH 104631, AMNH 1389), *Leipoa ocellata* (FMNH 105235, 1359); Craciidae: *Crax mitu* (AMNH 3815), *Nothocrax urumutum* (AMNH 6043); Anhimidae: *Anhima cornuta* (FMNH 105629, 105812, 104293).

2. *Molecular Analysis*

Tissue samples (muscle, liver) for DNA extraction were obtained from the frozen tissue collection at the Museum of Natural Science at Louisiana State University (LSUMZ; Baton Rouge, LA), Bernice P. Bishop Museum (BPBM; Honolulu, HI), and the American Museum of Natural History. These include *Tinamus major* (LSUMZ B15087), *Nothoprocta perdicaria* (AMNH 10558), *Struthio camelus* (LSUMZ B8610), *Rhea americana* (LSU B8608), *Dromaius novaehollandiae* (LSUMZ B8607), *Casuarius bennetti* (BPBM 109892), and *Apteryx australis mantelli* (LSUMZ B8606). Because Cooper *et al.* (1992) did not include *Tinamus* or *Nothoprocta* in their 12S rRNA study, it was necessary to use their tinamou species, *Eudromia elegans*, to complete the tandem alignment of the sequences.

B. DNA Extraction and Sequencing

DNAs were extracted by boiling minute quantities of muscle or liver in 250 μl of 5% (w/v) Chelex (Bio-Rad, Hercules, CA) suspension for 15 min, with occasional vortexing. The Chelex resin was pelleted by spinning it for 30 sec in a benchtop centrifuge, and the supernatant used as a template for polymerase chain reaction (PCR) amplifications. In the majority of cases, 10-μl reactions containing 1 μl of Chelex-extracted DNA, 1 μl of 2 mM dNTPs, 2 μl of Turbo buffer [250 m*M* Tris (pH 8.5), 10 m*M* $MgCl_2$, 100 m*M* KCl, bovine serum albumin (BSA, 2.5 mg/ml)], 0.75 U of *Taq* polymerase (Promega, Madison, WI), and 1 μl of each 10 μM primer, were sealed in glass capillary tubes for cycling in an Idaho Technologies air thermocycler. Reaction conditions were as follows: 94°C, 5 sec; 45–50°C, 2 sec; 72°C, 15 sec; for 35–40 cycles. Aliquots (5 μl) were separated on 2% low-melting-point agarose gels and visualized with ethidium bromide and ultraviolet (UV) light. Plugs were taken from appropriate-sized bands with sterile Pasteur pipettes and melted at 72°C for 15 min in 250 μl of H_2O. The gel-purified DNA was used as a template to prepare clean PCR products for direct sequencing. Reactions (40 μl) were assembled, containing 1.5 μl of DNA, 4 μl of 2 m*M* dNTPs, 8 μl of Turbo buffer, 0.75 U of *Taq* polymerase, and 2 μl of each primer. PCR conditions were as follows: 94°C, 10 sec; 48–52°C, 5 sec; 72°C, 25 sec; for 40 cycles. Aliquots (5 μl) were examined on agarose gels, and the remainder further purified with the BIO 101, Inc. (San Diego, CA). GeneClean II system, ultimately resuspending the total DNA from each PCR reaction in 6 μl of H_2O.

DNA was sequenced using dye-terminator chemistry in an Applied Biosystems, Inc. ABI; Foster City, CA) 377 automated sequencer. Reactions contained 0.6 μl of DNA from the procedure described above, 1.4 μl of H_2O, 1.1 μl of 10 μ*M* primer, and 2.9 μl of ABI Dye Terminator Cycle Sequencing Ready Reaction with AmpliTaq DNA polymerase, FS. Each DNA sample was sequenced in both directions. An initial denaturation step of 1 min at 95°C was followed by 32 cycles of 95°C, 10 sec; 50°C, 5 sec; 60°C, 3 min; and a 4°C soak cycle. Completed reactions were cleaned by passage through Sephadex G-50 spin columns for 3 min at 3000 rpm in a benchtop centrifuge. Samples were dried in a Speed-Vac (Savant, Hicksville, NY) for 30 min, and resuspended in loading buffer consisting of Blue

TABLE I. Primers Used to Amplify and Sequence Mitochondrial Genes Used in This Study[a,b]

Primer		Sequence	
16S rRNA			
L2311	5′	CAAAGCATTCAGCTTACACC	3′
L2313	5′	CAAAGCATTCAGCTTACACCTG	3′*
H2688	5′	CTCGGTAGGCTTTTCACCTCTAC	3′*
L2703	5′	AGCAGAGGTGAAAAGCC	3′
H2901	5′	TCTTTTGTTGGTGGCTGCTT	3′
L2909	5′	TGTAGACCTTCAAGCAGCCA	3′
H3287	5′	TTGATTGCGCTACCTTTGCACGGTTAGG	3′
H3620	5′	GGTCCATTGCTCAATTATATTGGG	3′
L3450	5′	GAAGACCCTGTGGAACTTTAA	3′
H4015	5′	GGAGAGGATTTGAACCTCTG	3′
L3183	5′	AAGGAACTCGGCA	3′
H3171	5′	TGCCGAGTTCCTT	3′
H3426	5′	AGGGTCTTCTCGTC	3′
Cytochrome *b*			
L14827	5′	CCACACTCCACACAGGCCTAATTAA	3′
H15298	5′	AAACTGCAGCCCCTCAGAATGATATTTGTCCTCA	3′**
L15068	5′	ACTAGCAATACACTACACAGCAGA	3′
H15505	5′	CTGCATGAATTCCTATTGGGTTGTTTGATCC	3′
L15311	5′	CTACCATGAGGACAAATATC	3′**
H15712	5′	GCGTATGCGAATAGGAAATA	3′
L15656	5′	AACCTACTAGGAGACCCAGA	3′
H16065	5′	GGAGTCTTCAGTCTCTGGTTTACAAGAC	3′**

(Continues)

Dextran–250 m*M* EDTA:formamide (1:6, v/v; ABI). Samples were run on 5% Long Ranger (FMC, Philadelphia, PA) gels in Tris–borate–EDTA (TBE) buffer at 3 V, 50°C for 3–4 hr. Sequences were edited and assembled with Sequencher 3.0 (Gene Codes, Ann Arbor, Michigan) software.

In the early phases of the study portions of the 16S rRNA gene sequences were obtained using nonautomated methods. The protocol of Allard *et al.* (1991) was followed for amplification of single-stranded DNA, which was then directly sequenced in both directions by the Sanger *et al.* (1977) dideoxy chain-termination method. Portions of the gene were sequenced using the Promega f-mol system, which utilizes a modified *Taq* polymerase to sequence double-stranded DNA in a thermal cycler under a high annealing temperature.

Primers used for PCR amplifications and cycle sequencing are listed in Table I. All sequences obtained in this study have been deposited in GenBank under the

TABLE I. (*Continued*)

Primer		Sequence	
COI:			
L6611	5′	TCGAACCTCTGTAAAAAGGACTAC	3′ *
L6615	5′	CCTCTGTAAAAAGGACTACAGCC	3′ ***
L6955	5′	CGCATGAACAACATAAGCTTCTG	3′ *
H7002	5′	CATCCTGTGCCGGCTCCAGCTTC	3′ *
H7032	5′	TTGCCAGCTAGTGGGGGGTA	3′ ***
L7318	5′	ACATTCTTCGACCCAGCCGGAGG	3′ *
H7350	5′	ACTTCGGGGTGTTCCGAAGAATCA	3′ *
H7662	5′	AGGAAGATGAATCCTAGGGCTCA	3′ *
COII:			
L8419	5′	TTCCACGACCACGCCCTAATAGT	3′ *
H8844	5′	TGGTTTAGTCGTCCAGGGATTGCGTC	3′ *
L8740	5′	GGCCACTTCCGACTACTAGAAGT	3′ *
H9085	5′	CAGGGGTTTGGGTTGAGTTGTGGCAT	3′ *
L8309	5′	CTGTCAAGACTAAATCACAGG	3′
H8907	5′	CCGCAGATTTCTGAGCATTGACC	3′
12S rRNA[c]			
L1264	5′	CAAACAAAGCATGGCACTGAAG	3′ *
H1861	5′	TCGATTATAGAACAGGCTCCTC	3′ *

[a]Numbers refer to base position on the *Gallus gallus* mitochondrial genome (Desjardin and Morais, 1990)

[b]All primers developed in our laboratory except for the following: *, primer courtesy of G. F. Barrowclough and J. Groth; **, primer courtesy of workers in A. C. Wilson laboratory (Berkeley, CA); ***, primer courtesy of D. Mindell.

[c]See also Cooper *et al.*, 1992, for 12S rRNA primers used in their study.

following accession numbers (16S rRNA, 12S rRNA, cytochrome *b*, COI, COII, and tRNA[Lys]): U76036–U76077.

C. Phylogenetic Analysis

Sequences of the 12S rRNA and all protein-coding genes were aligned by eye. The 16S rRNA gene sequences had hypervariable regions that could not be easily aligned by that method, therefore they were aligned using the MALIGN algorithm (Wheeler and Gladstein, 1992). Following this procedure, however, regions of both the 16S rRNA and 12S rRNA genes were considered to have such ambiguous alignments that homology statements about base positions were virtually arbitrary; these regions were eliminated from the phylogenetic analyses. These eliminated regions correspond to the aligned sequence between the following pairs of base positions (determined against the published *Gallus* sequence; Desjardin and Morais, 1990) for 16S rRNA and 12S rRNA, respectively: positions 2364–2385, 2396–2408, 2472–2477, 2514–2517, 2641–2644, 2758–2764, 2772–2774, 2779–2809, 2869–2890, 2934–2943, 2969–2971, 3260–3267, 3466–3478, 3502–3512, 3926–3951, 3962–3966; and positions 1383–1390, 1421–1426, 1520–1524, 1852–1860. In total, 225 base positions (characters) were eliminated, which also included gaps in the preceding sequence positions that were required by the alignment with the paleognath taxa.

Morphological character-states were polarized into primitive or derived states by the method of outgroup comparison. Character-states occurring in both ingroup and outgroup taxa were hypothesized to be primitive, whereas states restricted to ingroup taxa were considered derived (Hennig, 1966; Eldredge and Cracraft, 1980; Nelson and Platnick, 1981).

Outgroups were chosen on the basis of recent phylogenetic hypotheses for the basal clades of birds. Morphological (Cracraft, 1986, 1988) and molecular evidence (Prager *et al.*, 1976; Stapel *et al.*, 1984; Cracraft and Mindell, 1989) support the monophyly of the Neornithes. Within this clade, two monophyletic groups are supported by a diverse array of evidence: (1) the Paleognathae (Bock, 1963; Ho *et al.*, 1976; Prager *et al.*, 1976; Stapel *et al.*, 1984; Cracraft, 1986, 1988; Bock and Buhler, 1988; Cracraft and Mindell, 1989; Sibley and Ahlquist, 1990); and (2) the Neognathae (Ho *et al.*, 1976; Stapel *et al.*, 1984; Cracraft, 1986, 1988; Cracraft and Mindell, 1989; Sibley and Ahlquist, 1990). Within the Neognaths, the galliforms and anseriforms are postulated to be the basal clade (Cracraft, 1988; Cracraft and Mindell, 1989; Sibley and Ahlquist, 1990). The primary clades of the Anseriformes are the Anhimidae, the Anatidae, and the Anseranatidae. The Anhimidae are the sister-group of the other two lineages and are osteologically more primitive (Livezey, 1986; Sibley and Ahlquist, 1990). The main groups of the galliforms include the Cracidae, the Megapodiidae, the Odontophoridae, the Numididae, and the Phasianidae. The cracids and megapodes have been postulated as being basal lineages within the order (Cracraft, 1981). From these taxa, representatives from the Anhimidae, Megapodiidae, and Cracidae were chosen as outgroups for the morphological analysis. *Gallus gallus* was chosen to root the paleognath molecular tree (but see discussion below).

The two primary lineages of the paleognaths are the ratites and the tinamids. Various studies (Cracraft, 1974, 1981, 1986, 1988; Prager *et al.*, 1976; Stapel *et al.*, 1984; Bledsoe, 1988; Sibley and Ahlquist, 1990) have proposed that the tinamids are the sister-groups of the ratites. Using this finding as a working hypothesis, character-states that are present in basal neognaths and tinamous but not in ratites are here postulated to be primitive, whereas those restricted to one or more ratite taxa are derived.

Phylogenetic analyses of all data were performed using the exhaustive search algorithm of test version 4.0d46 of PAUPSTAR (Phylogenetic Analysis Using Parsimony; D. Swofford, 1996, personal communication). Different approaches to molecular data analysis were undertaken (Cracraft and Helm-Bychowski, 1991), as described below.

III. RESULTS

A. Morphological Analysis

A parsimony analysis of 58 skeletal characters (ordered as stated in Appendix I) resulted in a single most parsimonious tree of 80 steps, with a consistency index of 0.846 excluding uninformative characters (Fig. 7.2; see Appendix I for a detailed discussion of the characters and character-states supporting this hypothesis). The galliform and anseriform taxa were combined into one outgroup taxon since all of their character codings were identical in this analysis. Although the monophyly of the paleognaths is assumed here on the basis of other evidence not included in this study (Lowe, 1928; Bock, 1963; Ho *et al.*, 1976; Prager *et al.*, 1976; de Boer, 1980; Stapel *et al.*, 1984; Cracraft, 1974, 1981, 1986, 1988; Bledsoe, 1988; Cracraft and Mindell, 1989; Sibley and Ahlquist, 1990), two synapomorphies were identified in this data set that corroborate their monophyly: (17) shallow transverse ligamental sulcus of the humerus, and (34) internal cnemial crest of the tibiotarsus extended proximally beyond the articular surface.

Twenty characters support the monophyly of the ratites (Fig. 7.2). These characters include the following: (1) the loss of a keel, (6) the sternum essentially equal in width and length, (7) the loss of the sternal manubrium, (8) the fusion of the scapula and coracoid, (11) the humerus at least one-third longer than the ulna, (12) projection of the internal tuberosity of the humerus medially and proximally, (13) the deltoid crest of the humerus reduced to a small ridge, (14) proximal protrusion of both margins of the carpal trochleae greatly reduced so that both are flattened, (16) one metacarpal with phalangeal articulation, (18) reduced external epi-

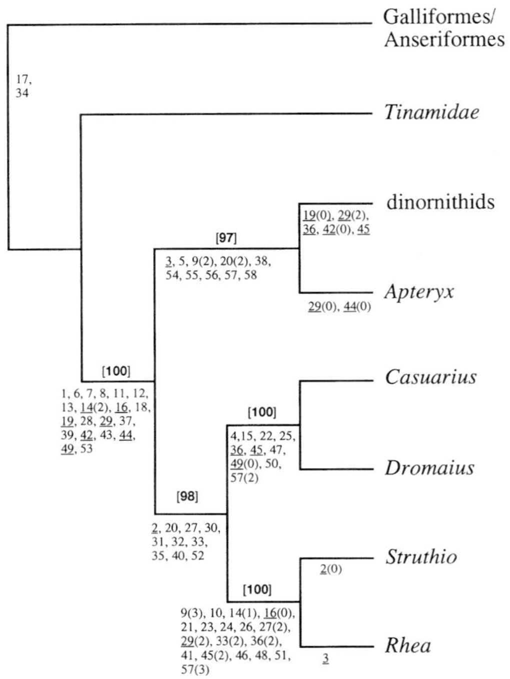

FIGURE 7.2 Most parsimonious solution (80 steps; CI= 0.846) using a revised morphological data set of 58 characters. All state changes in the characters represent transformations from 0 to 1 (Appendix I) unless noted in parentheses. Underlined characters represent parallelisms. Character transformations were ordered as discussed in Appendix I. Bootstrap percentages from 200 resamplings are shown in boldface within brackets.

condyle of the humerus, (19) narrowing of dorsal surface of synsacrum caudal to antitrochanter, (28) the intercotylar prominence equal in proximal extension with the hypotarsus on the tarsometatarsus, (29) external cotyla of tarsometatarsus slightly concave, and internal surface deeply concave with concurrent loss of sharp proximal protrusion on the medial margin, (37) a deep groove for the peroneus profundus muscle on the tibiotarsus, (39) a deep pit anteriorly and a groove posteriorly on the internal condyle of the tibiotarsus, (42) trochanteric crest of femur essentially on same level with the iliac facet, (43) the iliac facet of the femur with a rounded edge, loss of a lip, and convex to flattened surface, (44) external and fibular condyles of femur greatly enlarged and projecting distally beyond level of internal condyle, (49) posterior facet of internal condyle of femur triangular in shape, and (53) projection of zygomatic process of squamosal anterolaterally over at least two-thirds of the body of the quadrate.

The monophyly of the dinornithids and *Apteryx* (Fig. 7.2) is supported by 10 characters. The uniquely derived characters include the following: (5) a flattened sternum, (9) the coracoidal process of the scapulocorocoid in the form of a mediolaterally compressed ridge, (20) the anterior portion of the ilia more elongated than the posterior portion, (38) strong anterior projection of the internal condyle of the tibiotarsus, (54) the pterygoid divided into dorsal and ventral surfaces, forming a pterygoid fossa, (55) the intercondylar fossa of the quadrate with a deep, rounded pit, (56) mammillar tuberosites highly developed on the posterolateral margins of the basitemporal plate, (57) the antrum of the maxillopalatine present and ankylosed to the dorsal surface of the posterior maxillary, and (58) the third vestibule of the nasal region formed into an olfactory chamber. In addition, dinornithids and *Apteryx* share one derived character that is postulated to be homoplasious on the topology of Fig. 7.2: (3) the coracoidal sulci of the sternum laterally displaced (the sulci are also displaced in *Rhea*, but it is not as extreme, and its homology to the condition in dinornithids and *Apteryx* is questionable).

Ten characters support the monophyly of *Struthio*, *Rhea*, *Dromaius*, and *Casuarius* (Fig. 7.2), including nine that are postulated to be uniquely derived: (20) the posterior portion of the ilium longer than the anterior portion, (27) the internal ridge of the hypotarsus greatly reduced and a proximally protruding process on the external ridge, (30) the possession by the tarsometatarsus of a deep and narrow anterior metatarsal groove, (31) a sharp ridge present on the external side of the posterior shaft of the tarsometatarsus, (32) loss of digit I, (33) the base of the cnemial crests of the tibiotarsus compressed mediolaterally and the interarticular surface narrowed anteriorly, (35) supratendinal bridge of the tibiotarsus lost, (40) the anterior intercondylar fossa of the tibiotarsus widened, undercutting the condyles and forming a sharp ridge, and (52) the presence of an elongate supraorbital process on the lacrimal. One other synapomorphy, not uniquely derived, also supports the monophyly of the Struthionoidea: (2) loss of posterior lateral processes on the sternum.

The monophyly of *Dromaius* and *Casuarius* (Fig. 7.2) is supported by seven uniquely derived characters and three that are homoplastic. The former include the following: (4) posterior extension of the ventral lip of the coracoidal sulci on the sternum, (15) the presence of a phalangeal articulation for the os metacarpale alulare, (22) a club-shaped expansion of the distal ilium, (25) transverse processes of the sacral vertebrae broadened and fused to form a ventral plate of bone, (47) the external condyle of the femur elliptical in shape and projecting proximally, (50) a shallow pit on the external condyle of the femur cutting into the fibular condyle, (57) an antrum present as a large "pocket" formed from the maxillopalatines and also ankylosing anteriorly with the posterior margin of the maxillary. The three homoplasious synapomorphies include the following: (36) the external condyle of the tibiotarsus flattened along the distal margin with its anterior portion slightly undercut (parallel in dinornithids), (45) the fibular condyle of the femur level proximally with the external condyle and rounded posterolaterally (also found in dinornithids), and (49) posterior facet of internal condyle of femur ovoid, a reversal.

Finally, 18 characters support the monophyly of *Struthio* and *Rhea* (Fig. 7.2). Of these, 16 are uniquely derived: (9) coracoidal process of the scapulocoracoid pronounced and knoblike and projecting toward the glenoid facet, (10) glenoid facet oriented dorsolaterally, (14) internal and external margins of the carpal trochlea essentially level with each other and well rounded, (21) postacetabular ilium narrows dorsoventrally and mediolaterally and tapers to a conical shape, (23) pubis elongated beyond the ilium and fused to the ischium, (24) obturator process of the ischium fused to the pubis to form the obturator foramen, (26) presence of a puboischial bar, (27) internal ridge of the hypotarsus lost and external ridge knoblike proximally, (33) surface of anterior interarticular area greatly reduced as the base of the crests becomes sharply compressed, (36) external condyle of the tibiotarsus ovoid distally and sharply undercut anteriorly, (41) external condyle of the tibiotarsus possessing a moderate lateral extension, sharply undercut posteriorly, (45) posterior margin of the fibular condyle of the femur rounded and does not extend as far proximally as the internal condyle, (46) internal condyle of the femur flattened distally, (48) rotular groove of the femur narrow and deep, (51) popliteal fossa of the femur very deep, and (57) maxillopalatine antrum greatly reduced (see Appendix I). There are two homoplastic characters: (16) three metacarpals with phalangeal articulation, considered a reversal in Fig. 7.2, and (29) external and internal cotylar surfaces of the tarsometatarsus essentially level (also coded the same in dinornithids).

B. Sequence Analysis

The sequences reported here total 5444 bp from the mitochondrial genome, including 1682 bp of aligned sequence representing the entire 16S rRNA gene, 583 bp of the 12S rRNA gene determined in our laboratory, 361 bp of the 12S rRNA gene previously reported by Cooper *et al.* (1992), 1011 bp of the COI gene, 592 bp of the COII gene, 1143 bp of the entire cytochrome *b* gene, and 72 bp of $tRNA^{Lys}$. Because transition to transversion ratios between *Gallus* and the paleognaths, and even within most paleognath comparisons, are approaching unity (Table II), it appears that transition substitutions are nearing saturation. Therefore, the parsimony analysis reported here was performed using transversions only (analyses on all substitutions produced comparable results). In addition, because small "hypervariable" regions of the 16S rRNA and 12S rRNA genes were difficult to align, rendering homology statements among the taxa problematic at best, 225 bp of these genes were excluded; a total of 5219 bp was used in the analyses reported in this section. Finally, gaps in the sequences were treated as characters.

A transversion parsimony analysis of all 5219 bp yields a single most parsimonious tree of 1357 steps (Fig. 7.3A) using *Gallus* as a root for the tree. It is apparent from Tables II and III that *Gallus* is quite distant (8–10% transversion difference) from the paleognaths, and indeed extensive comparisons (see Section IV) suggest that *Gallus* is attaching to long branches within the paleognaths. Although both tinamous have very long branches, their clear synapomorphous sequence similarity unites them relative to the other taxa. The most parsimonious solution shows *Ap-*

TABLE II Pairwise Distances between Taxa Based on 5219 bp of Mitochondrial Sequences[a,b]

Taxon	*Gallus*	*Apteryx*	*Rhea*	*Struthio*	*Casuarius*	*Dromaius*	*Nothoprocta*	*Tinamus*
Gallus	—	0.170	0.157	0.161	0.163	0.158	0.188	0.184
Apteryx	471/414	—	0.129	0.132	0.118	0.114	0.165	0.168
Rhea	422/397	418/254	—	0.128	0.124	0.128	0.167	0.164
Struthio	421/417	444/246	398/272	—	0.118	0.122	0.164	0.167
Casuarius	451/398	437/180	403/245	387/231	—	0.085	0.159	0.162
Dromaius	431/395	419/178	417/249	416/219	346/95	—	0.153	0.159
Nothoprocta	481/500	474/389	470/403	435/419	443/388	410/386	—	0.159
Tinamus	456/502	479/400	446/410	450/423	450/397	428/400	467/365	—

[a] See materials and methods.
[b] Absolute transition/transversion distances below diagonal, mean uncorrected distances above diagonal.

teryx as the sister-group of *Casuarius* and *Dromaius*, *Struthio* as their sister-group, and then *Rhea* as theirs, with all groupings having strong bootstrap support (Fig. 7.3A).

To examine the influence that the long branch of the outgroup may be having on the topology, *Gallus* was eliminated and the two tinamous were used to root the tree. The resulting most parsimonious solution was a tree of 1068 steps (Fig. 7.3B), identical in topology to the tree of Fig. 7.3A, but now bootstrap support has weakened. Indeed, the second most parsimonious tree, only 5 steps longer at 1073 steps, has a topology identical to the morphological tree of Fig. 7.2.

C. Total Evidence Solution

A combined analysis was undertaken using all the sequence data, as above, and the 58 morphological characters previously discussed. With *Gallus* as the root of the tree, a single most parsimonious tree of 1450 steps was found (Fig. 7.4A) that was identical to the morphological tree (Fig. 7.2). In this solution morphological characters clearly influence the phylogenetic signal by uniting *Rhea* and *Struthio*, on the one hand, and *Casuarius* and *Dromaius*, on the other, and linking these clusters as sister-taxa, although bootstrap support for the latter relationship is not strong. The next most parsimonious tree was two steps away at 1452 steps and was identical in topology to the molecular tree of Fig. 7.3A.

When *Gallus* was excluded, however, and the tinamous used as the root, a single most parsimonious tree was found at 1144 steps that was identical to the tree including *Gallus*, but now bootstrap support increased substantially (Fig. 7.4B). The next most parsimonious tree was nine steps away at 1153 steps and had the kiwi as the sister-group of the emu and cassowary.

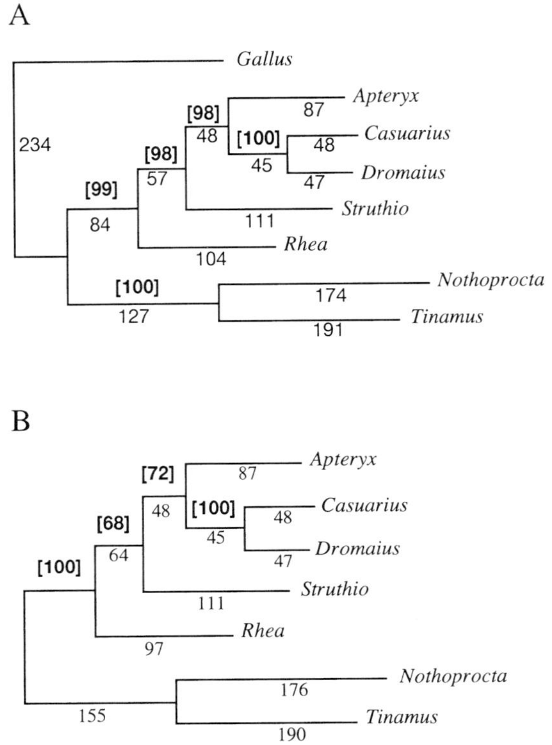

FIGURE 7.3 Transversion parsimony analysis of 5219 bp from the mitochondrial 16S rRNA, 12S rRNA, COI, COII, cytochrome *b*, and tRNA^{Lys} genes. (A) A single most parsimonious tree of 1357 steps resulted when using *Gallus gallus* as the root. (B) *Gallus gallus* has been excluded and two tinamous are used to root the tree. A single most parsimonious tree of 1068 steps was found. Bootstrap percentages from 200 resamplings are shown in boldface within brackets above branches, and branch lengths are shown below.

IV. DISCUSSION

A. Ratite Interrelationships: Morphological Data

The relationships proposed here are identical to those proposed by Cracraft (1974; see Fig. 7.1A) and contradict the hypothesis proposed by Bledsoe (1988; see our Fig. 7.1C). Bledsoe (1988) hypothesized that the kiwis are more closely related to the emus and cassowaries than to all remaining ratites. Our reevaluation of Bledsoe's characters, however, lead us to a different interpretation of the morphological data.

Six of Bledsoe's characters involve wing elements: humerus (Bledsoe's character

19), ulna-radius (26), and carpometacarpus (30, 31, 32, 34). Wing elements are greatly reduced in the kiwis, more so than in the other living ratites, the reduction presumably due in part to the much smaller size of the kiwis although it may be a secondary reduction shared with dinornithids. Because of this reduction in the size of the wing elements, most of the distinguishing processes and articulations are difficult to determine or differentiate (as was confirmed by an examination of considerable skeletal material), thus we conclude the forearm synapomorphies described by Bledsoe (1988) for *Apteryx*, *Casuarius*, and *Dromaius* are ambiguous. The other characters involve the femur (44, 48), the tibiotarsus (56, 64), and the tarsometatarsus (71). On reanalysis, each of these characters except one (44) was found to be uninformative either because it also occurred in the outgroup, it was too variable to distinguish a well-defined character-state, or because the character could not be determined (for more detail, see Appendix II). On the basis of the data, our reexamination of Bledsoe's characters (1988) shows weak support for Bledsoe's proposed sister-group relationship of *Apteryx* to *Dromaius* and *Casuarius* and instead supports the hypothesis of Fig. 7.2.

B. Why Do Molecular and Morphological Data Conflict?

The results of the morphological data (Fig. 7.2) confirm the topology found previously (e.g., Cracraft, 1974). This hypothesis, in which *Apteryx* is basal to the other ratites, is well supported as measured by high bootstrap values. The mitochondrial sequence data, on the other hand, come to a different conclusion—that *Apteryx* is linked with *Casuarius* and *Dromaius*; this hypothesis is also strongly supported by bootstrap procedures (Fig. 7.3A). Significantly, previous molecular studies cited earlier (Fig. 7.1) all fail to place kiwis at the base of the ratites. What might be the reasons for the conflict between morphological and molecular data sets?

One answer, of course, is that both trees are wrong, and that more data will be needed to discover the true tree. A second answer is that the molecular data are yielding the true tree. If this is the case, however, then either the morphological data have been seriously misinterpreted with respect to homology statements (which does not seem likely given different investigators having obtained roughly similar results) or there has been rampant homoplasy in morphology. That the morphological data are severely incongruent with the molecular tree is easily shown by optimizing the morphological data onto it (after excluding the dinornithids). This exercise results in a tree 25 steps longer relative to the morphological tree (without dinornithids) and also results in at least one internodal branch length of zero. A third alternative is that the morphological tree is correct, in which case one would infer that the molecular data possess some systematic error (see below). On face value, this third hypothesis might be considered more acceptable than the second inasmuch as the combined analysis (Fig. 7.4) indicates that a relatively small amount of

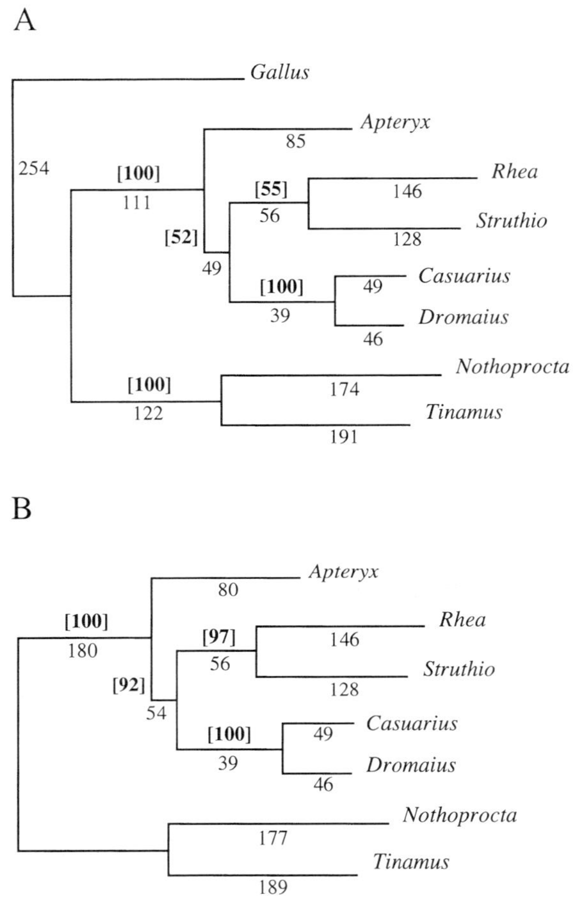

FIGURE 7.4 A total evidence solution based on 5219 bp from the mitochondrial 16S rRNA, 12S rRNA, COI, COII, cytochrome *b*, and $tRNA^{Lys}$ genes as well as 58 morphological characters. (A) A single most parsimonious tree of 1450 steps was found with *Gallus gallus* as the root of the tree. (B) With *Gallus gallus* excluded, and with two tinamous as the root, a single most parsimonious tree of 1144 steps was found. Both trees are identical to the morphological tree of Fig. 7.2. Bootstrap percentages from 200 resamplings are shown in boldface within brackets above branches, and branch lengths are shown below. All character transformations treated as in independent analyses shown in Figs. 7.2 and 7.3.

morphological data (58 characters) can outweigh a large amount of sequence data (5219 bp). The conclusion from the above analysis is that the molecular data fit the morphological hypothesis better than the morphological data fit the molecular hypothesis.

This last interpretation, if true, implies that the molecular topologies (Fig. 7.3A and B) are misleading with respect to the true relationships. The most obvious sus-

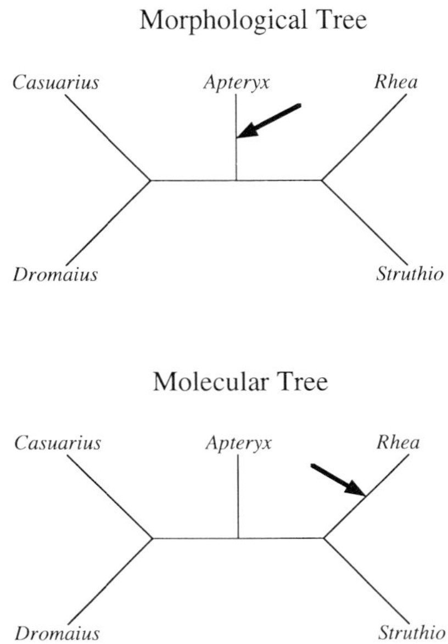

FIGURE 7.5 The conflict between morphological and molecular data sets seems to be a rooting problem because the ingroup topologies of the unrooted trees are identical. If the root joins to *Apteryx*, the resulting tree is like the morphological results of Fig. 7.2, whereas if the root joins to *Rhea*, the result is like the molecular tree of Fig. 7.3. It is suggested in text that long-branched outgroups such as *Gallus* and the tinamous are preferentially joining to a long-branched taxon of the ingroup, *Rhea*, owing to homoplasy generated by long-branch attractions. Morphological data are probably less susceptible to long-branch attractions, as discussed in text.

picion is that incorrect topologies are arising because of variation in evolutionary rate among the different lineages, that is, they exemplify a "long branches attract" problem or artifact (e.g., see Felsenstein, 1978; Penny *et al.*, 1987; Smith, 1994; Halanych, 1996; among many others). The proposition that the phylogenetic signal contained in both the morphological and molecular data sets is not actually different but that their apparent conflict is simply a rooting problem is shown in Fig. 7.5: if the root joins the lineage leading to *Apteryx*, one has the morphological tree; if it joins to *Rhea*, one obtains the molecular tree.

Examination of the molecular data and results is consistent with the above explanation. Tinamous, separately or together, have long branches relative to the ratites (see Sibley and Ahlquist, 1990, p. 810, Fig. 325; also Mindell *et al.*, 1996), and the long-branched outgroup, *Gallus*, not unexpectedly attaches to tinamous in the most parsimonious solution (Fig. 7.3A). Likewise, with *Gallus* excluded from the analysis, the tinamous join to another long branch, *Rhea* (Fig. 7.3B). This suggests the possibility that *Apteryx* is clustering with *Casuarius* and *Dromaius* merely as a consequence of *Rhea* and *Struthio* being "pulled down" to the base of the ratite tree by

TABLE III Pairwise Transversion Maximum Likelihood Distances for 5219 bp of Mitochondrial Sequences[a]

Taxon	*Gallus*	*Apteryx*	*Rhea*	*Struthio*	*Casuarius*	*Dromaius*	*Nothoprocta*
Gallus	—						
Apteryx	0.077	—					
Rhea	0.075	0.049	—				
Struthio	0.079	0.047	0.054	—			
Casuarius	0.073	0.032	0.047	0.044	—		
Dromaius	0.073	0.033	0.048	0.042	0.016	—	
Nothoprocta	0.096	0.075	0.080	0.084	0.075	0.075	—
Tinamus	0.096	0.078	0.080	0.083	0.076	0.077	0.070

[a] See materials and methods.

the outgroup, thereby precluding the placement of *Apteryx* at the base as suggested by the morphological and total evidence solutions (Figs. 7.2 and 7.4).

Other factors may be exacerbating the effect of long-branch attraction. First, it is apparent from the transversion distance matrices (Tables I and II), as well as from DNA hybridization distances (Sibley and Ahlquist, 1990, p. 810), that the radiation of the paleognaths is relatively ancient. Palaeognaths are at the base of the modern avian tree (Cracraft and Mindell, 1989; Sibley and Ahlquist, 1990; among others), and diversification among most of the paleognath lineages was likely in the Cretaceous. The mere fact that they represent ancient lineages creates certain difficulties for the systematic analysis of DNA hybridization distances and mitochondrial gene sequences because of the increased homoplasy associated with deep branching events.

An equally important factor, perhaps, is the observation that the major paleognath lineages are seemingly exemplary of a star phylogeny, that is, they diverged from one another relatively close in time. Transversion distances (Tables II and III), fitted DNA hybridization distances (Sibley and Ahlquist, 1990, p. 810, Fig. 325), and fitted maximum likelihood distances derived from the mitochondrial sequences (Fig. 7.6) all suggest that the lineages leading to *Casuarius*/*Dromaius*, *Struthio*, and *Rhea* diverged from one another relatively closely in time (but this picture may be distorted by rate differences among the lineages). Thus, internodal distances are generally much shorter than the branches leading to the terminal taxa.

V. CONCLUSIONS

The above arguments point to the morphological hypothesis as being best supported by all the data. Although the placement of the kiwi remains uncertain, both mor-

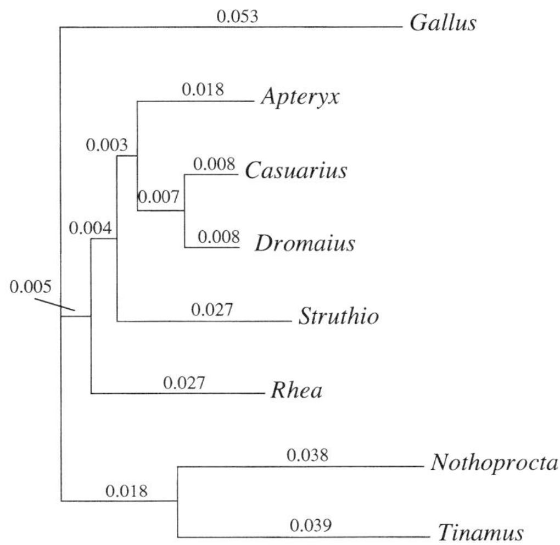

FIGURE 7.6 A maximum-likelihood tree of the molecular data (5219 bp; transversion parsimony; rates assumed to follow gamma distribution with shape parameter of 0.5). The fitted distances suggest short internodes among the major lineages of ratites, a result also found with DNA hybridization distances (Sibley and Ahlquist, 1990). The fact that the branches of *Gallus*, tinamous, *Rhea*, and *Struthio* are long relative to those of *Apteryx*, *Casuarius*, and *Dromaius* is also apparent.

phological and total evidence solutions place it at the base of the ratites, and a reevaluation of the morphological evidence confirms a sister-group relationship with moas (*contra* Cooper *et al.*, 1992).

A pertinent question is, what is the generality of these results? It was argued above (Fig. 7.5) that the debate over the placement of *Apteryx* on the tree seems dependent on where the outgroup joins to the ingroup. Because DNA hybridization distances (Sibley and Ahlquist, 1990) suggest that the relative rates in the nuclear genome essentially parallel those in the mitochondrial genome, there is at least some reason to predict that any long-branch artifacts of clustering arising within one data set might also be present in the other. Thus, the fact that there is congruence in results from DNA hybridization studies and those based on mitochondrial gene sequences does not necessarily constitute a sufficient reason for believing that molecular data are producing a robust estimate of relationships. Indeed, if internodal distances are short, it might be expected that in many instances morphology will actually have a better chance of resolving relationships than will molecular data. This follows from the observation that, even with short internodes, a suite of morphological markers (synapomorphies) can arise to characterize a clade, whereas insufficient molecular change will have accumulated during this time interval between branching events. The data presented here can be interpreted from this perspective, and a similar explanation has been proposed for corvine birds in which morphological data clearly unite manucodes with other birds of paradise but sequence data do not, except when close outgroups are chosen (Helm-Bychowski and Cracraft, 1993). Another factor is that phylogenetic reconstructions of molecular data, whether they be distances or discrete characters, might be expected to be more susceptible to rate effects than are morphological data merely because morphological data sets are rarely large enough to show sufficient "random" or parallel homoplasy to create long edges; usually the investigator consciously chooses characters that do not exhibit rampant variation—thus, parallelism—among the taxa; or the characters are chosen because they suggest synapomorphy rather than autapomorphy, the latter of which contributes to long edges. To our knowledge, no morphological study has invoked a long-edges-attract argument for a particular tree topology, although morphological systematists perhaps need to take this possibility more seriously.

One important conclusion from this study is that difficult problems in avian systematics will not be resolved by small amounts of DNA sequence or by inadequate taxon sampling. Although larger mitochondrial data sets have been shown to outperform smaller data sets in recovering relationships among relatively divergent taxa (Cummings *et al.*, 1995; Mindell and Thacker, 1996), it remains to be demonstrated whether any amount of sequence data will be able to resolve certain phylogenetic questions having characteristics similar to the ones surrounding the paleognaths: a distant outgroup, taxa having a deep history, a small number of taxa in the groups being compared, and/or short internodal distances relative to long terminals. Molecular data present many severe analytical difficulties, and thus morphological investigations will take on increasing importance for groups such as the paleognaths.

ACKNOWLEDGMENTS

We thank K. Helm-Bychowski for assistance in the early phases of this research. David Mindell, G. F. Barrowclough, Jeff Groth, and the laboratory of the late A. C. Wilson provided us with information on primers, for which we are grateful. As editor, David Mindell made numerous helpful suggestions, for which we are appreciative. Peter Houde and A. H. Bledsoe provided us with useful criticisms that improved the chapter. David Swofford kindly permitted use of a test version of PAUPSTAR. Tissues were kindly provided by the Lousiana State University Museum of Zoology (through Robert Zink) and the Bernice P. Bishop Museum (via Robert Fleischer). The Field Museum of Natural History provided access to its anatomical collections. Early phases of this work were supported by the Department of Anatomy and Cell Biology, University of Illinois at Chicago. The research reported in this chapter is a contribution from the Lewis B. and Dorothy Cullman Research Facility at the American Museum of Natual History and has received generous support from the Lewis B. and Dorothy Cullman Program for Molecular Systematics Studies, a joint initiative of The New York Botanical Garden and The American Museum of Natural History.

REFERENCES

Allard, M. W., Ellsworth, D. L., and Honeycutt, R. L. (1991). The production of single-stranded DNA suitable for sequencing using the polymerase chain reaction. *Biotechniques* **10**(1), 24–26.

Bledsoe, A. H. (1988). A phylogenetic analysis of postcranial skeletal characters of the ratite birds. *Ann. Carnegie Mus.* **57**, 73–90.

Bock, W. J. (1963). The cranial evidence for ratite affinities. *Proc. XIII Int. Ornithol. Congr.*, pp. 39–54.

Bock, W. J., and Bühler, P. (1990). The evolution and biogeographical history of the paleognathous birds. *In* "Proceedings of the International Centennial Meeting of the Deutsche Ornithologen-Gesellschaft," pp. 31–36.

Cooper, A., Mourer-Chauvire, C., Chambers, G. K., von Haeseler, A., Wilson, A. C., and Pääbo, S. (1992). Independent origins of New Zealand moas and kiwis. *Proc. Natl. Acad. Sci. U.S.A.* **89**, 8741–8744.

Cracraft, J. (1968). The lacrimal–ectethmoid complex in birds: A single character analysis. *Am. Midland Nat.* **80**(2), 316–359.

Cracraft, J. (1974). Phylogeny and evolution of the ratite birds. *Ibis* **116**, 494–521.

Cracraft, J. (1981). Toward a phylogenetic classification of the recent birds of the world (class Aves). *Auk* **98**, 681–714.

Cracraft, J. (1986). The origin and early diversification of birds. *Paleobiology* **12**(4), 383–399.

Cracraft, J. (1988). The major clades of birds. *In* "The Phylogeny and Classification of the Tetrapods," Vol. I: "Amphibians, Reptiles, Birds" (M. J. Benton, Ed.), pp. 339–361. Clarendon Press, Oxford.

Cracraft, J., and Helm-Bychowski, K. (1991). Parsimony and phylogenetic inference using DNA sequences: Some methodological strategies. *In* "Phylogenetic Analysis of DNA Sequences" (M. M. Miyamoto and J. Cracraft, Eds.), pp. 184–220. Oxford Press, New York.

Cracraft, J., and Mindell, D. P. (1989). The early history of modern birds: A comparison of molecular and morphological evidence. *In* "The Hierarchy of Life" (B. Frenholm, K. Bremer, and H. Jornvall, Eds.), pp. 389–403. Elsevier Science Publishers, Cambridge.

Cummings, M. P., Otto, S. P., and Wakeley, J. (1995). Sampling properties of DNA sequence data in phylogenetic analysis. *Mol. Biol. Evol.* **12**, 814–822.

de Boer, L. E. M. (1980). Do the chromosomes of the kiwi provide evidence for a monophyletic origin of the ratites? *Nature (London)* **287**, 84–85.

Desjardins, P., and Morais, R. (1990). Sequence and gene organization of the chicken mitochondrial genome. *J. Mol. Biol.* **212**, 599–634.

Eldredge, N., and Cracraft, J. (1980). "Phylogenetic Patterns and the Evolutionary Process." Columbia Univ. Press, New York.

Feduccia, A. (1980). "The Age of Birds." Harvard Univ. Press, Cambridge.

Feduccia, A. (1985). The morphological evidence for ratite monophyly: Fact or fiction. *Proc. XVIII Int. Ornithol. Congr.*, **1**, 184–190.

Felsenstein, J. (1978). Cases in which parsimony or compatibility methods will be positively misleading. *Syst. Zool.* **27**, 401–410.

Fürbringer, M. (1888). "Untersuchungen zur Morphologie und Systematik der Vogel," Vols. 1 and 2. Von Holkema, Amsterdam.

Fürbringer, M. (1902). Zur verglaichenden anatomie des Brustschulterapparates und der Schultermuskeln. Part 5. *Jena Z. Naturwiss.* **36**, 289–736.

Halanych, M. M. (1996). Testing hypotheses of chaetognath origins: Long branches revealed by 18S ribosomal DNA. *Syst. Biol.* **45**, 223–246.

Helm-Bychowski, K., and Cracraft, J. (1993). Recovering phylogenetic signal from DNA sequences: Relationships within the corvine assemblage (class Aves) as inferred from complete sequences of the mitochondrial cytochrome-*b* gene. *Mol. Biol. Evol.* **10**, 1196–1214.

Hennig, W. (1966). "Phylogenetic Systematics." University of Illinois Press, Champaign, Illinois.

Ho, C. Y.-K., Prager, E. M., Wilson, A. C., Osuga, D. T., and Feeney, R. E. (1976). Penguin evolution: Protein comparisons demonstrate phylogenetic relationships to flying aquatic birds. *J. Mol. Evol.* **8**, 271–282.

Houde, P. W. (1986). Ostrich ancestors found in the Northern Hemisphere suggest new hypothesis of ratite orgins. *Nature (London)* **324**, 563–565.

Houde, P. W. (1988). Paleognathous birds from the early Tertiary of the Northern Hemisphere. *Publ. Nuttall Ornithol. Club* **22**.

Houde, P. W. and Olson, S. L. (1981). Paleognathous carinate birds from the early Tertiary of North America. *Science* **214**, 1236–1237.

Huxley, T. H. (1867). On the classification of birds; and on the taxonomic value of the modifications of certain cranial bones observable in that class. *Proc. Zool. Soc. Lond.* **1867**, 415–472.

Jollie, M. T. (1957). The head skeleton of the chicken and remarks on the anatomy of this region in other birds. *J. Morphol.* **100**(3), 389–436.

Lesson, R.-P. (1831). "Traite d'Ornithologie," Vol. 1. Levrault, Paris.

Livezey, B. C. (1986). A phylogenetic analysis of recent anseriform genera using morphological characters. *Auk* **103**, 737–754.

Lowe, P. R. (1928). Studies and observations bearing on the phylogeny of the ostrich and its allies. *Proc. Zool. Soc. Lond.* **1928**, 185–247.

Marsh, O. C. (1880). Odontornithes: A monograph on the extinct toothed birds of North America. *In* "Report of the Geological Exploration of the Fortieth Parallel," Vol. 7, pp. 1–201. Washington D.C.

McDowell, S. (1948). The bony palate of birds. The Paleognathae. *Auk* **65**, 520–549.

Meise, W. (1963). Verhalten der straussatigen Vogel und Monophylie der Ratitae. *In* "Proceedings of the 13th International Ornithological Congress," pp. 115–125, Ithaca, New York.

Merrem, B. (1813). Tentamen systematis naturalis avium. *Abh. Konigel (Preussische) Akad. Wiss. Berlin (Physikal.)* 237–259.

Mindell, D. P., and Thacker, C. E. (1996). Rates of molecular evolution: Phylogenetic issues and applications. *Annu. Rev. Ecol. Syst.* (in press).

Mindell, D. P., Knight, A., Baer, C., and Huddleston, C. J. (1996). Slow rates of molecular evolution in birds and the metabolic rate and body temperature hypotheses. *Mol. Biol. Evol.* **13**, 422–426.

Mivart, G. (1877). On the axial skeleton of the Struthionidae. *Trans. Zool. Soc. Lond.* **10**, 1–52.

Muller, H. J. (1962). Die morphologie und entwicklung des craniums von *Rhea americana* Linne. *Sonderdruck Z. Wissensch. Zool.* **168**, 40–118.

Nelson, G., and Platnick, N. (1981). "Systematics and Biogeography." Columbia Univ. Press, New York.

Oliver, W. R. B. (1949). The moas of New Zealand and Australia. *Bull. Dominion Mus.* No. 15.

Olson, S. L. (1985). The fossil record of birds. *In* "Avian Biology" (Farner, D. S., King, J. R., and Parkes, K. C., Eds.), pp. 79–238. Academic Press, New York.

Parker, T. J. (1891). Observations on the anatomy and development of *Apteryx. Phil. Trans. R. Soc. Lond.* **182B**, 25–134.

Parker, T. J.(1895). On the cranial osteology, classification, and phylogeny of the Dinornithidae. *Trans. Zool. Soc. Lond.* **13**, 373–431.

Penny, D., Hendy, M. D., and Henderson, I. M. (1987). Reliability of evolutionary trees. *Cold Spring Harbor Symp. Quant. Biol.* **52**, 857–862.

Portmann, A. (1961). Olfaction. *In* "Biology and Comparative Physiology of Birds" (A. J. Marshall, Ed.). Academic Press, New York.

Prager, E. M., Wilson, A. C., Osuga, D. T. and Feeney, R. E. (1976). Evolution of flightless land birds on southern continents: Transferrin comparison shows monophyletic origin of ratites. *J. Mol. Evol.* **8**, 283–294.

Pycraft, W.P. (1900). On the morphology and phylogeny of the Paleognathae (Ratitae and Crypturi). *Trans. Zool. Soc. Lond.* **15**, 149–290.

Rich, P. V. (1979). The Dromornithidae, an extinct family of large ground birds endemic to Australia. *Bull. Aust. Bureau Min. Resources Geol. Geophys.* **184**, 1–194.

Sanger, F., Nicklen, S., and Coulsen, A. R. (1977). DNA sequencing with chain-terminating inhibitors. *Proc. Natl. Acad. Sci. U.S.A.* **74**(12), 5463–5467.

Sibley, C. G., and Ahlquist, J. E. (1981). The phylogeny and relationships of the ratite birds as indicated by DNA–DNA hybridization. *In* "Evolution Today, Proc. 2nd Int. Congr. for Systematic Evolutionary Biology" (G. G. E. Scudder and L. L. Reveal, Eds.), pp. 301–335. Hunt Inst. Botanic Document, Pittsburgh.

Sibley, C. G., and Ahlquist, J. E. (1990). "Phylogeny and Classification of Birds." Yale Univ. Press, New Haven, Connecticut.

Smith, B. B. (1994). Rooting molecular trees: Problems and strategies. *Biol. J. Linn. Soc.* **51**, 279–292.

Stapel, S. O., Leunissen, J. A. M., Versteeg, M., Wattel, J. and de Jong, W. W. (1984). Ratites as the oldest offshoot of avian stem-evidence from α-crystallin A sequences. *Nature (London)* **320**, 257–259.

Webb, M. (1957). The ontogeny of the cranial bones, cranial peripheral and cranial parasympathetic nerves, together with a study of the visceral muscles of *Struthio*. *Acta Zool.* **38**, 81–203.

Wheeler, W., and Gladstein, D. (1992). "MALIGN," version 1.01. Computer program distributed through the American Museum of Natural History, New York.

Morphological Character Descriptions for Hypothesis of Fig. 2

The primitive state is designated as (0), the derived states as (1, 2, 3) and unknown as (?); for multistate characters the order is listed under each character when necessary. The postcranial analysis is based on a reexamination of the characters described by Cracraft (1974) and Bledsoe (1988). Their analyses included *Aeypornis*, and Bledsoe's (1988) included the Dromornithidae, but because of a lack of available specimens, and the fact that dromornithids are probably not paleognaths, both groups were excluded from this analysis. With respect to the postcranial characters described by Cracraft (1974), four were modified before inclusion in this study. A reevaluation of Bledsoe (1988) resulted in the modification of 30 characters and the exclusion of 42 (these are discussed separately in Appendix II). Seventeen characters are unique to this analysis: 7 postcranial and 10 cranial.

Characters and character-states used for the morphological analysis of this chapter include the following:

	Character														
	1	2	3	4	5	6	7	8	9	10	11	12	13	14	15
Galliformes/ Anseriformes	0	0	0	0	0	0	0	0	?	0	0	0	0	0	0
Tinamidae	0	0	0	0	0	0	0	0	?	0	0	0	0	0	0
Dinornithids	1	1	1	?	1	1	1	1	2	0	?	?	?	?	?
Apteryx	1	1	1	?	1	1	1	1	2	0	1	1	1	2	?
Casuarius	1	1	0	1	0	1	1	1	1	0	1	1	1	2	1
Dromaius	1	1	0	1	0	1	1	1	1	0	1	1	1	2	1
Struthio	1	1	0	0	0	1	1	1	3	1	1	1	1	1	0
Rhea	1	1	1	?	0	1	1	1	3	1	1	1	1	1	0

	Character														
	16	17	18	19	20	21	22	23	24	25	26	27	28	29	30
Galliformes/ Anseriformes	0	0	0	0	0	0	0	0	0	0	0	0	0	0	0
Tinamidae	0	1	0	0	0	0	0	0	0	0	0	0	0	0	0
Dinornithids	?	?	?	0	2	0	0	0	0	0	0	2	1	2	0
Apteryx	1	1	1	1	2	0	0	0	0	0	0	2	1	0	0
Casuarius	1	1	1	1	1	0	1	0	0	1	0	1	1	1	1
Dromaius	1	1	1	1	1	0	1	0	0	1	0	1	1	1	1
Struthio	0	1	1	1	1	1	?	1	1	0	1	1	1	2	1
Rhea	0	1	1	1	1	1	?	1	1	0	1	1	1	2	1

	Character														
	31	32	33	34	35	36	37	38	39	40	41	42	43	44	45
Galliformes/ Anseriformes	0	0	0	0	0	0	0	0	0	0	0	0	0	0	0
Tinamidae	0	0	0	1	0	0	0	0	0	0	0	0	0	0	0
Dinornithids	0	0	0	1	0	1	1	1	1	0	0	0	1	1	0
Apteryx	0	0	0	1	0	0	1	1	1	0	0	1	1	0	1
Casuarius	1	1	1	1	1	1	1	0	1	1	0	1	1	1	1
Dromaius	1	1	1	1	1	1	1	0	1	1	0	1	1	1	1
Struthio	1	1	2	1	1	2	1	0	1	1	1	1	1	1	2
Rhea	1	1	2	1	1	2	1	0	1	1	1	1	1	1	2

	Character												
	46	47	48	49	50	51	52	53	54	55	56	57	58
Galliformes/ Anseriformes	0	0	0	0	0	0	0	0	0	0	0	0	0
Tinamidae	0	0	0	0	0	0	0	0	0	0	0	0	0
Dinornithids	0	0	0	1	0	0	0	1	1	1	1	1	1
Apteryx	0	0	0	1	0	0	0	1	1	1	1	1	1
Casuarius	0	1	0	0	1	0	1	1	0	0	0	2	0
Dromaius	0	1	0	0	1	0	1	1	0	0	0	2	0
Struthio	1	0	1	1	0	1	1	1	0	0	0	3	0
Rhea	1	0	1	1	0	1	1	1	0	0	0	?	0

1. Keel of the sternum (Cracraft, 1974, pp. 503, 506): (0) present, (1) absent. The loss of the keel is derived for all ratites.

2. Sternum, posterior lateral processes: (0) present, (1) absent. Cracraft (1974, pp. 503 and 506) took both the loss and reduction of the posterior lateral processes as being derived for all ratites. Only the presence/absence of the processes are recognized here as distinct character states. The complete loss of the posterior lateral processes in *Casuarius*, *Dromaius*, and *Rhea* is postulated to be derived.

3. Coracoidal sulci, lateral displacement: (0) sulci meet at or near the mid-

line, (1) sulci displaced laterally. Cracraft (1974, p. 503) described laterally displaced sulci as the derived condition, but did not state explicitly which taxa display this condition. Bledsoe (1988, character 3) described the cranial margin as being concave cranially (*Apteryx*, *Rhea*), straight (Dinornithidae), and convex cranially (remaining ratites) and further recognized the articular coracoidal sulci (Bledsoe's character 6) as having states (a) meeting or nearly meeting medially, and (b) restricted to the lateral margins. Bledsoe considered *Apteryx*, *Rhea,* and the dinornithids to have the derived condition. In this study only the lateral displacement of the sulci is described because the curvature of the cranial margin is largely related to the location of the coracoidal sulci. *Apteryx*, *Rhea*, and the Dinornithidae display the derived condition, although the displacement in *Rhea* is not as extreme as it is in the other taxa.

4. Coracoidal sulci, ventral lip: (0) extends laterally toward the base of the scapulocoracoid process, (1) directed more strongly posteriorly. The Tinamidae, Anhimidae, and *Struthio* are characterized by the primitive condition. In the derived condition found in *Dromaius* and *Casuarius*, the ventral lip of the sulci angles more posteriorly toward the midline of the body of the sternum (Mivart, 1877, pp. 25, 33, Figs. 19 and 27). *Rhea*, *Apteryx*, and the Dinornithidae cannot be scored for this condition because of the lateral displacement of the entire sulci (see character 3 above).

5. Sternum, lateral view: (0) moderate to highly curved, (1) flattened. Bledsoe (1988, character 5) described this condition as four character-states: (a) highly curved, carnia present, (b) highly curved, low medial ridge present, (c) moderately curved, no distinct medial ridge, and (d) flattened. Character-state (b) was unique to *Rhea*, (c) was described for *Casuarius*, *Dromaius*, and *Struthio*, and (d) for *Apteryx* and the dinornithids. The presence or absence of a keel is described above (See character 1). Regarding the curvature of the sternum, it is either curved (ancestral) or flattened (derived) and only *Apteryx* and the Dinornithidae display a flattened sternum.

6. Proportions of the body of the sternum: (0) much longer anteroposteriorly than wide mediolaterally, (1) essentially equal in width and length. Bledsoe (1988, character 1) coded unique proportions for *Apteryx* and *Aepyornis* whereas here *Apteryx* is coded as being the same as the other ratites, which display the derived condition.

7. Sternum, manubrium: (0) present, (1) absent. The loss of the sternal manubrium is a synapomorphy for all of the ratites.

8. Scapula and coracoid (Cracraft, 1974, pp. 503, 506): (0) not fused, (1) fused. The fusion of the scapula and coracoid is derived for all ratites.

9. Coracoidal process (Cracraft, 1974, pp. 505, 507–508; Bledsoe, 1988, character 9): (1) absent, (2) in the form of a ridge located anteriorly to the glenoid facet and compressed mediolaterally, (3) very pronounced and knoblike, projects toward the glenoid facet. This is an unordered character because the primitive condition is unknown. The fusion of the scapula and coracoid is

unique to the ratites (see character 9), and thus the outgroups have no homologous condition to polarize this character. The process, absent in *Casuarius* and *Dromaius*, is in the form of a mediolaterally compressed ridge in *Apteryx* and the moas, and very knoblike and pronounced in *Struthio* and *Rhea*.

10. Glenoid facet (Bledsoe, 1988, character 7): (0) oriented laterally, (1) oriented dorsolaterally. The dorsolateral orientation of the glenoid facet on the scapulocoracoid is derived for *Struthio* and *Rhea*.

11. Humerus, length relative to the ulna (Cracraft, 1974, pp. 505–506): (0) shorter than the ulna, (1) at least one-third longer than the ulna. This character combines characters 11 and 29 of Bledsoe (1988). In character 11 Bledsoe described the proportions of the humerus as being (a) slender, length moderate, (b) slender, length elongate, (c) stout, short, and (d) slender, short. Character-states (c) and (d) are unique for *Casuarius* and *Dromaius*, respectively, but essentially the elements of both are short. Although the humerus of these two taxa is short, it is at least one-third longer than the ulna. In character 29 Bledsoe described the lengths of the ulna–radius relative to the humerus as being (a) subequal, (b) ulna–radius longer than the humerus, and (c) humerus longer than the ulna–radius. For this latter character, state (c) inexplicably did not appear in Bledsoe's data matrix, (a) was ancestral, and (b) was derived for all of the ratites. The primitive condition is coded here as a humerus that is shorter than the ulna–radius, and all of the ratites display the derived condition. Because dinornithids cannot be scored for wing elements, characters 11–18 are coded as unknown.

12. Humerus, internal tuberosity: (0) protrudes slightly medially, does not extend proximally to the level of the head of the humerus, (1) knoblike with great medial protrusion, extends proximally to the level or slightly beyond the humeral head. Bledsoe (1988, character 14) described three character states: (a) head protrudes beyond tuberculum (tuberosity), (b) subequal, (c) tuberosity protrudes beyond head. Bledsoe indicated that *Apteryx*, *Rhea*, and *Struthio* have the tuberculum protruding beyond the head of the humerus and that *Casuarius* and *Dromaius* are subequal. Given the variation observed in the specimens examined for this study, *Struthio*, *Rhea*, *Dromaius*, *Casuarius*, and *Apteryx* are all coded as having the tuberosity more or less protruding to the level of the humeral head.

13. Deltoid crest: (0) greatly flares laterally from the base of the external tuberosity and tapers to a prominent ridge as it continues distally, (1) slightly raised ridge beginning from the base of the external tuberosity. Bledsoe (1988, character 18) described this character in terms of the absence or presence of a ridge at the base of the pectoral crest, and described the ridge as being present in *Apteryx*, *Struthio*, and *Rhea*. In this study, the reduction of the crest to a barely perceptible ridge is postulated as being derived for all ratites.

14. Carpal trochleae, carpometacarpus: (0) strong proximal protrusion of the external margin relative to the internal margin, with both margins well

rounded, (1) proximal protrusion of the external margin reduced so as to be essentially level with the internal margin, and with both margins well rounded, (2) proximal protrusion of both margins greatly reduced so that they are flattened. Bledsoe (1988, character 30) described the carpal trochlea as (a) highly curved, (b) moderately curved, or (c) flattened. *Rhea* and *Struthio* were scored as primitive (a), *Casuarius* as moderately curved, and *Apteryx* and *Dromaius* as flattened. For the specimens examined here, *Struthio* and *Rhea* are distinguishable from the ancestral state and display an intermediate condition in which the external margin of the trochlea is greatly reduced relative to the level of the internal margin, and both margins are well rounded. *Casuarius*, *Dromaius*, and *Apteryx* are considered to have the derived condition in which the trochlea are so reduced as to appear flattened. This character is ordered 1–0–2.

15. Phalangeal articulation for os metacarpale alulare (Bledsoe, 1988, character 31): (0) present, (1) absent. The loss of this articulation is derived for *Casuarius* and *Dromaius*. *Apteryx* is coded as questionable because the articulation has been found present in some specimens and absent in others.

16. Number of metacarpals with phalangeal articulation (Bledsoe, 1988, character 34): (0) three, (1) one. *Dromaius*, *Casuarius*, and *Apteryx* are considered to display the derived condition.

17. Humerus, transverse ligamental sulcus: (0) deep, (1) shallow. Bledsoe (1988, character 12) described the sulcus as (a) deep, (b) shallow, and (c) absent. The Dromornithidae were the only taxon coded as lacking a sulcus. The condition in the Tinamidae (a shallow transverse ligamental sulcus) is here considered to be homologous to that of the ratites, which display the derived condition.

18. Humerus, external epicondyle: (0) well developed and pronounced, (1) reduced to slightly raised surface. Bledsoe (1988, character 19) described the epicondyle as (a) well developed, (b) moderately developed, and (c) highly reduced. Bledsoe considered *Struthio* to be well developed, *Rhea* to be moderately developed, and *Casuarius*, *Dromaius*, and *Apteryx* to be highly reduced. In this analysis the external epicondyle is taken to be greatly reduced and derived for all ratites. Because the humerus of ratites is so modified, and skeletons exhibit considerable variation in what might be recognized as an epicondyle, it is not always easy to partition variation into clear character-states.

19. Dorsal surface of synsacrum caudal to antitrochanter (Cracraft, 1974, pp. 503–504; Bledsoe 1988, character 35): (0) broad, (1) narrow. The narrowing of the dorsal, posterior synsacrum is derived for all ratites except the Dinornithidae.

20. Relative lengths of the anterior and posterior ilia (relative to the acetabulum): (0) approximately equal to each other in length, (1) posterior portion longer than the anterior portion, (2) anterior portion longer than the posterior portion. Apparently Bledsoe (1988, character 37) miscoded this character because the anterior portion is described as being more elongate in *Rhea* and *Stru-*

thio; the posterior portions are clearly more elongate than the anterior portions (for *Rhea*, see Mivart, 1877, p. 2, Fig. 2). *Casuarius* and *Dromaius* also display this derived condition (Mivart, 1877, pp. 16 and 27, Figs. 13 and 22). *Apteryx* and the dinornithids display the other derived condition in which the anterior portion is more elongate than the posterior portion (for the dinornithids, see Oliver, 1949; for *Apteryx*, Parker, 1891, plate 18, Figs. 278–279). The character is ordered 1–0–2.

21. Postacetabular ilium, lateral view (Cracraft, 1974, pp. 503–504, Fig. 9): (0) essentially the same height dorsoventrally and width mediolaterally, dorsal margin curved, (1) narrows dorsoventrally and mediolaterally, tapering to a conical shape, dorsal margin flattened. Bledsoe (1988, character 40) described this condition as the dorsoventral width of the postacetabular ilium: (a) narrow, (b) moderately wide, and (c) very wide. Character-state (a) was ascribed to *Struthio* and *Rhea*, (b) was unique for *Apteryx*, and (c) was described for *Casuarius* and *Dromaius*. On the basis of an examination of *Hesperornis* (Marsh, 1880, plate X, Fig. 2) and the relative proportions of the other outgroups, the condition in which the posterior ilium remains approximately the same height and width for its entire length is considered ancestral. The condition found in *Rhea* and *Struthio* in which the posterior ilium narrows in height and width, tapering to a conical shape, is derived.

22. Club-shape expansion of the distal ischium: (0) absent, (1) ischium long, straight and narrow, ending in a hammer-like expansion that fuses or abuts against the posterior end of the ilium. In the outgroups, *Apteryx*, and the dinornithids the ischium is broad dorsoventrally for almost the entire length and the distal end is free (Cracraft, 1974, p. 504, Fig. 9; Oliver, 1949, Figs. 90 and 104). In *Struthio* and *Rhea* the ischium fuses with the pubis (see character 23 below), and so are coded as (?) since this character does not apply to these taxa. The ischium of *Dromaius* and *Casuarius* is narrow dorsoventrally and the distal portion ends in a hammer-like expansion (Mivart, 1877, Figs. 13 and 22).

23. Elongation and fusion of the pubis (Cracraft, 1974, pp. 503–504, 508): (0) essentially equal in length to the ilium and unfused, (1) elongated beyond the ilium and fused to the ischium. Bledsoe (1988, character 38) described the caudal protrusion of the ilium, ischium, and pubis as (a) pubis and ischium extend beyond ilium, (b) pubis extends beyond ischium, which extends beyond ilium, (c) protrusion subequal, and (d) ischium protrudes beyond subequal pubis and ilium. Character-state (d) was unique to *Aeypornis*. Character-state (b) occurred in *Rhea*, *Struthio*, and the Dinornithidae, and character-state (c) was found in *Dromaius* and *Casuarius*. For this analysis, the elongation of the pubis beyond the ilium and fusion to the ischium is found to occur only in *Struthio* and *Rhea*. In the other ratites and the outgroups, the posterior extensions of the pubis, ilium, and ischium are essentially the same and they are unfused.

24. Obturator process of the ischium: (0) not fused with the pubis, (1) fused

with the pubis to form a complete obturator foramen. This is derived for *Struthio* and *Rhea*.

25. Transverse processes of sacral vertebrae, ventral view: (0) processes separate and not broadened to fuse with each other, narrowly fused to anterior and posterior ilia, (1) processes broadened and fuse with each other and with the ilia to form a ventral plate of bone. The derived condition is found only in *Dromaius* and *Casuarius* (Mivart, 1877, Figs. 16 and 26).

26. Puboischial bar (Cracraft, 1974, pp. 503–504, 508): (0) absent, (1) present. A puboischial bar is present only in *Struthio* and *Rhea*. Bledsoe (1988, character 42) described this character as the dorsoventral width of pubis and ischial bar; it appears Bledsoe is referring to the same character as a puboischial bar.

27. Hypotarsus: (0) broad with two ridges, the internal ridge projecting more posteriorly than the external, (1) internal ridge greatly reduced, external ridge with sharp process protruding proximally, (2) internal ridge completely lost, external ridge knoblike at proximal end (modified from Cracraft, 1974, pp. 502, 506–508 and from Bledsoe, 1988, characters 66, 67, 68, and 73). Cracraft (1974) described four character-states. Cracraft coded a broad hypotarsus with two ridges (internal being larger) as primitive, and both ridges being pronounced more posteriorly as an intermediate condition. Cracraft then recognized two derived states, one in which there is a reduction of the external ridge and consequent development toward a single ridge, and another being a single ridge located along the external side of the bone. The intermediate state united *Apteryx* with the Dinornithidae, the first derived state united *Casuarius* and *Dromaius*, and the second derived state united *Rhea* and *Struthio*. Bledsoe's character 66 described *Aeypornis* and the Dromornithidae as differing in the shape of the hypotarsus from the other ratites. Bledsoe's character 67 was autapomorphic for the Dromornithidae, character 68 united all of the ratites (excluding *Aeypornis* and the Dromornithidae), and character 73 united *Struthio*, *Rhea*, *Casuarius*, and *Dromaius*. This analysis postulates *Apteryx* and the Dinornithidae to resemble the primitive condition, *Casuarius* and *Dromaius* to display an intermediate condition, and *Rhea* and *Struthio* to display a further derived condition. For the latter character-state, the complete loss of the internal ridge results in the hypotarsus being located more laterally than in the other taxa where it remains near the midline. The character order is 0–1–2.

28. Tarsometatarsus, intercotylar prominence (Cracraft, 1974, pp. 503, 506–507; Bledsoe, 1988, character 70): (0) proximally extended beyond hypotarsus, (1) essentially equal to the hypotarsus in proximal extension. The derived condition occurs in all ratites.

29. Tarsometatarsus, depth of the internal and external cotylar surfaces (modified from Bledsoe, 1988, characters 69 and 72.): (0) external cotylar surface relatively flat, internal surface concave with sharp proximal protrusion on its medial margin, (1) external cotylar surface slightly concave, internal surface deeply concave with concurrent loss of sharp proximal protrusion on the me-

dial margin, (2) external and internal cotylar surfaces concave and essentially equal in depth. The intermediate condition occurs in *Dromaius* and *Casuarius* and the derived condition is found in the Dinornithidae, *Struthio*, and *Rhea*. *Apteryx* shows the primitive condition. The character is ordered 0-1-2.

30. Anterior metatarsal groove (Cracraft, 1974, pp. 503, 507): (0) shallow and broad proximally, flattens out distally, (1) deep and narrow for the entire length. A deep and narrow metatarsal groove on the tarsometatarsus is derived for *Casuarius*, *Dromaius*, *Struthio*, and *Rhea*.

31. Posterior view of shaft of the tarsometatarsus (Cracraft, 1974, p. 502, Bledsoe, 1988, character 74): (0) generally smooth, lacking significant ridges, (1) sharp ridge present on external side running proximodistally. The derived condition is found in *Casuarius*, *Dromaius*, *Struthio*, and *Rhea*.

32. Trochleae/digits of the tarsometatarsus (Cracraft, 1974, p. 502): (0) four present, (1) loss of digit I. The loss of digit I is derived for *Casuarius*, *Dromaius*, *Struthio*, and *Rhea*.

33. Cnemial crests, tibiotarsus: (0) surface of anterior interarticular area and base of the crests wide, (1) surface of the anterior interarticular area narrows as the base of the crests becomes mediolaterally compressed, (2) surface of anterior interarticular area greatly reduced as the base of the crests becomes sharply compressed. Cracraft (1974, pp. 502, 506–508, Fig. 7D–F) recognized only a difference between *Rhea* and *Struthio* and the other ratites. Bledsoe (1988) saw variation in terms of three separate character descriptions:

> Character 51. Mediolateral compression of cranial and lateral crests: (a) slight or no compression, (b) moderate compression, and (c) substantial compression. Character-state (b) was unique for the Dromornithidae and state (c) united *Rhea* and *Struthio*.
>
> Character 52. Extent of cnemial crest and remaining articular surface in proximal view: (a) cnemial crest equal in extent to remaining articular surface, (b) cnemial crest less extensive than remaining articular surface, and (c) cnemial crest more extensive than remaining articular surface. Character-state (b) was described for dinornithids, *Apteryx*, and *Aepyornis*, and state (c) united *Struthio*, *Rhea*, *Casuarius*, and *Dromaius*.
>
> Character 54. Lateral margin between lateral cnemial crest and lateral articular surface: (a) shallowly concave, and (b) deeply concave. The derived condition was restricted to *Casuarius* and *Dromaius*.

In this analysis, reduction of the anterior interarticular surface and the development of a deeply concave lateral margin are considered to be due to the mediolateral constriction at the base of the cnemial crests. *Apteryx* and the dinornithids resemble the ancestral condition. *Dromaius* and *Casuarius* display an intermediate condition in which the mediolateral compression is not as extreme as it is in *Struthio* and *Rhea*. The character is ordered 0–1–2.

34. Tibiotarsus, proximal extension of internal cnemial crest: (0) flat to slightly extended above level of interarticular surface, (1) moderate to greatly extended beyond articular surface. In describing the proximal extension of the internal cnemial crest, Bledsoe (1988, character 53) recognized three character states: (a) moderately beyond the articular surface, displayed by *Apteryx* and the

dinornithids, (b) slightly beyond the articular surface, described for *Rhea* and *Struthio*, and (c) far beyond the articular surface, found in *Dromaius* and *Casuarius*. The ancestral character-state found in outgroups shows the internal cnemial crest flattened or only slightly extended above the level of the articular surface. Enough variation occurred among the specimens examined that the derived condition is best described as moderate to greatly extended above the articular surface (Cracraft, 1974, p. 501, Fig. 7A–C) and is found in all of the ratites and the Tinamidae.

35. Supratendinal bridge of the tibiotarsus (Cracraft, 1974, p. 501; Bledsoe, 1988, character 57): (0) present, (1) absent. The loss of the supratendinal bridge on the tibiotarsus is derived for *Casuarius*, *Dromaius*, *Struthio*, and *Rhea.*

36. Tibiotarsus, external condyle (Cracraft, 1974, pp. 500–501, 507–508, Fig. 6A): (0) rounded along distal margin, anterior margin grades into the shaft smoothly (not undercut), (1) flattened along distal margin, anterior portion slightly undercut, (2) ovoid along distal margin, anterior portion sharply undercut. When Bledsoe (1988, character 61) described this, Bledsoe only recognized a difference between the primitive condition and the derived state occurring in *Struthio* and *Rhea.* Here a derived condition is also hypothesized to occur in *Dromaius*, *Casuarius*, and the dinornithids (character-state 1). The other derived condition (character-state 2) is postulated for *Struthio* and *Rhea. Apteryx* resembles the primitive condition. The character is ordered 1–0–2.

37. Tibiotarsus, depression on the lateral surface of external condyle: (0) shallow, (1) deep (Bledsoe, 1988, character 63). The depression is presumed to be the groove for the peroneus profundus muscle, and a deep groove is derived for all ratites.

38. Tibiotarsus, internal condyle distal view (Cracraft, 1974, pp. 501, 507–508, Fig. 6B): (0) essentially level or slightly projected anteriorly relative to external condyle, (1) projects strongly anteriorly relative to external condyle. The derived condition is found only in *Apteryx* and the dinornithids.

39. Tibiotarsus, medial side of the internal condyle (Cracraft, 1974, pp. 501, 507–508, Fig. 6B): (0) slight depression near anterior margin, (1) deep pit in anterior margin and a groove along the posterior margin. The derived condition occurs in all ratites.

40. Tibiotarsus, anterior intercondylar fossa (Cracraft, 1974, pp. 498, 506, Fig. 5): (0) narrow and does not undercut the condyles, (1) widens and undercuts the condyles at the proximal margin, forming a slight ridge that distinguishes the articular surface from the fossa. The derived condition is found in *Dromaius*, *Casuarius*, *Struthio*, and *Rhea.*

41. Tibiotarsus, posterior margin of the external condyle (Cracraft, 1974, pp. 500–501, 508, Fig. 6A; Bledsoe, 1988, character 62): (0) rounded, smoothly grades into shaft, (1) moderate lateral extension, sharply undercuts base of shaft. Only *Struthio* and *Rhea* display the derived condition.

42. Femur, trochanteric crest: (0) extends proximally beyond the level of

the trochanteric fossa (iliac facet), (1) crest essentially on the same level with the iliac facet. Bledsoe (1988, character 44) recognized four character-states, but three of those four are each unique to *Aepyornis*, *Struthio*, and the dromornithids. Bledsoe described *Struthio* as having the trochanteric fossa extended greatly beyond the level of the trochanteric crest, whereas here this condition is considered to be homologous to that of *Rhea*, *Casuarius*, *Dromaius*, and *Apteryx*.

43. Femur, margin of the iliac facet: (0) curved sharply to form a lip that faces medially, surface highly concave, (1) rounded edge, no lip present, surface flattened to slightly convex. A similar condition was described by Bledsoe (1988, character 43) for the caudal margin of the proximal antitrochanteric articular surface: (a) highly concave, (b) moderately concave, (c) straight or nearly so. Bledsoe coded character-state (b) for *Struthio* and *Rhea*, and character-state (c) for *Apteryx*. In this study no distinction could be made between moderately concave and straight. The derived condition occurs in all of the ratites.

44. Femur, external and fibular condyles (Cracraft, 1974, pp. 499, 500–501, 507–508, Fig. 3): (0) essentially equal in size and in their distal extension relative to the internal condyle, (1) greatly enlarged and project distally beyond level of internal condyle. Bledsoe (1988, character 46) recognized three character-states: (a) subequal distally, (b) lateral condyle extended slightly distally beyond medial condyle, and (c) lateral condyle extended distally far beyond medial condyle. Character-state (b) united dinornithids with *Casuarius* and *Dromaius*, and character-state (c) united *Struthio* and *Rhea*. The specimens examined exhibited so much variation that it is difficult to distinguish between Bledsoe's character-states (b) and (c). Therefore, *Struthio* and *Rhea* are coded for the same derived condition as that in *Dromaius*, *Casuarius*, and the dinornithids.

45. Femur, fibular condyle, posterior view: (0) relatively sharp posterolateral margin, proximal margin essentially level with external condyle, (1) rounded posterolateral margin, proximal margin essentially level with external condyle, (2) rounded posterolateral margin, proximal margin not extended as far proximally as external condyle. Cracraft (1974, pp. 498–499, 507, Fig. 3) recognized only one derived condition, which Cracraft described as occurring in *Dromaius* and *Casuarius*. Bledsoe (1988, character 48) partitioned the variation into two character-states: (a) lateral and fibular condyles extend distally about the same distance, (b) fibular condyle extends less distally (three-quarters or less) than the lateral condyle. Bledsoe coded *Apteryx*, *Casuarius*, *Dromaius*, and *Rhea* as displaying character-state (b). The dinornithids are considered here to have the same derived condition as *Dromaius* and *Casuarius* : a rounded (not sharp) posterolateral margin and a proximal margin level with the external condyle. *Struthio* and *Rhea* display a second derived condition in which the proximal margin of the external condyle extends proximally beyond the fibular condyle. *Apteryx* resembles the ancestral condition. The character order is 1–0–2.

46. Femur, internal condyle, medial view (Cracraft, 1974, pp. 498–499, 508, Fig. 3): (0) rounded along distal margin, (1) flattened along distal margin.

Bledsoe (1988, character 50) recognized three character-states: (a) semicircular, (b) triangular or elliptical, and (c) flattened. Bledsoe considered the primitive condition (a) to occur in the dinornithids and *Apteryx*, the derived state (b) to occur in *Dromaius* and *Casuarius*, and state (c) to occur in *Struthio* and *Rhea*. Specimens examined in this study displayed no distinguishable difference between semicircular and triangular or elliptical, so only *Rhea* and *Struthio* were considered to display the derived condition.

47. Femur, external condyle viewed laterally (Cracraft, 1974, pp. 498, 507): (0) appears more rounded and projects more posteriorly, (1) appears more elliptical and projects more proximally. The derived condition is found in *Casuarius* and *Dromaius*.

48. Femur, rotular groove (Cracraft, 1974, pp. 498, 508): (0) broad and shallow, (1) narrow and deep. The derived character-state is found to occur in *Struthio* and *Rhea*.

49. Femur, internal condyle: (0) posterior facet ovoid in shape, (1) posterior facet triangular in shape. The derived condition occurs in *Rhea*, *Struthio*, *Apteryx*, and the dinornithids.

50. Femur, distal end of external condyle: (0) pit for the tibialis anticus deeply excavated and narrow mediolaterally, (1) pit shallow and wide, cutting into the fibular condyle laterally. The pit for the attachment for the tibialis anticus becomes shallow and widens so as to slightly excavate laterally into the fibular condyle. The derived condition is found in *Dromaius* and *Casuarius*.

51. Femur, popliteal fossa (Bledsoe, 1988, character 49): (0) shallow, almost flat, (1) very deep, extending anteriorly. *Struthio* and *Rhea* display the derived condition.

52. Lacrimal, elongate supraorbital process, projecting posterolaterally over the orbit: (0) absent, (1) present. The process is absent in the Tinamidae, Anhimidae, Megapodiidae, dinornithids, *Apteryx* and is present in *Struthio*, *Rhea*, *Casuarius*, and *Dromaius*.

53. Squamosal, projection of zygomatic process: (0) projects slightly anterolaterally just over the articulation with the quadrate, (1) projects anterolaterally over at least two-thirds of the body of the quadrate. The Anhimidae, Megapodiidae, and Tinamidae have a zygomatic process that just slightly projects anterolaterally over the articulation with the quadrate, and this condition is here postulated to be primitive. All the ratites have the derived condition.

54. Pterygoid: (0) not divided, simple rodlike structure, (1) divided into dorsal and ventral surfaces. The primitive condition occurs in the outgroups, the Tinamidae, *Struthio*, *Rhea*, *Casuarius*, and *Dromaius*. The body of the pterygoid is found as a simple rodlike unit that articulates posteriorly with the quadrate and the basipterygoid processes and anteriorly with the vomer and the palatines. Variation in the shape of the articulating surfaces of these elements is autapomorphic for each taxon. In *Apteryx* and the dinornithids the pterygoid is divided anteriorly, forming dorsal and ventral surfaces. This saddle-shaped por-

tion on the medial side of the bone is henceforth called the pterygoid fossa. The dorsal fork appears to be the main body of the pterygoid and homologous to the body of the pterygoid in the other paleognaths. This fork extends anteriorly and ankyloses to the dorsal surface of the vomer. The ventral fork is the unique structure and ankyloses with the palatine anterolaterally and extends anteriorly to ankylose to the ventral surface of the vomer. Parker (1891, pp. 55–56) described the pterygoid of *Apteryx* as divided anteriorly into medial and lateral processes, with the medial process articulating with the lateral border of the vomer and the lateral process articulating with the dorsolateral border of the palatine. Jollie (1957, p. 420) and McDowell (1948, pp. 527–528) also described the pterygoid in essentially the same manner. This divided pterygoid and the resultant pterygoid fossa are apparently not found in any other birds except perhaps the "Lithornithidae" (Houde, 1988, pp. 20, 47, Fig. 5).

55. Quadrate, intercondylar fossa: (0) shallow, (1) deep, rounded pit. The fossa is shallow in the outgroups, *Struthio, Rhea, Casuarius*, and *Dromaius*. In *Apteryx* and the dinornithids the fossa is deeply marked, partially as a consequence of the posterior articulating surface having a strong ventral elevation.

56. Basitemporal plate and mammillar tuberosities: (0) posterolateral margins of the basitemporal plate flattened or just slightly developed into mammillar tuberosities, (1) posterolateral margins of the plate with well-developed mammillar tuberosities. The tuberosities have been shown to occur at the junction of the basioccipital, exoccipital, and prootic bones (Parker, 1895, p. 384) in dinornithids and *Apteryx*. The tuberosities are poorly developed or absent in the outgroups and the other ratites.

57. Maxillopalatine antrum: (0) absent, (1) present as a "large pocket" formed from the maxillopalatines with the anterior portion ankylosed to the dorsal surface of the posterior maxillary, (2) present as a "large pocket" formed from the maxillopalatines, the anterior margin ankylosing with the posterior margin of the maxillary, (3) greatly reduced to a "small pocket" that ankyloses anteriorly with the posterior maxillary. The antrum is present in all ratites (except *Rhea*, see below), but in different character-states. The antrum in all taxa (Parker, 1895, plate LXII, Figs. 59–63) is located on the dorsolateral surface of the maxillopalatine. The antrum consists of a "pocket," the floor of which is formed by the maxillopalatine, and is covered by a thin lamina of bone with a posterior opening. Pycraft (1900, pp. 185–187) described the antrum as the interior half of the maxillopalatine having the outer and inner borders turned upward to meet in the middorsal line to form a long, thin-walled tunnel. McDowell (1948, pp. 527–528) described the antrum as a dorsal arched lamina of the maxillopalatine joining the ventral lamina at its margins, forming hollow cones with posterior openings. In the dinornithids and *Apteryx*, the dorsal and ventral surfaces are entirely formed from maxillopalatine bone, but the anterior portion ankyloses to the dorsal surface of the posterior maxillary (character-state 1). In the dinornithids the posterior openings are wide, whereas in *Apteryx*

the posterior openings are small foramina. In some *Apteryx* specimens the posterior foramina could not be found. The antrum of *Dromaius* and *Casuarius* is also formed entirely from maxillopalatine bone, but ankyloses anteriorly with the maxillary (character-state 2). *Struthio* has the antrum reduced to a small, essentially open, anterior portion ankylosing with the posterior maxillary (character-state 3). *Rhea* displays a structure similar to that in *Struthio*, but it is so reduced that the homologous condition is uncertain. For the purposes of this analysis, *Rhea* has been coded as questionable. The presence of an antrum is derived; in this analysis the polarity of the character transformation beyond the primitive state is unordered.

58. Olfactory chamber/tubercle: (0) ossification of chamber relatively poorly developed, (1) ossification of chamber well developed. The third, posteriormost vestibule of the nasal cavity, when present, is the olfactory chamber (Portmann, 1961, pp. 42–43). The Tinamidae, Anhimidae, and Megapodiidae, and all ratites except *Apteryx* and the dinornithids have a chamber that is poorly ossified. In *Apteryx* and the dinornithids, the olfactory chamber is enlarged and fused to the ectethmoid complex (Parker, 1891, pp. 48–50; 1895, p. 389). In *Apteryx* the chamber contains a complex of ossified turbinals, whereas in dinornithids the chamber is empty.

Comments on Morphological Characters of Bledsoe (1988)

Bledsoe's (1988) characters 10, 16, 17, 20, 23, 24, 25, 36, 39, 58, 59, 65, 75, 76, 77, 79, 81, 82, and 83 were excluded from this analysis because they were autapomorphies and thus phylogenetically uninformative. There are 23 additional characters in Bledsoe's analysis that have also been excluded from this study. The reasons for this warrant further discussion (the character numbers and descriptions correspond to those in Bledsoe, 1988).

2. Craniolateral process (sternum): (a) elongate, (b) shortened, (c) very short. Character-state (b) was described only for the Dromornithidae. Bledsoe (1988) described a short craniolateral process (scapulocoracoid) to occur in *Casuarius*, *Dromaius*, and *Struthio*. In the specimens examined for this study only *Casuarius* was found to have shortened processes. The processes for *Dromaius* and *Struthio* did not differ from the ancestral condition, so this character is an autapomorphy and uninformative within ratites.

4. Number of incisures (sternum): (a) two, (b) none, (c) four. Bledsoe (1988) refers to the posterior margin of the sternum and the presence or absence of posterior medial processes. Bledsoe described four incisures unique to *Struthio* and no incisures as derived for *Casuarius*, *Dromaius*, and *Rhea*. The homologous condition could not be determined among taxa due to variation among specimens and the loss of the posterior lateral processes in *Dromaius*, *Casuarius*, and *Rhea* (Appendix I, character 2).

8. Medial groove of glenoid cavity (scapulocoracoid): (a) absent, (b) present. Bledsoe (1988) described the groove as present in *Struthio* and *Dromaius*, but in the specimens examined here, the groove was found to vary in occurrence across all ratite taxa and as such is considered uninformative.

13. Pneumatic foramen (humerus): (a) present, (b) absent. The presence of this character is inconsistent across taxa. Bledsoe (1988) described the foramen as being absent in the dinornithids, *Apteryx*, *Struthio*, and *Rhea*, but it was also found to be absent in several specimens of *Dromaius* and *Casuarius*.

15. Position of the head (humerus): (a) near midline, (b) shifted dorsally, (c) shifted ventrally. Character-state (c) is unique for *Aeypornis*. Bledsoe (1988) described the humeral head as being shifted dorsally in *Dromaius*, *Rhea*, and *Struthio*. In the specimens examined, the humeral head was essentially near the midline for all taxa.

21. Shape of shaft in cross-section (humerus): (a) elliptical proximally and distally, (b) circular proximally and distally, (c) triangular proximally and distally, (d) triangular proximally, elliptical distally. Variation in the specimens examined was sufficiently great to blur distinction among the character-states.

22. Distal end (humerus): (a) widest cranially, (b) widest caudally, (c) cranial, medial, and caudal widths subequal, (d) widest medially. The specimens examined in this analysis displayed insufficient distinct variation from character-state (a), the primitive condition.

26. Ulna, proximal end: (a) broad in proximal view, (b) narrow in proximal view. Bledsoe (1988) described *Casuarius*, *Dromaius*, and *Apteryx* as being narrow in proximal view, but the specimens examined here all appeared to be broad in proximal view and not distinct from the other taxa.

27. Ulna and radius: (a) unfused, (b) fused. Bledsoe (1988) described *Casuarius* as polymorphic and *Dromaius* as having character-state (b). All of the specimens examined for this analysis have unfused ulnae and radii.

28. Width of shafts of ulna and radius: (a) ulna broader, (b) subequal. Bledsoe (1988) described *Struthio* and *Casuarius* as having the primitive condition in which the ulna is broader than the radius. The specimens examined in this study displayed an ulna that varied along the length of the bone, thus the distinctness of the character-states is unclear.

32. Fusion of os metacarpale majus and minus: (a) incomplete, (b) complete. Bledsoe (1988) described *Apteryx*, *Casuarius*, and *Dromaius* as having the fusion complete. This study found the fusion, reduction, or loss of these elements to be unique for each taxon and thus highly variable across ratites. Both the majus and minus are present in *Struthio* and articulate at the distal end, but are often not entirely fused. In *Casuarius*, the majus and minus frequently are not extended completely distally in which case they are neither fused nor articulating; however, in some specimens they are bridged by a small piece of bone and they become "fused." In *Dromaius* the minus cannot be found owing to a loss of the bone or fusion with the majus, and in *Apteryx* the minus is present, but it is not fused with the majus distally. The condition in *Rhea* resembles that in the Tinamidae and Megapodiidae in which the majus and minus are fused distally, although in some specimens fusion is essentially lacking.

33. Os metacarpale majus: (a) wide dorsoventrally, (b) compressed dorsoventrally. Too much variation occurred across the specimens examined to distinguish character-states.

41. Preacetabular tuberculum: (a) elongate, (b) short, (c) absent. Character-state (c) is autapomorphic for the dromornithids. Bledsoe (1988) coded *Casua-*

rius, *Dromaius*, and *Rhea* as having a short tuberculum, but so much variation occurred among the specimens examined that no distinct character-state could be determined.

43. Caudal margin of proximal antitrochanteric articular surface (femur): (a) highly concave, (b) moderately concave, (c) straight or nearly so. The specimens examined had variation that was indistinguishable from the ancestral condition.

45. Relationship of longest axis of shaft of femur to longest axes of medial and lateral condyles: (a) parallel or nearly so, (b) divergent by 15 degrees or greater. The two character-states could not be distinguished.

47. Dorsal margin of lateral condyle (femur): (a) straight or nearly so, (b) moderately concave dorsally, (c) highly concave dorsally. Because of so much variability no distinct character-states could be distinguished.

55. Channeling at margins of intercondylar eminence (tibiotarsus): (a) present, (b) absent. Because none of the specimens examined seemed to display an intercondylar eminence (character 56, see below), the presence of channeling at the margins was considered to be ambiguous.

56. Intercondylar eminence (tibiotarsus): (a) present, (b) absent. None of the specimens in this analysis displayed a structure that might be distinguished as an intercondylar eminence.

60. Craniodistal margin of lateral condyle (tibiotarsus): (a) semicircular, (b) elliptical. In the specimens examined, variation was too great to discern character-states.

64. Width and length of medial condyle (tibiotarsus): (a) moderate, (b) moderate in width, short in length, (c) very wide, very short. This study revealed too much variation to individuate discrete character-states.

71. Depression between intercotylar area and hypotarsus (tarsometatarsus): (a) present, (b) absent. Bledsoe (1988) described the depression as being absent in *Struthio* and the Dinornithidae, whereas in this analysis, the depression was found to be present in those taxa so there is no difference from the ancestral condition.

78. Medial and lateral margins of trochlea III (tarsometatarsus): (a) not parallel, (b) parallel. No discrete states could be recognized in the specimens examined.

80. Proximodistal lengths of proximal phalanges: (a) III longest, IV shortest, or III and IV subequal, (b) III longest, (c) II longest, (d) II and III subequal, IV shortest. No distinct character-states could be determined owing to high variability in the specimens examined.

CHAPTER 8

Phylogenetic Relationships among and within Select Avian Orders Based on Mitochondrial DNA

DAVID P. MINDELL,[*][†] MICHAEL D. SORENSON,[†] CHRISTOPHER J. HUDDLESTON,[‡] HECTOR C. MIRANDA, JR.,[§] ALEC KNIGHT,[||] STEVEN J. SAWCHUK,[**] TAMAKI YURI[*][†]

[*]*Department of Biology, University of Michigan, Ann Arbor, Michigan 48109;* [†]*Museum of Zoology, University of Michigan, Ann Arbor, Michigan 48109;* [‡]*Laboratory of Molecular Systematics, National Museum of Natural History, Smithsonian Institution, Washington, DC 20560;* [§]*Department of Biological Sciences, University of Cincinnati, Cincinnati, Ohio 45221;* [||]*Department of Biology, Sul Ross State University, Alpine, Texas 79832;* [**]*School of Veterinary Medicine, Ohio State University, Columbus, Ohio 43210*

I. INTRODUCTION

Well-corroborated phylogenies are crucial to understanding the evolution of the molecular, morphological, and behavioral differences among modern birds. However, our understanding of phylogenetic relationships among and within avian orders is incomplete. Progress has been made, but our knowledge remains limited, owing in part to the apparent rapid radiation of modern birds. As a result, there are few shared derived characters, still conserved, delineating basal groups. Among the many challenges facing avian systematists is the need for additional discrete character data sets and knowledge of the constraints influencing character change over time. Our objective in this chapter is to present phylogenetic analyses of new molecular sequence data for select avian lineages, to place these analyses in the context of existing phylogenetic hypotheses, and to discuss pertinent issues regarding methods of phylogenetic inference.

Most researchers recognize two primary groups of extant birds: (1) Paleognathae, which usually includes ratites (*Struthio*, *Rhea*, *Casuarius*, *Dromaius*, *Apteryx*) and nine genera of tinamous, and (2) Neognathae, which includes all other birds (Cracraft, 1981; Cracraft and Mindell, 1989; Olson, 1985; Sibley and Ahlquist, 1990). Galliformes (pheasants, megapodes, curassows) and Anseriformes (waterfowl, screamers) are supported as sister taxa by diverse data sets, although their placement relative to Paleognathae and Neognathae remains unclear. Beyond this basal split between paleognaths and neognaths, however, there is little agreement among data sets or researchers on relationships among avian orders.

Ordinal phylogenies based on morphological characters have many polytomies deep within them (e.g., Cracraft, 1988) because few derived morphological characters have been recognized that unite particular avian orders in sister-group relationships. The only comprehensive ordinal phylogeny based on molecular data is from DNA–DNA hybridization analyses (Sibley and Ahlquist, 1990), and although significant insights have been gained, methodological and empirical problems remain. Many of the branches uniting orders are short (reflecting small $\Delta T_{50}H$ values), and in many instances differences in $\Delta T_{50}H$ values between replicate experiments (with identical taxa) exceed differences between experiments involving different taxa. This suggests that the DNA–DNA hybridization data are not well suited for divergences as large as those among many avian orders. Other difficulties with the Sibley and Ahlquist analyses pertain to violated assumptions of equivalence in genome size, gene arrangement, and AT:GC ratios; undesired hybridization and comparison of paralogous sequences, owing to mixing of single-copy and low-copy number genes; significant technical difficulties in measuring "normalized percent hybridization"; unenumerated corrections for evolutionary rate heterogeneity across taxa; and extrapolation of $\Delta T_{50}H$ values to an unseen portion of a curve whose shape must be estimated. The lack of complete data matrices ($\Delta T_{50}H$ values for pairwise comparisons of all taxa), understandable given the labor involved, is also problematic. Many species are compared to a single reference taxon (a radiolabeled tracer) and these many species comparisons are then "chained" together according to their increasing distance from the reference taxon. Thus, all species are not directly compared to each other as is usual in phylogenetic analyses, and heterogeneity in rates of change can have additional significant effects on the inferred clustering. Also, analyses of $\Delta T_{50}H$ values yield unrooted trees that may, in the absence of a specified outgroup, have multiple phylogenetic interpretations consistent with them (Cracraft, 1987; Houde, 1987; Gill and Sheldon, 1991; Mindell, 1992; Lanyon, 1992).

Existing phylogenetic hypotheses for our current study taxa (Table I), based on both molecular and morphological data, do indicate a close relationship between Procellariiformes (albatrosses, petrels, shearwaters) and Pelecaniformes (gannets, cormorants, pelicans); however, the relative position of Ciconiiformes (herons, storks, ibises) varies (e.g., Cracraft, 1988; Sibley and Ahlquist, 1990). Phylogenetic placement of "Turniciformes" (or Turnicidae; buttonquails or hemipodes) is poorly known. Falconiformes (falcons, hawks, eagles) and Strigiformes (owls) are variously placed as sisters or as distant relatives, with the latter requiring independent evolution of a general raptorial morphology and natural history in the two groups. Placement of Passeriformes (songbirds), Cuculiformes (cuckoos), Caprimulgiformes (goatsuckers, nightjars), Apodiformes (swifts), Trogoniformes (trogons), Coraciiformes (rollers, kingfishers), and Charadriiformes (shorebirds, gulls, auks) relative to other orders is similarly uncertain. Discussion of the 20–30 commonly recognized avian orders is complicated by the fact that monophyly for the orders as generally configured (e.g., Peters, 1931–1951; Wetmore, 1960; Mayr and Cottrell, 1979) cannot be presumed. Monophyly for Ciconiiformes, Gruiformes, Falconiformes, Pelecaniformes and Cuculiformes has been particularly contentious (Cracraft, 1981, 1982; Olson, 1985; Sibley and Ahlquist, 1990; Sheldon and Bledsoe, 1993; Hedges and Sibley, 1994; Sibley, 1994; Hedges *et al.*, 1995; Hackett *et al.*, 1995). Discussion of ordinal relationships here is intended to focus on noncontroversial members of those orders, unless otherwise indicated.

II. ISSUES IN PHYLOGENETIC ANALYSES

A. Constraints on Molecular Evolution

If we presume the existence of (1) a natural, divergent hierarchy of species based on common descent and (2) identifiable shared–derived characters (homologies) for taxa within the hierarchy, then discovery of monophyletic groups appears to be a straightforward task. Preference for a parsimony criterion in this discovery process derives from the general scientific practice of minimizing ad hoc assumptions (of homoplasy in the case of phylogenetic analyses) and the notion that parsimony correctly determines which phylogenetic hypothesis is best supported by the character evidence (Farris, 1983; Nelson, 1994). Ideally, one need not invoke any specifics of evolutionary process to estimate genealogy.

However, putative homologies can be difficult to identify. This may be attrib-

TABLE I Study Species and Orders Represented[a]

Order	Species	Order	Species
Struthioniformes	*Rhea americana***	**Gruiformes**	*Fulica atra*
	Struthio camelus		(*Turnix varia*)
Tinamiformes	*Crypturellus undulatus*	**Charadriiformes**	*Scolopax minor*
Procellariiformes	*Diomedea nigripes*		*Pterocles coronatus*
Pelecaniformes	*Phalacrocorax pelagicus*	**Cuculiformes**	(*Tauraco hartlaubi*)*
Ciconiiformes	*Mycteria americana*		*Coccyzus erythopthalmus*
	Nyctanassa violacea		*Opisthocomus hoazin*
	(*Phoenicopterus ruber*)	**Strigiformes**	*Tyto alba*
Falconiformes	*Accipiter superciliosis**		*Nyctea scandiaca**
	*Circus aeruginosus**		*Otus longicornis*
	*Circaetus gallicus**		*Otus mirus*
	*Gyps fulvus**		*Otus mindorensis*
	*Buteo buteo**		*Otus megalotis everetti**
	*Buteo jamaicensis**		*Otus megalotis nigrorum*
	*Milvus migrans**		*Mimizuku gurneyi**
	*Haliaeetus leucocephalus**		*Bubo virginianus**
	Gampsonyx swainsonii		*Asio flammeus**
	*Pernis apivorus**		*Aegolius acadicus**
	*Pandion haliaetus**		*Ninox philippensis**
	*Sagittarius serpentarius**	**Caprimulgiformes**	*Chordeiles minor***
	*Falco peregrinus***	**Apodiformes**	*Chaetura cinereiventris**
Anseriformes	*Anhima cornuta*	**Trogoniformes**	*Trogon melanurus*
	Chauna chavena	**Coraciiformes**	*Coracias caudata*
	Anseranas semipalmata	**Passeriformes**	*Vidua chalybeata***
	Cygnus buccinator		*Motacilla cinerea*
	Cygnus atratus		*Junco hyemalis*
	Anas formosa		*Sturnella magna*
	*Anas platyrhynchos**		*Cardinalis phoeniceus*
	Aix sponsa		*Lanius collurio*
	*Aythya americana***		*Sayornis phoebe*
	Anser rossi	**(Nonavian reptiles, outgroups)**	*Alligator mississippiensis***
	Branta sandvicensis		*Crocodylus porosus*
	Dendrocygna arcuata		
	Dendrocygna bicolor		
	Somateria fisheri		
	Thalassornis leuconotus		
Galliformes	*Gallus gallus***		
	Coturnix coturnix		
	Bonasa umbellus		
	Meleagris gallopavo		
	Phasianus colchicus		

[a] Ordinal assignments for species in parentheses are poorly known. The entire mt 12S gene was analyzed for all species listed. Single asterisks denote species having an additional 518 bases of mt COI available for analyses. Double asterisks denote species having 12 mt protein-coding genes (all but ND6) and both mt rDNAs available for analyses. Mitochondrial 12S and COI sequences have been deposited in GenBank with accession numbers U83709–U83787 and U86138–U86142. Sequences for *Gallus gallus* and *Coturnix coturnix* are from Desjardins and Morais (1990 and 1991, respectively).

uted in part to the existence of a finite number of character states (four in the case of DNA) and rates of change sufficient to yield independent expressions of the same state. Putative homologies, like putative relationships among taxa, are products of phylogenetic analyses; attempts to improve them must, therefore, involve refinements in phylogenetic analyses. Such refinement has long been sought in the use of conservative characters. By giving greater weight in phylogenetic analyses to characters changing less frequently, confounding effects of homoplastic similarity can be reduced. Thus, improved understanding of rates of molecular evolution can potentially improve phylogenetic analyses.

Primary influences on rates of molecular evolution may be viewed as constraints on mutation and fixation (Mindell and Thacker, 1996). Variation in mutation rate can stem from differences in replication frequency, replication repair (different mechanisms and different enzymes), replication fidelity, exposure to mutagens (especially DNA-damaging oxygen free radicals), and the initial conditions of differential codon and nucleotide base composition for genes and taxa (see Britten, 1986; Shigenaga *et al.*, 1989; Ohta, 1993). These factors and others may influence variation in both the overall rate and the rate for particular kinds of change. The rate at which one or more mutations become fixed is influenced by a set of continuous and overlapping constraints, including the genetic code, secondary or tertiary structure, gene function, population size, frequency of cladogenesis, and natural selection. Variation in the nature of these constraints across taxa and over time contributes to patterns of variation in rates of molecular sequence evolution for different taxa and different kinds of character changes. Such patterns, however, must be considered as hypotheses to be judged on their individual merits for any set of taxa or genes in which they are proposed. Patterns should not be assumed *a priori*.

Given information on relative rates of change and variability for different characters, unequal weighting may be used to reduce levels of homoplasy (convergent similarity) in the data (Swofford *et al.*, 1996; Mindell and Thacker, 1996). Emphasizing change in relatively conserved, or slowly changing, characters is one of the oldest principles in systematics (Darwin, 1859; Hennig, 1966; Farris, 1966), although the means for doing this remain controversial. Character weights based on phylogenetically determined measures of homoplasy (Farris, 1969; Goloboff, 1993) make no claims independent of phylogeny about the information content of characters. Other weighting schemes, whether based on measures of character compatibility (Penny and Hendy, 1985; Sharkey, 1989) or, as is more common, various estimates of comparative absolute rate (e.g., Mindell *et al.*, 1991; Honeycutt and Adkins, 1993) or frequency of cooccurrence of alternative states at homologous sites (Knight and Mindell, 1993; Wheeler, 1990a), do make claims that are not derived from character distributions on a phylogenetic hypothesis. However, the effects of fundamental physical–chemical constraints, such as the genetic code, Watson-Crick base pairing, and the initial base composition, do not rely on any particular evolutionary theory. Rather, they circumscribe physical limitations on sequence character change across organisms, such that not all character changes are equally probable. Several studies indicate the success of unequal character weighting in recovering "known" phylogenies, whether those phylogenies are well corroborated and "noncontroversial," simulated, or result from manipulation of populations in a laboratory (Allard and Miyamoto, 1992; Atchley and Fitch, 1991; Hillis *et al.*, 1994; Huelsenbeck, 1995; Miyamoto *et al.*, 1994; Smith, 1994). Mindell and Thacker (1996) demonstrated that for a set of diverse vertebrate taxa, relative rates of evolution estimated for various characters were similar whether based on discrete character branch lengths from a phylogenetic tree or pairwise comparisons.

B. Sampling of Characters and Taxa

Considerable debate has focused on two issues related to sampling in phylogenetic analyses: (1) the relative merits of partitioning and combining data sets, and (2) taxonomic versus character congruence (Miyamoto, 1985; Cracraft and Mindell, 1989; Kluge, 1989; Bull *et al.*, 1993; Eernisse and Kluge, 1993; Chippindale and Wiens, 1994; deQueiroz *et al.*, 1995). What can be called a "total data" approach accords equal weight to all characters regardless of their degree of homoplasy. An alternative "total evidence" approach (in our usage of that term) considers all characters as potentially informative, but can use successive approximations (Farris, 1969) and character analyses yielding rate estimates as sources of evidence in determining the relative information content of character sets. Weights may be applied in alternative phylogenetic analyses accordingly. In this view, total evidence includes the characters themselves and our improving understanding of their history of change. This is in keeping with the view that theory and practice of systematics can be mutually informing and that informed unequal weighting can improve phylogenetic hypotheses. The total data approach with equal weighting of all characters may be appropriate in the absence of evolutionary rate heterogeneity, or if homoplasy can be demonstrated to be randomly distributed for a particular set of taxa and characters. However, the constraints on rates of molecular evolution discussed above, varying across taxa and characters sets, tend to impose decidedly nonrandom effects on patterns of homoplasy accumulation.

Numerous studies suggest that larger data sets perform better in recovery of a well-corroborated tree or a known tree based on simulation or laboratory manipulation, and a threshold for number of DNA characters may exist at which phylogenetic analyses recover well-corroborated, "noncontroversial" sister relationships with equal weighting of most characters. Parsimony trees from random samples of 7000 equally weighted mitochondrial DNA characters (excluding the rapidly evolving D-loop region) yielded the same tree for 10 taxa, as did the whole mitochondrial genome about 90% of the time (Cummings *et al.*, 1995). In a numerical simulation study of four taxa with equal rates of evolution, Hillis *et al.* (1994) found that equally weighted parsimony analyses yielded the known tree 100% of the time with about 1500 bases. Parsimony analyses with unequal weighting required even fewer characters in recovering the known tree 100% of the time. These studies also suggest that larger numbers of taxa and (nonidealized) unequal rates of sequence change, as well as the presence of short internodes, will tend to require larger amounts of sequence data. Larger sets of DNA characters are also preferable to smaller sets because they provide a more comprehensive test of congruence among characters and often provide phylogenies that are less sensitive to alternative character weighting schemes, alternative sequence alignments (particularly of ribosomal DNA), and alternative inclusion/exclusion of study taxa, including outgroups (e.g., Kluge, 1989; Wheeler, 1995; Mindell and Thacker, 1996). This is reflected in phylogenies having high levels of statistical support based on mammalian whole mitochondrial genomes and lower levels of support based on a single or small number of genes (see Cao *et al.*, 1994a,b; Graur and Higgins, 1994; Janke *et al.*, 1994).

Similarly, more taxa are generally preferable to fewer taxa. Additional taxa can reduce the incidence of long branches within a tree, and so reduce the potential for attraction among them owing to convergent similarity. As more closely related taxa are included, more of the multiple substitutions at individual nucleotide positions may be recovered. However, the addition of single representatives of distantly related taxa, with long branches, may have the opposite effect of increasing the level of convergent similarity within the data set. These issues are particularly relevant to selection of outgroup taxa. Analyses by Smith (1994) demonstrate that parsimony recovers a well-corroborated tree more often when multiple outgroup taxa representing a single sister group, with relatively few long branches, are used, rather than multiple distantly related outgroup lineages.

III. METHODS

A. Study Characters and Taxa

We provide analyses of three overlapping mt DNA sequence data sets. Our largest character data set consists of 13,298 nucleotide bases from 12 mitochondrial (mt) protein-coding genes (all but ND6) and both ribosomal (rDNA) genes from each of six taxa: greater rhea (*Rhea americana*), domestic chicken (*Gallus gallus*), redhead duck (*Aythya americana*), peregrine falcon (*Falco peregrinus*), village indigobird (*Vidua chalybeata*; Passeriformes), and *Alligator mississippiensis*. We present analyses of mt 12S rDNA, comprising 859 aligned nucleotide positions, for 72 species of birds representing 18 different orders as traditionally configured (Table I). We also present analyses of 518 nucleotide positions of mt COI for a subset of the study taxa. Our 12S rDNA analyses include multiple species within eight orders, as well as several taxa whose ordinal relationships have been particularly difficult to resolve: the hoatzin (*Opisthocomus hoazin*), a buttonquail (*Turnix varia*), and a flamingo (*Phoenicopterus ruber*).

Mitochondrial 12S rDNA and COI sequences for most falconiform and strigiform taxa ($n=20$) were determined using standard polymerase chain reaction (PCR) amplification techniques and manual sequencing with CircumVent (New England BioLabs, Beverly, MA), and direct incorporation of ^{35}S-labeled dATPs (see Knight and Mindell, 1993). Sequences for all other taxa in Table I, including the six with the largest set of sequence characters, were determined using an Applied Biosystems (ABI; Foster City, CA) 377 automated sequencer. To maximize efficiency and to generate greater overlap between contiguous sequences for the six taxa having 13 kb sequenced, long PCR products were generated with an r*Tth* DNA polymerase-based XL-PCR kit (Perkin-Elmer, Norwalk, CT) and, after gel purification, sequenced directly with both the PCR primers and multiple internal primers. The PCR products were gel purified and then extracted from agarose with a QIAquick gel extraction kit (Qiagen, Chatsworth, CA). An aliquot of the purified PCR product was run on a minigel and compared to marker bands with known quantities of DNA to estimate DNA concentration. Approximately 30–150 ng of PCR product (depending on the size of the PCR fragment) was used in each sequencing reaction. Sequencing reactions used the ABI Prism dye terminator cycle sequencing ready reaction kit with AmpliTaq DNA polymerase FS (Perkin-Elmer). Unincorporated fluorescent labeled ddNTPs were removed using a Sephadex (G-50-Fine; Pharmacia, Piscataway, NJ) spin column. Sequencing reactions were run on a Long Ranger (FMC, Philadelphia, PA) acrylamide gel. Careful interpretation of the machine output, sequencing and reconciling of both DNA strands, and repetition of reactions when necessary yield accurate sequence. Interpretation and checking of machine output was completed by use of the ABI Sequence Navigator program, which allows chromatograms to be aligned, compared, and edited on screen.

Insertions of mtDNA sequences into the nuclear genome have been documented in a wide variety of taxa and now appear to be a common phenomenon (Zhang and Hewitt, 1996; Quinn, Chapter 1 in this volume). We are keenly aware of the potential problems nuclear homologs may cause in the accurate determination of mtDNA sequences as well as the potential errors introduced by their unwitting inclusion in phylogenetic analyses. Extracts of total genomic DNA from blood (birds have nucleated red blood cells) were used in previous published reports of nuclear insertions in birds (Quinn, 1992; Arctander, 1995; Sorenson and Fleischer, 1996), and no blood samples were used in the current study. Additional precautions in our study include (1) the amplification of long fragments (although a continuous nuclear ex-mtDNA sequence of 7.9 kb has been documented in the cat, most are much smaller in length; Zhang and Hewitt, 1996), (2) the inclusion of degenerate positions in primers complementary to protein-coding regions (this reduces the possibility that primers will preferentially amplify a low copy number nuclear sequence owing to primer mismatch with the mtDNA), (3) broadly overlapping PCR products (it is unlikely that two different sets of PCR pairs will both amplify a nuclear sequence from an mtDNA-rich sample), and (4) careful examination of sequences for double peaks (resulting from coamplification of both mtDNA and nuclear sequences), unexpected insertions or deletions (indels), and frameshifts or stop codons.

Alignments for all mitochondrial genes were done using CLUSTAL W (Thompson *et al.*, 1994) and adjustments by eye. For 12S rDNA the initial alignments were overlaid on a secondary structural model (see Fig. 8.1), and adjustments were sought

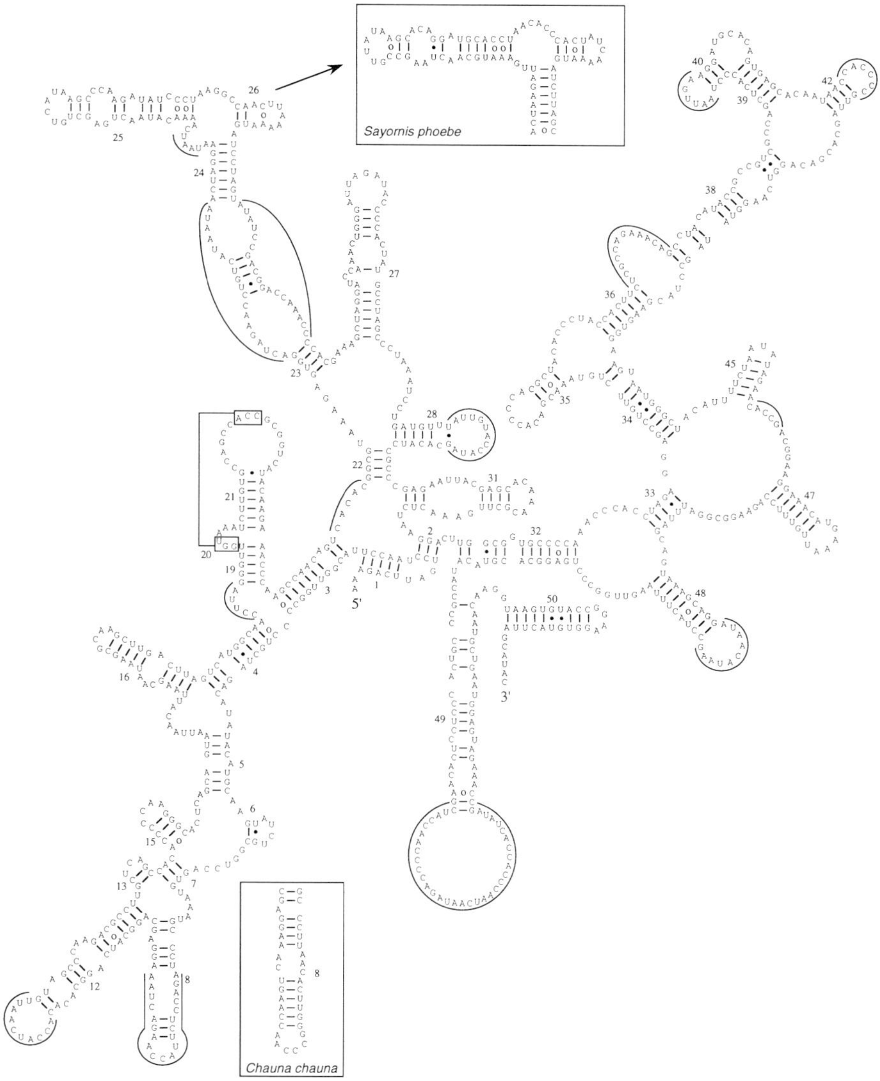

FIGURE 8.1 Hypothetical mt 12S rRNA secondary structure for peregrine falcon (*Falco peregrinus*). The structure is based on previously published secondary structure models (Gutell, 1994; Van de Peer *et al.*, 1994; Sullivan *et al.*, 1995; Hickson *et al.*, 1996) with identification of structures aided by compensatory base changes across helices among the birds in our data set. Helices are numbered according to Van de Peer *et al.* (1994). Unconventional base pairings are indicated by filled circles for (G–U) and unfilled circles for (A–C). Regions of substantial length variation among birds are indicated by solid curved lines adjacent to the *Falco peregrinus* sequence, and correspond roughly to regions omitted from parsimony analyses.

that would (1) maintain alignment of both stems and nonstems across taxa, and (2) maintain base pairing within stems (see Kjer, 1995). Hypervariable regions, appearing to be randomly aligned, which could not be improved by eye, were omitted from phylogenetic analyses below. For mt 12S rDNA, 242 base positions were omitted from the total alignment length of 1101 positions for the study taxa (Table I), yielding 859 base positions for analyses. For the mt 16S rDNA available for six taxa, 171 base positions were omitted from the total alignment length of 1671 positions, based on comparisons to published secondary structural hypotheses (Gutell *et al.*, 1993). We have not refined a 16S rDNA model for birds.

The hypothetical mt 12S rRNA secondary structure for *Falco peregrinus* in Fig. 8.1 is based on previously published secondary structure models (Gutell, 1994; Van de Peer *et al.*, 1994; Sullivan *et al.*, 1995; Hickson *et al.*, 1996) with identification of structures aided by compensatory base changes across helices among the birds in our data set. Helix 8 is particularly variable in the number of base pairs in its distal portion. Base pairing for helix 26 and the adjacent ends of helices 24 and 25 were difficult to determine in many taxa, suggesting structural variability among species. *Falco* has four extra bases between helices 24 and 25, unlike any of the other birds in our data set. Additional avian examples for helices 8 and 24–26 are shown. The proposed 5-bp helix between helices 23 and 24 is also unusual in *Falco*. Nonetheless, most of the birds in our data set have at least two or three bases of complementary sequence in positions homologous to the distal 2-bp in *Falco*. In most other taxa, this region has been represented as two long single-stranded sequences (e.g., Houde, Chapter 5 in this volume). We have drawn helix 38 as in Hickson *et al.* (1996): an alternative structure is given by Houde (Chapter 5 in this volume). Increasingly refined models of secondary structure as in Fig. 8.1 are useful as templates for informing alignments and character homology hypotheses for conserved structural features.

Although a complete discussion of patterns of sequence variation in relation to secondary structure is beyond the scope of this study, we suggest that a weighting scheme based on a simple categorization of 12S rDNA sequence characters into helices, bulges, loops, and other unpaired regions on the premise that these structures evolve at different rates would make unrealistic assumptions. Whereas the most variable regions are generally terminal loops, other loops are highly conserved. Likewise, certain helices have high rates of base substitution, while others show little if any variation across taxa. In addition, some helices (e.g., helix 8) appear to vary in length and to shift slightly in position.

B. Phylogenetic Analyses

Phylogenetic analyses were conducted using the criteria of parsimony and congruence among characters. Given the large numbers of taxa and characters, heuristic searches for shortest trees were conducted with 1000 replicates, using starting topologies based on random addition of taxa to reduce the possibility of finding a

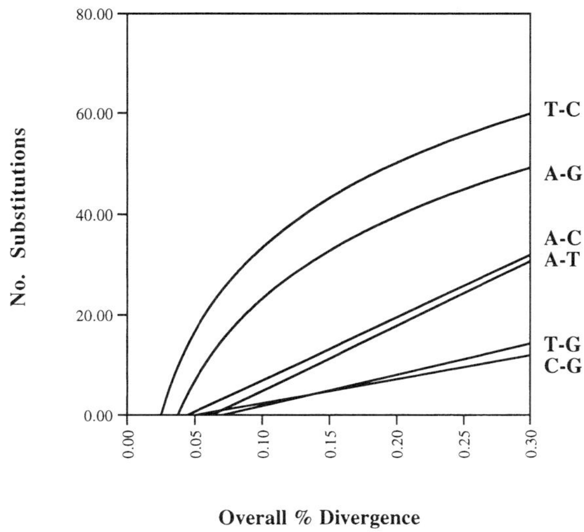

FIGURE 8.2 Number of substitutions for nucleotide pairs versus overall percent divergence for the mitochondrial 12S rDNA gene from 50 birds (same species as in Fig. 8.5), plus *Crocodylus porosus*, *Homo sapiens*, *Mus musculus*, and *Didelphis virginiana*. Lines were drawn using logarithmic curves for T–C and A–G substitutions and straight lines for the other substitution types, based on their fit to plotted points (not shown). Substitution tallies have been standardized to account for differences in base composition, by dividing the observed number of substitutions for nucleotide pairs by the mean ratio of observed to expected (25%) frequencies.

local parsimony optimum rather than the universal optimum, using PAUP (Swofford, 1993). Exhaustive searches for the most parsimonious tree were conducted for smaller subsets of taxa to focus on particular taxa and test alternative existing hypotheses regarding their phylogenetic relationships. Support indices are calculated for nodes within most parsimonious trees to denote their degree of character support (Bremer, 1988) where all characters have been equally weighted. They indicate the number of additional steps required for the shortest tree lacking the particular clade.

To inform character weighting in mt 12S rDNA analyses, we compare inferred rates of substitution among the 6 possible nucleotide pairs in all pairwise comparisons among 50 birds, a crocodilian, and 3 mammals. As expected, we find transitions to be more frequent than transversions and a decreasing frequency for T–C, A–G, A–C, A–T, T–G, and C–G substitution types (Fig. 8.2). We also plot mean percent divergence among these same 12S sequences against estimated time since divergence for the taxa compared (Fig. 8.3). Divergence time estimates based on fossils include dates of 310 million years ago (MYA) for the split between mammals and reptiles/birds and 245 MYA for the split between birds and crocodilians (Benton, 1990). Divergence times among avian orders are poorly known, although they are likely similar for most orders given an apparent rapid radiation of forms. Current

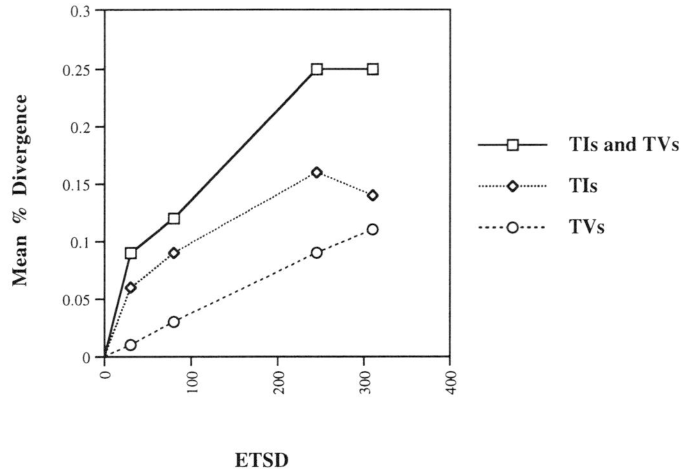

FIGURE 8.3 Mean percent divergence in mitochondrial 12S rDNA sequence versus estimated time since divergence (ETSD) for all species pairs for 50 birds (same species as in Fig. 8.5), *Crocodylus porosus*, *Homo sapiens*, *Mus musculus*, and *Didelphis virginiana*. Divergence time estimates based on fossils include dates of 310 million years ago for the split between mammals and reptiles/birds and 245 MYA for the split between birds and crocodilians (Benton, 1990). Divergences for avian orders were set to 80 million years ago and divergences among Accipitridae and Strigidae species were set to 30 million years ago (see text). Substitution tallies have been standardized to account for differences in base composition as in Fig. 8.2. TIs and TVs, Transitions and transversions, respectively.

estimates for extant avian ordinal ages range from more than 100 MYA based on molecular data (Sibley and Ahlquist, 1990; Hedges *et al.*, 1996) to less than 60 MYA based on fossil evidence (see Feduccia, 1995), and we have used an estimate of 80 MYA for divergences among avian orders in Fig. 8.3. We use an estimate of 30 MYA for divergences among species within the avian families Accipitridae and Strigidae, consistent with the oldest known fossils for these taxa (Brodkorb, 1964; Olson, 1985; Carroll, 1988).

Figures 8.2 and 8.3 indicate that transitions are more likely to entail convergent similarities than are transversions for analyses of avian orders in our study set, and for our analyses of single representatives from diverse orders we use transversions only. For analyses mixing orders with both single and numerous representative taxa we use two sets of weights in alternative analyses, (1) transversions and transitions weighted equally and (2) a weighting ratio of 5:1, respectively. Mitochondrial DNA exhibits diverse rates of evolution, varying both among and within genes, gene regions, and across taxa. Thus, statements regarding rate must be viewed as simplified hypotheses specific to the data set on which they are based.

Differences among taxa in base composition can contribute to increased amounts of convergent similarity and confound phylogenetic analyses. However, this is ameliorated by focusing on relatively conserved, nonsaturated types of character change. 12S rDNA mean percent base composition for 71 avian species (Table I) is A: 31.1,

TABLE II Parsimony-Based Relative Rate Tests among Eight Birds[a]

	1	2	3	4	5	6	7
1. *Struthio*							
2. *Crypturellus*	8/23**						
3. *Anser*	7/10	26/14*					
4. *Meleagris*	12/11	29/13**	14/10				
5. *Turnix*	6/14	20/13	6/11	10/19			
6. *Motacilla*	9/26**	20/22	11/25*	10/28**	12/21		
7. *Sagittarius*	5/19**	20/19	9/20*	9/24**	10/16	19/17	
8. *Otus*	11/27**	25/26	11/24*	16/33*	14/22	26/25	20/22

[a]Tests are based on mt 12S rDNA transversions only using *Crocodylus acutus* as an outgroup (Mindell and Honeycutt, 1990). Numbers denote unambiguous, autapomorphic character changes (column taxon/row taxon) based on branch lengths for a series of three-taxon trees using MacClade (Maddison and Maddison, 1992). *, $P < 0.05$; **, $P < 0.01$. See Table I for full species names (*Otus megalotis everetti* is used here).

C: 26.9, G: 22.5, T: 19.5. A chi-square test of homogeneity of base frequencies across taxa found no significant differences. Similar analyses of relative rate were conducted for mt COI sequences from a subset of the avian study taxa, and the observed patterns, with higher rates of change at third-codon positions compared to first and second positions, are consistent with studies of other protein-coding mt genes in vertebrates (see Fig. 2 in Mindell and Thacker, 1996). Mitochondrial COI mean percent base composition for 27 avian species (Table I) is A: 25.9, C: 31.8; G: 16.5, T: 25.8. Again, a chi-square test of homogeneity of base frequencies across taxa found no significant differences.

We compared relative rates of mt 12S rDNA sequence change among the study taxa, using a parsimony-based approach (Mindell and Honeycutt, 1990; Mindell *et al.*, 1996), and present a subset of the significant comparisons in Table II. Taxa showing faster rates of change relative to others include a tinamou (*Crypturellus*), a passeriform (*Motacilla*), secretary bird (*Sagittarius*), and a strigiform (*Otus*). Other passeriform and strigiform study taxa also show relatively fast rates, although the comparisons vary among species. Relative rate comparisons help in identifying potential long branch attraction problems, particularly for lineages represented by single species.

IV. RESULTS AND DISCUSSION

A. Phylogenetic Placement of Anseriformes and Galliformes

Our largest character data set includes 13,298 mtDNA characters for 5 birds and a crocodilian outgroup. Single most parsimonious trees based on all characters equally

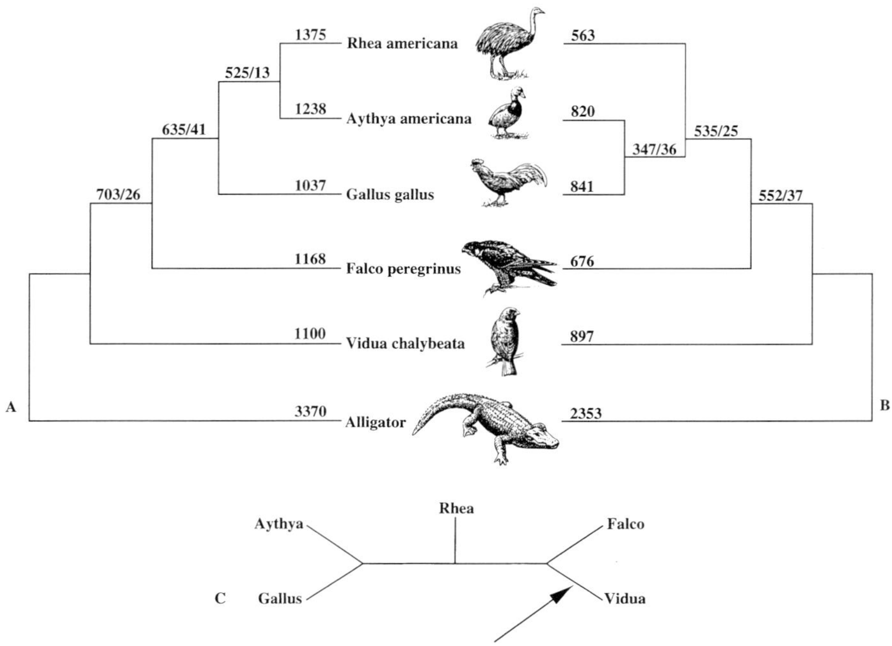

FIGURE 8.4 Single most parsimonious trees based on exhaustive searches using 13,298 nucleotide base positions from 12 mitochondrial protein-coding genes (only ND6 is missing) and the 2 rDNA genes for each of 6 taxa, with (A) all characters given equal weight (11,151 steps) and (B) all characters weighted equally at codon positions 1 and 2 and with third codon position and rDNA transitions given zero weight (7584 steps). *Alligator mississippiensis* was designated as the outgroup. Numbers on branches denote: branch lengths/support indices (Bremer, 1988). The topology shown in (B) was also found to be most parsimonious in analysis of the corresponding amino acid sequences for 12 mt protein-coding genes using the PROTPARS weight matrix, and found to be optimal in maximum-likelihood analysis of amino acid sequences using protML and the JTT model in the MOLPHY set of programs (Adachi and Hasegawa, 1992). The basal position of a passeriform (*Vidua*) among primary lineages of extant birds is unconventional (see also Fig. 8.5). (C) shows the ingroup topology from (B) as an unrooted network, illustrating sensitivity of topology to the root placement (indicated by an arrow). Placement of the root along the lineage leading to *Rhea* in (C) would yield the more conventional topology (Cracraft and Mindell, 1989; Sibley and Ahlquist, 1990) with the *Aythya/Gallus* clade being sister to the neognaths (*Falco/Vidua*). Addition of sequences from more taxa within the avian ingroup (including passeriforms), and from more diverse crocodilians, may help reduce potential attraction among long branches, and further test the position of Passeriformes in future analyses.

weighted (Fig. 8.4A) and codon positions 1 and 2 characters equally weighted with rDNA transversions only (Fig. 8.4B) differ only in the relative position of *Aythya*, an anseriform, being sister to *Rhea* in the first analysis and sister to *Gallus* in the second. We believe the second analysis to be a better phylogenetic estimate, in light of greater saturation of third-codon positions and rDNA transitions. Although the character set is smaller in Fig. 8.4B, the support index for the *Gallus/Aythya* node

(36) is substantially larger than for the *Rhea/Aythya* node (13) in Fig. 8.4A. Figure 8.4B unites representative galliform and anseriform species as sisters, most recently sharing a common ancestor with *Rhea*, a paleognath. A falconiform is sister to that group, and a passeriform is sister to the clade including all four other birds. We also found the Fig. 8.4B topology in analyzing just the 12 mt protein-coding genes based on (1) both DNA (weighted as in Fig. 8.4B) and amino acids (using the PROTPARS weight matrix) in parsimony analyses with PAUP, and (2) using a maximum-likelihood approach with protML (version 2.2) and the JTT model (Jones *et al.*, 1992) in the MOLPHY set of programs (Kishino and Hasegawa, 1990; Adachi and Hasegawa, 1992). Analysis of only the two mt rDNA genes placed the root (*Alligator*) along the *Gallus* branch, yielding a tree with *Gallus* as basal among birds and with *Rhea* sister to Falconiformes/Passeriformes.

The sister relationship for Galliformes and Anseriformes (Fig. 8.4B) is in agreement with both morphological and molecular studies; however, the sister relationship between a Galliformes/Anseriformes clade and paleognaths conflicts with previous molecular and morphological analyses (Stapel *et al.*, 1984; Cracraft, 1988; Cracraft and Mindell, 1989; Sibley and Ahlquist, 1990). Curiously, summary DNA–DNA hybridization distance analyses actually do indicate a sister relationship between paleognaths and Galliformes/Anseriformes (Sibley and Ahlquist, 1990, Fig. 357); however, this was considered as (Sibley and Ahlquist, 1990, p. 255) "misplaced," and an arrangement with Galliformes/Anseriformes as the oldest neognath clade was said to be (Sibley and Ahlquist, 1990, p. 288) "the best representation of all the data, morphological and molecular." Stapel *et al.* (1984) found evidence placing Galliformes and Anseriformes in a clade with 14 neognath species based on 173 amino acids from the nuclear-encoded α-crystallin A gene.

Placement of the root is critical in resolving the position of Anseriformes/Galliformes as paleognaths (as in our best estimate) or as neognaths. Root placement must be considered carefully, as distant outgroups in particular can lead to spurious rooting on the longest internal branch of the ingroup (Wheeler, 1990b; Smith, 1994). Considering the ingroup topology in Fig. 8.4B as an unrooted network (Fig. 8.4C), we find sister relationships for Anseriformes and Galliformes and for the two neognaths, with *Rhea* attached to their internode. Attachment of the root (*Alligator*) to the passeriform (*Vidua*), as in Fig. 8.4B, yields a sister relationship for the Anseriformes/Galliformes clade and a paleognath, as well as the unexpected position of Passeriformes as basal to other birds. Attachment of the root to *Rhea* yields the conventional placement of Anseriformes/Galliformes as sister to all other neognaths (Cracraft and Mindell, 1989; Sibley and Ahlquist, 1990). However, if the root were placed along the galliform or anseriform lineages *Rhea* would be sister to the two neognaths. Long branch attraction involving a distantly related outgroup, as in our case, is a strong possibility, and we look for inclusion of additional ingroup and outgroup taxa to help bisect long branches in future studies. Nonetheless, the basal position of Passeriformes and sister relationship for Anseriformes/Galliformes and a paleognath (Fig. 8.4B) is the most parsimonious explanation for this large and relatively conserved mt data set, and it is not our intention at present to dismiss this topology as a rooting anomaly. Mitochondrial 12S rDNA analyses discussed below strongly support Passeriform monophyly and a basal position for Passeriformes among an expanded set of neognath taxa.

B. Relationships among Neognath Orders

Rooting is also a critical issue for resolving relationships among traditional neognath orders. If Passeriformes is basal among birds, as suggested by Fig. 8.4B, then crocodilians are the appropriate outgroup. However, use of such an early divergent (distant) outgroup appears to be pushing the limits of historical informativeness for 12S rDNA characters alone. 12S rDNA is better suited for analysis of relationships among traditional neognath orders if an avian outgroup is used, and we provide such analyses using two anseriform taxa as an outgroup (Fig. 8.5). Use of anseriform taxa as an outgroup is inconsistent with our Fig. 8.4B topology, however, it is consistent with the conventional placement of Anseriformes/Galliformes as sister to all other neognaths (Cracraft and Mindell, 1989; Sibley and Ahlquist, 1990) and with the unrooted ingroup topology in Fig. 8.4C.

Parsimony analyses of mt 12S rDNA for all characters using two anseriforms as an outgroup for 48 birds representing 15 traditional orders were conducted using 1000 replicate searches with random addition of taxa in each replicate. Equal weighting for all characters yielded 10 equally parsimonious trees whose strict consensus resolved only 4 sets of ordinal relationships. *Opisthocomus* is sister to *Tauraco*, *Trogon* is sister to *Chordeiles*, *Falco* is sister to *Tyto*, and Passeriformes are sister to all the other neognath orders combined (Fig. 8.5A). Weighting transversions:transitions as 5:1 yielded a single tree also showing Passeriformes as basal to the other neognath orders, but showing differences from the equally weighted tree for the other taxa mentioned above (Fig. 8.5B).

We consider the 5:1 tree a better representation of the phylogenetic signal within the 12S character set, given greater levels of convergent similarity in transitions (Figs. 8.2 and 8.3). The small number of resolved nodes in Fig. 8.5A further indicates reduced phylogenetic informativeness. Relationships among falconiform and strigiform taxa and within the families Accipitridae and Strigidae (similar in both trees) are similar to the few previous studies considering those taxa, and are discussed in following sections. Separate analyses focusing on buttonquail, hoatzin, and flamingo are also discussed below.

Placement of Passeriformes as basal to the other neognath taxa, based on both the large 14-gene mt data set (Fig. 8.4) and the 12S rDNA data set including more taxa (Fig. 8.5), has not been indicated in previous analyses, and was unexpected. The fossil record for passeriform birds dates back only to the upper Oligocene, about 25 MYA (e.g., Mourer-Chauviré *et al.*, 1989), and the earlier appearance in the fossil record of other modern bird lineages (see Olson, 1985; Feduccia, 1996)

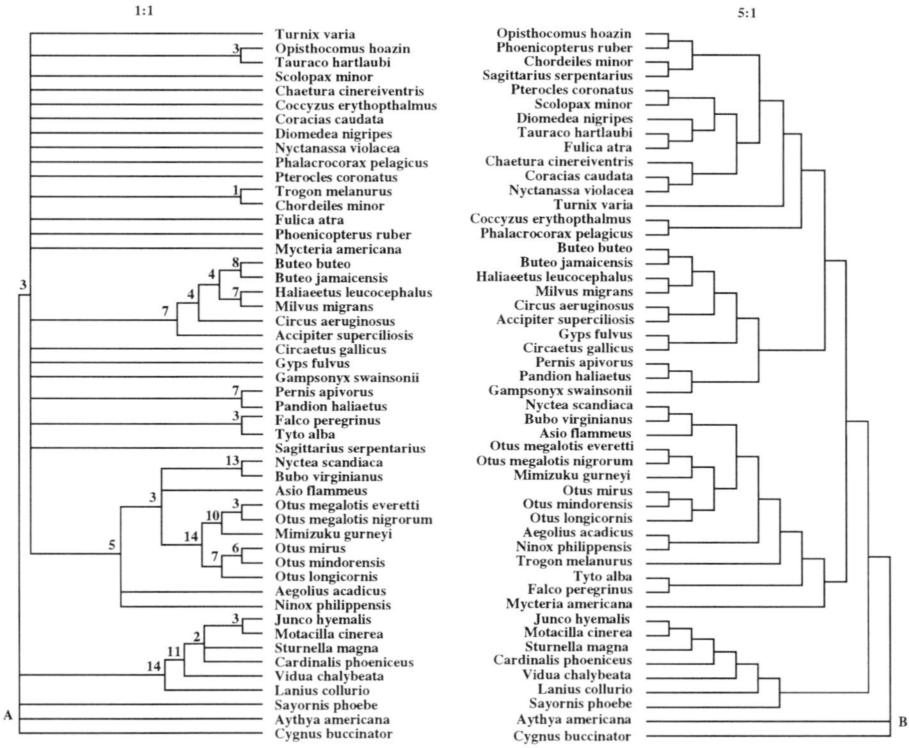

FIGURE 8.5 Phylogenetic hypotheses based on 859 12S mt rDNA sequence characters from 48 (in-group) bird species representing 15 traditional orders, using *Cygnus* and *Aythya* as outgroup taxa. (A) Strict consensus of 10 equally parsimonious trees of 2020 steps based on equal weighting for all characters in 1000 replicate searches with random addition of taxa. (B) Single most parsimonious tree of 3760 steps based on a weighting transversion-to-transition ratio of 5 : 1 in 1000 replicate searches with random addition of taxa. A maximum of 10 trees were saved and branch-swapped per replicate search. Numbers on branches denote support indices.

argues against a basal position for Passeriformes among extant avian lineages. There remains, however, a lack of fossil evidence for the passerine radiation, and Boles (1995) has described two bones, dating back to 54 MYA, that closely resemble those of Passeriformes. Passeriform placement in our analyses may be influenced by faster rates of mtDNA sequence evolution among them. We found a tendency toward faster rates of 12S mt rDNA evolution in some Passeriformes compared to other birds (Table II), and this could result in a long-branch attraction (Felsenstein, 1978) with Passeriformes drawn basally in phylogenetic hypotheses owing to greater convergent similarity at numerous sites with crocodilian (Fig. 8.4) or anseriform (Fig. 8.5) outgroups. Emphasis on relatively conserved, slowly evolving characters for both data sets, however, works to reduce long-branch attraction, and we cannot assume that the basal position indicated for Passeriformes is an artifact of rate differences. The position of Passeriformes in DNA–DNA hybridization analyses (Sibley and Ahlquist, 1990) well within their "Neoaves," and sister to the group including their Columbiformes, Gruiformes, and Ciconiiformes, is based on a short internode and is less than conclusive.

Phylogenetic placement of sandgrouse (represented by *Pterocles*) has been controversial, with most debate focusing on their alternative placement with Columbiformes (pigeons and doves; e.g., Wetmore, 1960; Mayr and Amadon, 1951) or Charadriiformes (shorebirds, gulls, and alcids; e.g., MacLean, 1969; Sibley and Ahlquist, 1990). We find *Pterocles* placed as sister to a charadriiform (*Scolopax*) in Fig. 8.5B (and sister to a clade including *Scolopax* in Fig. 8.6); however, we do not have sequence data for a columbiform and cannot test a sister relationship directly for *Pterocles* and Columbiformes relative to Charadriiformes. Lack of resolution for *Pterocles* (and many other taxa) in Fig. 8.5A suggests that the *Pterocles* lineage is a relatively old one in which multiple substitutions at individual nucleotide positions have diminished historical signal based on all characters equally weighted.

Various higher level relationships based on single species representatives (*Diomedia*, *Tauraco*, *Fulica*, *Chaetura*, *Coracias*, *Nyctanassa*, *Coccyzus*, and *Phalacrocorax*) of different orders, indicated in Fig. 8.5B, are largely unexpected on the basis of previous studies. Use of single species representatives for divergent orders is particularly subject to confounding effects of convergent similarity and misdiagnosis of derived characters as ancestral, and we are skeptical of the indicated relationships. More sampling of taxa is needed. The nonsister relationship of *Tauraco* (family Musophagidae) and *Coccyzus* (family Cuculidae) in Fig. 8.5 is consistent with the view of polyphyly for traditional Cuculiformes and the findings of Sibley and Ahlquist (1990).

C. Placement of Buttonquail, Hoatzin, and Flamingoes

Phylogenetic relationships for buttonquail and for the hoatzin (*Opisthocomus hoazin*) have been controversial. They overlap in their combined set of hypothesized relatives, so will be considered concomitantly here. Buttonquail (Turnicidae), also known as hemipodes, are small, running birds, similar in appearance to Old World quail. They exhibit several unique features, including reversed sexual dimorphism, a tendency toward polyandrous nesting habits, and, in some species, lack of a hind toe. They have variously been placed as Galliformes, Gruiformes, or as the sole family within Turniciformes. Analyses by Sibley and Ahlquist placed them as basal to all neognaths. They viewed this as an artifact of evolutionary rate differences among taxa and suggested that buttonquail are as likely to be members of Gruiformes as of any other order, although they placed the buttonquail in a separate order in their classification.

The hoatzin is a slender, pheasant-like bird, with bare, blue skin on the face, that nests communally (with the aid of nest-helpers). Morphological features include a large two-part crop used in fermentation and digestion of cellulose, and the presence of two claws on the wings of nestlings. The hoatzin has been variously placed together with turacos, cuckoos, and Galliformes (reviewed in Cracraft, 1981; Sibley and Ahlquist, 1990; Hedges *et al.*, 1995).

Our analyses of mt 12S rDNA emphasizing relatively conserved characters indicate that the hoatzin is most closely related to cuckoos (as represented by *Coccyzus*), and not turacos or Galliformes (Fig. 8.6). This is in agreement with other molecular analyses (Hedges *et al.*, 1995; Sibley and Ahlquist, 1972, 1990). We find this inference based on both transversion parsimony and a transversions:transitions weighting ratio of 5:1. Equal weighting for all characters yielded the following strict consensus topology for three equally parsimonious trees with 589 steps:

(((((*Scolopax*, *Fulica*), *Pterocles*), *Turnix*, *Opisthocomus*, *Tauraco*), *Gallus*), *Coccyzus*).

Our analyses show a buttonquail (*Turnix*) to be more closely related to a gruiform (*Fulica*) than to a galliform. However, *Turnix* is sister to an *Opisthocomus*/*Coccyzus* clade, and not to the gruiform. Relative rate comparisons suggest that *Turnix* mt 12S rDNA has not been changing at a rate significantly different from that of other birds in our sample (e.g., Table II). *Turnix* appears to represent a distinctive group of uncertain affinity. *Turnix* does not appear basal among other neognath orders (Fig. 8.5), which is inconsistent with DNA–DNA hybridization findings. Based on 12S rDNA transversions and the taxa set in Fig. 8.6, nine additional steps are required for the shortest tree uniting *Turnix* and *Gallus* as sisters, and four additional steps are required to unite *Turnix* and *Fulica* as sisters.

Flamingoes (Phoenicopteridae) present a mosaic of morphological features seen in disparate orders including webbed feet, lamellate bills, long legs, and long necks. Flamingoes have been considered variously as Anseriformes, Ciconiiformes, Charadriiformes, or equally closely related to some subset of these three groups (reviewed in Sibley and Ahlquist, 1990; Olson and Fedducia, 1980). Mitochondrial 12S rDNA analyses suggest that *Phoenicopterus* is more closely related to cormorants (*Phalacrocorax*), herons (*Nyctanassa*), and storks (*Mycteria*) than to Anseriformes and Charadriiformes based on both transversions only, and a transversions:transitions weighting ratio of 5:1 (Fig. 8.7). Based on transversions only, three additional steps are required to place *Scolopax* as sister to *Phoenicopterus*, and to place *Aythya* as sister to *Phoenicopterus*. Exhaustive searches based on transversions and on a weighting ratio of 5:1, having *Diomedea*, *Nyctanassa*, and *Phalacrocorax* excluded, placed *Phoenicopterus* and *Mycteria* as sisters, with *Scolopax* sister to them. The transversion parsimony support indices and branch lengths are small; however, they represent our most conservative set of characters for application to this question.

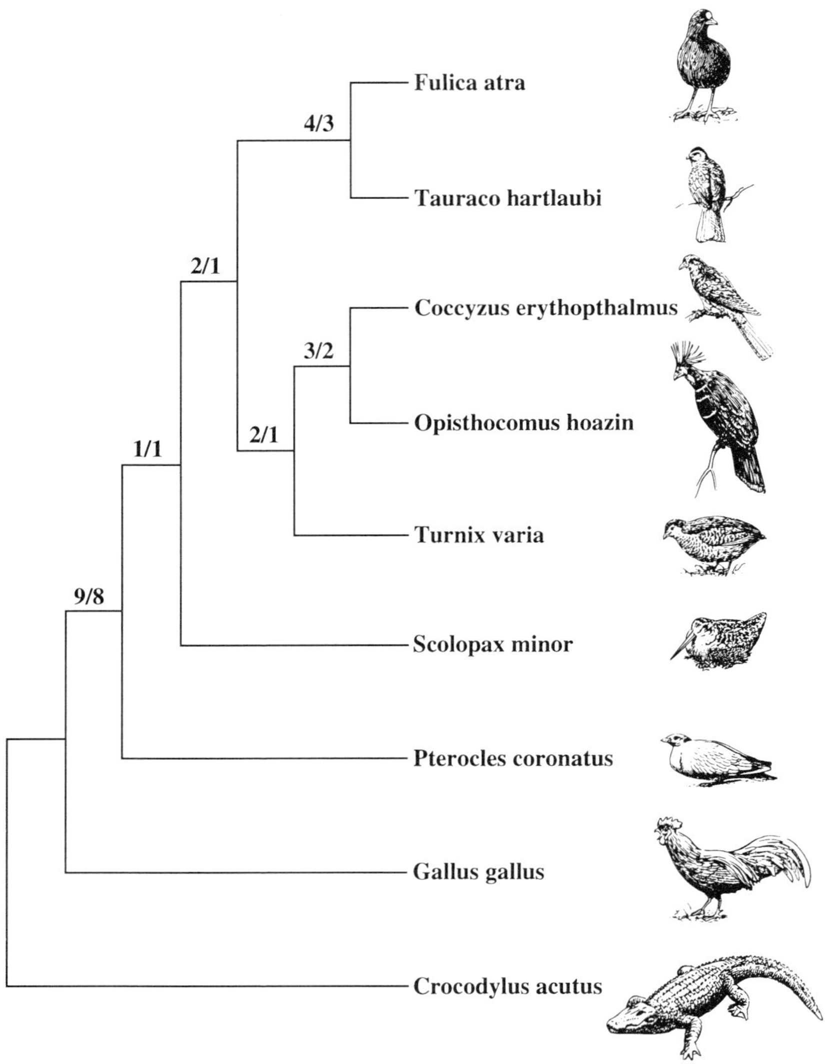

FIGURE 8.6 Single most parsimonious tree (163 steps) based on an exhaustive search of mitochondrial 12S rDNA transversions only, indicating phylogenetic position of a buttonquail (*Turnix varia*) and the hoatzin (*Opisthocomus hoazin*) among traditionally hypothesized relatives. Numbers on branches denote branch lengths/support indices. Analysis using a transversions:transitions weighting ratio of 5:1 yielded a single most parsimonious tree (1257 steps) with the same topology as above, except *Pterocles* is placed basal to *Turnix*.

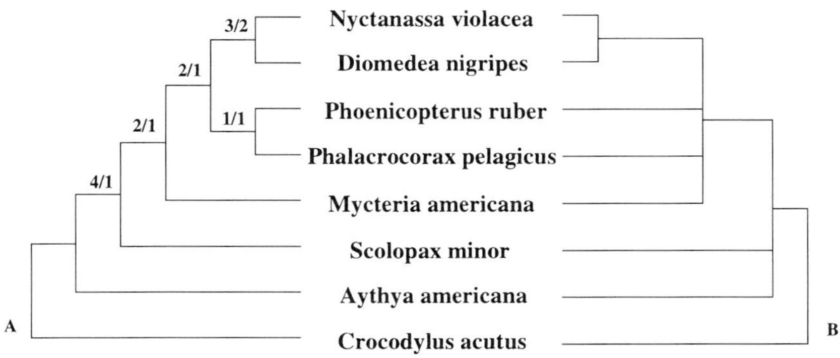

FIGURE 8.7 Phylogenetic position of a flamingo (*Phoenicopterus ruber*) relative to disparate taxa traditionally proposed as relatives, based on exhaustive parsimony searches of mt 12S rDNA (A) transversions only (126 steps), and (B) a transversions:transitions weight ratio of 5 : 1 (consensus of two trees requiring 1011 steps). Numbers in (A) denote branch lengths/support indices.

D. Relationships among Passeriformes

Our study taxa include seven passeriform species, and hypothesized relationships among them based on mt 12S rDNA (Fig. 8.5) are consistent with many previous analyses, showing an early divergence between suboscines (represented by *Sayornis*) and oscines and a basal divergence for *Lanius* relative to the other study oscines. Placement of *Motacilla* and *Junco* as sisters is unexpected, as *Junco* and *Cardinalis* are considered closer based on morphology and are members of the New World nine-primaried oscine family Emberizidae (Paynter, 1970) or subfamily Emberizinae (Sibley and Ahlquist, 1990). *Motacilla*, an Old World oscine in the family Motacillidae (Mayr and Greenway, 1960) or Passeridae (Sibley and Ahlquist, 1990), shows a significantly faster rate of mt 12S rDNA evolution compared to five of the six other oscines (comparisons not shown), and this may have influenced phylogenetic placement. A sister relationship between *Motacilla* and *Vidua* would be consistent with DNA–DNA hybridization analyses and existing classifications (Sibley and Ahlquist, 1990).

E. Relationships among Anseriformes and Galliformes

Our 12S rDNA analyses indicate monophyly for Galliformes and Anseriformes based on the set of 20 taxa analyzed (Fig. 8.8). The phylogenetic hypothesis shows sister relationships for swans and geese, and for swans and geese with an unresolved set of five diverse ducks (from the traditional subfamily Anatinae). Sister to all of

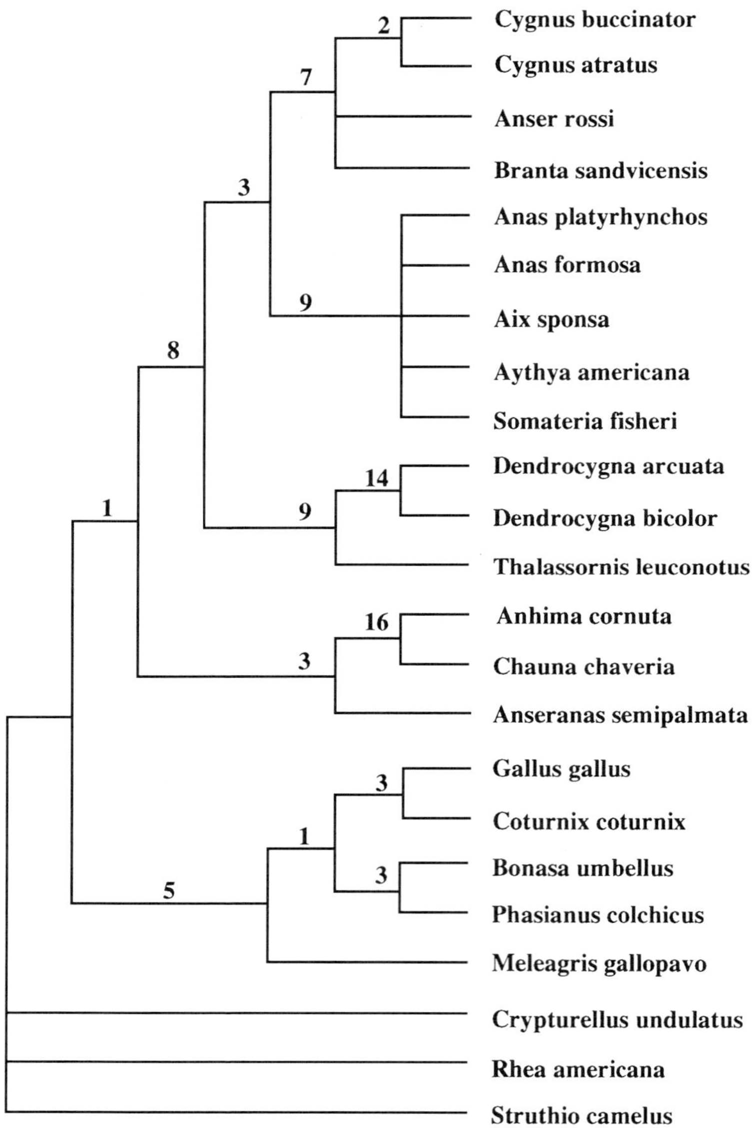

FIGURE 8.8 Strict consensus of four equally parsimonious trees of 625 steps based on equal weights for all mt 12S rDNA characters for 20 anseriform and galliform species, using 3 paleognaths as an outgroup. Random addition sequences for taxa were used in 1000 replicate searches. Numbers on branches denote support indices. Analysis using a transversions:transitions weighting of 5 : 1 (not shown) yielded a strict consensus of 5 trees requiring 1149 steps with the same topology shown, except that the only node resolved within galliforms was a sister relationship for *Bonasa* and *Phasianus*.

these is a clade including two whistling ducks (*Dendrocygna*) and the white-backed duck (*Thalassornis*), and sister to all the above anseriforms is a clade including two screamers (*Anhima, Chauna*; family Anhimidae) and the magpie goose (*Anseranas*). The whistling ducks have traditionally been considered sister to the geese and swans within the subfamily Anserinae (Delacour and Mayr, 1945; Sibley and Ahlquist, 1990); however, our analysis suggests a basal position to both geese and swans and the Anatinae, as found by Livezey (1986). *Thalassornis* was traditionally considered as one of the stiff-tailed ducks (tribe Oxyurini; Delacour and Mayr, 1945). Behavioral and morphological characters were later identified that indicated a closer relationship between *Thalassornis* and whistling ducks (Johnsgard, 1967; Livezey, 1986), although they were still not considered as sister taxa. Our 12S analyses do place *Thalassornis* and whistling ducks as sisters. *Anseranas* has traditionally been considered as within the Anatidae (e.g., Delacour and Mayr, 1945; Sibley and Ahlquist, 1972), although Verheyen (1955) postulated a sister relationship for screamers and *Anseranas* based on skeletal morphology as we have found, and as indicated by DNA–DNA hybridization analyses of Sibley and Ahlquist (1990). DNA–DNA hybridization analyses of Madsen *et al.* (1988) and morphological analyses of Livezey (1986), however, did not support this sister relationship.

12S rDNA analyses suggest that *Meleagris* is sister to a clade including the other four galliform species, in which *Gallus* and *Coturnix* are sisters and *Bonasa* and *Phasianus* are sisters (Fig. 8.8). A basal divergence for *Meleagris* within the group is consistent with morphological classifications (e.g., Wetmore, 1960), DNA–DNA hybridization (Sibley and Ahlquist, 1990), and mt cytochrome *b* (cyt*b*) analyses (Kornegay *et al.*, 1993). Placement of *Bonasa* is unexpected given its apparent close relationship to *Meleagris* in other analyses (Sibley and Ahlquist, 1990; Ellsworth *et al.*, 1996). An exhaustive search including the five phasianid species and *Cygnus buccinator* and *Aythya americana* as an outgroup, using all characters, yielded two equally parsimonious trees, one identical to the topology in Fig. 8.8, and the other with *Meleagris* sister to a *Bonasa, Phasianus* clade.

F. Relationships among Falconiformes and Strigiformes

Phylogenetic relationships among the primary groups of predatory birds [including owls, hawks, eagles, Old and New World vultures, falcons, caracaras, and the secretary bird (*Sagittarius serpentarius*)] have been controversial. Morphological analyses have led some to suggest that certain of these primary groups (particularly owls, falcons, hawks, and secretary bird) are no more closely related to each other than to various other orders of birds (Hudson, 1948; Jollie, 1976–1977). Strigiformes (owls) have been considered closely related to the Falconiformes (hawks, eagles, vultures, falcons, caracaras, and secretary bird) by a few researchers (i.e., Reichenow,

1913–1914; Cracraft, 1981), but only distantly related by most others (Chandler, 1916; Wetmore, 1960; Brown and Amadon, 1968; Sibley and Ahlquist, 1990; Griffiths, 1994). Correspondingly, behavioral similarities between falcons and owls, such as an absence of nest-building, killing of prey by severing neck vertebrae, and holding of food in one claw have been variously considered as either shared derived or convergent traits. The cursorial, snake-hunting secretary bird is most often considered a specialized accipitrid (hawks and eagles), although character support for this relationship is limited. Several researchers have noted superficial similarities between secretary bird and cariamids within the Gruiformes (e.g., Mayr and Amadon, 1951). The osprey (*Pandion haliaetus*) traditionally comprises the family Pandionidae and is placed closest to Accipitridae. Considerable evidence indicates that New World vultures are actually Ciconiiformes (e.g., Friedmann, 1950; Ligon, 1967; Sibley and Ahlquist, 1990), although this is not supported by analyses of syringeal morphology (Griffiths, 1994; see also Seibold and Helbig, 1995). We have no sequence data for New World vultures at present, and cannot address this last issue.

In their analyses of falconiform lineages, Sibley and Ahlquist (1990; p. 486) say the "positions of these [melting] curves are probably due as much to the different rates of DNA evolution as they are to times of divergence." In light of this, their phylogenetic placement of these taxa rests entirely on the unspecified "corrections" applied, and it is difficult to assess the evidential basis of their hypotheses.

Our analyses of mt 12S rDNA are inconsistent with traditional configurations of Falconiformes and Strigiformes. For analyses using both 1:1 and 5:1 weighting ratios of transversions:transitions, *Falco* and *Tyto* are sister taxa, joined basally to a clade of 11 owls in the family Strigidae (Fig. 8.9). A sister relationship for Strigiformes and Caprimulgiformes (represented by *Chordeiles*), as suggested by some (e.g., Sibley and Ahlquist, 1990) is not supported by the 12S characters. 12S analyses do not place secretary bird any closer to the accipitrids or *Falco* than to taxa representing other orders. Nor does 12S mt rDNA support a sister relationship between secretary bird and a gruiform (*Fulica*). Exhaustive parsimony searches (not shown) for *Fulica*, *Sagittarius*, *Buteo*, and *Falco* with *Aythya* as an outgroup yielded the ingroup topology:

(((*Buteo*, *Falco*) *Sagittarius*) *Fulica*) for all characters and an unresolved trichotomy
((*Buteo*, *Falco*, *Sagittarius*) *Fulica*) for weight ratios of 5:1 and 1:0.

Placement of *Sagittarius* may be influenced by relatively fast rates of 12S sequence change (Table II) and potential long-branch attraction.

We have 518 bases of mt COI available for 22 of the 29 taxa in the above 12S analyses (Fig. 8.9) and combined them in further parsimony searches. The total evidence analysis for 22 ingroup taxa based on equal weighting for all characters indicates a sister relationship for *Falco* and *Sagittarius*, and an unresolved polytomy for those two taxa and the other primary lineages (Fig. 8.10A). Use of unequal weighting (5:1 for 12S and codon positions 1 and 2 only for COI) indicates an unresolved polytomy for three lineages: *Falco*, *Sagittarius*, and accipitrids plus *Pan-*

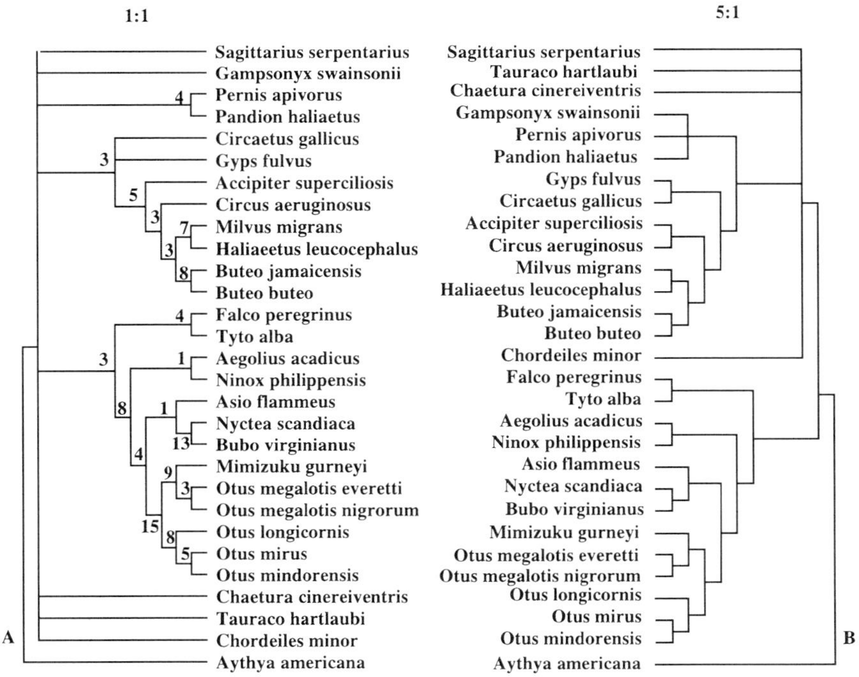

FIGURE 8.9 Phylogenetic hypotheses based on 1000 replicate searches with random addition of taxa for 12S mt rDNA for 28 (ingroup) bird species focusing on falconiform and strigiform taxa. *Aythya americana* was the designated outgroup. (A) Strict consensus of 3 equally parsimonious trees of 1119 steps based on a transversions:transitions weighting ratio of 1 : 1. Numbers denote support indices. (B) Strict consensus of 2 equally parsimonious trees of 2092 steps, based on a transversions:transitions weighting ratio of 5 : 1.

dion (Fig. 8.10B). This unites the lineages traditionally included in Falconiformes and we consider this our current best estimate of their phylogenetic relationships.

1. *Relationships within Accipitridae*

The 208 species in the cosmopolitan family Accipitridae represent the largest radiation of diurnal birds of prey. Subgroups that have been recognized based on osteology, myology, plumage, and behavior include milvine and nonmilvine kites, sea eagles, Old World vultures, snake eagles, accipiters, chanting goshawks, harriers, booted eagles, buteos (or buzzards), subbuteos, and harpy eagles (Brown and Amadon, 1968). However, phylogenetic relationships among the 60 or so accipitrid genera are little known (Amadon, 1982). Jollie (1976–1977) professed an inability to identify either derived features or character transformation series from his detailed anatomical studies. Jollie and others have attributed this to the extreme specializations found in most avian predators and subsequent difficulty in distinguish-

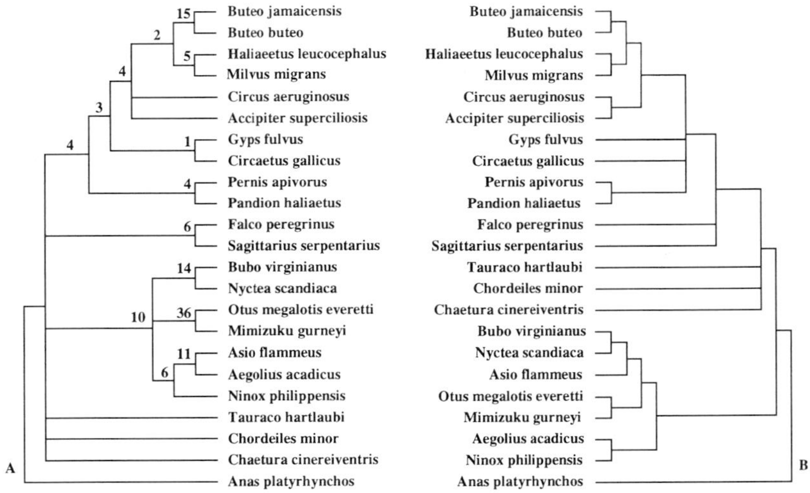

FIGURE 8.10 Phylogenetic hypotheses based on combined 12S mt rDNA (859 characters) and mt COI sequences (518 characters) for 22 (ingroup) bird species, focusing on falconiform and strigiform taxa. *Anas platyrhynchos* was the designated outgroup. One thousand replicate searches with random addition of taxa were conducted for (A) all characters equally weighted (strict consensus of 6 trees requiring 1908 steps) and (B) 12S characters weighted with a transversions:transitions ratio of 5:1 and COI third-codon positions given a weight of 0 (strict consensus of 4 trees requiring 1930 steps). Numbers on branches in (A) denote support indices.

ing results of common ancestry and convergence. The only previous molecular study has been based on DNA–DNA hybridization and included a small number of accipitrid groups (Sibley and Ahlquist, 1990).

We present two analyses focusing on accipitrid relationships using *Sagittarius* and *Falco* as outgroup taxa in exhaustive searches. The first is based on 12S rDNA and includes 11 ingroup taxa (Fig. 8.11), and the second is based on 12S rDNA and COI for 10 ingroup taxa (Fig. 8.12). These analyses are largely consistent with those based on all falconiform and strigiform taxa (Figs. 8.9 and 8.10) and may be summarized as follows. The two *Buteo* species are sisters as expected. A sister relationship for *Milvus* and *Haliaeetus* is consistent with the hypothesis of close relationship between the sea eagles and the milvine kites postulated by others based on aspects of morphology and behavior (Brown and Amadon, 1968; Amadon, 1982; Olson, 1982). We found consistent support for a sister relationship between the *Buteos* and the *Milvus/Haliaeetus* clade (Figs. 8.9–8.12), suggesting that the milvine kites and sea eagles may not be basal among accipitrids as they have been considered to be based on morphology and behavior (Amadon, 1982; but see Griffiths, 1994). Moving toward the base of the combined data set topology (Fig. 8.10), *Circus* and *Accipiter* are weakly supported as sisters, followed by a *Gyps/Circaetus* clade and a

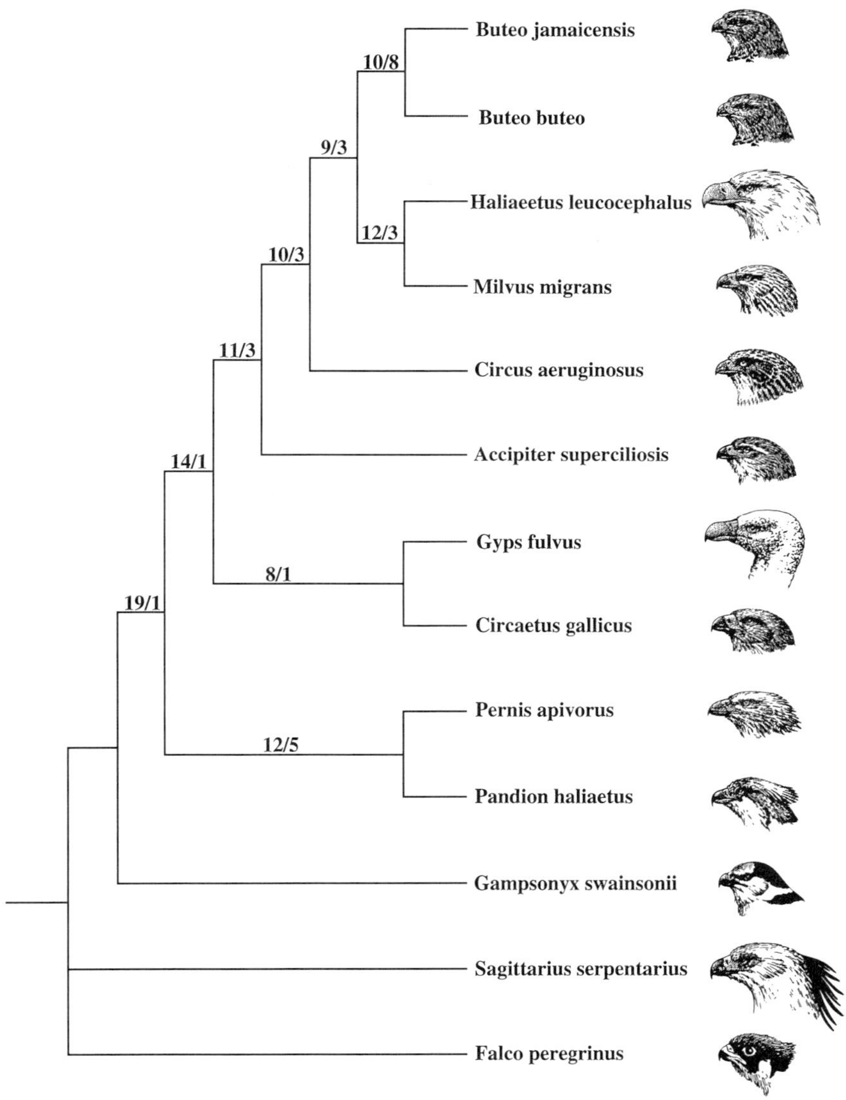

FIGURE 8.11 Single most parsimonious tree (475 steps) based on mitochondrial 12S rDNA using a transversions:transitions weighting ratio of 1:1 for 11 Accipitridae (ingroup) species with *Falco* and *Sagittarius* as an outgroup. Numbers on branches denote branch lengths/support indices.

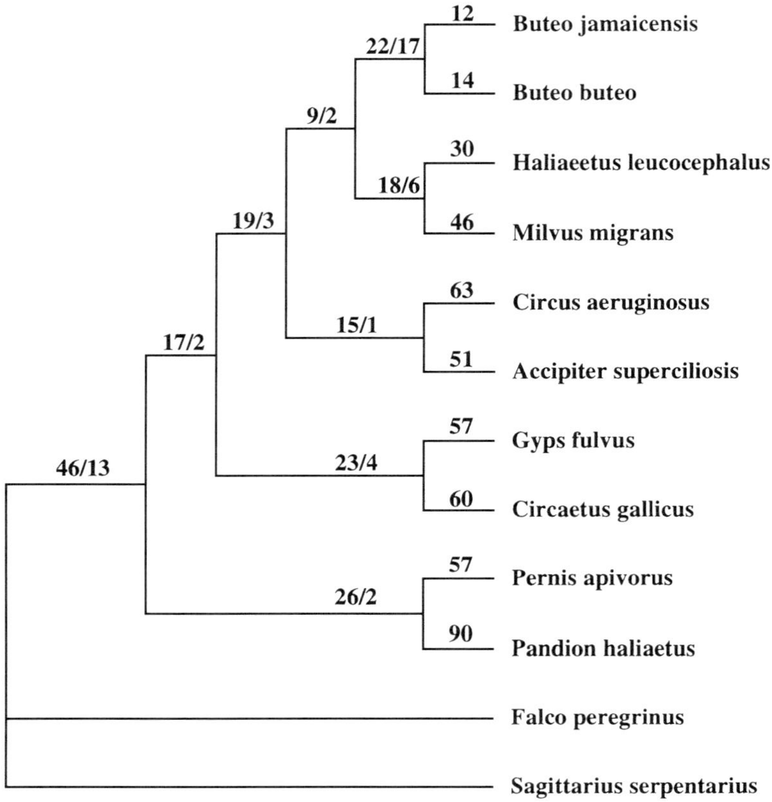

FIGURE 8.12 Single most parsimonious tree (858 steps) based on mt 12S rDNA and mt COI sequences combined with all characters equally weighted for 10 (ingroup) Accipitridae species with *Falco* and *Sagittarius* as an outgroup. Numbers on branches denote branch lengths/support indices.

Pernis/Pandion clade. Based on 12S rDNA characters *Gampsonyx* appears as either a basal unresolved lineage (Fig. 8.9A), together with *Pernis* and *Pandion* (Fig. 8.9B), or basal among accipitrids and *Pandion* (Fig. 8.11).

The kites comprise a diverse set of accipitrids, and suspected primitive features, thought to be retained from less predatory ancestors, include mild predatory habits and reduced sexual dimorphism in many species (Brown and Amadon, 1968). Friedmann (1950) divided kites into three groups, represented by *Milvus, Pernis,* and *Gampsonyx* in our study taxa, which are clearly polyphyletic in our analyses (Figs. 8.9–8.12). Analyses of syringeal morphology also indicate polyphyly of the kites (Griffiths, 1994). *Pandion*, the osprey, although generally considered as comprising a separate family (Pandionidae) based on distinctive morphological features, is consistently shown as sister to *Pernis* (Figs. 8.9–8.12).

Greater topological similarity of Fig. 8.11 (12S, 1:1) with Fig. 8.9B (12S, 5:1) than with Fig. 8.9A (12S, 1 : 1) is likely due to reduction of the confounding effects of homoplasy in using outgroups that are more closely related to the ingroup taxa.

2. Relationships within Strigidae

The Strigidae, or recent owls (Amadon and Bull, 1988), includes two primary clades. The barn owls and bay owls (subfamily Tytoninae) comprise one, and all the other owls (the "typical owls"; subfamily Striginae) comprise the second. A variety of classifications (e.g., Ford, 1967) have been presented for Striginae taxa, although few phylogenetic analyses have been conducted.

Our hypothesized relationships for strigids are identical in Figs. 8.9A and B, and 8.10B. The only conflicting topology (Fig. 8.10A) involves a different placement for *Asio* which may be influenced by inclusion of rapidly evolving COI third-codon positions. We found monophyly of the scops owls (*Otus*, *Mimizuku*), a close relationship between the scops owls and a *Nyctea/Bubo* clade, and a sister relationship for *Aegolius* and *Ninox*, which are, in turn, sister to the others. *Tyto* is consistently placed outside the Striginae taxa. Relationships for the seven Striginae genera common to our analyses and DNA–DNA hybridization analyses (Sibley and Ahlquist, 1990) are congruent with the exception that our analyses show *Aegolius* and *Ninox* as sisters, rather than unresolved, and *Asio* as sister to the *Nyctea/Bubo* clade rather than unresolved.

V. CONCLUSIONS

Phylogenetic analyses presented here based on mitochondrial DNA characters address a number of controversial issues. Anseriformes and Galliformes are supported as sister taxa that are more closely related to a paleognath (*Rhea*) than to a set of neognaths. Placement of the root is critical in this latter determination, and addition of sequences from more taxa within the avian ingroup and within the crocodilian outgroup clade (including alligatorids, crocodilids, and gavialids) may help reduce potential attraction among long branches. It is doubtful, however, that greater sampling of extant forms will ever eliminate the problem entirely. Molecular and morphological characters from extinct lineages arising from the phylogenetic internode between crocodilians and birds could be useful in rooting phylogenies, if any such characters become available. Use of duplicated gene sequences as outgroups, where duplications occur prior to diversification events, has been demonstrated by Iwabe *et al.* (1989). This approach could also be useful in studies of avian phylogeny, if gene duplications (or ex-mt nuclear genes) are found that predate divergences among extant avian lineages and postdate the split between birds and crocodilians.

In analyses of two different data sets, Passeriformes are indicated as basal (1) among five lineages representing the oldest divergences among extant birds, and (2) among a set of neognaths. Basal placement among neognaths uses an avian outgroup, which helps reduce the potential effect of a long branch introduced with a crocodilian outgroup. However, this makes the assumption that Passeriformes are not basal to the outgroup Anseriformes. Our rooted phylogeny in Fig. 8.4B based on more than 13 kb of mitochondrial sequence is inconsistent with the sequence of appearance of fossil forms and with previous molecular analyses of Sibley and Ahlquist (1990). It should be remembered, however, that DNA–DNA hybridization distance analyses by Sibley and Ahlquist are entirely unrooted, with the earliest divergence among extant birds being based on a form of midpoint rooting and unsupported assumptions of evolutionary rate homogeneity.

Our evidence indicates that a buttonquail is more closely related to a gruiform than to a galliform. However, the buttonquail is sister to a cuckoo/hoatzin clade, and not sister to the gruiform. Hoatzin appears most closely related to cuckoos (as represented by *Coccyzus*), and not turacos or Galliformes. A flamingo (*Phoenicopterus*) appears more closely related to cormorants, herons, and storks than to either Anseriformes or Charadriiformes. The magpie goose (*Anseranas*) is supported as being sister to screamers (Anhimidae) rather than sister to the other waterfowl (Anatidae). Based on conserved mt 12S rDNA and COI characters combined, we found Falconiformes to be monophyletic with a polytomy for three primary lineages: Accipitridae species, secretary bird, and a representative of Falconidae. Kites are found to be polyphyletic, with osprey (*Pandion*) being placed within an accipitrid clade (as sister to *Pernis*) rather than sister to all accipitrids.

Limitations of our analyses of avian phylogeny are common to many others involving higher level taxa and rapid radiations of forms. Greater sampling of taxa within diverse lineages (different orders) of birds is needed, as inclusion of only one or a few taxa from distantly related lineages may work to increase prevalence of long branches. We suspect that if all orders within our study taxa were as well represented as those of Falconiformes and Strigiformes there would be fewer polytomies denoting ordinal relationships and greater stability of topology based on alternative character weighting approaches. Despite more complete and more balanced sampling of taxa, some relationships are likely to remain poorly resolved, particularly where time elapsed between divergences is small or where divergences are concomitant. Ideally, systematists would like to find characters that experienced significant change during the radiations, but little thereafter, and such a pattern may occur more often for morphological characters associated with selection and cladogenetic events (Lanyon, 1988; Olmstead *et al.*, 1990). Collection of additional morphological and molecular characters for use in combined analyses may help in this regard. Numerous studies indicate increased resolving power of larger character data sets (e.g., Hillis *et al.*, 1994; Charleston *et al.*, 1994; Cummings *et al.*, 1995; Mindell and Thacker, 1996). Further analyses of the sequence data presented here are needed, using alternative 12S rDNA alignments, and iterative weighting approaches, and these are likely to yield some differences in phylogenetic inference.

Ultimately, enhanced phylogenetic resolution requires comprehensive analyses of increased numbers of taxa and characters. The primary challenge for systematists, beyond data collection, is determining how to conduct the comprehensive analyses in light of increased understanding of the varied constraints on character evolution.

ACKNOWLEDGMENTS

We thank Jennifer Ast, Derek Dimcheff, Emily Malvitz, and Karen Ober for assistance with laboratory work. Helpful discussion or comments on earlier drafts were provided by Jennifer Ast, Derek Dimcheff, and Robert Payne. We thank the University of Tel Aviv Research Zoological Garden, Louisiana State University Museum of Natural Science, University of Minnesota Raptor Rehabilitation Program, the University of Michigan Museum of Zoology Bird Division, and Robert Payne for help in obtaining tissue samples for analysis. David Swofford kindly provided us with a prerelease version of PAUP (versions 4d48–52) for testing. We thank John Megahan for his bird illustrations used in the figures. This work was funded by grants from the National Science Foundation to D.P.M. (DEB-9496343, 9019669). and by contributions from the Swales and Fargo Funds of the University of Michigan Museum of Zoology Bird Division. M.D.S. was supported by a National Science Foundation grant (ISBN-9412399) to R. B. Payne.

REFERENCES

Adachi, J., and Hasegawa, M. (1992). MOLPHY: Programs for molecular phylogenetics. I. PROTML: maximum likelihood inference of protein phylogeny. *In* "Computer Science Monographs," No. 27. Institute of Statistical Mathematics, Tokyo.

Allard, M. W., and Miyamoto, M. M. (1992). Testing phylogenetic approaches with empirical data, as illustrated with the parsimony method. *Mol. Biol. Evol.* **9**, 778–786.

Amadon, D. (1982). A revision of the sub-buteonine hawks (Accipitridae, Aves). *Am. Mus. Novitates* **2741**, 1–20.

Amadon, D., and Bull, J. (1988). Hawks and owls of the world. *Proc. West. Found. Vert. Zool.* **3**, 295–357.

Arctander, P. (1995). Comparison of a mitochondrial gene and a corresponding nuclear pseudogene. *Proc. R. Soc. Lond. B* **262**, 13–19.

Atchley, W. R., and Fitch, W. M. (1991). Gene trees and the origins of inbred strains of mice. *Science* **254**, 554–558.

Benton, M. J. (1990). Phylogeny of the major tetrapod groups: Morphological data and divergence dates. *J. Mol. Evol.* **30**, 409–424.

Boles, W. E. (1995). A preliminary analysis of the Passeriformes from Riversleigh, northwestern Queensland, Australia, with the description of a new species of lyrebird. *Courier Forsch. Inst. Senckenberg* **181**, 163–170.

Bremer, K. (1988). The limits of amino-acid sequence data in angiosperm phylogenetic reconstruction. *Evolution* **42**, 795–803.

Britten, R. J. (1986). Rates of DNA sequence evolution differ between taxonomic groups. *Science* **231**, 1393–1398.

Brodkorb, P. (1964). Catalogue of fossil birds. Part 2 (Anseriformes through Galliformes). *Bull. Flor. State Mus. Biol. Sci.* **8**, 195–335.

Brown, L., and Amadon, D. (1968). "Eagles, Hawks, and Falcons of the World." McGraw-Hill, New York.

Bull, J. J., Huelsenbeck, J. P., Cunningham, C. W., Swofford, D. L., and Waddell, P. J. (1993). Partitioning and combining data in phylogenetic analysis. *Syst. Biol.* **42**, 384–397.

Cao, Y., Adachi, J., and Hasegawa, M. (1994a). Eutherian phylogeny as inferred from mitochondrial sequence data. *Jpn. J. Genet.* **69**, 455–472.

Cao, Y., Adachi, J., Janke, A., Pääbo, S., and Hasegawa, M. (1994b). Phylogenetic relationships among eutherian orders estimated from inferred sequences of mitochondrial proteins: instability of a tree based on a single gene. *J. Mol. Evol.* **39**, 519–527.

Carroll, R. L. (1988). "Vertebrate Paleontology and Evolution." W. H. Freeman, New York.

Chandler, A. C. (1916). A study of the structure of feathers, with reference to their taxonomic significance. *Univ. Calif. Publ. Zool.* **13**, 243–446.

Charleston, M. A., Hendy, M. D., and Penny, D. (1994). The effects of sequence length, tree topology and number on the performance of phylogenetic methods. *J. Comput. Biol.* **1**, 133–151.

Chippindale, P. T., and Wiens, J. J. (1994). Weighting, partitioning, and combining characters in phylogenetic analysis. *Syst. Biol.* **43**, 278–287.

Cracraft, J. (1981). Toward a phylogenetic classification of birds of the world (class Aves). *Auk* **98**, 681–714.

Cracraft, J. (1982). Phylogenetic relationships and transatlantic biogeography of some gruiform birds. *Geobios, mémoire spécial.* **6**, 393–402.

Cracraft, J. (1987). DNA hybridization and avian systematics. *In* "Evolutionary Biology" (M. K. Hecht, B. Wallace, and G. T. Prance, Eds.), Vol. 21, pp. 47–96. Plenum, New York.

Cracraft, J. (1988). The major clades of birds. *In* "The Phylogeny and Classification of the Tetrapods" (M. J. Benton, Ed.), Systematics Assoc. Special Vol. No. 35A, pp. 333–355. Clarendon Press, Oxford.

Cracraft, J., and Mindell, D. P. (1989). The early history of modern birds: A comparison of molecular and morphological evidence. *In* "The Hierarchy of Life" (B. Fernholm, K. Bremer, and H. Jörnvall, Eds.), Proc. of Nobel Symposia, pp. 389–403. Elsevier Science Publishers, Amsterdam.

Cummings, M. P., Otto, S. P., and Wakeley, J. (1995). Sampling properties of DNA sequence data in phylogenetic analysis. *Mol. Biol. Evol.* **12**, 814–822.

Darwin, C. (1859). "On the Origin of Species by Means of Natural Selection, or the Preservation of Favoured Races in the Struggle for Life." Murray, London.

Delacour, J., and Mayr, E. (1945). The family Anatidae. *Wilson Bull.* **57**, 3–55.

deQueiroz, A., Donoghue, M. J., and Kim, J. (1995). Separate versus combined analysis of phylogenetic evidence. *Annu. Rev. Ecol. Syst.* **26**, 657–681.

Desjardins, P., and Morais, R. (1990). Sequence and gene organization of the chicken mitochondrial genome. *J. Mol. Biol.* **212**, 599–634.

Desjardins, P., and Morais, R. (1991). Nucleotide sequence and evolution of coding and noncoding regions of a quail mitochondrial genome. *J. Mol. Evol.* **32**, 153–161.

Eernisse, D. J., and Kluge, A. G. (1993). Taxonomic congruence versus total evidence, and amniote phylogeny inferred from fossils, molecules, and morphology. *Mol. Biol. Evol.* **10**, 1170–1195.

Ellsworth, D. L., Honeycutt, R. L., and Silvy, N. J. (1996). Systematics of grouse and ptarmigan determined by nucleotide sequences of the mitochondrial cytochrome-*B* gene. *Auk* **113**, 811–822.

Farris, J. S. (1966). Estimation of conservatism of characters by constancy within biological populations. *Evolution* **20**, 587–591.

Farris, J. S. (1969). A successive approximations approach to character weighting. *Syst. Zool.* **18**, 374–385.

Farris, J. S. (1983). The logical basis of phylogenetic analysis. *In* "Advances in Cladistics" (N. W. Platnick and V. A. Funk, Eds.), pp. 7–36. Columbia Univ. Press, New York.

Feduccia, A. (1995). Explosive evolution in Tertiary birds and mammals. *Science* **267**, 637–638.

Feduccia, A. (1996). "The Origin and Evolution of Birds." Yale Univ. Press, New Haven, Connecticut.

Felsenstein, J. (1978). Cases in which parsimony or compatibility methods will be positively misleading. *Syst. Zool.* **27**, 401–410.

Ford, M. L. (1967). "A Systematic Study of the Owls Based on Comparative Osteology." Unpublished Ph.D. dissertation. University of Michigan, Ann Arbor, Michigan.

Friedmann, H. (1950). Birds of North and Middle America, Falconiformes. *U.S. Natl. Mus. Bull.* **50**(part XI), 1–793.

Gill, F. B., and Sheldon, F. H. (1991). The birds reclassified. *Science* **252**, 1003–1005.

Goloboff, P. A. (1993). Estimating character weights during tree search. *Cladistics* **9**, 83–91.

Graur, D., and Higgins, D. G. (1994). Molecular evidence for the inclusion of cetaceans within the order Artiodactyla. *Mol. Biol. Evol.* **11**, 357–364.

Griffiths, C. S. (1994). Monophyly of the Falconiformes based on syringeal morphology. *Auk* **111**, 787–805.

Gutell, R. R. (1994). Collection of small subunit (16S- and 16S-like) ribosomal RNA structures: 1994. *Nucleic Acids Res.* **22**, 3502–3507.

Gutell, R. R., Gray, M. W., and Schnare, M. N. (1993). Compilation of large subunit (23S- and 23S-like) ribosomal RNA structures: 1993. *Nucleic Acids Res.* **21**, 3055–3074.

Hackett, S. J., Griffiths, C. S., Bates, J. M., and Klein, N. K. (1995). A commentary on the use of sequence data for phylogeny reconstruction. *Mol. Phylogenet. Evol.* **4**, 350–356.

Hedges, S. B., and Sibley, C. G. (1994). Molecules vs. morphology in avian evolution: The case of the "pelecaniform" birds. *Proc. Natl. Acad. Sci. U.S.A.* **91**, 9861–9865.

Hedges, S. B., Simmons, M. D., van Dijk, M. A. M., Caspers, G.-J., de Jong, W. W., and Sibley, C. G. (1995). Phylogenetic relationships of the hoatzin, an enigmatic South American bird. *Proc. Natl. Acad. Sci. U.S.A.* **92**, 11662–11665.

Hedges, S. B., Parker, P. H., Sibley, C. G., and Kumar, S. (1996). Continental breakup and the ordinal diversification of birds and mammals. *Nature (London)* **381**, 226–229.

Hennig, W. (1966). "Phylogenetic Systematics." Univ. of Illinois Press, Urbana, Illinois.

Hickson, R. E., Simon, C., Cooper, A., Spicer, G. S., Sulliman, J., and Penny, D. (1996). Conserved sequence motifs, alignment and secondary structure for the third domain of animal 12S rRNA. *Mol. Biol. Evol.* **13**, 150–169.

Hillis, D. M., Huelsenbeck, J. P., and Cunningham, C. W. (1994). Application and accuracy of molecular phylogenies. *Science* **264**, 671–677.

Honeycutt, R. L., and Adkins, R. M. (1993). The higher level systematics of eutherian mammals: An assessment of molecular characters and phylogenetic hypotheses. *Annu. Rev. Ecol. Syst.* **24**, 279–305.

Houde, P. (1987). Critical evaluation of DNA hybridization studies in avian systematics. *Auk* **104**, 17–32.

Hudson, G. E. (1948). Studies on the muscles of the pelvic appendage in birds. II. The heterogeneous order Falconiformes. *Am. Midl. Nat.* **39**, 102–127.

Huelsenbeck, J. P. (1995). The performance of phylogenetic methods in simulation. *Syst. Biol.* **44**, 17–48.

Iwabe, N., Kuma, K., Hasegawa, M., Osawa, S., and Miyata, T. (1989). Evolutionary relationship of archaebacteria, eubacteria, and eukaryotes inferred from phylogenetic trees of duplicated genes. *Proc. Natl. Acad. Sci. U.S.A.* **86**, 9355–9359.

Janke, A., Feldmaier-Fuchs, G., Thomas, W. K., von Haeseler, A., and Pääbo, S. (1994). The marsupial mitochondrial genome and the evolution of placental mammals. *Genetics* **137**, 243–256.

Johnsgard, P. A. (1967). Observations on the behavior and relationships of the white-backed duck and the stiff-tailed ducks. *In* "Wildfowl Trust, 18th Annual Report (1965–1966)," pp. 98–107.

Jollie, M. (1976–1977). A contribution to the morphology and phylogeny of the Falconiformes. *Evol. Theory* **1**, 285–298; **2**, 285–300; **3**, 1–141.

Jones, D. T., Taylor, W. R., and Thornton, J. M. (1992). The rapid generation of mutation data matrices from protein sequences. *Comput. Appl. Biosci.* **8**, 25–282.

Kishino, H., and Hasegawa, M. (1990). Maximum likelihood inference of protein phylogeny and the origin of chloroplasts. *J. Mol. Evol.* **31**, 151–160.

Kjer, K. M. (1995). Use of secondary structure in phylogenetic studies to identify homologous positions: An example of alignment and data presentation from the frogs. *Mol. Phylogenet. Evol.* **4**, 314–330.

Kluge, A. G. (1989). A concern for evidence and a phylogenetic hypothesis for relationships among Epicrates (Boidae, Serpentes). *Syst. Zool.* **38**, 7–25.

Knight, R. A., and Mindell, D. P. (1993). Substitution bias, a priori weighting of DNA sequence change, and the phylogenetic position of *Azemiops feae*. *Syst. Biol.* **42**, 18–31.

Kornegay, J. R., Kocher, T. D., Williams, L. A., and Wilson, A. C. (1993). Pathways of lysozyme evolution inferred from the sequences of cytochrome *b* in birds. *J. Mol. Ecol.* **37**, 367–379.

Lanyon, S. M. (1988). The stochastic mode of molecular evolution: What consequences for systematic investigation? *Auk* **105**, 565–573.

Lanyon, S. M. (1992). Review. *Condor* **94**, 304–310.

Ligon, J. D. (1967). Relationships of the cathartid vultures. *Occas. Papers Univ. Michigan Mus. Zool.* **651**, 26 pp.

Livezey, B. C. (1986). A phylogenetic analysis of recent anseriform genera using morphological characters. *Auk* **103**, 737–754.

MacLean, G. L. (1969). The sandgrouse-doves or plovers? *J. Ornithol.* **110**, 104–107.

Maddison, W. P., and Maddison, D. R. (1992). "MacClade: Analysis of Phylogeny and Character Evolution." Sinauer, Sunderland, Massachusetts.

Madsen, C. S., McHugh, K. P., and de Kloet, S. R. (1988). A partial classification of waterfowl (Anatidae) based on single-copy DNA. *Auk* **105**, 452–459.

Mayr, E., and Amadon, D. (1951). A classification of recent birds. *Am. Mus. Novit.* **1496**, 1–42.

Mayr, E., and Cottrell, G. W. (1979). "Revision of the Work of J. L. Peters: Check-List of Birds of the World," Vol. 1. Museum of Comparative Zoology, Cambridge.

Mayr, E., and Greenway, J. C. (1960). "Check-List of Birds of the World," Vol. 9. Museum of Comparative Zoology, Cambridge.

Mindell, D. P. (1992). DNA–DNA hybridization and avian phylogeny. *Syst. Biol.* **41**, 126–134.

Mindell, D. P., and Honeycutt, R. L. (1990). Ribosomal RNA in vertebrates: Evolution and phylogenetic applications. *Annu. Rev. Ecol. Syst.* **21**, 541–566.

Mindell, D. P., and Thacker, C. E. (1996). Rates of molecular evolution: Phylogenetic issues and applications. *Annu. Rev. Ecol. Syst.* **27**, 279–303.

Mindell, D. P., Dick, C. W., and Baker, R. J. (1991). Phylogenetic relationships among megabats, microbats and primates. *Proc. Natl. Acad. Sci. U.S.A.* **88**, 10322–10326.

Mindell, D. P., Knight, A., Baer, C., and Huddleston, C. J. (1996). Slow rates of molecular evolution in birds and the metabolic rate and body temperature hypotheses. *Mol. Biol. Evol.* **13**, 422–426.

Miyamoto, M. M. (1985). Consensus cladograms and general classification. *Cladistics* **1**, 186–189.

Miyamoto, M. M., Allard, M. W., Adkins, R. M., Janecek, L. L., and Honeycutt, R. L. (1994). A congruence test of reliability using linked mitochondrial DNA sequences. *Syst. Biol.* **43**, 236–249.

Mourer-Chauviré, C., Hugueney, C., and Jonet, P. (1989). Découverte de Passeriformes dans l'Oligocene supérieur de France. *Acad. Sci. C. R. Ser. 2A* **309**, 843–849.

Nelson, G. (1994). Homology and systematics. *In* "Homology: The Hierarchical Basis of Comparative Biology" (B. Hall, Ed.), pp. 101–149. Academic Press, San Diego, California.

Ohta, T. (1993). An examination of the generation-time effect on molecular evolution. *Proc. Natl. Acad. Sci. U.S.A.* **90**, 10676–10680.

Olmstead, R. G., Jansen, R. K., Michaels, H. J., Downie, S. R., and Palmer, J. D. (1990). Chloroplast DNA and phylogenetic studies in the Asteridae. *In* "Biological Approaches and Evolutionary Trends in Plants" (S. Kawano, Ed.), pp. 119–134. Academic Press, San Diego, California.

Olson, S. L. (1982). The distribution of fused phalanges of the inner toe in the Accipitridae. *Bull. Br. Ornithol. Club* **102**, 8–12.

Olson, S. L. (1985). The fossil record of birds. *In* "Avian Biology" (D. S. Farner, J. R. King, and K. C. Parkes, Eds.), Vol. 8, pp. 79–238. Academic Press, New York.

Olson, S. L., and Fedducia, A. (1980). Relationships and evolution of flamingoes (Aves: Phoenicopteridae). *Smithsonian Contrib. Zool.* **316**, 1–73.

Paynter, R. A. (1970). "Check-List of Birds of the World," Vol. 13. Museum of Comparative Zoology, Cambridge.

Penny, D., and Hendy, M. D. (1985). Testing methods of evolutionary tree construction. *Cladistics* **1**, 266–278.

Peters, J. L. (1931–1951). "Check-List of Birds of the World," Vols. 1–7. Museum of Comparative Zoology, Cambridge.

Quinn, T. W. (1992). The genetic legacy of Mother Goose—Phylogeographic patterns of lesser snow goose *Chen caerulescens caerulescens* maternal lineages. *Mol. Ecol.* **1**, 105–117.

Reichenow, A. (1913–1914). "Die Vögel: Handbuch der systematischen Ornithologie," Vols. 1 and 2. F. Enke, Stuttgart, Germany.

Seibold, I., and Helbig, A. J. (1995). Evolutionary history of New and Old World vultures inferred from nucleotide sequences of the mitochondrial cytochrome *b* gene. *Philos. Trans. Royal Soc. Land., B, Biol. Sci.* **350**, 163–178.

Sharkey, M. J. (1989). A hypothesis-independent method of character weighting for cladistic analysis. *Cladistics* **5**, 63–86.

Sheldon, F. H., and Bledsoe, A. H. (1993). Avian molecular systematics 1970s to 1990s. *Annu. Rev. Ecol. Syst.* **24**, 243–278.

Shigenaga, M. K., Gimeno, C. J., and Ames, B. N. (1989). Urinary 8-hydroxy-2′-deoxyguanosine as a biological marker of *in vivo* oxidative DNA damage. *Proc. Natl. Acad. Sci. U.S.A.* **86**, 9697–9701.

Sibley, C. G. (1994). On the phylogeny and classification of living birds. *J. Avian Biol.* **25**, 87–92.

Sibley, C. G., and Ahlquist, J. E. (1972). A comparative study of the egg-white proteins of non-passerine birds. *Bull. Peabody Mus. Nat. Hist.* **39**, 1–276.

Sibley, C. G., and Ahlquist, J. A. (1990). "Phylogeny and Classification of Birds: A Study in Molecular Evolution," Yale University Press, New Haven, Connecticut.

Smith, A. B. (1994). Rooting molecular trees: Problems and strategies. *Biol. J. Linnean Soc.* **51**, 279–292.

Sorenson, M. D., and Fleischer, R. (1996). Multiple independent transpositions of mitochondrial DNA control region sequences to the nucleus. *Proc. Natl. Acad. Sci. U.S.A.* (in press).

Stapel, S. O., Leunissen, J. A. M., Versteeg, M., Wattel, J., and de Jong, W. W. (1984). Ratites as oldest offshoot of avian stem—evidence from alpha-crystallin A sequences. *Nature (London)* **311**, 257–259.

Sullivan, J., Holsinger, K. E., and Simon, C. (1995). Among-site rate variation and phylogenetic analysis of 12S rRNA in sigmodontine rodents. *Mol. Biol. Evol.* **12**, 988–1001.

Swofford, D. L. (1993). "PAUP: Phylogenetic Analysis Using Parsimony," version 3.1.1. Illinois Natural History Survey, Champaign, Illinois.

Swofford, D. L., Olsen, G. J., Waddell, P. J., and Hillis, D. M. (1996). Phylogenetic inference. *In* "Molecular Systematics" (D. M. Hillis, C. Moritz, and B. K. Mable, Eds.), pp. 407–514. Sinauer, Sunderland, Massachusetts.

Thompson, J. D., Higgins, D. G., and Gibson, T. J. (1994). CLUSTAL W: Improving the sensitivity of progressive multiple sequence alignment through sequence weighting, positions-specific gap penalties and weight matrix choice. *Nucleic Acids Res.* **22**, 4673–4680.

Van de Peer, Y., Van den Broeck, I., de Rijk, P., and de Wachter, R. (1994). Database on the structure of small ribosomal subunit RNA. *Nucleic Acids Res.* **22**, 3488–3494.

Verheyen, R. (1955). La systematique des Anseriformes basée sur l'osteologie comparée. *Bull. Inst. R. Sci. Nat. Belgique* **31**(35–38), 72 pp.

Wetmore, A. (1960). A classification for the birds of the world. *Smiths. Misc. Coll.* **139**, 1–37.

Wheeler, W. C. (1990a). Combinatorial weights in phylogenetic analysis: A statistical parsimony procedure. *Cladistics* **6**, 269–275.

Wheeler, W. C. (1990b). Nucleic acid sequence phylogeny and random outgroups. *Cladistics* **6**, 363–368.

Wheeler, W. C. (1995). Sequence alignment, parameter sensitivity, and the phylogenetic analysis of molecular data. *Syst. Biol.* **44**, 321–331.

Zhang, D.-X., and Hewitt, G. M. (1996). Nuclear integrations: Challenges for mitochondrial DNA markers. *TREE* **11**, 247–251.

PART II

Applying Phylogeny and Population Genetics to Broader Issues

CHAPTER 9

Relevance of Microevolutionary Processes to Higher Level Molecular Systematics

SCOTT V. EDWARDS
Department of Zoology and Burke Museum, University of Washington, Seattle, Washington 98195

I. INTRODUCTION

The schism between microevolutionary and macroevolutionary research erected in the 1970s is considered by many biologists today to be an artificial one. Whereas the pattern of evolution and the relative strength of various evolutionary forces may well differ markedly at different hierarchical levels (Gould, 1994), the restriction of specific mechanisms to individual levels and the independence of these levels from one another is considered by many a nonissue. Evolutionary processes of drift, se-

Avian Molecular Evolution and Systematics

lection, and mutation acting within populations (microevolution) can explain large-scale evolutionary trends between species and higher taxa (macroevolution) without recourse to additional phenomena (Charlesworth *et al.*, 1982). Most of the early debates focused on phenotypic evolution as a forum for thrashing out these issues, and molecular evolution likely received far less attention in this context because of the ease with which gene trees spanning multiple hierarchical levels can be drawn (Avise, 1994).

While the bridges between microevolutionary and macroevolutionary studies of molecular evolution may be more obvious than those for phenotypic evolution, they are far from complete. Enormous conceptual shifts in this direction have been brought about by efforts to build and interpret large-scale phylogenetic trees of DNA sequences and alleles within species (Avise *et al.*, 1987; Cann *et al.*, 1987; Vigilant *et al.*, 1991; Edwards, 1993a; Baker *et al.*, 1993; Bowen *et al.*, 1994), the advent of "genealogical" models in population genetics (Slatkin and Maddison, 1989, 1990; Hudson, 1991), and the inculcation of "tree thinking" and the comparative method, both above the species level (Felsenstein, 1985; O'Hara, 1988; Harvey and Pagel, 1991; Brooks and McLennan, 1990) and below (O'Hara, 1993; Edwards and Kot, 1995). But Felsenstein's (1988, p. 445) statement that "systematists and evolutionary geneticists don't often talk to each other" was both retrospective and prospective, and the problem is particularly acute in molecular evolutionary ornithology. Although much of the conceptual framework that justifies ignoring the species boundary in molecular evolution has been laid, the flood of sequence data in higher level avian systematics has not been accompanied by an equivalently enthusiastic concern for the population genetic bases underlying the patterns of sequence variability observed among higher taxa. To the extent that this dialogue between levels in the hierarchy is not pursued, the potential richness of the interactions between levels will not be realized.

This chapter reviews studies of molecular evolution in birds that illustrate how molecular and population phenomena observable within species can affect both the analysis and outcome of patterns observed when comparing sequences from representatives of higher taxa. This theme will be explored with examples from five avenues of current research by this author and others: (1) the implications of intraspecific polymorphism for higher level molecular systematics, (2) the application of patterns of nucleotide substitution inferred from lower level comparisons to phylogenetic analyses of higher taxa, (3) the effect of incomplete lineage sorting and hybridization gene trees among higher taxa, (4) the effect of population structure on the dating of vicariant biogeographic splits, and (5) the impact of selection occurring within species on patterns of diversity and phylogenetic analysis of long-diverged sequences. One may conclude that molecular evolutionary studies of birds below the species level have more than passing relevance for higher level systematics and that useful insights can be gained by simultaneous analysis at both levels.

II. MOLECULAR VARIABILITY

A. The Specter of Polymorphism

Intraspecific polymorphism at the molecular level is a ghostly specter, an ever-present, inescapable shadow looming over the shoulder of the higher level systematist. For either sequence or allozyme data, the problem has been raised repeatedly and is greatest when the variability of characters is observable only on examination of taxonomic levels lower than the sampling scheme employed. Characters that may appear to unite members of particular clades may in fact represent parallelisms or convergences on denser sampling. For DNA sequence data, sites that appear to serve as synapomorphies linking clusters of taxa may in fact be polymorphic within taxa, or undergo further change, at the tips of these clusters, rendering their actual use as synapomorphies, or their inferred number of changes on the tree, dubious (Fig. 9.1). In this context, "polymorphism" is a problem at any level in the hierarchy, within or between species, so long as it occurs at a level lower than that being analyzed.

Intraspecific polymorphism in DNA sequences can have two sources. One is when alleles in an ancestral species do not sort completely between the time of population splitting and the time of sampling; here, even in the absence of mutation, there is a possibility of reaching incorrect phylogenetic conclusions when a single allele from a species is sampled (see Section II,B). The other source is mutations arising within species whose allelic lineages have sorted completely. The widespread use of mitochondrial DNA (mtDNA) in higher level avian systematics has positive and negative attributes with respect to these sources of polymorphism. Because the effective population size of mtDNA is usually about one-quarter that of a nuclear gene (barring large deviations from equal sex ratios and high rates of paternal leakage), the problem of incomplete lineage sorting prior to sampling is minimized compared to an average nuclear gene (Moore, 1995). However, while the high rate of substitution in mtDNA coding and noncoding sequences renders them useful at a variety of hierarchical levels, the possibility of ignoring unseen polymorphism due to mutations is great, especially if the time between population splitting and sampling is long, i.e., even long after reciprocal monophyly of descendent lineages has been achieved. (*Reciprocal monophyly* for a given locus refers to the condition in which all allelic lineages within each of two species descending from a single ancestral gene pool form monophyletic groups.) This latter situation characterizes studies of cytochrome *b* in babblers of the Australo–Papuan songbird genus *Pomatostomus*. Sequencing of a 282-bp segment of this gene revealed 17 variable sites among 16 individuals from the 2 lineages (eastern and western, gray-crowned and red-breasted, respectively) within the gray-crowned babbler (*Pomatostomus temporalis*; Edwards and Wilson, 1990). Phylogenetic analysis, however showed that the mtDNAs within either lineage were monophyletic, i.e., stemmed from a common ancestor within those lineages. Thus, whatever polymorphism is missed in

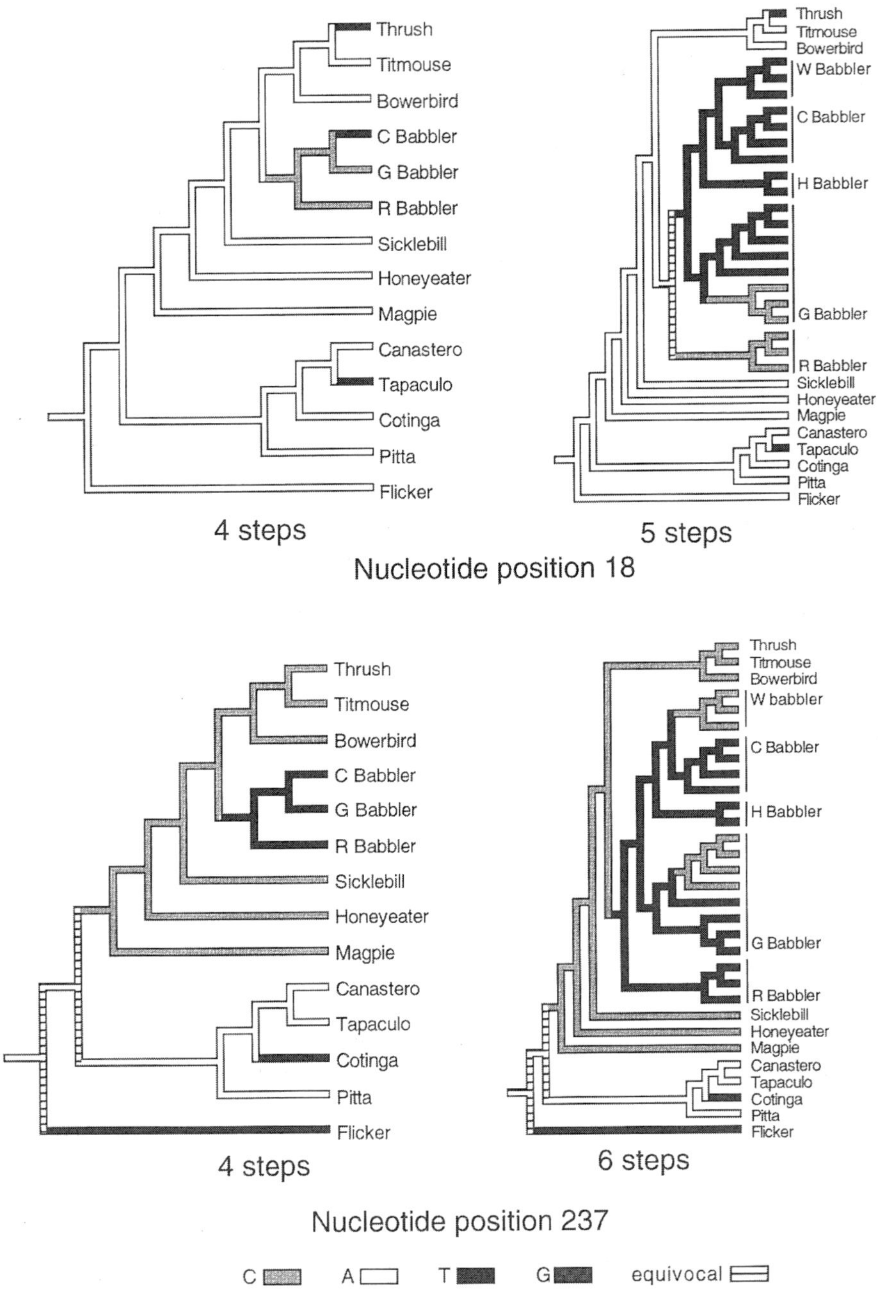

FIGURE 9.1 Examples from two sites in the mitochondrial cytochrome *b* gene illustrating the discovery of hidden polymorphism on denser taxonomic sampling. Illustrated here is the reconstruction of variability in two sites (positions 18 and 237 of the 282-bp cytochrome *b* segment sequenced in Edwards and Wilson, 1990) along a phylogenetic tree consisting of 14 passerine birds (left; Edwards *et al.*, 1991; with reanalysis in Edwards and Arctander, 1996) and a second tree (right) consisting of the above taxa plus 17 additional sequences within babblers (*Pomatostomus*). Common names of babblers are as follows:

comparisons of single individuals at this or higher levels likely stems from recurrent mutation, not incomplete sorting (Fig. 9.1).

We can estimate the magnitude of the problem of polymorphism by measuring changes in the phylogenetic signal brought about by changing the level of taxon sampling in phylogenetic analysis (Table I). For example, for 3 babblers, 10 other perching birds (Passeriformes), and a woodpecker, Edwards *et al.* (1991) compared cytochrome *b* sequences (the "exemplar tree") that spanned those determined within each of the babbler species earlier (the "babbler tree"). Despite the larger number of sequences in the babbler tree (Table I) the phylogenetic information in the sequences at this level, as measured by consistency and retention indexes, increases. By contrast, the strength of the branch leading to the babblers drops slightly on more intense sampling at low taxonomic levels (exemplar tree vs full tree, Table I). Thus, in this example the phylogenetic level of sampling as well as the number of sequences used can influence the perceived phylogenetic signal in the data (cf. Sanderson and Donoghue, 1989). Although the number of characters inferred to have changed unambiguously on this branch remains the same, the list of characters changes (Table I): three of these sites are polymorphic either within the gray-crowned babbler or among the five babbler species (Edwards and Wilson, 1990; Table I) and are interpreted as convergences or homoplasy only on denser taxon sampling. This effect reflects the fact that the sampling of taxa can influence the inferred reconstruction of events by parsimony (Wilson *et al.*, 1991).

We can further use the combination of low taxonomic level and high taxonomic level data sets to test the adequacy of certain methods of correcting for unseen substitutions due to unsampled nodes in a phylogeny. Fitch and Bruschi (1987) pointed out that there is a positive correlation between the number of nodes passed through from tips of a tree to the ancestor of the ingroup ("penultimate ancestor") and the number of substitutions inferred by parsimony. Such a correlation, which is also evident in the cytochrome *b* data for passerines (Fig. 9.2A), suggests that inferred substitutions are being missed along lineages with fewer branches. The method they propose for correcting branch lengths for such unseen substitutions leaves the lengths of those lineages passing through the largest number of nodes in the tree (the "trunk" of the tree) uncorrected. We can test the adequacy of this aspect of their method by asking how the branch lengths of such lineages change when we add the 17 additional babbler sequences to the exemplar tree. The increase in branch length inferred by parsimony in the full tree (Fig. 9.2B; Table I) suggests that, as Fitch and Bruschi (1987) suggest, even the lengths of those lineages passing through the largest number of nodes are likely underestimated by parsimony. In summary, although noise in the higher level phylogenetic analysis was increased by

---

G Babbler, gray-crowned (*temporalis*); C Babbler, chestnut-crowned (*ruficeps*); H Babbler, Hall's (*halli*); W Babbler, white-browed (*superciliosus*); R Babbler, rufous (*isidori*). The two sets of trees illustrate the discovery of nucleotide polymorphism (both sites) and nucleotide convergence (site 237) at lower taxonomic levels, and the extra steps revealed by denser sampling. Both sites are third positions of codons.

TABLE I Perceived Information Content and Substitution Dynamics of Perching Bird Mitochondrial Cytochrome *b* Sequences under Different Intensities of Taxon Sampling[a]

Parameter	Babbler tree	Exemplar tree	Full tree
Taxonomic level of sequence comparison	Within genus	Between families	Within genus, between families
Number of sequences	20	14	34
Tree length	106	373	429
Perceived information content			
Consistency index	0.61	0.52	0.46
Retention index	0.83	0.30	0.61
g_1 statistic	−0.78	−0.58	−0.51
Support for branch leading to babbler sequences			
Number of unambiguous sites supporting branch[b]	—	5 (sites 18, 49, 105, 111, 237)	5 (sites 49, 105, 111, 237, 279)
Bootstrap value of branch	—	96	88
Number of inferred unambiguous changes within babbler clade	106	40	63
Transition/transversion ratio			
Parsimony	14.14	1.34	1.69
Maximum likelihood[c]	29.1	3.0	3.8
Variation in substitution rate among sites (Wakeley, 1993a)			
Mean number of steps per site	0.38	1.32	1.52
Variance in number of steps per site	0.67	3.46	4.53
f value[d]	5.2	19.1	28.7
α[e]	0.48	0.82	0.77

[a] The "exemplar" tree consists of 12 perching bird sequences from Edwards *et al.* (1991) and using the thrush (*Catharus guttatus*) sequence from Helm-Bychowski and Cracraft (1993) as analyzed in Edwards and Arctander (1996). The "full" tree consists of the above tree plus 17 additional sequences from within the "babbler" genus *Pomatostomus*, whose phylogenetic relationships are presented in Edwards and Wilson (1990). The babbler tree consists of the latter 17 sequences only. All analyses were performed in MacClade (version 3.0; Maddison and Maddison, 1992) and PAUP (version 3.0s; Swofford, 1991).

[b] Site numbers correspond to those listed for the 282-bp segment in Edwards and Wilson (1990), beginning with 1 for the first site.

[c] The maximum likelihood estimate of the transition/transversion ratio was obtained using the program NUCML in the phylogenetics package MOLPHY by Adachi and Hasegawa (1995).

[d] Test statistic for rate nonuniformity based on the mean and variance of the number of parsimony steps per site (Wakeley, 1993).

[e] Inverse of the coefficient of variation of the substitution rate among sites. Smaller values of α indicate greater perceived variation in rate among sites relative to larger values.

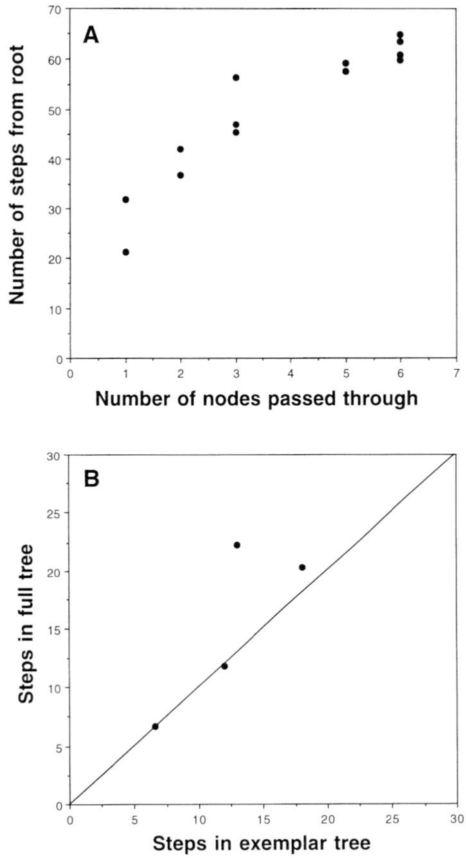

FIGURE 9.2 Effect of number of nodes and taxonomic sampling on the inferred number of changes along lineages in trees for passerine cytochrome *b* sequences. (A) Fitch-Bruschi (1987) plot illustating the dependence of number of inferred parsimony steps on the number of nodes passed through from root of the perching bird tree (Edwards *et al.*, 1991; Edwards and Arctander, 1995). The slope of the line is 6.5. (B) Capture of extra parsimony steps along four branches in the babbler portion of the perching bird tree on denser taxonomic sampling. The "exemplar" tree represents the 14 sequences in Edwards *et al.* (1991); the "full" tree represents these plus 17 additional babbler sequences (Edwards and Wilson, 1990). The diagonal line represents no change in number of steps along these branches in the two trees.

including the polymorphism found in the low taxonomic level sequences, the parsimony reconstructions are more realistic. More importantly, what appear in the exemplar tree to be sites clearly delineating a particular lineage turn out to be noisier than this coarse sampling scheme would suggest (Fig. 9.1; Table I).

In practice, the problem of polymorphism is usually addressed by ignoring or downweighting those sites or types of nucleotide change that are likely to be polymorphic at lower levels. Several higher level systematic analyses of avian DNA sequence data have dropped variability in the third positions of codons (Edwards *et al.*, 1991) or transition changes in all positions (Helm-Bychowski and Cracraft, 1993). Alternatively, step matrices, in which there is an increased cost in number of steps for particular substitution types (e.g., transversions), can be utilized. When such an approach is taken, the increase in consistency index at lower phylogenetic levels disappears, making these results more consistent with the findings of Sanderson and Donoghue (1989). Implementation of all of these methods entails an *a priori* model, as in the maximum likelihood method of phylogenetic inference (Felsenstein, 1981). Maximum likelihood essentially accounts for polymorphism and sites known to change frequently by downweighting the contribution of these sites to the total likelihood. As outlined in Section II,B, these methods of dealing with unseen polymorphism are connected to another type of reliance on data from lower taxonomic levels—inferring the actual pattern of nucleotide substitution.

B. Pattern of Nucleotide Substitution

It was not until the polymerase chain reaction (PCR) was used to obtain multiple closely related sequences that the high transition bias originally observed in primate mitochondrial DNA (Brown *et al.*, 1982) was confirmed for birds (Kocher *et al.*, 1989; Edwards and Wilson, 1990). This observation, which has had important consequences for the analysis of sequence data in higher level avian systematics (Helm-Bychowski and Cracraft, 1993; Lanyon and Hall, 1994; see also Cracraft and Helm-Bychowski, 1991), was determined almost wholly by observing patterns of nucleotide change among close relatives (sequences between which there have been few if any multiple changes at single sites).

That the transition bias in avian mtDNA is best observed among close relatives can be illustrated again with the cytochrome *b* data from babblers and other passerines (Table I). Edwards and Wilson (1990) observed that the most closely related sequences within babbler species differed from one another solely by transition changes, and the skew toward C/T changes was high in most comparisons; a maximum likelihood method yields a ratio as high as 29 (Table I). By contrast, were our knowledge of avian mtDNA dynamics drawn solely from comparisons of distantly related sequences, the bias would appear much less extreme (Table I). Parsimony reconstructions along these same trees confirm this trend (Table I).

It is becoming increasingly clear that some sort of weighting of rarely occurring substitution types (e.g., transversions) can considerably improve phylogenetic accuracy of mtDNA analyses (*sensu* Mindell and Honeycutt, 1990; Cracraft and Helm-Bychowski, 1991; Hillis *et al.*, 1993; Miyamoto *et al.*, 1994). An elegant way of visualizing the dynamics of substitution for use in weighting schemes is simply to plot observed numbers of transitions on numbers of transversions for pairs of sequences (Hasegawa *et al.*, 1985); here transversions serve as an approximate time scale along which the dynamics of transitions can be plotted, and the correlation between the axes of the plot is minimized.[1] Edwards and Wilson (1990) used this approach to estimate the transition bias in cytochrome *b* among close relatives; a crude estimate of the bias at this level (about 20 : 1, primarily in third positions of codons) was then assumed *a priori* to apply to the set of distantly related sequences later analyzed (Edwards *et al.*, 1991). The appeal of this approach is that the model of nucleotide substitution assumed in the analysis of diverged sequences is based on real data taken from a subset of the taxa under consideration; other approaches either assume an arbitrary model of substitution or one based on unrelated species. Since the transition bias in animal mtDNA likely differs between major taxonomic groups, the former approach may not make use of all available information, whereas the latter approach may be somewhat misleading.

Sequences for multiple gene regions determined for the same set of close relatives can then be used to compare the dynamics of substitution among those regions. Figure 9.3 plots the dynamics of substitutions among 27 babbler mtDNAs for portions of 2 regions: cytochrome *b* and the control region (Edwards, 1992). The comparison reveals that the approach to saturation for sites in region I (Fig. 9.3A) is much slower and less steep than that for third positions of cytochrome *b* (Fig. 9.3B). The higher rate of substitution and transition bias in third positions is also suggested by a maximum likelihood analysis of base substitutions (Table II; Hasegawa *et al.*, 1991; Edwards, 1992), although the standard errors of most of the estimates in this analysis are quite large owing to short DNA sequences. This initial result, however, is surprising, as one might expect some parts of the control region, a noncoding region not subject to the same constraints of coding sequences, to reflect more faithfully the underlying mutational bias toward transitions (Brown *et al.*, 1982; Thomas and Beckenbach, 1989; Quinn and Wilson, 1993). The difference in base compositional bias between the two regions (Fig. 9.3C and D) suggests a similar pattern: the more even base composition of region I appears not to faithfully reflect the composition expected on the basis of the directional mutation pressure on the L strand for chordate mtDNA (Jermiin *et al.*, 1995). The fact that the segment termed here "region I" contains at its 3′ end 45 bp of regulatory sequences (including the F box; Southern *et al.*, 1988) may partly explain this pattern. On the other hand, one could argue that the conservative evolution of cytochrome *b* at the amino

[1] Authors have applied a number of graphical methods for visualizing the magnitude of the transition bias, and hence the model to be employed in a phylogenetic weighting scheme for vertebrate mtDNA. The problem to be overcome is that the numbers of observed transitions and transversions between two sequences are not independent of one another, and that visualization of a change in observed bias over time requires an estimate of time when there is none. One common method plots the observed transition bias between pairs of sequences on the *y* axis and an estimate of corrected divergence on the *x* axis (for nonavian examples see Moritz *et al.*, 1992). Although this method appears to fulfill the second criterion by using distance as a proxy for time, it actually introduces even more correlations between the axes of the plot, since the estimate of divergence depends on the numbers of both transitions and transversions. The simpler graphical method of Hasegawa *et al.* (1985) accomplishes the same goal with less conflation of the axes.

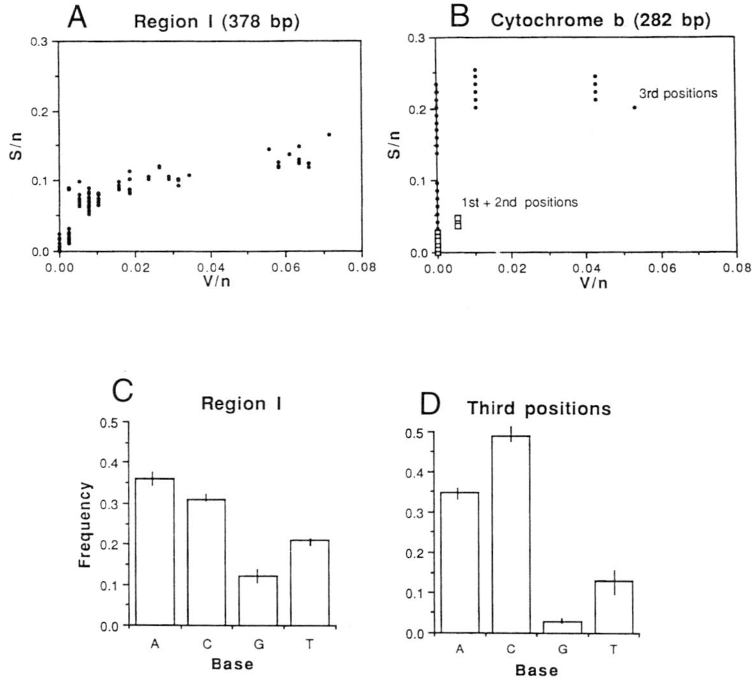

FIGURE 9.3 Dynamics of base substitution and base composition in region I of the control region and cytochrome *b* from 27 babbler sequences (Edwards and Wilson, 1990; Edwards, 1992). (A and B) Plots of observed numbers of transitions (*S*) per site (*n*) and transversions (*V*) per site for all pairwise comparisons of region I and cytochrome *b* sequences, respectively. (C and D) Observed base compositions for region I and third positions of cytochrome *b*, respectively.

acid level (Meyer, 1994) permits few transversions in the third positions, and that the mutational bias at these sites is actually much lower. The sequencing strategy suggested by this analysis, i.e., determining third positions instead of region I, is not only cumbersome but might also result in a data set with reduced signal even within species (Edwards, 1992). Furthermore, these comparisons could be confounded with differences in the extent of among-site rate variation between the two regions (Wakeley, 1993b).

In theory, a single point sample taken at any point along a unique transition/transversion curve (i.e., determined by comparing close or distant relatives) would appear sufficient to estimate the transition bias; in practice, however, curves corresponding to different biases are most distinct at small to intermediate distances, implying that comparisons at these levels will be most fruitful (Wakeley, 1993b). Determining the pattern of nucleotide substitution is difficult since the trees on which substitutions are traced are themselves based on some model of change, although

TABLE II Maximum Likelihood Estimates of Rates of Change (and Standard Errors) in Region I and Third Positions of Cytochrome *b* for 20 Pairs of Sequences from Babblers (*Pomatostomus*)[a]

Parameter being estimated	Estimates based on two subsets of data			
	Region 1		Third positions	
f	1	0.41	1	0.87
AIC	110.84	106.08	52.96	53.69
α	0.055 (0.006)	0.364 (0.110)	2.37 (2.59)	5.08 (12.65)
β	0.006 (0.001)	0.021 (0.004)	0.006 (0.003)	0.007 (0.003)
ν_s (Myr^{-1})	0.012 (0.001)	0.079 (0.024)	0.35 (0.38)	0.75 (1.87)
ν_v (Myr^{-1})	0.003 (0.001)	0.010 (0.002)	0.003 (0.001)	0.003 (0.002)
ν_s/ν_v	4	7.9	116.7	250
ν_t	0.015 (0.002)	0.089 (0.025)	0.35 (0.38)	0.76 (1.87)

[a]The maximum likelihood method of Hasegawa *et al.* (1990) was used to calculate all values. This method is based on the tree of the sequences and incorporates information on base composition into the model. *f*, Fraction of variable sites assumed; AIC, Akaike information criterion, a measure of the explanatory power of a model given the number of parameters in the model; α and β, parameters determining transition and transversion rates, ν_s and ν_v, respectively. To obtain absolute rates per million years a divergence time of 9 million years ago (MYA) for *P. isidori* from the other species was assumed (Sibley and Ahlquist, 1985).

some recently proposed methods appear to yield results that are less sensitive to tree topology (Yang *et al.*, 1994). Either way, the empirical estimation of patterns of nucleotide substitution depends critically on the range of degrees of sequence divergence employed, perhaps more so than on the particular method of estimation method used (Table I). In this regard it is unfortunate there are not more data sets in birds consisting of sequences from protein-coding genes sampled from within species or between recently diverged species (Edwards and Wilson, 1990; Birt-Friesen *et al.*, 1992; Moum and Johansen, 1992). Since these regions are among the more popular choices for use in higher level avian systematics, knowledge of their dynamics gleaned from analysis at lower levels can be applied in additional contexts (Edwards *et al.*, 1991; Helm-Bychowski and Cracraft, 1993). Most studies at or below the species level to date, however, make use of gene regions that would not be utilized in higher level systematics (e.g., the control region; e.g., Quinn, 1992; Wenink *et al.*, 1993, 1994).

Just as in the estimation of the transition bias, the extent to which different sites in a sequence change at different rates is also best observed among close or intermediate relatives (Table I). As sequences become more diverged the increase in number of saturated sites (reflected in the increase in mean number of parsimony changes) obscures the variability in rates among sites. Although in practice it is difficult to distinguish transition bias from among-site rate variation (Wakeley, 1993b), this again underscores the potential of lower level studies to yield information of critical importance to the interpretation and analysis of highly diverged sequences.

III. POPULATION PROCESSES

A. Gene and Species Trees

Sections IIA and B dealt with the enhanced ability to detect and measure mutational processes from analyses at lower taxonomic levels. In these cases, phylogenetic analysis of data sampled at low taxonomic levels can improve the phylogenetic accuracy (*sensu* Hillis and Bull, 1993) of analyses performed at higher levels. Sections III,A and B describe cases in which the trees inferred from DNA sequences are assumed to be completely accurate; here, lower level processes better inform analyses at higher levels not by improving our ability to recover the true tree, but by drawing attention to processes affecting the concordance between gene trees and trees of higher taxa.

1. Neutrality

Although the assumption of neutrality of DNA sequences is a prerequisite for using many new models appropriate for population-level processes (such as for gene flow and genetic drift; Slatkin and Maddison, 1989; Hudson 1991), this assumption is rarely tested in more recent avian studies (but see Barrowclough *et al.*, 1985). Neutrality tests have been applied to the noncoding mitochondrial region I data from *Pomatostomus* babblers (Edwards, 1993a,b). Using Tajima's (1989) test, selective neutrality was not rejected for sequence variability observed in 11 of 12 babbler populations (Fig. 9.4). Neutrality could also not be rejected for control region variation in Eurasian finches (Marshall and Baker, 1997). These results augur well for future studies in birds aimed at using neutral mtDNA sequences to infer population histories.

2. Incomplete Lineage Sorting

The possibility of discordances between the trees of alleles sampled from species and the historical sequence of separation of those species is well known (Nei, 1987). This discordance occurs not because gene trees have been incorrectly reconstructed, but because ancestral populations leave their imprint on descendent popu-

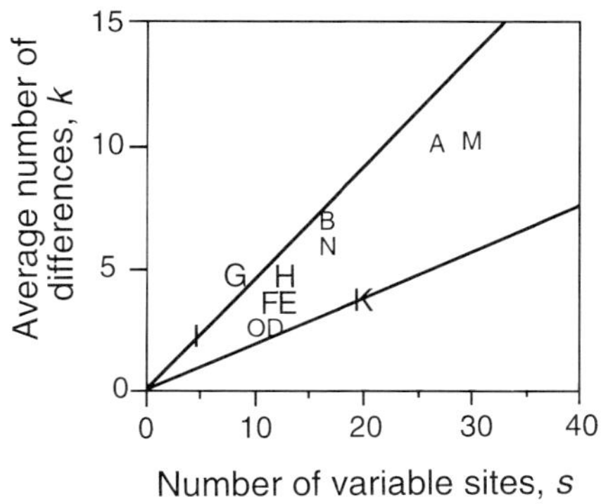

FIGURE 9.4 Tests of neutrality of mtDNA region I sequences in 12 populations of gray-crowned babblers throughout Australia and New Guinea. Letters denote individual populations as described in Edwards (1993a,b). Tajima's (1989) test compares the estimate of $\theta = 4N\mu$ provided by the number of variable sites (s; x axis) and the number of pairwise differences (k; y axis) for sequences sampled from a given population. When the estimate of θ from these two sources differs, selection is a possible interpretation. Thick lines indicate approximate 95% confidence limits on the joint values of s and k under neutrality for a sample size of 15.

lations for an extended time after population separation. The completion of "lineage sorting," the genealogical term applied to simple genetic drift, comes about when the genetic lineages of two descendent populations trace back to ancestors within each population; the average time required for this process is $4N_e$ generations, where N_e is the effective population size of the gene or organelle in question (Neigel and Avise, 1986). When lineages have not completely sorted, as will occur with high probability if the time since separation is short ($\ll 4N$ generations), there is a substantial probability that (1) if single alleles are chosen to represent each populations, the tree relating them to one another will not reflect the tree of populations; or (2) if multiple alleles per population are sampled, the resulting allele tree will be scrambled, with little evidence for monophyly of lineages within populations (Fig. 5A). The problem with inferring recent historical scenarios of this sort is that gene flow (migration) between long-separated populations can produce gene trees that mimic exactly those produced by incomplete sorting (Takahata and Slatkin, 1990).

The chances of inferring the population history correctly from the gene tree increase as the ratio of time between population splits (t) and effective population size (N_e) of the gene in question increases. Moore (1995) has reminded systematists that the relatively small effective population size of mitochondrial DNA makes it a much more likely candidate for tracking the population or species tree than an average nuclear gene. In fact, Moore (1995) suggests that it would take 16 nuclear genes to provide the same level of confidence of concordance between gene and species trees as the single locus provided by mtDNA! This conclusion does not bode well for ornithologists interested in improving the picture of population history

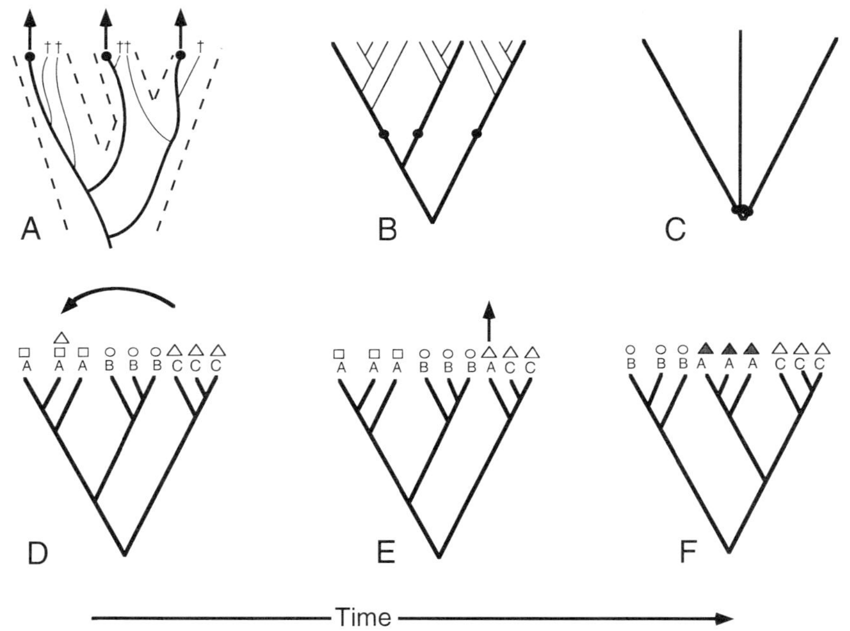

FIGURE 9.5 Summary and permanence of demographic effects creating discordances between gene and species trees. (A) Discordance between gene and species trees created by incomplete lineage sorting. Thick lines represent allelic lineages whose phylogeny is discordant with that of the three populations shown with dashes. Dots at tips represent ancestors of subsequent lineages in (B) and (C). Arrows indicate that these alleles are eventually fixed in their respective populations. (B) Shape of the gene tree in (A) soon after population splitting. Dots indicate ancestors that have given rise to new lineages. (C) Expected shape of discordant gene tree after much elapsed time. The tree is still discordant with the species tree but the short internode relative to the terminal branches makes resolution of an observed trichotomy difficult. (D) Effect of hybridization on gene trees. Letters represent localities or populations, shapes represent major allelic types. Curved arrow represents flow of an allele from population C into population A. (E) Observed gene tree immediately after hybridization. Arrow indicates eventual fixation of the new allele in (A) in that population. (F) Shape of gene tree after much time has elapsed from (D). Shaded triangles indicate divergence of allelic lineages from an ancestral unshaded state (D). The gene tree at this stage is still discordant with the species tree, even though all lineages have sorted in their respective populations.

with variable nuclear loci (such as microsatellites) when mtDNA lineages appear not to have sorted completely; although high variability will always aid in tracing lineages, for any given time frame, it is small N_e for a locus, not high mutation rates, that helps guarantee concordance of gene and species trees.

Population samples of control region sequences from the gray-crowned babblers provide a way to test models consistent with the observation of incomplete lineage sorting. Edwards (1993b) sequenced 400 bp of region I from 44 babblers sampled from 3 geographically close localities in the Northern Territory, Australia. Phylogenetic analyses suggested that none of the populations were monophyletic with

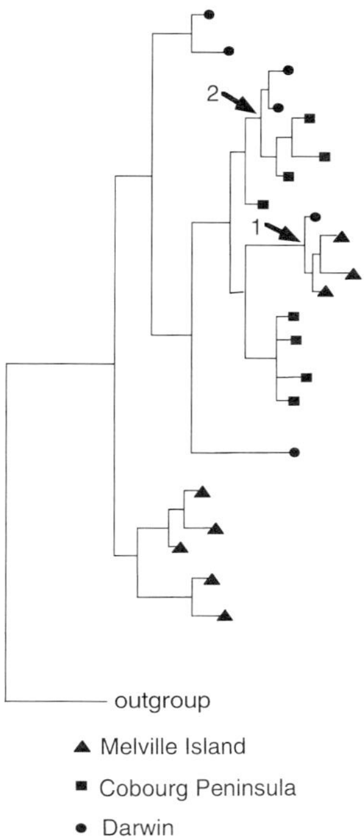

FIGURE 9.6 Fitch–Margoliash tree of 22 types of region I sequences from 3 populations of babblers (*Pomatostomus temporalis*) from the Northern Territory, Australia (Edwards, 1993a). Sequences sampled from the Melville Island, Cobourg Peninsula, and Darwin populations are represented by triangles, squares, and circles, respectively. The arrows indicate the first occurrence of an interpopulational coalescent event in the sample (1, between lineages from Melville Island and Darwin) and the second occurrence (2, between Cobourg Peninsula and Darwin). The eastern form of *P. temporalis* was used as an outgroup (see Edwards, 1993b).

respect to the sampled mitochondrial lineages (Edwards, 1993b; Fig. 9.6). Times of most recent Pleistocene sea level rises were used as proxies for times of population splitting (about 8000–10,000 years ago) and suggested that a historical model of genetic drift in the absence of gene flow between the isolated populations could be ruled out if the long-term N_e of the populations was less than about 13,000. Because this value seemed large for a passerine bird, particularly a social one, it seemed unlikely that the lack of monophyly of lineages within populations was due to persistence of alleles in large populations, and statistical tests of neutrality (Tajima, 1989) ruled out some sort of balancing selection. Rather, migration between populations across water barriers (approximately 50–150 km wide) seemed a better explanation of the data, even though this scenario might seem unlikely for a sedentary species. However, Milligan *et al.* (1994), using a completely different approach (Kuhner *et al.*, 1995), estimated from these same data that the N_e could have been much larger. If the gene trees are reasonably accurate, the possibility exists that the pattern is due to incomplete lineage sorting in isolated populations. If this is the case, we can still use the trees to determine the order of population splitting, even though they appear hopelessly scrambled and far from monophyly. Takahata (1989) showed that the phylogeny of closely related populations could nonetheless be extracted from such trees. The key observation is that populations must have split after the common ancestor of any allelic lineages found in those populations has split. This restriction requires that descendent populations have split after the most recent split between genetic lineages found in two different populations.

The key event for this problem is the most recent node connecting lineages found in two different populations in the trees, in this case the Darwin and Melville Island populations (Fig. 9.6). When the probability that at least one observed interpopulational coalescent event occurred after the third (Cobourg) population split but before the Darwin and Melville split is large (as it is with the sample sizes used in the study), the condition specifying high consistency of the allele and organismal trees is met, and the temporal order of interpopulational coalescent events is the same as that for populations. The scenario implied by this analysis is one in which the island population was not the first to bud off, which might suggest a less important role for water barriers in the diversification of these populations.

3. Hybridization

Hybridization is another important way in which population-level processes can influence higher level systematics (reviewed in Moore, 1995). Flow of nuclear or mitochondrial genes between taxa can, with inadequate sampling of populations and loci, prevent accurate reconstruction of evolutionary history; although the gene tree may be correctly inferred, the species tree will not (Fig. 5D). Organelle genomes are particularly susceptible to flow between species during hybridization because they will flow across taxa unconstrained by physical linkage to nuclear loci, although epistatic interactions with nuclear loci may impede flow (Barton and Jones, 1983; Harrison, 1989). Discordances between gene and species phylogenies via hybridization appear to be much less common in animals than in plants, where interspecific capture of foreign organelles (i.e., "chloroplast capture"; Reiseberg and Soltis, 1991) has led to major phylogenetic conflicts with other evidence. The best example in birds is likely Degnan's (Degnan and Moritz, 1992; Degnan, 1993) nuclear and mtDNA surveys in silvereyes (*Zosterops*); Degnan convincingly demonstrates that the mitochondrial tree can misrepresent the tree of nuclear loci, warning against sole reliance on this molecule for systematic purposes.

B. Permanent Effects of Incomplete Lineage Sorting and Hybridization

There is a suspicion among systematists that effects such as incomplete lineage sorting and hybridization are temporary, and that sequences sampled at highly diverged phylogenetic levels will be immune to them. This is not the case. Alleles with a history that is discordant with that of their respective populations nonetheless have just as much chance of reaching fixation as concordant alleles. This scenario would cause the gene tree to misrepresent the species tree permanently. Such effects are particularly important in studies of closely related species; here the length of the internode comprises a substantial fraction of the length of the entire tree (Fig. 5B); under some rare conditions, this branch might actually be deemed significant by a bootstrap (Felsenstein, 1985) or other test. If such a scenario occurs, however, it is never erased. No matter how long the three populations have been diverged from one another, in principle, the gene tree will always misrepresent the species tree. However, as time increases, the length of the internode becomes trivial compared to those leading to the tips, and, in practice, the recovered phylogeny resembles a star phylogeny (Fig. 5C). Thus, if one wished to determine the phylogeny of a monophyletic trio of species that had been diverging for millions of years, contrary to intuition, there is a chance that the true gene tree would misrepresent the true species tree even in this case. There is such risk in all single gene trees of higher taxa (Edwards *et al.*, 1991; Helm-Bychowski and Cracraft, 1993), but our ability to distinguish the true gene tree from an observed trichotomy is vanishingly small (DeSalle *et al.*, 1994). Furthermore, in order for the discordant gene tree to be realized, sampling must take place prior to any extinction of species, as extinction can restore concordance. Using multiple loci to assess the higher level tree will nearly always improve resolution and ability to infer the species tree.

Like lineage sorting, the imprint of hybridization on higher level phylogenetic trees can be long lasting but can also be erased by extinction of the species into which mtDNA has flowed. If foreign mtDNA somehow invades and takes over mtDNA "native" to a particular species, the mtDNA of that species will continue to yield atypical results (Fig. 5E and F) until hybridization with closer relatives or extinction of the lineage might possibly restore concordance. Several higher level molecular trees in birds are suspected to have been influenced by past hybridization events (Crow *et al.*, 1992; Avise *et al.*, 1990), and the lingering possibility of this phenomenon, even when working far above the species level, should compel workers to score nuclear loci simultaneously, either indirectly via the phenotype, or more directly, via nuclear markers (Moore, 1995).

C. Dating Biogeographic Events: Coalescence in Subdivided Populations

Population structure can leave its imprint on analyses above the species level through effects other than lineage sorting and hybridization. It has been known for more than 10 years that the total divergence between two species, as measured by, say, the average pairwise divergence between lineage tips in the two species (δ_{xy}), includes both the divergence of alleles between the species that has accumulated after lineage splitting as well as the divergence of alleles within the common ancestral species. A suggested measure of interspecific divergence (δ) employing a correction for that component of divergence caused solely by diversity within the ancestor is

$$\delta = \delta_{xy} - 0.5(\delta_x + \delta_y) \quad (1)$$

where δ_x and δ_y are the average divergence of alleles within species x and y (Stephens and Nei, 1985; Wilson *et al.*, 1985). As the equation implies, the correction becomes more important as diversity within the ancestor (as estimated by observed diversity within the two descendent species) increases. Although this correction becomes less important as the divergence time of species becomes large, for recently diverged species there is a risk of significantly overestimating the divergence time when species divergence (δ) is equated with gene divergence (δ_{xy}).

The magnitude of this error for biogeographic studies employing a molecular clock of mtDNA, such as those testing models of Pleistocene speciation (Bermingham *et al.*, 1992; Zink and Slowinski, 1995), can be estimated by examining the range of within-species mtDNA diversity and comparing this to typical interspecific divergences. Moore (1995) compiled such data (primarily for North American species assayed via restriction enzymes) and showed that the average maximum branch length within species was 0.007 substitutions per site, yielding an average depth for intraspecific trees of 0.0035 substitutions per site, or approximately 350,000 years. Thus, estimates of interspecific divergence will in general be overestimates by about this amount without correction for ancestral diversity. A list of minimum values for δ_{xy} (for nominate species and subspecies) for 18 genera of North American birds has a range of <0.001 (subspecies of red-winged blackbirds, *Agelaius phoeniceus*) to 0.09 (species of sandpipers, *Calidris*). Although many estimates of δ_{xy} for North American birds are large, suggesting gene divergence long before the Pleistocene (e.g., >2 million years ago), the mtDNA of a variety of avian species pairs reveals smaller δ_{xy} values (Bermingham *et al.*, 1992; Zink and Slowinski, 1995). If any of the ancestors of these species pairs had coalescence times approaching 350,000 years, then the fraction of δ_{xy} attributable to ancestral diversity could be quite large.

The effect of intraspecific diversity on estimated dates of interspecific divergence will increase as the ancestral species becomes more structured. This is because the effective population size of a species structured into many semiisolated demes is greater than the sum of the effective sizes of those demes. Hence the within-species diversity of a structured species will also be disproportionately large. Nei and Takahata (1993), building on Wright (1943), Maruyama (1970), and Slatkin (1991), rederived the quantity for the effective size of a species (N_e) subdivided into n semi-isolated demes, each of size N. It is

$$N_e = Nn\left[1 + \frac{(n - 1)^2}{4Nmn^2}\right] \quad (2)$$

where Nm is the level of gene flow (in migrants per generation) between subpopulations in a finite island model. When the ancestral species is structured, for example when Nm is low (0.001) and the number of demes high (32), the coalescence time is on average more than 400 times the coalescence time of a similarly sized panmictic species (Nei and Takahata, 1993)! This makes intuitive sense, since the demes comprising a highly structured species are nearly independent of one another; the low level of gene flow just prevents them from diverging *ad infinitum*. Values for δ_x in such a species will be similarly inflated over those values in a panmictic species of the same total size. Thus, possible structuring in the species from which contemporary species diverged will exacerbate the failure to correct interspecific distances for divergence within the ancestor.

A more intuitive way of expressing the increase in coalescence time in a structured species is in terms of F_{st}, since this is what is typically measured. We can do this by noting an equation for N_e of a structured species nearly identical to Eq. (2) for a large number of demes (n; Felsenstein, 1992):

$$N_e = Nn + \frac{(n - 1)}{4m} \tag{3}$$

$$= Nn\left[1 + \frac{1}{4Nm}\frac{(n - 1)}{n}\right] \tag{4}$$

$$= Nn\left[\frac{4Nm + \frac{(n - 1)}{n}}{4Nm}\right] \tag{5}$$

When n is large, $(n - 1)/n$ approaches 1. So, we can substitute in for F_{st} by noting that F_{st} in an island model is often expressed as $1/(1 + 4Nm)$. This leaves us with

$$N_e = Nn\frac{1}{(1 - F_{st})}$$

This means that the effective size of a structured species is inflated by a factor $1/(1 - F_{st})$ over a panmictic one of similar total size (Wright, 1943).

This formulation provides a new perspective on uses of a continent-wide survey of control region sequences in the gray-crowned babbler (Edwards, 1993a; Fig. 9.7) to date an important vicariant barrier for this and other species in Australia. The eastern and western lineages of gray-crowned babblers diverged across a well-known biogeographic barrier in northeastern Australia known as the Carpenterian barrier. While this barrier is considered quite old on the basis of the relative timing of taxa diverging across it (Cracraft, 1986), there have been no estimates of its relative or absolute age from molecular data. Average sequence divergence in region I between the eastern and western lineages across this barrier is 8.3% (Edwards and Kot, 1995). If the rate of divergence between the two babbler lineages in region I is 11–17% per million years as in humans (Nei, 1993; but see Mindell *et al.*, 1996), this suggests a divergence time of about 440,000–680,000 years ago (= genetic dis-

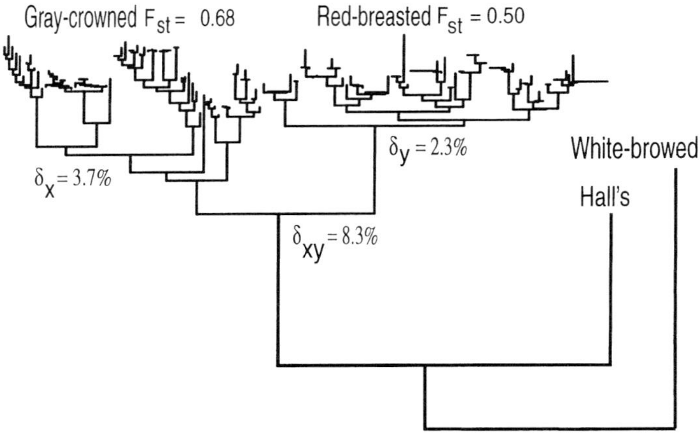

FIGURE 9.7 Gene genealogy of region I (control region) sequences from 12 populations of gray-crowned babblers (*P. temporalis*) across Australia (Edwards, 1993a). The common names of the two major lineages within *temporalis* are indicated, along with estimates of F_{st} for this region for each group. The estimated average pairwise sequence divergence within (δ_x, δ_y) and between these groups (δ_{xy}) is indicated. White-browed (*P. superciliosus*) and Hall's (*P. halli*) sequences were used as outgroups (Edwards, 1992). The tree was constructed by the neighbor-joining method (Saitou and Nei, 1987) using gamma distances with $\alpha = 0.5$.

tance/rate/2), without any correction for possible diversity in the ancestral population. However, if diversity and genetic structure of the ancestor were similar to that of contemporary lineages, then it would be high: F_{st} for region I within the eastern and western lineages of 0.68 and 0.50, respectively (Fig. 9.7). This means that δ_x and δ_y, for the eastern and western lineages, are about three and two times, respectively, what they would be were these lineages panmictic, and the differences in coalescence times within each lineage appear to reflect this pattern (Fig. 9.7). The average intraspecific divergences are 3.7 and 2.3%, respectively (Edwards and Kot, 1995), yielding a corrected between-lineage divergence of 5.3%, or only 275,000–425,000 years. These times differ from the uncorrected value by more than 35%. Clearly, estimating the times of recent biogeographic events such as the Carpenterian break analyzed here requires correction for intraspecific diversity.

The logic employed in corrections for ancestral diversity is that population structure and diversity levels have stayed roughly the same during the evolution of the two descendent species. This is highly unlikely for many species pairs, and may well be misleading. Indeed, the short distance between the base of each babbler lineage and their common ancestor implies that the corrected divergence time would have occurred after all alleles had coalesced within the gray-crowned and red-breasted lineages—highly implausible unless a bottleneck or some other departure from demographic stasis occurred (J. Felsenstein, personal communication). In addition, in many cases the overestimate of interspecific divergence by gene divergence might be swamped out by the errors associated with the molecular clock itself. The number of intraspecific gene trees that could be useful for such purposes in birds is growing (e.g., Zink, 1993; Zink and Dittmann, 1993a,b; Joseph and Moritz, 1994), and in the future, multiple population samples of sequences will considerably improve our picture of the evolution of intraspecific diversity as well as better refine our estimates of divergence times between species (e.g., Wood and Krajewski, 1996).

D. Selection

It is sometimes stated that the use of markers that are under the influence of natural selection, whether morphological or molecular, is taboo for systematists interested in retrieving the true organismal tree. Admittedly, however, in the past this has been a vague rule of thumb: rarely is there any quantitative method applied to the decision to invoke selection for a particular character (much less what type of selection), and it is often unknown exactly what the phylogenetic consequences of that selection will be. These are some of the reasons why the habit of disregarding characters under selection *a priori* has been questioned. Generally, it is possible that a systematist might avoid characters influenced by positive Darwinian selection because the phylogenetic signal might be obscured: selection may accelerate the rate of change for a character, resulting in excess homoplasy, and selection may increase homoplasy by directly causing convergence or parallelism in unrelated lines.

Although there are abundant examples in birds of false phylogenetic trails being left by morphological characters under selection (Mayr, 1963), there have been few similar claims for molecular characters. Nonetheless, with more diversified molecular and statistical techniques (Tajima, 1989; Golding and Felsenstein, 1990; Golding, 1994), there are an increasing number of examples of natural selection at the molecular level. In particular, polymorphisms in genes of the major histocompatibility complex (MHC) of vertebrates suggest that systematists might avoid using characters under selection for yet another reason: balancing selection (heterozygote advantage) can create a situation in which the true gene tree (allelic genealogy), even if correctly reconstructed, often would not reflect the tree of species splitting ("organismal" tree).

MHC molecules are glycoproteins that bind antigenic peptides from bacteria and pathogens and present these to T cells for initiation of the immune response. Several decades of intense molecular and comparative research have yielded a detailed picture of the causes and consequences of variability in those portions of MHC molecules that specifically bind foreign antigens—the antigen-binding sites (ABSs). MHC loci are now considered the most extreme example of natural selection at the molecular level in vertebrates (reviewed in Klein *et al.*, 1993; Hedrick, 1994).

The particular type of selection at MHC loci, namely balancing selection (heterozygote advantage or frequency-dependent selection), has the effect of maintaining alleles for extremely long periods of time in populations; it makes alleles more resistant to extinction via genetic drift. The result is not only an increase in intraspecific allelic diversity but a long life span of alleles and a higher incidence of alleles shared between long-separated species; this maintenance of ancestral polymorphism at MHC loci across speciation events is predicted to occur if balancing selection is intense enough (Klein *et al.*, 1993). Many examples of maintained ancestral polymorphism have been documented at mammalian MHC loci (e.g., Edwards *et al.*, 1997), with some alleles apparently having been maintained up to 40 million years (Klein *et al.*, 1993). Some of this apparent ancestral polymorphism is more likely convergence (i.e., incorrect phylogenetic reconstruction) in long-separated lineages (Hughes *et al.*, 1994; Takahata, 1994), and there are reasons other than selection (recombination, intra- and interlocus gene conversion) explaining why MHC phylogenies often appear so scrambled (She *et al.*, 1991; Gyllensten *et al.*, 1991); thus MHC loci are excellent examples of characters that systematists interested in an organismal phylogeny would want to avoid both because natural selection produces phylogenetic results that are inconsistent with other data as well as because the phylogenetic signal embedded in them is obscured. They are also excellent examples of lower level processes influencing higher level phylogenetic trees.

The cloning and comparative analysis of MHC class II genes in birds (Edwards *et al.*, 1995a,b) illustrates these points. One popular way of detecting selection at the ABS of MHC loci is to find evidence for elevated rates of nonsynonymous (amino acid changing) substitutions (Hughes and Nei, 1988). Such rates are expected to increase under balancing selection because alleles with new amino acid sequences, and hence with new capacities for binding foreign peptides, will be advantageous when rare or when individuals are heterozygous. Edwards *et al.* (1995b) tested for the action of balancing selection at the ABS of MHC sequences amplified from three songbirds; the results suggested that, as in mammals, MHC loci in birds are subject to the action of balancing selection. To extend these results, phylogenies of three functional domains of the MHC sequences amplified from songbirds were built using a chicken sequence as an outgroup. At face value, the results suggest that ancestral polymorphisms may have been maintained for portions of the antigen-binding site, but that other portions of the molecule, such as the anchor exons (exon 3), exhibited no evidence for this effect (Fig. 9.8). Assuming the gene trees are correct, such a result can be explained by invoking recombination between functional domains of the MHC. The phylogenetic signal in the domains subject to selection (α helix and β sheet) was strong enough to reject the topology of the non-trans-species tree exhibited by exon 3 (Fig. 9.8). However, the divergence times of the species in that study, which included representatives of two major songbird groups, namely Passerida and Corvida (Sibley and Ahlquist, 1990), were quite large, suggesting that possibly the phylogenies reflected more convergence in the ABS than maintained alleles. That the variability in the third positions of codons also reflects the conflict between phylogenies of subdomains argues against the convergence hypothesis; on the other hand, the extremely skewed base composition of

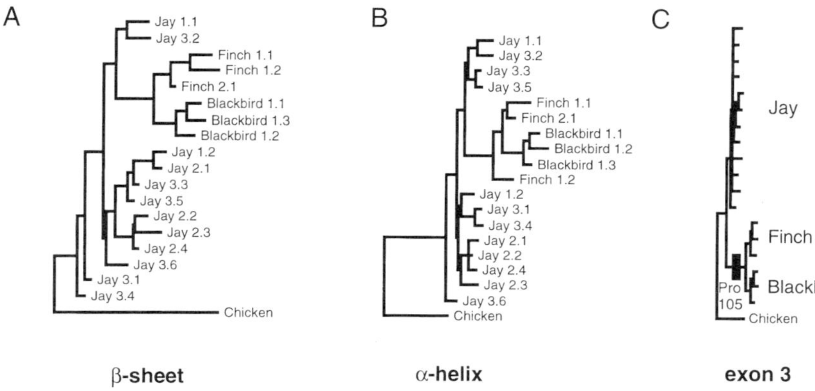

FIGURE 9.8 Phylogenetic trees (Swofford, 1991; Saitou and Nei, 1985) of MHC class II B (β chain) sequences amplified from red-winged blackbirds (*Agelaius phoeniceus*), western scrub jays (*Aphelocoma coerulescens californica*), and house finches (*Carpodacus mexicanus*). Each sequence is designated by a species; the number represents the individual and clone (e.g., 3.3 indicates clone 3 from individual 3). (A) Tree of the β sheet, exon 2; (B) Tree of the α helix, exon 2; (C) Tree of exon 3. Pro 105, Insertion of a proline codon at position 105 in the blackbird and finch sequences relative to the jay and chicken sequences. See Edwards *et al.* (1995b) for details.

third positions of avian MHC class II genes makes convergence even at these positions more likely (Edwards *et al.*, 1995b). The uncertainty as to the gene and locus relationships of the amplified sequences further complicates the picture. Thus, although MHC genes are undoubtedly excellent examples of the tight link between population level processes and higher level systematics, the precise nature of the link, at least for birds, is still obscure.

IV. CONCLUSION

An eminent molecular evolutionist of birds once challenged his upper level molecular evolution class with the question, "Can systematists working far above the species level effectively ignore processes operating below the species level?" As usual with such pop questions, the class squirmed uneasily, looking away or at each other or any place other than the front of the room, even though the instructor had spent most of the lecture championing the affirmative. This chapter has attempted to review the evidence and marshal new perspectives championing the negative. Not only can higher level systematists not ignore population processes in their quest for the major branches of avian phylogenetic trees—they are effectively studying such processes at the same time. Of course, the instructor (Wilson *et al.*, 1985), as well as numerous other geneticists (Avise *et al.*, 1987; Moritz *et al.*, 1987; Crozier, 1990), had proffered the same perspective in their writings. For this reason I believe A. C.

Wilson's question was primarily meant to challenge and stimulate, as was so often the case.

It is likely that the reversed question—Can avian population geneticists effectively ignore patterns occurring far above the species level?—can be answered just as forcefully in the negative. For example, embedding the focal species in its appropriate phylogenetic context is often the only way of determining the direction of evolutionary trends and of adding rigor to the writing of "evolutionary chronicles" (O'Hara, 1988). Molecular data, particularly those bearing on questions of selection and gene genealogies in structured populations, will likely continue to play an important role in fusing these two levels.

ACKNOWLEDGMENTS

I thank Joe Felsenstein, Mary Kuhner, Masami Hasegawa, David Mindell, Tom Quinn, Monty Slatkin, and John Wakeley for helpful discussions and assistance.

REFERENCES

Adachi, J., and Hasegawa, M. (1995). "MOLPHY: Programs for Molecular Phylogenetics," version 2.3. Institute of Statistical Mathematics, Tokyo.

Avise, J. C. (1994). "Molecular Markers, Natural History and Evolution." Chapman & Hall, New York.

Avise, J. C., Arnold, J., Ball, R. M., Bermingham, E., Lamb, T., Neigel, J. C., Reeb, C. A., and Saunders, N. C. (1987). Intraspecific phylogeography: The mitochondrial DNA bridge between population genetics and systematics. *Annu. Rev. Ecol. Syst.* **18**, 489–522.

Avise, J. C., Ankey, C. D, and Nelson, W. S. (1990). Mitochondrial gene trees and the evolutionary relationship of mallard and black ducks. *Evolution* **44**, 1109–1119.

Baker, C. S., Perry, A., Bannister, J., Weinrich, M., Abernathy, R., Calambolcidis, J., Lien, J., Lambertsen, R., Urbán Ramírez, J., Vasquez, O., Clapham, P., Alling, A., O'Brien, S., and Palumbi, S. R. (1993). Abundant mitochondrial DNA variation and world-wide population structure in humpback whales. *Proc. Natl. Acad. Sci. U.S.A.* **90**, 8239–8243.

Barrowclough, G. F., Johnson, N. K., and Zink, R. M. (1985). On the nature of genic variation in birds. *Curr. Ornithol.* **2**, 135–154.

Barton, N. H., and Jones, J. S. (1983). Mitochondrial DNA: New clues about evolution. *Nature* (*London*) **306**, 317–318.

Bermingham, E., Rohwer, S., Freeman, S., and Wood, C. (1992). Vicariance biogeography in the Pleistocene and speciation in North American wood warblers: A test of Mengel's model. *Proc. Natl. Acad. Sci. U.S.A.* **89**, 6624–6628.

Birt-Friesen, V. L., Montevecchi, W. A., Gaston, A. J., and Davidson, W. S. (1992). Genetic structure of thick-billed murre (*Uria lomvia*) populations examined using direct sequence analysis of amplified DNA. *Evolution* **46**, 267–272.

Bowen, B. W., Kamezaki, N., Limpus, C. J., Hughes, G. R., Meylan, A. B., and Avise, J. C. (1994). Global phylogeography of the loggerhead turtle (*Caretta caretta*) as indicated by mitochondrial DNA haplotypes. *Evolution* **48**, 1820–1828.

Brooks, D. R., and McLennan, D. A. (1990). "Phylogeny, Ecology and Behavior." Univ. of Chicago Press, Chicago.

Brown, W. M., Prager, E. M., Wang, A., and Wilson, A. C. (1982). Mitochondrial DNA sequences of primates: Tempo and mode of evolution. *J. Mol. Evol.* **18**, 225–239.

Cann, R. L., Stoneking, M., and Wilson, A. C. (1987). Mitochondrial DNA and human evolution. *Nature (London)* **325**, 31–36.

Charlesworth, B., Lande, R., and Slatkin, M. (1982). A Neodarwinian commentary on macroevolution. *Evolution* **36**, 474–498.

Cracraft, J. (1986). Origin and evolution of continental biotas: Speciation and historical congruence in the Australian avifauna. *Evolution* **40**, 977–996.

Cracraft, J., and Helm-Bychowski, K. (1991). Parsimony and phylogenetic inference using DNA sequences: Some methodological strategies. "Phylogenetic Analysis of DNA Sequences" (M. Miyamoto and J. Cracraft, Eds.), pp. 184–220. Oxford University Press, London.

Crow, T. M., Harley, E. H., Jakutowicz, M. B., Komen, J., and Crowe, A. A. (1992). Phylogenetic, taxonomic and biogeographical implications of genetic, morphological and behavioral variation in francolins (Phasianidae: *Francolinus*). *Auk* **109**, 24–42.

Crozier, R. H. (1990). From population genetics to phylogeny: Uses and limits of mitochondrial DNA. *Aust. Syst. Bot.* **3**, 111–124.

Degnan, S. M. (1993). The perils of single gene trees—mitochondrial versus single-copy nuclear DNA variation in white-eyes (Aves: Zosteropidae). *Mol. Ecol.* **2**, 219–225.

Degnan, S. M., and Moritz, C. (1992). Phylogeography of mitochondrial DNA in two species of white-eye in Australia. *Auk* **109**, 800–811.

DeSalle, R., Absher, R., and Amato, G. (1994). Speciation and phylogenetic resolution. *Trends Ecol. Evol.* **9**, 297–298.

Edwards, S. V. (1992). Gene flow and mitochondrial DNA evolution in social babblers (*Pomatostomus*). Ph.D. dissertation. The University of California, Berkeley.

Edwards, S. V. (1993a). Long-distance gene flow in a cooperative breeder detected in genealogies of mitochondrial DNA sequences. *Proc. R. Soc. Lond. B.* **252**, 177–185.

Edwards, S. V. (1993b). Mitochondrial gene genealogy and gene flow among island and mainland populations of a sedentary songbird, the grey-crowned babbler (*Pomatostomus temporalis*). *Evolution* **47**, 1118–1137.

Edwards, S. V., and Arctander, P. (1996). Detecting scrambled mtDNA sequences using intragenic tests of congruence: Retraction and reanalysis of a thrush (*Catharus guttatus*) cytochrome *b* sequence. (Submitted).

Edwards, S. V., and Kot, M. (1995). Comparative methods at the species level: Geographic variation in morphology and group size in grey-crowned babblers (*Pomatostomus temporalis*). *Evolution* **47**, 1134–1146.

Edwards, S. V., and Wilson, A. C. (1990). Phylogenetically informative length polymorphism and sequence variability in mitochondrial DNA of Australian songbirds (*Pomatostomus*). *Genetics* **126**, 695–711.

Edwards, S. V., Arctander, P., and Wilson, A. C. (1991). Mitochondrial resolution of a deep branch of the genealogical tree for perching birds. *Proc. R. Soc. Lond. B.* **243**, 99–107

Edwards, S. V., Chestnut, K., Satta, Y., and Wakeland, E. K. (1997). Ancestral polymorphism of *Mhc* class II genes in mice: implications for balancing selection and the mammalian molecular clock. *Genetics, in press.*

Edwards, S. V., Grahn, M., and Potts, W. K. (1995a). Dynamics of MHC evolution in birds and crocodilians: Amplification with degenerate primers. *Mol. Ecol.* **4**, 719–729.

Edwards, S. V., Wakeland, E. K., and Potts, W. K. (1995b). Contrasting histories of avian and mammalian *Mhc* genes revealed by class II B sequences from songbirds. *Proc. Natl. Acad. Sci. U.S.A.* **92**, 12200–12204.

Felsenstein, J. (1981). Evolutionary trees from DNA sequences: A maximum likelihood approach. *J. Mol. Evol.* **17**, 368–376.

Felsenstein, J. (1985). Phylogenies and the comparative method. *Am. Naturalist* **125**, 1–15.

Felsenstein, J. (1988). Phylogenies and quantitative characters. *Annu. Rev. Ecol. Syst.* **19**, 445–71.

Felsenstein, J. (1992). "Lecture Notes in Population Genetics." University of Washington, Seattle, Washington.

Fitch, W. M., and Bruschi, M. (1987). The evolution of prokaryotic ferredoxins—with a general method correcting for unobserved substitutions in less branched lineages. *Mol. Biol. Evol.* **4**, 381–94.

Golding, B., Ed. (1994). "Non-Neutral Evolution: Theories and Molecular Data." Chapman & Hall, New York.

Golding, B., and Felsenstein, J. (1990). A maximum-likelihood approach to the detection of selection from a phylogeny. *J. Mol. Evol.* **31**, 511–523.

Gould, S. J. (1994). Tempo and mode in the macroevolutionary reconstruction of Darwinism. *Proc. Natl. Acad. Sci. U.S.A.* **91**, 6764–6769.

Gyllensten, U., Sundvall, M., and Erlich, H. (1991). Allelic diversity is generated by intra-exon exchange at the DRB locus of primates. *Proc. Natl. Acad. Sci. U.S.A.* **88**, 3686–3690.

Harrison, R. G. (1989). Animal mitochondrial DNA as a genetic marker in population and evolutionary biology. *Trends Ecol. Evol.* **4**, 6–11.

Harvey, P., and Pagel, M. (1991). "The Comparative Method in Evolutionary Biology." Oxford University Press, London.

Hasegawa, M., Kishino, H., and Yano, T. (1985). Dating of the human–ape splitting by a molecular clock of mitochondrial DNA. *J. Mol. Evol.* **22**, 160–174.

Hasegawa, M., Kishino, H., Hayasaka, K., and Horai, S. (1991). Mitochondrial DNA evolution in primates: Transition rate is extremely low in lemur. *J. Mol. Evol.* **31**, 113–121.

Hedrick, P. W. (1994). Evolutionary genetics of the major histocompatibility complex. *Am. Nat.* **143**, 945–964.

Helm-Bychowski, K., and Cracraft, J. (1993). Recovering phylogenetic signal from DNA sequences: Relationships within the corvine assemblage (class Aves) as inferred from complete sequences of the mitochondrial DNA cytochrome-*b* gene. *Mol. Biol. Evol.* **10**, 1196–1214.

Hillis, D. M., and Bull, J. J. (1993). An empirical test of bootstrapping as a method for assessing confidence in phylogenetic analysis. *Syst. Biol.* **42**, 182–192.

Hillis, D. M., Allard, M. W., and Miyamoto, M. M. (1993). Analysis of DNA sequence data: Phylogenetic inference. *Methods Enzymol.* **224**, 456–487.

Hudson, R. R. (1991). Gene genealogies and the coalescent process. *Oxford Surv. Evol. Biol.* **7**, 1–44.

Huelsenbeck, J. P. (1991). Tree-length distribution skewness: An indicator of phylogenetic information. *Syst. Zool.* **40**, 257–270.

Hughes, A., and Nei, M. (1988). Pattern of nucleotide substitution at class I major histocompatibility complex loci reveals overdominant selection. *Nature* (*London*) **335**, 167–170.

Hughes, A. L., Hughes, M. K., Howell, C. Y., and Nei, M. (1994). Natural selection at the class II major histocompatibility loci of mammals. *Phil. Trans. R. Soc. Lond. B* **345**, 359–367.

Jermiin, L. S., Graur, D., and Crozier, R. H. (1995). Evidence from analyses of intergenic regions for strand-specific directional mutation pressure in metazoan mitochondrial DNA. *Mol. Biol. Evol.* **12**, 558–563.

Joseph, L., and Moritz, C. (1994). Mitochondrial DNA phylogeography of birds in eastern Australian rainforests: First fragments. *Aust. J. Zool.* **42**, 385–403.

Klein, J., Satta, Y., O'hUigin, C., and Takahata, N. (1993). The molecular descent of the major histocompatibility complex. *Annu. Rev. Immunol.* **11**, 269–295.

Kocher, T. D., Thomas, W. K., Meyer, A. Edwards, S. V., Pääbo, Villablanca, F. X., and Wilson, A. C. (1989). Dynamics of mitochondrial DNA evolution in animals: amplification and sequencing with conserved primers. *Proc. Natl. Acad. Sci. USA* **86**, 6196–6200.

Kuhner, M. K., Yamato, J., and Felsenstein, J. (1995). Estimating effective population size and mutation rate from sequence data using Metropolis–Hastings sampling. *Genetics* **140**, 1421–1442.

Lanyon, S. M., and Hall, J. G. (1994). Reexamination of barbet monophyly using mitochondrial-DNA sequence data. *Auk* **111**, 389–397.

Maddison, W. P., and Maddison, D. R. (1992). "MacClade: Analysis of Phylogeny and Character Evolution," version 3.0. Sinauer Associates, Sunderland, Massachusetts.

Marshall, H. D., and Baker, A. J. (1997). Structural variation and conservation of the mitochondrial control region of Fringilline finches (*Fringilla* spp.) and the Greenfinch (*Carduelis chloris*). *Mol. Biol. Evol.*, in press.

Maruyama, T. (1970). Distribution of gene frequencies in a geographically structured population. I. Distribution of neutral genes and genes with small effect. *Ann. Hum. Genet.* **35**, 411–423.

Mayr, E. (1963). "Animal Species and Evolution." Belknap Press, Cambridge, Massachusetts.

Meyer, A. (1994). Shortcomings of the cytochrome *b* gene as a molecular marker. *Trends Ecol. Evol.* **9**, 278–280.

Milligan, B. G., Mack-Leebens, J., and Strand, A. E. (1994). Conservation genetics: Beyond maintenance of marker diversity. *Mol. Ecol.* **3**, 423–435.

Mindell, D. P., and Honeycutt, R. L. (1990). Ribosomal RNA in vertebrates: Evolution and phylogenetic applications. *Annu. Rev. Ecol. Syst.* **21**, 541–566.

Mindell, D. P., Knight, A., Baer, C., and Huddleston, C. J. (1996). Slow rates of molecular evolution in birds and the metabolic rate and body temperature hypotheses. *Mol. Biol. Evol.* **13**, 422–426.

Miyamoto, M. M., Allard, M. W., Adkins, R. M., Janecek, L. L., and Honeycutt, R. L. (1994). A congruence test of reliability using linked mitochondrial DNA sequences. *Syst. Biol.* **43**, 236–249.

Moore, W. S. (1995). Inferring phylogenies form mtDNA variation: Mitochondrial-gene trees versus nuclear-gene trees. *Evolution* **49**, 718–726.

Moritz, C., Dowling, T. E., and Brown, W. M. (1987). Evolution of animal mitochondrial DNA: Relevance for population biology and systematics. *Annu. Rev. Ecol. Syst.* **18**, 269–292.

Moritz, C., Schneider, C. J., and Wake, D. B. (1992). Evolutionary relationships within the *Ensatina escholtzii* complex confirm the ring species interpretation. *Syst. Biol.* **41**, 273.

Moum, T., and Johansen, S. (1992). The mitochondrial NADH dehydrogenase subunit 6 (ND6) gene in murres: Relevance to phylogenetic and population studies among birds. *Genome* **35**, 903–906.

Nei, M. (1987). "Molecular Evolutionary Genetics." Columbia University Press, New York.

Nei, M. (1993). Age of the common ancestor of human mitochondrial DNA. *Mol. Biol. Evol.* **9**, 1176–1178.

Nei, M., and Takahata, N. (1993). Effective population size, genetic diversity, and coalescence time in subdivided populations. *J. Mol. Evol.* **37**, 240–244.

Neigel, J. E., and Avise, J. C. (1986). Phylogenetic relationships of mitochondrial DNA under various demographic models of speciation. *In* "Evolutionary Processes and Theory" (S. Karlin and E. Nevo, Eds.), pp. 515–534. Academic Press, New York.

O'Hara, R. J. (1988). Homage to Clio, or, toward an historical philosophy of evolutionary biology. *Syst. Zool.* **37**, 142–155.

O'Hara, R. J. (1993). Systematic generalization, historical fate, and the species problem. *Syst. Biol.* **42**, 231–246.

Quinn, T. W. (1992). The genetic legacy of Mother Goose—phylogeographic patterns of lesser snow goose *Chen caeurlescens caerulescens* maternal lineages. *Mol. Ecol.* **1**, 105–117.

Quinn, T. W., and Wilson, A. C. (1993). Sequence evolution in and around the mitochondrial control region in birds. *J. Mol. Evol.* **37**, 417–425.

Reiseberg, L. H., and Soltis, D. E. (1991). Phylogenetic consequences of cytoplasmic gene flow in plants. *Trends Plants* **5**, 65–84.

Saitou, N., and Nei, M. (1985). The neighbor-joining method: A new method for reconstructing phylogenetic trees. *Mol. Biol. Evol.* **4**, 406–425.

Sanderson, M. J., and Donoghue, M. J. (1989). Patterns of variation in levels of homoplasy. *Evolution* **43**, 1781–1795.

She, J.-X., Boehm, S., Wang, T. W., Bonhomme, F., and Wakeland, E. K. (1991). Amplification of MHC class II gene polymorphism by intra-exonic recombination. *Proc. Natl. Acad. Sci. U.S.A.* **88**, 453–457.

Sibley, C. G., and Ahlquist, J. E. (1985). The phylogeny and classification of the Australo–Papuan passerine birds. *Emu* **85**, 1–14.

Sibley, C. G., and Ahlquist, J. E. (1990). "The Phylogeny and Classification of Birds: A Study in Molecular Evolution." Yale Univ. Press, New Haven, Connecticut.

Slatkin, M. (1991). Inbreeding coefficients and coalescence times. *Genet. Res.* **58**, 167–175.

Slatkin, M., and Maddison, W. P. (1989). A cladistic measure of gene flow inferred from the phylogenies of alleles. *Genetics* **123**, 603–613.

Slatkin, M., and Maddison, W. P. (1990). Detecting isolation by distance using phylogenies of genes. *Genetics* **126**, 249–260.

Southern, S. O., Southern, P. J., and Dizon, A. E. (1988). Molecular characterization of a cloned dolphin mitochondrial genome. *J. Mol. Evol.* **28**, 32–42.

Stephens, J. C., and Nei, M. (1985). Phylogenetic analysis of polymorphic DNA sequences at the *Adh* locus in *Drosophila melanogaster* and its sibling species. *J. Mol. Evol.* **22**, 289–300.

Swofford, D. L. (1991). "Phylogenetic Analysis Using Parsimony (PAUP)," version 3.0s. Illinois Natural History Survey, Champaign, Illinois.

Tajima, F. (1989). Statistical method for testing the neutral mutation hypothesis by DNA polymorphism. *Genetics* **123**, 585–595.

Takahata, N. (1989). Gene genealogy in three related populations: Consistency probability between gene and population trees. *Genetics* **122**, 957–966.

Takahata, N. (1994). Comments on the detection of reciprocal recombination or gene conversion. *Immunogenetics* **39**, 146–149.

Takahata, N., and Slatkin, M. (1990). Genealogy of neutral genes in two partially isolated populations. *Theor. Pop. Biol.* **38**, 331–350.

Thomas, W. K., and Beckenbach, A. T. (1989). Variation in salmonid mitochondrial DNA: Evolutionary constraints and mechanisms of substitution. *J. Mol. Evol.* **29**, 233–245.

Vigilant, L., Stoneking, M., Harpending, H., Hawkes, K., and Wilson, A. C. (1991). African populations and the evolution of human mitochondrial DNA. *Science* **253**, 1503–1507.

Wakeley, J. (1993a). Substitution rate variation among sites in hypervariable region I of human mitochondrial DNA. *J. Mol. Evol.* **37**, 613–623.

Wakeley, J. (1993b). Substitution-rate variation among sites and the estimation of transition bias. *Mol. Biol. Evol.* **11**, 436–442.

Wenink, P. W., Baker, A. J., and Tilanus, M. G. (1993). Hypervariable-control-region sequences reveal global population structuring in a long-distance migrant shorebird, the dunlin (*Calidris alpinus*). *Proc. Natl. Acad. Sci. U.S.A.* **90**, 94–98.

Wenink, P. W., Baker, A. J., and Tilanus, M. G. (1994). Mitochondrial control-region sequences in two shorebird species, the turnstone and the dunlin, and their utility in population genetic studies. *Mol. Biol. Evol.* **11**, 22–31.

Wilson, A. C., Cann, R. L., Carr, S. M., George, M., Gyllensten, U. B., Helm-Bychowski, K. M., Higuchi, R. G., Palumbi, S. R., Prager, E. M., Sage, R. D., and Stoneking, M. (1985). Mitochondrial DNA and two perspectives on evolutionary genetics. *Biol. J. Linn. Soc.* **26**, 375–400.

Wilson, A. C., Stoneking, M., and Cann, R. L. (1991). Ancestral geographic states and the peril of parsimony. *Syst. Zool.* **40**, 363–365.

Wood, T. C., and Krajewski, C. (1996). Mitochondrial DNA sequence variation among the subspecies of Sarus Crane (*grus antigone*). *Auk* **113**, 655–663.

Wright, S. (1943). Isolation by distance. *Genetics* **28**, 114–138.

Yang, Z., Goldman, N., and Friday, A. (1994). Comparison of models for nucleotide substitution used in maximum likelihood phylogenetic estimation. *Mol. Biol. Evol.* **11**, 316–324.

Zink, R. M. (1993). Geography of mitochondrial DNA variation, population structure, and hybridization in the fox sparrow (*Passerella iliaca*). *Evolution* (in press).

Zink, R. M., and Dittmann, D. L. (1993a). Gene flow, refugia, and evolution of geographic variation in the song sparrow (*Melospiza melodia*). *Evolution* **47**, 717–729.

Zink, R. M., and Dittman, D. L. (1993b). Population structure and gene flow in the chipping sparrow (*Spizella passerina*) inferred from variation in mitochondrial DNA, and a hypothesis for the evolution of the genus *Spizella*. *Wilson Bull.* **105**, 399–413.

Zink, R. M., and Slowinski, J. B. (1995). Evidence from molecular systematics for decreased avian diversification in the Pleistocene epoch. *Proc. Natl. Acad. Sci. U.S.A.* **92**, 5832–5835.

CHAPTER 10

Phylogeny in Studies of Bird Ecology, Behavior, and Morphology

FREDERICK H. SHELDON AND LINDA A. WHITTINGHAM

Museum of Natural Science, Louisiana State University, Baton Rouge, Louisiana 70803

I. INTRODUCTION

The theory and methods of phylogenetic reconstruction have improved dramatically in the last 20 years, leading to the next logical step in evolutionary biology: the use of phylogenies to interpret historical trends in ecology, behavior, and morphology. This new field is termed "historical ecology" (e.g., Brooks, 1985). It encompasses two interrelated areas: *ecomorphology*, in which morphological evolution is interpreted in terms of ecology, and *ecophylogenetics*, in which ecology and behavior are interpreted in light of phylogeny. Historical ecology relies on the *comparative method* (e.g., Lorenz, 1950; Tinbergen, 1964; Ridley, 1983; Pagel and Harvey, 1988), which may be divided into two main approaches in this context. The first is the *homology approach* (Coddington, 1994), in which characters are optimized on (or in) phylogenetic trees and their patterns of change are observed for insight into evolutionary processes (summarized in Brooks and McLennan, 1991). The second, the *convergence approach*, began with statistical efforts to remove phylogenetic bias in comparisons of ecological characteristics among taxa (summarized in Harvey and Pagel, 1991), and now encompasses a wide range of methods for studying repeated evolutionary patterns in a phylogenetic context (e.g., Pagel, 1994). In this chapter we emphasize a combined approach, in which hypotheses of evolutionary processes (such as adaptation and phylogenetic constraint) are tested by optimization of apparently convergent characters in multiple phylogenetic settings.

The development of historical ecology may be divided into three stages. The first was the conception, in principle, of the value of phylogeny in interpreting the evolution of ecological and behavioral characteristics. Brooks and McLennan (1991) provide a review of this stage. It began with Darwin and reached a particularly rich period in the 1940s and 1950s with the development of modern ethology (Lorenz, 1941, 1950; Tinbergen, 1953, 1964). However, emphasis on historical interpretation waned in the 1960s and 1970s as ecologists began to emphasize local processes (e.g., physical environment, resource distribution, competition, and predation) as the determining forces in community development (e.g., Ricklefs, 1987) and systematists questioned the identification of behavioral homology (e.g., Atz, 1970). Ultimately, this "eclipse of history" (Brooks and McLennan, 1991) stemmed from a lack of accurate phylogenies for ecological inference.

The refinement and promulgation of cladistics and the introduction of quantitative molecular methods in the 1970s permitted movement to the second stage of historical ecology in the 1980s. Rigorously constructed, increasingly accurate, phylogenies became available to interpret ecological and behavioral patterns. During this period, the introduction of microcomputer parsimony programs, namely, MacClade (Maddison and Maddison, 1992) and PAUP (Swofford, 1993), facilitated the process of mapping and observing character change and provided simple methods for testing null models of phylogenetic effects (e.g., Maddison and Slatkin, 1991). The period also saw phylogenetic rationale, and consequently greater rigor, applied to the definition or understanding of previously slippery concepts, especially homology (e.g., Patterson, 1982, 1988), adaptation (e.g., Gould and Vrba, 1982; Coddington, 1988; Baum and Larson, 1991), and phylogenetic constraint (e.g., McKitrick, 1993). Most importantly, during this period the number of empirical studies of historical ecology increased dramatically.

The explosive growth of historical ecology has been a source of opportunity and frustration for ornithologists. Because more ecological and behavioral data are available for phylogenetic interpretation in birds than in any other major group of organisms, ornithologists have been leaders in empirical ecophylogenetics and ecomorphology (e.g., Höglund, 1989; Prum, 1990, 1994; Björklund, 1991; Lanyon, 1992; McKitrick, 1992; Richman and Price, 1992; Edwards and Naeem, 1993; Moreno and Carrascal, 1993a,b; Winkler and Sheldon, 1993; Höglund and Sillén-Tullberg, 1994). However, the number of avian phylogenetic studies has not kept pace with ecological studies, and this disparity has created a temporary crisis. It has led, in particular, to a range of standards concerning the importance of phylogenetic accuracy to ecological interpretation, with the consequence that at least a few ornithologists have undertaken ecophylogenetic studies using unsubstantiated or obviously inaccurate phylogenies (e.g., McKitrick, 1992; Harvey and Nee, 1994; Møller and Birkhead, 1994). In addition, as a consequence of initial enthusiasm, there has been a tendency among historical ecologists to overinterpret the significance of patterns of character change. We often hear, for example, that such-and-such a change is "adaptive," or that a given case of stasis is the result of "phylogenetic constraint." But the demonstration of such processes is extremely difficult. We are at a point where the rationale, methods, and statistics of historical ecology lag far behind our enthusiasm or empirical capabilities.

This brings us to the beginning of the third stage of historical ecology. The field has been through an initial period of theoretical and empirical development, and now it is time to reassess where it has been and where it needs to go. Such an assessment is difficult because it requires a consideration of all elements of historical ecology, from the gathering of phylogenetic and ecological data, to the detection of patterns in the data, to the explanation of evolutionary processes responsible for those patterns. Each of these elements has a large and often contradictory literature, and the participants in historical ecology come from widely differing backgrounds in systematics and ecology. Nevertheless, there are some basic themes running through historical ecology that permit the practical and philosophical issues to be sorted and assessed. Among these are the tenets that (1) no phylogenetic approach or statistical method is universally correct or appropriate to a problem, (2) all terms must be defined and understood equally by all participants in historical ecology, and (3) patterns and processes must be demonstrated rigorously.

Working within these guidelines, our aim is to help set the stage for the next period in historical ecology. Using examples mainly from projects on which we have worked, we explore three fundamental issues: phylogenetic accuracy, adaptation, and phylogenetic constraint. These issues are central to historical ecology and enable us to cover a broad range of subjects in the field. In the end, we hope to identify areas in which historical ecology has made substantial advances, areas where there is hope of discovery, and areas where progress will be more difficult.

II. ACCURATE PHYLOGENETIC ESTIMATES AND HISTORICAL ECOLOGY

The most important step to successful historical ecology is the collection of appropriate, accurate, ecological and phylogenetic data. Although obvious, this assertion is remarkably underemphasized in ecophylogenetic studies. Emphasis is placed primarily on the quality of ecological data, which are viewed as dependent variables, instead of on the accuracy of phylogeny, which is generally viewed as an independent variable without error (Lanyon, 1993). However, phylogenetic estimates have error distributions (Lanyon, 1993; Miyamoto and Fitch, 1995), but these are commonly ignored because they are too complicated to quantify.

When phylogenetic error is recognized in historical ecological studies, it is often rationalized or discounted. Some historical ecologists feel that if comparative studies include enough ecological data and phylogenetic estimates for a wide array of taxa evolutionarily informative patterns will emerge regardless of errors in specific trees. Other historical ecologists, notably cladists, feel that the accuracy of a tree is moot because a tree is a hypothesis and simply the best that one can do with given data. Thus, all historical analyses are, in some sense, preliminary, and the accuracy of the hypothesis will increase with the collection and comparison of additional data. This is a reasonable view. However, when a preliminary phylogenetic hypothesis is clearly inaccurate, although rigorously constructed, it is not an acceptable premise to ecophylogenetic analysis.

This disparity in views on the importance of phylogenetic accuracy is complicated by arguments among systematists who support different methods to assess accuracy. These arguments center around two alternative approaches: *taxonomic congruence* (the *consensus approach*), in which trees derived from distinct phylogenetic data sets are compared for agreement in branching patterns (e.g., Cracraft and Mindell, 1989; Bledsoe and Raikow, 1990; Miyamoto and Cracraft, 1991; Sheldon and Bledsoe, 1993), and *character congruence* (the *combined approach*), in which all data sets are combined to produce a single best estimate of phylogeny based on "total evidence" (e.g., Miyamoto, 1985; Cracraft and Mindell, 1989; Kluge, 1989; Kluge and Wolf, 1993).

Both approaches have appeal, but both have limitations. In taxonomic congruence analysis, if two data sets produce trees that concur, there is strong probabilistic support of phylogeny (e.g., Miyamoto and Fitch, 1995). However, if branching patterns disagree, then a decision must be made as to which data set provides the better estimate of phylogeny, and this decision is subjective. In character congruence, the single best estimate of phylogeny is based on evidence, as opposed to inference (Kluge, 1989; Kluge and Wolf, 1993). However, if one (or all) of the combined data sets are "positively misleading" (Felsenstein, 1978), an incorrect tree may be produced, and faith in such a tree will be misplaced (Bull *et al.*, 1993; de Queiroz, 1993; Miyamoto and Fitch, 1995). Of course, when one (or more) of the alternative data sets consists of obligate distances (as produced by microcomplement fixation or DNA hybridization), the test of congruence must be taxonomic congruence, because distances cannot be combined with character data. Thus, taxonomic congruence is an important method in avian systematics because Sibley and Ahlquist (1990) compared a wide variety of groups by DNA hybridization and many historical ecologists employ their phylogenetic estimates (e.g., Moreno and Carrascal, 1993a,b; Harvey and Nee, 1994; Møller and Birkhead, 1994).

We believe that historical ecologists must identify strong and weak parts of phylogenetic estimates before attempting ecophylogenetic analyses. Otherwise they will not be able to differentiate between likely and only possible scenarios of character evolution. The assessment of branch support requires an understanding of tree-building and testing methods and a reevaluation of all pertinent data. However, it does not require that ecologists collect phylogenetic data, as suggested by Kluge and Wolf (1993). To do so would be a highly inefficient use of time and skills in this age of specialization.

For single data sets, branch robustness may be determined by approaches such as bootstrapping (Felsenstein, 1985), jackknifing (Lanyon, 1985), and decay analysis (e.g., Bremer, 1988); in addition, overall phylogenetic information content can be assessed via such tests as Hillis's (1991) *g* test. Multiple data sets should be analyzed both for character and taxonomic congruence. Because data are lumped in character congruence, tests of tree quality are the same as for a single data set. Taxonomic congruence may be assessed descriptively by simple comparison of trees produced by different data sets, or statistically using randomized data [e.g., component analysis (Page, 1993)] or data error distributions (e.g., Templeton, 1983; Kishino and Hasegawa, 1989). In any event, if substantially different trees are produced by different data sets, then the single tree produced by character congruence should be viewed cautiously because one (or all) of the data sets has problems (Bull *et al.*, 1993; de Queiroz, 1993). Lanyon (1993) suggested that in this situation it is useful to produce a strict or majority-rule consensus tree for each data set, in which only strongly supported branches are resolved. Consensus trees from different data sets should then be compared to one another for congruence. Although the consensus trees from individual data sets may not be well resolved, in many cases when they are compared to one another, resolved parts in one tree may complement unresolved parts in other trees. The result is a reliable "phylogenetic framework" for historical ecological study.

A. An Example of the Phylogenetic Framework Approach

In Fig. 10.1, we provide a simple example of how published data sets may be compared for taxonomic congruence to produce a phylogenetic framework (Lanyon, 1993). Figure 10.1A pairs two alternative trees that estimate the phylogeny of day herons (Ardeidae: Ardeinae). Tree (a) is the best-fit DNA hybridization tree from Sheldon (1987a,b); tree (b) is the most parsimonious Wagner tree based on the osteological study of Payne and Risley (1976). By simple congruence analysis, the two trees disagree in several respects. For example, they conflict in the placement of the cattle egret (*Bubulcus ibis*). From the perspective of an historical ecologist, this is unfortunate because the cattle egret is an interesting heron in that it is an upland feeder with apparent adaptations to that life style (e.g., several osteological characters associated with its short legs, neck, and bill; Payne and Risley, 1976). To understand the evolution of these characters requires that they be compared to those of

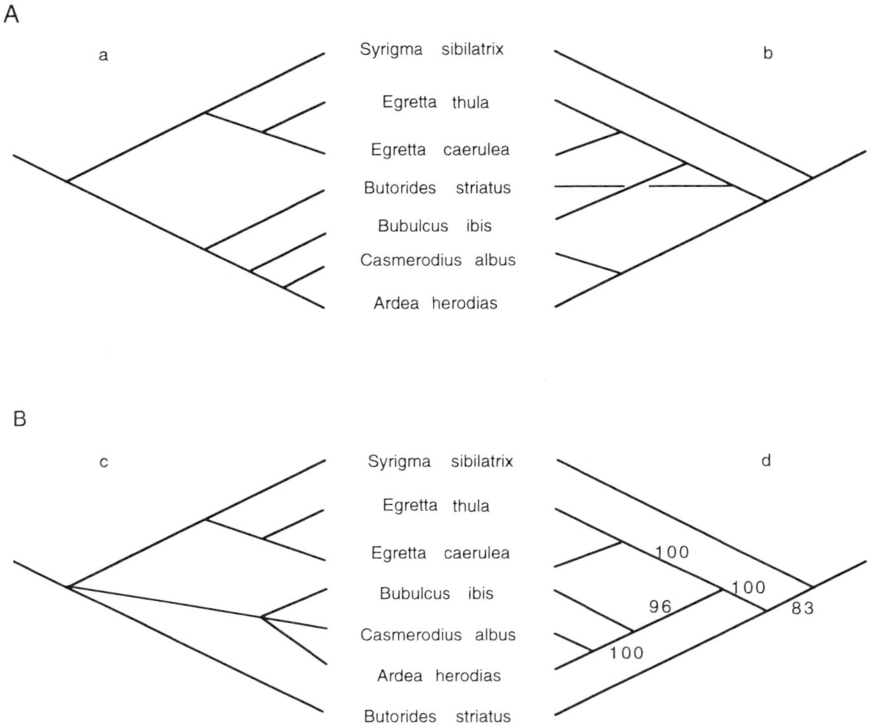

FIGURE 10.1 (A) A taxonomic congruence assessment of day heron phylogeny featuring (a) the best-fit DNA hybridization tree of Sheldon (1987a, Fig. 1) and (b) the most parsimonious Wagner tree of Payne and Risley (1976, Figs. 34 and 35), based on osteological characters. (B) A congruence assessment following reanalysis of the original data sets: (c) DNA hybridization data have been assessed by jackknifing taxa and bootstrapping distances to produce a strict consensus tree (as described in Sheldon and Winkler, 1993); (d) osteological data were subjected to modern analysis using global parsimony in PAUP (Swofford, 1993), multiple (instead of composite) outgroups, and bootstrapping (Felsenstein, 1985) to produce a majority rule tree (K. McCracken and F. H. Sheldon, unpublished analysis). Values on tree (d) represent percentage bootstrap support of branches.

the cattle egret's sister taxon and closest relatives (Sheldon and Gill, 1996). The identification of these relatives, in turn, requires the knowledge of day heron phylogeny. However, from the available studies, we do not have a good picture of day heron phylogeny.

In Fig. 10.1B, we compare trees from the same data sets after having tested the robustness of their branching patterns. To construct the DNA hybridization tree (c), we jackknifed (Lanyon, 1985) and bootstrapped (Krajewski and Dickerman, 1990) the data, and for the morphological tree (d), we used PAUP (Swofford, 1993) and bootstrapping (Felsenstein, 1985). The result of the reanalysis is that the morphological tree is much more congruent with, and helps to resolve, the DNA hybridization tree. Conversely, the DNA hybridization tree fortifies several key branches of the morphological tree. The only disagreement that remains is over the position of the whistling heron (*Syrigma sibilatrix*); one or both trees incorrectly place this species. Most importantly, we have resolved the sister relationship of the cattle egret and, thus, have a firmer phylogenetic framework for a comparative morphological study of its upland habits.

III. APPLICATION OF THE COMPARATIVE APPROACH TO CLASSIC PROBLEMS OF EVOLUTION

Given the problems inherent in phylogenetic estimation and the energy and effort required to gather complete ecological, behavioral, or morphological data for any sizable set of taxa, it would seem that the most difficult part of historical ecology is the initial accumulation of basic natural history data. However, as we have already suggested, it is equally difficult to detect meaningful patterns of character change from the basic data and far more difficult to demonstrate the causes of these patterns. To illustrate these difficulties, we discuss them in terms of the two evolutionary questions most frequently addressed via the comparative method: adaptation and phylogenetic constraint.

A. Adaptation

Gould and Vrba (1982), Coddington (1988, 1994), Baum and Larson (1991), and others have helped to develop a definition of adaptation in the context of phylogeny. An adaptation is an apomorphic feature that evolved in response to an apomorphic function (Coddington, 1994). It has current utility and was generated historically through the action of natural selection for its current biological role (Baum and Larson, 1991). By extension, convergent characters are adaptations for the same function in distinct lineages (i.e., not synapomorphies and not accidentally similar).

Despite (or perhaps because of) the increasing rigor in definition, it is extremely difficult to demonstrate adaptation through the comparative methods of historical ecology. Several authors suggest that hypotheses of adaptive character evolution can be tested directly through a phylogenetic examination of the character trait in relation to the selective environment. Coddington (1988) and Miles and Dunham (1993) argued that adaptation is evident when a trait change occurs at the same location within a phylogeny as the environmental change (Fig. 10.2a), whereas Baum and Larson (1991) suggested that adaptation is evident only when the environmental change precedes the trait change (Fig. 10.2b). However, identifying a correlation between environmental and trait change is only part of the process of testing adaptation. A causal relationship must be demonstrated between environ-

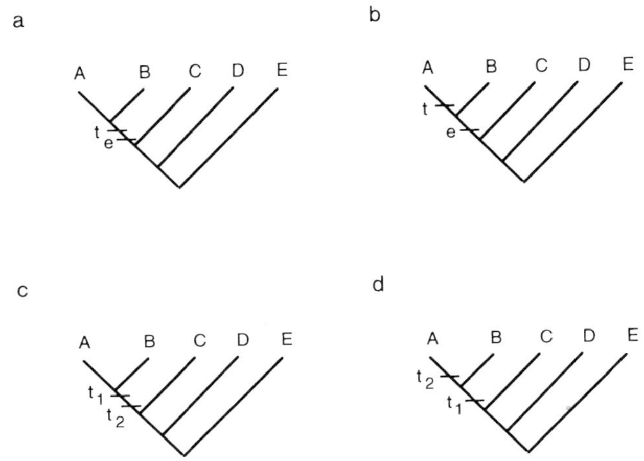

FIGURE 10.2 Scenarios used to test adaptation and phylogenetic constraint. (a) The origin of a novel trait *t* coincides with the origin of the novel environment *e*. This scenario is evidence of adaptation according to Coddington (1988) and Miles and Dunham (1993). (b) Trait *t* evolves subsequent to the environmental change *e*. This scenario is evidence of adaptation according to Baum and Larson (1991). (c) The origin of a novel trait t_1 coincides with the origin of another trait t_2. This is evidence that one character is constraining another (McKitrick, 1993) or that the development (adaptation) of one character depends on the appearance of another, as in (a). (d) The origin of t_1 precedes the origin of t_2. This scenario refutes phylogenetic constraint of character t_1 by t_2 (McKitrick, 1993).

mental and trait changes; that is, that natural selection is responsible for the establishment and maintenance of a character.

The difficulty in demonstrating adaptation stems from meeting all these requirements. A causal relationship implies that the environmental change preceded the trait change. Thus, reliable paleoecological data must be available to characterize the environment, and a reliable phylogeny must be available to demonstrate timing. The phylogeny must have at least two lineages appearing after the environmental change, one with the putative adaptation and one without (e.g., Fig. 10.2b). Otherwise, it is impossible to determine unambiguously by parsimony whether the environmental change occurred before or concurrently with the trait change. Not only must data be available on paleoenvironmental conditions and timing, but character and environmental changes in intervening time must meet certain assumptions. It must be shown that the adaptive trait was more beneficial than alternative character states that never spread or that disappeared through extinction (Dobson, 1985). In addition, the trait must have been favored because it conferred its current adaptive function, i.e., is not an exaptation (Gould and Vrba, 1982). This implies that the selective environment has not changed substantially between the time when the trait arose and the present. Some biologists feel that this assumption is unlikely to hold in most cases (Frumhoff and Reeve, 1994). Others feel that it is reasonable.

Pagel (1994, p. 38), for example, noted it is a peculiar view of evolution to suppose "... that traits are labile and evolve for various functions until the organism chances upon using the trait for its current function, at which time there is no further modification of the trait."

Given the difficulty of determining past environmental conditions and organismal interactions, most inferences of adaptation quite reasonably rely on correlations of character changes in a phylogenetic context (e.g., Donoghue, 1989; Maddison, 1990). A change in morphology, behavior, or ecology may be related (1) to a change in physical environment or (2) to a previous change in morphology, behavior, or ecology, which changes the functional environment. In principle, adaptation is demonstrated when it can be shown repeatedly that a particular character appears following the development of a particular environment. Similarly, if it can be shown repeatedly that a particular trait appears following the development of another trait, then selection of the second trait has been shown to be favored by the appearance of the first trait (e.g., Donoghue, 1989; Prum, 1990). Unfortunately, most examples of character "correlation" consist of a single historical association between a trait and an environmental change, or between two trait changes. A single coincident event provides little evidence of a correlative relationship, let alone a causal relationship. In contrast, cases in which character transformations occur in a predicted sequence in multiple independent lineages provide more substantial evidence of causal relationships (Frumhoff and Reeve, 1994; Leroi *et al.*, 1994; Pagel, 1994). In such circumstances, statistical tests based on null models of chance occurrence (e.g., Ridley, 1983; Maddison, 1990) assume real power. Thus, to investigate hypotheses concerning the evolutionary sequence of traits, it is necessary to focus on traits that have multiple origins (i.e., convergent or homoplastic characters). Although such characters are a nuisance to phylogeneticists, they hold a wealth of information about evolutionary processes.

Although repeated independent correlations among traits provide strong evidence for adaptation, selection is only one of several evolutionary mechanisms that can produce such patterns (Frumhoff and Reeve, 1994; Leroi *et al.*, 1994). They can also occur as the result of interactive genetic processes such as pleiotropy or genetic linkage (see descriptions of how these genetic systems cause false correlations in Section III,B). Most studies offering adaptive explanations do so in the absence of direct evidence about selection or the genetic interaction among traits. Direct evidence of evolutionary mechanisms can be obtained only if (1) natural selection is assessed experimentally and (2) the phenotypic and genetic covariance structure among traits is measured. While such detailed investigations may provide secure conclusions about adaptation, they set a high standard and require a long-term commitment of energy and money.

The comparative method can begin the process of distinguishing between correlations caused by natural selection, drift, and genetic interactions, if the nature of these forces is considered carefully. Drift is the easiest factor to discount in correlation analyses because it is not expected to cause consistent convergent changes in multiple lineages. Selection and genetic interactions are more difficult to distinguish, but the two may be teased apart by considering fundamental differences in how they produce environment/trait and trait/trait correlations. Selection moves relatively slowly; the environment changes first and subsequently the trait changes, or one trait changes and then subsequently the other changes. When genetic interactions are at play in trait/trait correlations, the effect is instantaneous because one gene acts directly on another (e.g., by pleiotropy). Thus, if trait/trait correlations appear simultaneously (i.e., on the same branch), genetic interactions are possible. If trait/trait correlations appear sequentially, e.g., following cladogenetic events, selection is a more likely explanation. A more difficult situation concerns the distinction between selection and genetic interactions when the correlations are of the environment/trait variety. In such cases, genetic interactions can be viewed as a chain reaction, in which the environment affects an unrecognized trait that, in turn, affects the recognized trait. Thus, the timing of genetic interactions in environment/trait cases is likely to be sequential and may be indistinguishable from the pattern produced by selection. Even so, it may still be possible to differentiate between the two forces using the logic of Simpson (1944). Simpson noted that if the dependent (second) trait is highly consistent among clades, genetic effects are suggested because selection is a more haphazard process and would be expected to produce substantial variation.

Prum (1990, 1994) discovered and outlined a convincing example of a causal relationship between two traits: multiple origins of elaborate plumage ornaments and display behaviors in the manakins (Pipridae). Prum predicted that if the derived display behaviors were distributed more generally among taxa than the plumage novelty, then the behavior evolved prior to the plumage, and the hypothesis that plumage has evolved as a consequence of the display would be corroborated. If the plumage were more generally distributed, then the opposite hypothesis would be supported. In manakins, derived male plumage traits have evolved subsequently to the behavioral novelties in which they are prominently featured. This implies that behavioral diversification is driving some aspects of morphological diversification within the family.

Adaptations and Key Innovations: Parus Example

In their review of the subject, Heard and Hauser (1995, p. 152) defined a key innovation as ". . . an evolutionary change in individual trait(s) that is causally linked to an increased diversification rate in the resulting clade (for which it is a synapomorphy)." A key innovation, therefore, would be a special case of adaptation: an environmental condition exists, a taxon acquires a trait that is selectively advantageous in that environment, and radiation ensues. The difference between an adaptation and a key innovation is radiation; the key innovation must create circumstances in which diversification increases in those lineages having the trait relative to those that lack the trait. The number of species would be expected to increase if

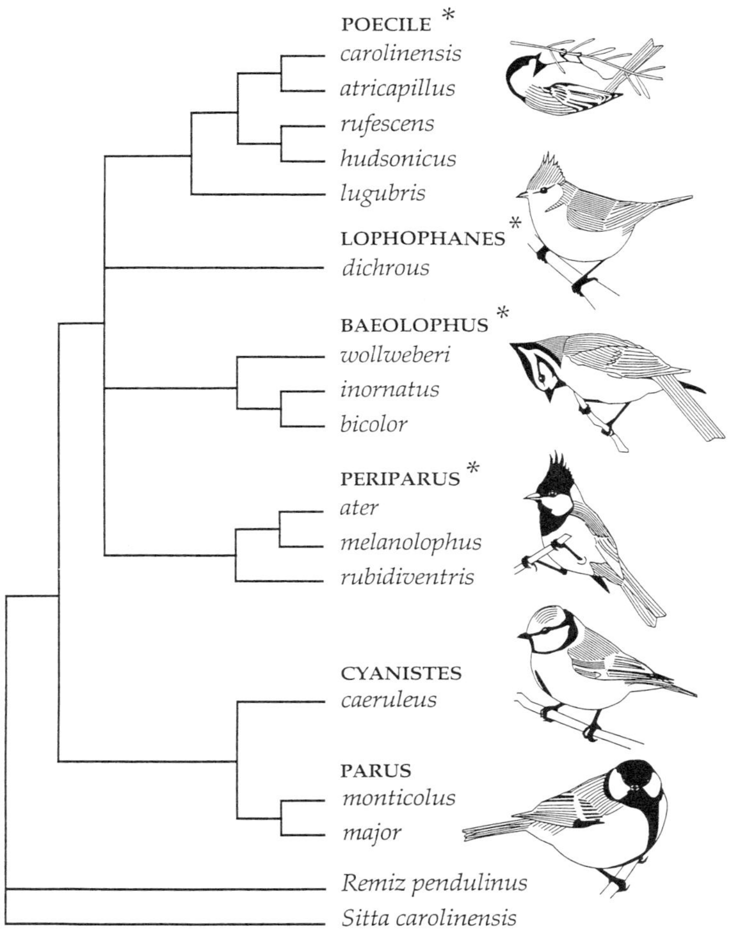

FIGURE 10.3 The phylogeny of *Parus* estimated by Sheldon *et al.* (1992) and Slikas *et al.* (1996). Subgenera that cache seeds are marked with asterisks.

the fitness conferred by the trait increased the longevity or range of individual species, thereby creating opportunities for speciation by vicariance or dispersal, or diminishing the likelihood of extinction (Heard and Hauser, 1995).

An example of a putative adaptation that may also be a key innovation is seed caching in the genus *Parus*, chickadees and titmice (Sheldon and Gill, 1996). This hypothesis is based on the observation that *Parus* is divided into two lineages (Fig. 10.3). One of these (blue and great tits) consists of seven species (Eck, 1988), none of which is known to cache seeds. The other lineage consists of 23 species (Eck, 1988), all of which apparently cache seeds (e.g., Ekman, 1989). Thus, seed

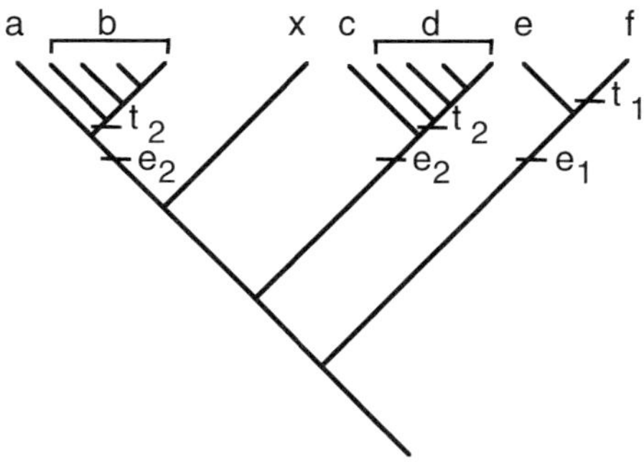

FIGURE 10.4 Scenario required to support a hypothesis of adaptation, as in the case of seed caching by *Parus*. *e* is an environmental change hypothesized to be responsible for the selection of *t*. There must be multiple independent instances in which a trait arises in a given environment and results in radiation. Moreover, lineages in the same environment that lack the trait must be relatively species poor, as may lineages with the trait arising in different environments.

caching is a synapomorphy for the speciose clade, and it seems to be largely responsible for the radiation of parid species in that clade. By ensuring a predictable supply of food in winter, seed caching would permit extensive exploitation of coniferous and deciduous forests.

The test for this hypotheses is the same as any comparative test of adaptation, except that the trait in question must be associated with a relative increase in the number of species. Therefore, we need to examine caching and noncaching lineages in other groups that live in the same environment (Fig. 10.4). The environmental criterion is essential because the function for which the caching develops must be the same in all taxa. If it were not, we would not be examining the same phenomenon, and the hypothesis would likely be falsified. For example, if we referred to bowerbirds, some of which are known to cache (Pruett-Jones and Pruett-Jones, 1985), we might not find that caching is correlated with diversification. But bowerbirds live in tropical and subtropical forest, where the supposed advantages of caching to parids (i.e., winter food) do not apply. Similarly, we probably would not include cachers such as woodpeckers or squirrels in our comparisons, even though they may live in the same environment, because they are phylogenetically remote and many elements of their biology are bound to differ (Pagel, 1994). Instead, we should look to other temperate coniferous and deciduous forest passerines that cache: namely nuthatches (Sittidae) and crows, jays, and nutcrackers (Corvidae). These are both oscine groups, but neither is sister taxon to the Paridae (Sibley and Ahlquist, 1990; Sheldon and Gill, 1996). Thus, they are relatively close genetically, but have acquired seed caching independently of the Paridae (i.e., seed caching is not a synapomorphy). Moreover, they share a similar range of nuclear DNA divergence with the titmice (ca. 1–4%; Sibley and Ahlquist, 1990), suggesting a temporally coincident radiation (given an approximate molecular clock) that is possibly driven by the same environmental forces as that of the Paridae.

Unfortunately, the phylogenies of sittids and pertinent corvids have not been well studied, and although knowledge of caching in certain taxa is extensive (e.g., nutcrackers; Balda and Kamil, 1989), information on the patterns and specifics of caching for many taxa is unknown. Moreover, the influence of habitat on the evolution of caching in all three families is also poorly known. Thus, considerably more work is required to test the hypothesis. However, the elements of a truly rigorous analysis of seed caching as an adaptation are in place. Similar possibilities for the rigorous study of adaptations and key innovations exist throughout the Passeriformes, as this order is rife with examples of multiple convergent evolution (e.g., to seed eating, nectivory, trunk probing, leaf gleaning, and flycatching) and subsequent radiation (e.g., Bledsoe, 1988; Sibley and Ahlquist, 1990; Sheldon and Gill, 1996).

B. Phylogenetic Constraints

In contrast to adaptive change, one may find that a certain trait occurs throughout a monophyletic group and varies little among members of that group. This character uniformity may be attributed to *phylogenetic constraint*, which has been defined as "any result or component of the phylogenetic history of a lineage that prevents an anticipated course of evolution in that lineage" (McKitrick, 1993). That is, phylogenetically constrained traits are expected, *a priori*, to vary in response to variable selection, but they resist adaptive modification as a result of inherited genetic conditions.

The difficulty that phylogenetic constraint presents to historical ecologists is that phenotypic variation may be limited for several reasons, some of which may be considered phylogenetic constraint and some of which cannot (Edwards and Naeem, 1993; Frumhoff and Reeve, 1994; Leroi *et al.*, 1994). Phenotypic variation may be limited because of (1) a lack of genetic variation, (2) pleiotropy, in which a single gene underlies the expression of several traits and the evolution of one of those traits can be constrained by selection on others, (3) gene linkage, in which a change in one gene may be restricted by its proximity to other genes on the same chromosome, and (4) stabilizing selection, in which a trait is maintained by selection against alternative phenotypes.

The first three limitations may be considered phylogenetic constraints; the lack of variation is caused by inherited genetic conditions whose momentum or complexity make evolutionary change unlikely or difficult to induce (*phylogenetic inertia*). In contrast, stabilizing selection maintains stasis on the basis of current ecological interactions and not common ancestry; hence we refer to it as an *ecological constraint*. We emphasize one further distinction. Although a trait may be correlated strongly with phylogeny (e.g., Winkler and Sheldon, 1993), such a correlation is not a demonstration of phylogenetic constraint. Instead, it is a *phylogenetic effect* (e.g., Miles and Dunham, 1993) because (1) there is no *a priori* expectation of variation or directionality and (2) the cause may be phylogenetic constraint, ecological constraint, or a combination of the two.

Some authors have observed that all species of gulls in the genus *Larus* lay three eggs (Graves *et al.*, 1984; McLennan *et al.*, 1988), even though there may be selective disadvantages to doing so (e.g., insufficient or overabundant food to feed three nestlings). Similarly, all species in the family Megapodidae incubate their eggs in a variety of habitats using heat sources other than body temperature (Ligon, 1993). On the basis of consistency in the face of expected variation, these authors conclude that these traits are phylogenetically constrained within these groups. The idea that character evolution is constrained by phylogeny, however, is only a hypothesis about evolutionary processes based on a pattern. To establish constraint, the hypothesis must be tested.

Although Frumhoff and Reeve (1994) suggested that it is not possible to assess whether a character's presence in extant taxa results from phylogenetic constraint on adaptive evolution, other workers have felt that a hypothesis of phylogenetic constraint is testable. McKitrick (1993) proposed an examination of the relative timing of the historical sequence of character evolution to test the hypothesis of phylogenetic constraint (e.g., Fig. 10.2). McKitrick maintained that a hypothesis of phylogenetic constraint is supported when two interrelated, mutually constraining traits arise at the same point within the phylogeny (Fig. 10.2c). Alternatively, McKitrick argued that a hypothesis of constraint is falsified when the evolution of the supposed constrained trait precedes the evolution of the trait thought to constrain it (Fig. 10.2d). Interestingly, these analyses of trait-appearance patterns are analogous to the tests that Coddington (1988) and Baum and Larson (1991) promoted to support hypotheses of adaptation (Fig. 10.2a and b, respectively).

Such an approach to testing phylogenetic constraint is extremely difficult, if not impossible. The same test is used to demonstrate adaptation (an ecological phenomenon) and constraint (a phylogenetic phenomenon). This dilemma demonstrates the hierarchical interrelationship between adaptation and constraint. As Ligon (1993) pointed out, adaptation at one level is constraint at another. Indeed, the identification of a "key" adaptation, i.e., one that underlies a radiation, rests on the phylogenetic conservativeness of that trait at a lower hierarchical level (see the swallow example in the next section).

Because of these problems, hypotheses of phylogenetic constraint may be examined more effectively if restricted variation in a trait is associated with fluctuations in the selective environment (Edwards and Naeem, 1993; Ligon, 1993; Miles and Dunham, 1993). Edwards and Naeem (1993) proposed that cooperative breeding in Australian birds persists even though the various species occupy different ecological regimes. They note that the lack of a relationship between cooperative breeding and habitat refutes the hypothesis that cooperative breeding necessarily reflects responses to current environmental or ecological conditions. The rejection of this hypothesis is an important advance in understanding the evolution of cooperative breeding. Of course, once an adaptive scenario is disproved, another may be erected in its place for testing. For example, in this instance, habitat may not be the primary selective force in the evolution of cooperative breeding; perhaps the force is some other selective agent, such as the availability of mates (e.g., Pruett-Jones and Lewis, 1990). However, a careful consideration of the natural history of the study group should make it possible to limit the number of ad hoc scenarios that require testing.

The first step in indentifying phylogenetic constraint is, thus, to show that the lack of phenotypic variation is more likely to be the result of ancestral genetic effects than selection. Eventually, we may be able to determine the genetic basis of particular traits (through breeding experiments or molecular analyses) and mapping those genetic attributes onto phylogenies. The strength of the hypothesis then must be assessed by documenting significant selection differences among species. If phylogenetic constraint is the likely explanation for stasis, the analytical focus should then shift to identifying the genetic mechanism of constraint.

An Avian Example of Issues in Phylogenetic Constraint

In Fig. 10.5, we present an example of difficulties encountered when arguing for phylogenetic constraint. Figure 10.5 presents an estimate of swallow intergeneric phylogeny (Sheldon and Winkler, 1993). When nest structure is mapped onto the tree and randomized data sets are compared (via the randomization function in MacClade; Maddison and Maddison, 1992), a strong phylogenetic effect for nest

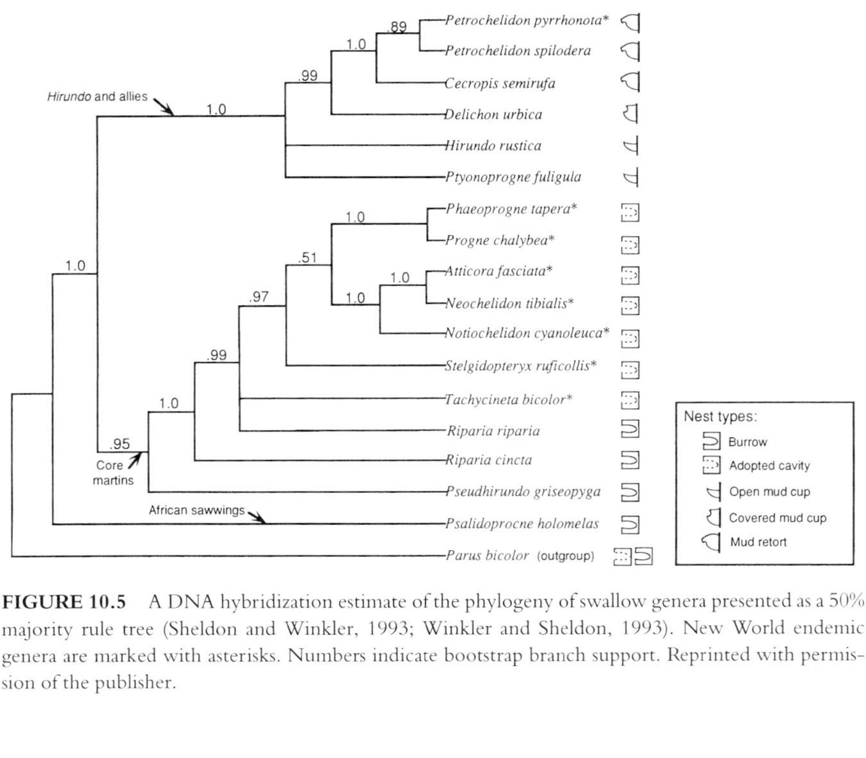

FIGURE 10.5 A DNA hybridization estimate of the phylogeny of swallow genera presented as a 50% majority rule tree (Sheldon and Winkler, 1993; Winkler and Sheldon, 1993). New World endemic genera are marked with asterisks. Numbers indicate bootstrap branch support. Reprinted with permission of the publisher.

type is indicated (Winkler and Sheldon, 1993). For example, all members of one speciose clade (*Hirundo, sensu lato*) use mud to build their nests. Members of its sister group either dig their nests in sandy soil or adopt holes or niches in trees and cliffs. Despite this consistency, substantial variation exists within the major nest-construction themes. The mud nesters, for example, construct a variety of nest types, from simple cups (e.g., barn swallows, *Hirundo rustica*) to enclosed globes with entrance tunnels [e.g., cliff swallows (*Petrochelidon pyrrhonota*)]. The Australian tree martin (*Petrochelidon nigricans*) adopts a hole in a tree, but lines or dams it with mud.

Are the swallows in these clades phylogenetically constrained, or does selection determine the structure of nests in the major clades? Most mud nesters live in sub-Saharan Africa, where tree holes are at a premium and mud is freely available. Most hole adopters live in the New World tropics, where mud nests may be less adaptive. Emlen (1954), for example, noted that mud nests of swallows can crumble in conditions of high humidity, even without being directly moistened. If habitat controls nest type, the basic nesting strategy (mud building, adoption, burrowing) may be considered a key adaptation (Winkler and Sheldon, 1993). If genetics controls nest type, then nesting strategy is constrained.

The problem is that variations on the three themes could be simple modifications within the confines of stabilizing selection, phylogenetic constraint, or a combination of the two (Winkler and Sheldon, 1994). Moreover, the outwardly simple pattern is confused by exceptions. The purple martin (*Progne subis*), for example, is undoubtedly a member of the large core martin clade (Fig. 10.5). It is a well-known hole adopter and a New World endemic (like all hole adopters in the core martin clade). Even so, purple martins are known occasionally to build open nests with mud walls (F. H. Sheldon and L. A. Whittingham, personal observation). Does this mean that all swallows have the genetic capacity to build mud nests, and it is only globally manifested in one major clade? Perhaps hole nesting is the phylogenetic constraint, and burrowing, adopting a hole, and building a mud nest with an entrance tunnel are variations on that theme. The point is that any number of stories that invoke phylogenetic constraint can be formulated, and the demonstration of constraint in this instance will require a careful study of nest types in variable environments.

IV. SUMMARY

Historical ecology is burgeoning because it provides structure to the study of ecological patterns and evolutionary processes. However, in early efforts to apply historical methods, some ecologists and systematists have neglected the importance of accurate phylogenies to ecophylogenetic analysis, and others have been too hasty to invoke adaptation and phylogenetic constraint to explain patterns of character development. We have emphasized the use of accurate phylogenetic data because the

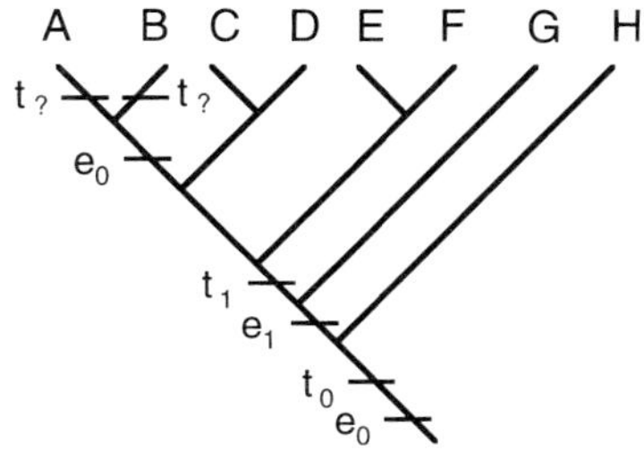

FIGURE 10.6 Use of phylogenies to predict character distributions. The state of trait $t_?$ is unknown in taxa A and B. However, the pattern of trait (*t*) covariation with environmental change (*e*) suggests that $t_?$ is likely to be t_0. By studying A and B, the hypothesis that $t_?$ is t_0 can be tested and additional evidence may be gathered to support or refute the influence of *e* on *t*.

interpretation of evolutionary patterns obviously changes as relationships among taxa change. We have also emphasized the difficulty in demonstrating adaptation and constraint in the hope that more rigor will be applied to the study of these phenomena. In doing so, we have outlined approaches that investigators might take to examine trait evolution and interaction. Although progress will be limited without a knowledge of the quantitative genetics of specific traits, we think that initial hypothesis testing is possible with prudent use of the phylogenetic approach, provided that multiple examples of potential evolutionary phenomena are examined. For example, a hypothesis of adaptation is strengthened substantially if it can be shown that the putative adaptation has evolved convergently in several distinct lineages under highly similar environmental conditions (e.g., Pagel, 1994). For phylogenetic constraint, alternative explanations for stasis should be ruled out by testing for trait constancy in a variety of selective regimes.

An important benefit of the historical approach is that it may be used to generate hypotheses and focus future research. For example, we can identify traits that occur multiple times within a phylogeny, and thus may provide productive grist for quantitative genetic studies. We can also predict the characteristics of unstudied taxa. Fig. 10.6 is a hypothetical example showing that trait *t* appears to be related to an environmental condition *e*. A and B may be unstudied taxa where the state of *t* is unknown, but their environment is known (e_0). We can predict that A and B will have trait t_0, if this trait evolves in response to e_0. Phylogenies can also be used to test evolutionary models, such as sexual section models of female mate choice. For example, the sensory bias hypothesis predicts that female preferences for exaggerated male secondary sexual traits evolved before the male trait (e.g., Basolo, 1990; Hill, 1994). This hypothesis can be tested by observing in the phylogeny the sequence in which the male trait and the female preference evolved. In short, the predictive strength and usefulness of the historical approach are limited only by the availability of accurate ecological and phylogenetic data.

ACKNOWLEDGMENTS

We thank P. Dunn, F. Gill, D. Ligon, D. Mindell, R. Prum, and B. Slikas for comments on the manuscript. *Proceeding of the National Academy of Sciences, USA* and the *Journal of Avian Biology* kindly gave permission to reproduce figures. This research was supported by Grants NSF/BSR 90–20183, NSF/BSR-92-07991, and NSF/LaSER 1992-96-ADP-02.

REFERENCES

Atz, J. W. (1970). The application of the idea of homology to behavior. *In* "Development and Evolution of Behavior" (L. R. Aronson, E. Tobach, D. S. Lehrman, and J. S. Rosenblatt, Eds.), pp. 51–74. Freeman and Co., San Francisco.

Balda, R. P., and Kamil, A. C. (1989). A comparative study of cache recovery by three corvid species. *Anim. Behav.* **38**, 486–495.

Basolo, A. L. (1990). Female preference predates the evolution of the sword in swordtail fish. *Science* **250**, 808–810.

Baum, D. A., and Larson, A. (1991). Adaptation reviewed: A phylogenetic methodology for studying character macroevolution. *Syst. Zool.* **40**, 1–18.

Björklund, M. (1991). Evolution, phylogeny, sexual dimorphism and mating system in the grackles (*Quiscalus* spp.: Icterinae). *Evolution* **45**, 608–621.

Bledsoe, A. H. (1988). Nuclear DNA evolution and phylogeny of the New World nine-primaried oscines. *Auk* **105**, 504–515.

Bledsoe, A. H., and Raikow, R. J. (1990). A quantitative assessment of congruence between molecular and nonmolecular estimates of phylogeny. *J. Mol. Evol.* **30**, 247–259.

Bremer, K. (1988). The limits of amino acid sequence data in angiosperm phylogenetic reconstruction. *Evolution* **42**, 795–802.

Brooks, D. R. (1985). Historical ecology: A new approach to studying the evolution of ecological associations. *Ann. Missouri Bot. Garden* **72**, 660–680.

Brooks, D. R., and McLennan, D. A. (1991). "Phylogeny, Ecology, and Behavior." Univ. Chicago Press, Chicago.

Bull, J. J., Huelsenbeck, J. P., Cunningham, C. W., Swofford, D. L., and Waddell, P. J. (1993). Partitioning and combining data in phylogenetic analysis. *Syst. Biol.* **42**, 384–397.

Coddington, J. A. (1988). Cladistic tests of adaptational hypotheses. *Cladistics* **4**, 3–22.

Coddington, J. A. (1994). The roles of homology and convergen in studies of adaptation. *In* "Phylogenetics and Ecology" (P. Eggleton and R. I. Vane-Wright, Eds.), pp. 53–78. Academic Press, London.

Cracraft, J., and Mindell, D. P. (1989). The early history of modern birds: A comparison of molecular and morphological evidence. *In* "The Hierarchy of Life" (B. Fernholm, K. Bremer, and H. Jornvall, Eds.), pp. 389–403. Exerpta Medica, Amsterdam.

de Queiroz, A. (1993). For consensus (sometimes). *Syst. Biol.* **42**, 368–372.

Dobson, F. S. (1985). The use of phylogeny in behavior and ecology. *Evolution* **39**, 1384–1388.

Donoghue, M. J. (1989). Phylogenies and the analysis of evolutionary sequences, with examples from seed plants. *Evolution* **3**, 1137–1156.

Eck, S. (1988). Gesichtspunkte zur Art-Systematik der Meisen (Paridae). *Zoologische Abhandlungen Staatliches Museum fur Tierkunde Dresden* **43**, 101–134.

Edwards, S. V., and Naeem, S. (1993). The phylogenetic component of cooperative breeding in perching birds. *Am. Nat.* **141**, 754–789.

Ekman, J. (1989). Ecology of non-breeding social systems of *Parus. Wilson Bull.* **101**, 263–288.

Emlen, J. T. (1954). Territory, nest building, and pair formation in the cliff swallow. *Auk* **71**, 16–35.

Felsenstein, J. (1978). Cases in which parsimony or compatibility methods will be positively misleading. *Syst. Zool.* **27**, 401–410.

Felsenstein, J. (1985). Confidence limits on phylogenies: An approach using the bootstrap. *Evolution* **39**, 783–791.

Frumhoff, P. C., and Reeve, H. K. (1994). Using phylogenies to test hypotheses of adaptation: A critique of some current proposals. *Evolution* **48**, 172–180.

Gould, S. J., and Vrba, E. S. (1982). Exaptation—a missing term in the science of form. *Paleobiology* **8**, 4–15.

Graves, J., Whiten, A., and Henzi, P. (1984). Why does the herring gull lay three eggs? *Anim. Behav.* **32**, 798–805.

Harvey, P. H., and Nee, S. (1994). Comparing real with expected patterns from molecular phylogenies. *In* "Phylogenetics and Ecology" (P. Eggleton and R. I. Vane-Wright, Eds.), pp. 219–231. Academic Press, London.

Harvey, P. H., and Pagel, M. D. (1991). "The Comparative Method in Evolutionary Biology." Oxford Univ. Press, New York.

Heard, S. B., and Hauser, D. L. (1995). Key evolutionary innovations and their ecological mechanisms. *Hist. Biol.* **10**, 151–173.

Hill, G. E. (1994). Geographic variation in male ornamentation and female mate preference in the house finch: A comparative test of models of sexual selection. *Behav. Ecol.* **5**, 64–73.

Hillis, D. M. (1991). Discriminating between phylogenetic signal and random noise in DNA sequences. *In* "Phylogenetic Analysis of DNA Sequences" (M. M. Miyamoto and J. Cracraft, Eds.), pp. 278–294. Oxford Univ. Press, New York.

Höglund, J. (1989). Size and plumage dimorphism in lek-breeding birds: A comparative analysis. *Am. Nat.* **134**, 72–87.

Höglund, J., and Sillén-Tullberg, B. (1994). Does lekking promote the evolution of male-biased size dimorphism in birds? On the use of comparative approaches. *Am. Nat.* **144**, 881–889.

Kishino, H., and Hasegawa, M. (1989). Evaluation of the maximum likelihood estimate of the evolutionary tree topologies from DNA sequence data, and the branching order in Hominoidea. *J. Mol. Evol.* **29**, 170–179.

Kluge, A. G. (1989). A concern for evidence and a phylogenetic hypothesis of relationships among *Epicrates* (Boidae, Serpentes). *Syst. Zool.* **38**, 7–25.

Kluge, A. G., and Wolf, A. J. (1993). Cladistics: What's in a word? *Cladistics* **9**, 183–199.

Krajewski, C., and Dickerman, A. W. (1990). Bootstrap analysis of phylogenetic trees derived from DNA hybridization distances. *Syst. Zool.* **39**, 383–390.

Lanyon, S. M. (1985). Detecting internal inconsistencies in distance data. *Syst. Zool.* **34**, 397–403.

Lanyon, S. M. (1992). Interspecific brood parasitism in blackbirds (Icterinae): A phylogenetic perspective. *Science* **255**, 77–79.

Lanyon, S. M. (1993). Phylogenetic frameworks: Towards a firmer foundation for the comparative approach. *Biol. J. Linn. Soc.* **49**, 45–61.

Leroi, A. M., Rose, M. R., and Lauder, G. V. (1994). What does the comparative method reveal about adaptation? *Am. Nat.* **143**, 381–402.

Ligon, J. D. (1993). The role of phylogenetic history in the evolution of contemporary avian mating and parental care systems. *Curr. Ornithol.* **10**, 1–46.

Lorenz, K. (1941). Vergleichende Bewegungsstudien an Anatiden. *J. Ornithol. Suppl.* **89**, 194–293.

Lorenz, K. (1950). The comparative method in studying innate behaviour patterns. *Symp. Soc. Exp. Biol.* **4**, 221–268.

Maddison, W. P. (1990). A method for testing the correlated evolution of two binary characters: Are gains or losses concentrated on certain branches of a phylogenetic tree? *Evolution* **44**, 539–557.

Maddison, W. P., and Maddison, D. R. (1992). "MacClade: Analysis of Phylogeny and Character Evolution," version 3.0. Sinauer, Sunderland, Massachusetts.

Maddison, W. P., and Slatkin, M. (1991). Null models for the number of evolutionary steps in a character on a phylogenetic tree. *Evolution* **45**, 1184–1197.

McKitrick, M. C. (1992). Phylogenetic analysis of avian parental care behavior. *Auk* **109**, 828–846.
McKitrick, M. C. (1993). Phylogenetic constraint in evolutionary theory: Has it any explanatory power? *Annu. Rev. Ecol. Syst.* **24**, 307–330.
McLennan, D. A., Brooks, D. R., and McPhail, J. D. (1988). The benefits of communication between comparative ethology and phylogenetic systematics: A case study using gasterosteid fishes. *Can. J. Zool.* **66**, 2177–2190.
Miles, D. B., and Dunham, A. E. (1993). Historical perspectives in ecology and evolutionary biology: The use of phylogenetic comparative analyses. *Annu. Rev. Ecol. Syst.* **24**, 587–619.
Miyamoto, M. M. (1985). Consensus cladograms and general classifications. *Cladistics* **1**, 186–189.
Miyamoto, M. M., and Cracraft, J. (1991). Phylogenetic inference, DNA sequence analysis, and the future of molecular systematics. *In* "Phylogenetic Analysis of DNA Sequences" (M. M. Miyamoto and J. Cracraft, Eds.), pp. 3–17. Oxford Univ. Press, New York.
Miyamoto, M. M., and Fitch, W. M. (1995). Testing species phylogenies and phylogenetic methods with congruence. *Syst. Biol.* **44**, 64–76.
Møller, A. P., and Birkhead, T. R. (1994). The evolution of plumage brightness in birds is related to extra pair paternity. *Evolution* **48**, 1089–1100.
Moreno, E., and Carrascal, L. M. (1993a). Leg morphology and feeding postures in four *Parus* species: An experimental ecomorphological approach. *Ecology* **74**, 2037–2044.
Moreno, E., and Carrascal, L. M. (1993b). Ecomorphological patterns of aerial feeding in oscines (Passeriformes: Passeri). *Biol. J. Linn. Soc.* **50**, 147–165.
Page, R. D. M. (1993). "COMPONENT 2.0. Tree Comparison Software for Use with Microsoft Windows." The Natural History Museum, London.
Pagel, M. D. (1994). The adaptationist wager. *In* "Phylogenetics and Ecology" (P. Eggleton and R. I. Vane-Wright, Eds.), pp. 29–52. Academic Press, London.
Pagel, M. D., and Harvey, P. H. (1988). Recent developments in the analysis of comparative data. *Q. Rev. Biol.* **63**, 413–440.
Patterson, C. (1982). Morphological characters and homology. *In* "Problems of Phylogenetic Reconstruction" (K. A. Joysey and A. E. Friday, Eds.), pp. 22–74. Academic Press, London.
Patterson, C. (1988). Homology in classical and molecular biology. *Mol. Biol. Evol.* **5**, 603–625.
Payne, R. B., and Risley, C. J. (1976). Systematics and evolutionary relationships among the herons (Ardeidae). *Misc. Publ. Univ. Michigan Mus. Zool.* **150**, 1–115.
Pruett-Jones, M. A., and Pruett-Jones, S. G. (1985). Food caching in the tropical frugivore, MacGregor's bowerbird (*Amblyornis macgregoriae*). *Auk* **102**, 334–341.
Pruett-Jones, S. G., and Lewis, M. J. (1990). Sex ratio and habitat limitation promote delayed dispersal in superb fairy-wrens. *Nature (London)* **348**, 541–542.
Prum, R. O. (1990). Phylogenetic analysis of the evolution of display behavior in the neotropical manakins (Aves: Pipridae). *Ethology* **84**, 202–231.
Prum, R. O. (1994). Phylogenetic analysis of the evolution of alternative social behavior in the manakins (Aves: Pipridae). *Evolution* **48**, 1657–1675.
Richman, A. D., and Price, T. (1992). Evolution of ecological differences in the Old World leaf warblers. *Nature (London)* **355**, 817–821.
Ricklefs, R. E. (1987). Community diversity: Relative roles of local and regional processes. *Science* **235**, 167–171.
Ridley, M. (1983). The Explanation of Organic Diversity: The Comparative Method and Adaptations for Mating. Clarendon Press, Oxford.
Sheldon, F. H. (1987a). Phylogeny of herons estimated from DNA–DNA hybridization data. *Auk* **104**, 97–108.
Sheldon, F. H. (1987b). Rates of single-copy DNA evolution in herons. *Mol. Biol. Evol.* **4**, 56–69.
Sheldon, F. H., and Bledsoe, A. H. (1993). Avian molecular systematics, 1970s to 1990s. *Annu. Rev. Ecol. Syst.* **24**, 243–278.
Sheldon, F. H., and Winkler, D. W. (1993). Intergeneric phylogenetic relationships of swallows estimated by DNA–DNA hybridization. *Auk* **110**, 798–824.

Sheldon, F. H., and Gill, F. B. (1996). A reconsideration of songbird phylogeny, with emphasis on the evolution of titmice and their sylvioid relatives. *Syst. Biol.* **45**, 471–493.

Sheldon, F. H., Slikas, B., Kinnarney, M., Gill, F. B., Zhao, E., and Silverin, B. (1992). DNA–DNA hybridization evidence of phylogenetic relationships among major lineages of *Parus*. *Auk* **109**, 173–185.

Sibley, C. G., and Ahlquist, J. E. (1990). "Phylogeny and Classification of Birds." Yale Univ. Press, New Haven, Connecticut.

Simpson, G. G. (1944). "Tempo and Mode in Evolution." Columbia Univ. Press, New York.

Slikas, B., Sheldon, F. H., and Gill, F. B. (1996). Phylogeny of titmice (Paridae): I. Estimate of relationships among subgenera based on DNA–DNA hybridization. *J. Avian Biol.* **27**, 70–82.

Swofford, D. L. (1993). "PAUP: Phylogenetic Analysis Using Parsimony," version 3.1. Illinois Natural History Survey, Champaign, Illinois.

Templeton, A. R. (1983). Phylogenetic inference from restriction endonuclease cleavage site maps with particular reference to the evolution of humans and the apes. *Evolution* **37**, 221–244.

Tinbergen, N. (1953). "Social Behaviour in Animals: With Special Reference to Vertebrates." Methuen, London.

Tinbergen, N. (1964). On aims and methods of ethology. *Z. Tierpsychol.* **20**, 410–433.

Winkler, D. W., and Sheldon, F. H. (1993). The evolution of nest construction in swallows (Hirundinidae): A molecular phylogenetic perspective. *Proc. Natl. Acad. Sci. U.S.A.* **90**, 5705–5707.

Winkler, D. W., and Sheldon, F. H. (1994). Phylogenetic hierarchy in character variability and its causes: Lessons from character-state distributions in swallows, Hirundinidae. *J. Ornithol.* **135**, 342. [Abstract]

CHAPTER 11

Phylogeographic Studies of North American Birds

Robert M. Zink
J. F. Bell Museum, University of Minnesota, St. Paul, Minnesota 55108

I. INTRODUCTION

Patterns of geographic variation in morphological characteristics of organisms reveal a variety of evolutionary processes (James, 1970). External phenotypic characters of American robins (*Turdus migratorius*) tend to covary with geographic temperature–humidity gradients, illustrating the potential of local adaptation to effect geographic variation patterns (Aldrich and James, 1991; James, 1970). The difficulty of establishing the genetic basis of polygenic fitness traits (e.g., James, 1983), and therefore the evolutionary interpretation of spatial patterns, has led investigators to document the geography of genetic variation by various indirect methods. These methods include protein electrophoresis, restriction fragment length polymorphisms (RFLPs) in mitochondrial DNA (mtDNA), and most recently the direct sequencing of DNA itself. Such methods are of interest because the way in which genetic variation is apportioned within and among populations reveals signatures left by various evolutionary processes.

In this chapter I review ways in which molecular techniques have been applied to the study of geographic variation in birds. I concentrate on studies of mtDNA variation because this provides the largest database for inference. After a brief review of allozymic studies, which set the stage for subsequent work, I evaluate how studies of mtDNA have contributed to understanding the evolution of geographic variation, population structure, and gene flow. A relatively new field, comparative phylogeography, is presented as a way of testing causal mechanisms of geographic differentiation as well as the historical stability of community composition.

II. ALLOZYMIC STUDIES OF AVIAN GEOGRAPHIC VARIATION

Several authors (e.g., Barrowclough, 1983; Corbin, 1987; Evans, 1987; Barrowclough and Johnson, 1988) reviewed allozymic studies of bird species, and they summarized resultant estimates of genetic variability. Although there were reports of natural selection influencing gene loci surveyed with electrophoresis (Redfield, 1974; Gyllensten *et al.*, 1979), most studies used allozymic variants as neutral genetic markers for estimating levels of genetic variability, and for investigating demographic structures and processes (Barrowclough *et al.*, 1985). Data on heterozygosity suggested that avian populations generally had levels of genetic variation consistent with those observed in other vertebrates. Another genetic estimate of interest was the fraction of genetic variation distributed among populations. Some studies found significant allelic frequency differences (e.g., Johnson and Marten, 1988), but in some of these cases comparisons might have been interspecific. In general the amount of genetic variation distributed among populations of temperate North American birds was low compared to many other vertebrates (Barrowclough, 1983; Avise, 1983). In fact, levels of among-population differentiation seemed anomolously low. For example, Barrowclough (1980a) described patterns of allelic variation in the yellow-rumped warbler (*Dendroica coronata*), and despite considerable geographic distance and morphological variation among population samples once ascribed to different biological species, no diagnostic or frequency differences were observed. Barrowclough concluded that population sizes are large, and connected by either ongoing or recently ceased gene flow. This view of population structure was the typical one inferred for North American birds from allozyme studies. The situation in the tropics might be different (Capparella, 1991).

Two hypotheses were offered to explain the general lack of population structure in birds, but were not resolved with allozymic data: avian populations have high effective population sizes and levels of gene flow (Barrowclough, 1980b), or avian molecular evolution proceeds at a slow rate relative to molecular evolution in other vertebrates, and to plumage evolution in the birds themselves. Avise (1983) suggested that a molecular rate slowdown might be due to high avian body temperatures that exert a stringent selective environment. A variant of the molecular slowdown explanation exists, which suggests that allozymic evolution proceeds at a slower pace than other avian genomic regions (Zink, 1991). For whatever reason(s), a problem in interpreting allozyme evidence is that currently segregating allozyme alleles probably have common allele ancestors that predated the fragmentation of populations and in some cases species (Zink and Remsen, 1986), and are therefore poor in information about recent population history. Mindell *et al.* (1996) discuss evidence for a rate slowdown at the DNA level, and suggest that relatively high avian body temperature might account for reduced differentiation (Kessler and Avise, 1985) via lower rates of change.

The notion that avian populations were largely unstructured prompted James (1991) to suggest that strong selection at the morphological level must overcome high levels of gene flow. Needed was an independent set of molecular markers to test the nature of avian population structure and gene flow deduced from allozyme evidence. The DNA revolution provided a new set of markers that has captured the attention of researchers even more than allozyme electrophoresis (Lewontin, 1974).

III. THE DNA REVOLUTION IN INTRASPECIFIC STUDIES

Surveys of restriction fragment length polymorphisms in mtDNA rapidly supplanted allozymic studies as a source of markers for studying both population-level and higher level systematic questions (Avise, 1994). The maternally inherited, nonrecombining, rapidly evolving mtDNA genome is rich in information about population-level processes. An advantage of mtDNA RFLP (and sequence) data is that the variant individual patterns (haplotypes) can be analyzed phylogenetically. In this approach, one infers the phylogenetic history of haplotypes (effectively alleles at the haploid mtDNA "locus") in the same way that one infers, for example, the phylogeny of species in a genus. Haplotype phylogenies are superimposed over geography, an approach aptly termed *phylogeography* by Avise *et al.* (1987). At one extreme, all haplotypes at all localities would trace to single common ancestors, themselves geographically arrayed in the phylogenetic tree, which would signal a highly substructured population and would offer a hypothesis for the history of isolation events. The haplotypes themselves could be similar or different in percentage sequence divergence, which would correspond to strong and weak phylogeographic divisions. The "depth" of the structured haplotype trees could be an index

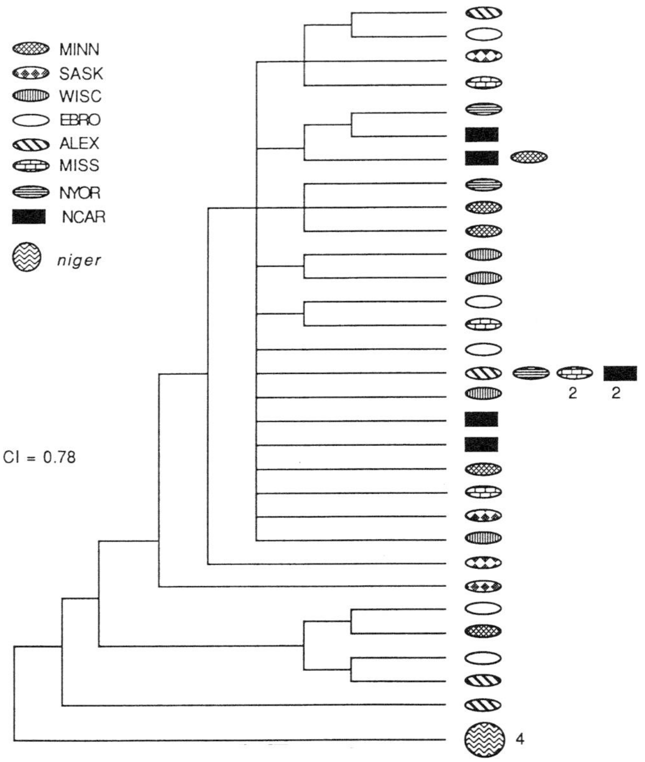

FIGURE 11.1 Geographic pattern of haplotype relationships in the common grackle. (From Zink *et al.*, 1991a.) Reprinted with permission of the publisher.

to the age of the population fragmentation recorded by the mtDNA. At the other extreme, haplotypes would appear geographically "scrambled," which would suggest either recent expansion of a species' range or high levels of current gene flow. Recent range expansions can result in haplotype phylogenies that are not geographically structured, a process termed *lineage sorting* (Avise, 1994) or *retained ancestral polymorphism*. Geographically unstructured haplotypes could be similar or divergent, each of which would yield different interpretations. For instance, similar haplotypes not showing a geographic pattern might suggest a recent bottleneck followed by extensive gene flow. Alternatively, some (e.g., Rand *et al.*, 1994) have proposed "selective sweeps" where one mtDNA haplotype is rapidly substituted because it is highly favored. Divergent haplotypes that are geographically unstructured might indicate previous allopatric divergence followed by a breakdown of barriers and population admixture, such as in snow geese (*Chen caerulescens*; Avise *et al.*, 1990). These are four more or less extreme phylogeographic structures with a large variety of intermediate possibilities (Avise *et al.*, 1987). Each reveals information about the

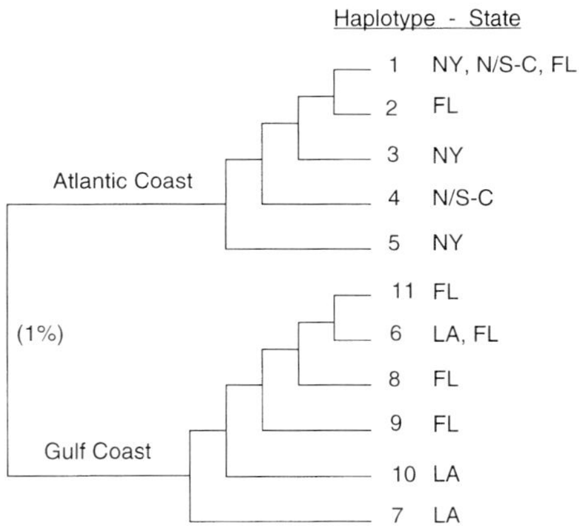

FIGURE 11.2 Phylogenetic relationships among haplotypes in the seaside sparrow (from Avise and Nelson, 1989), indicating a lack of phylogeographic structure within the two major clades corresponding to Atlantic and Gulf coasts. Abbreviations are for states in the U.S., and "N/S-C" refers to either North or South Carolina, which could not be deduced from Avise and Nelson's (1989) paper. Degree of haplotype divergence not indicated except for the distance between the two principal clades (1%).

history of mtDNA lineages, which likely reflect the history of populations, even though the mtDNA genome is but a single "gene" embedded in the organismal phylogeny (Avise, 1994).

IV. PHYLOGEOGRAPHIC STUDIES IN BIRDS

A. Nature of Variation

The first major study of avian mtDNA phylogeography (Ball *et al.*, 1988) showed that a phenotypically variable species, the red-winged blackbird (*Agelaius phoeniceus*), was essentially unstructured geographically. Another species that lacked phylogeographic structure was the common grackle (*Quiscalus quiscula*; Fig. 11.1), which was once considered two species. An alternative was the seaside sparrow (*Ammodramus maritimus*), which Avise and Nelson (1989) found consisted of two discrete parapatric geographic units (Fig. 11.2).

I compiled studies of mtDNA differentiation (Table I) in a variety of North American birds to determine general correlates of differentiation. I ascertained whether populations sampled on either side of mountains, deserts, Beringia, those on islands, or those separated by long distances without barriers, tend to be differentiated (i.e., haplotypes at a locality form a clade); some species figured multiple times in the tests if they were relevant to more than one type of barrier. The data set was limited by the idiosyncratic nature of individual studies, and by the difficulty in defining barriers for vagile organisms such as birds.

TABLE I Mitochondrial DNA Differentiation in North American Birds[a]

Species	Ref.[b]	Barrier	Phylogeographic variation	Method
Pipilo erythrophthalmus	1	Distance (E/W)	S?/S	RFLP
Geothlypis trichas	1	Distance (E/W)	S?/S	RFLP
Melospiza melodia	11	Distance	N	RFLP
Molothrus ater	1	Distance	N	RFLP
Zenaida macroura	1	Distance	N	RFLP
Picoides pubescens	1	Distance	N	RFLP
Agelaius phoeniceus	3	Distance	N	RFLP
Dendroica petechia	9	Distance (E/W)	W?/G	RFLP
Calidris alpina	4	Distance (E/W)	S/S?	Seq
Agelaius phoeniceus	3	Distance (N/S)	W/S	RFLP
Chen caerulescens	8	Distance (E/W)	S	Seq
Ammodramus caudacutus	12	Distance (E/W)	S/S	RFLP
Spizella passerina	14	Distance	N	RFLP
Quiscalus quiscula	15	Distance	N	RFLP
Parus bicolor	13	Distance	N	RFLP
Parus wollweberi	13	Distance	N	RFLP
Ammodramus maritimus	21	Distance	S/S	RFLP
Gulf coast	21	Distance	N	RFLP
Atlantic coast	21	Distance	N	RFLP
Colaptes auratus	23	Distance	W/G?	RFLP
Parus carolinensis	24	Distance (E/W)	S	RFLP
Parus hudsonicus	24	Distance	N	RFLP
Parus atricapillus	24	Distance	N	RFLP
Tympanuchus sp.	25	Distance	N	RFLP
Branta canadensis	26	Distance (N/S)	S	RFLP
Passerella iliaca[c]	7	Distance	S/S	RFLP
Passerella iliaca iliaca	7	Distance	N	RFLP
Passerella iliaca megarhyncha	7	Distance	N	RFLP
Passerella iliaca unalaschcensis	7	Distance	N	RFLP
Passerella iliaca schistacea	7	Distance	N	RFLP
Branta bernicla	27	Distance	N	RFLP
Geothlypis trichas	2	Cascades, S. Nevada	W/S?	Seq
Passerella iliaca	7	Cascades, S. Nevada	S	RFLP
Picoides pubescens	1	Cascades, Rockies	N	RFLP
Amphispiza belli	5	Sierra Nevada	W?/S	Seq
Agelaius phoeniceus	3	Cascades, Rockies	N	RFLP
Melospiza melodia	11	Cascades, Rockies	N	RFLP
Zenaida macroura	1	Cascades, Rockies	N	RFLP
Parus inornatus	13	Sierra Nevada	S	RFLP
Spizella passerina	14	Cascades, S. Nevada	N	RFLP
Agelaius phoeniceus	3	Cascades, S. Nevada	W?/G?	RFLP
Molothrus ater	1	Cascades, Rockies	N	RFLP
	22	Sierra Nevada	W/S	RFLP
Parus atricapillus	24	Cascades, Rockies	N	RFLP
Phalocrocorax pelagicus	6	Beringia	W/S	RFLP

(*Continues*)

TABLE I (*Continued*)

Species	Ref.[b]	Barrier	Phylogeographic variation	Method
Anas crecca	6	Beringia	N	RFLP
Gallinago gallinago	6	Beringia	W	RFLP
Numenius phaeopus	6	Beringia	S	RFLP
Larus canus	6	Beringia	S	RFLP
Sterna hirundo	6	Beringia	W	RFLP
Brachyramphus marmoratus	6	Beringia	S	RFLP
Picoides tridactylus	6	Beringia	S	RFLP
Hirundo rustica	6	Beringia	W	RFLP
Pica pica	6	Beringia	S	RFLP
Anthus spinoletta	6	Beringia	S	RFLP
Calcarius lapponicus	6	Beringia	N	RFLP
Leucosticte arctoa	6	Beringia	S	RFLP
Calidris alpina	4	Beringia	S	Seq
Arenaria interpres	4	Beringia	N	Seq
Branta bernicla	27	Island (N)	S	RFLP
Dendroica petechia	9	Island (S)	S	RFLP
Coereba flaveola	10	Island (S)	S	RFLP
Melospiza melodia	11	Island (N)	N	RFLP
Passerella iliaca	7	Island (N)	W	RFLP
Saltator albicollis	16	Island (S)	S	RFLP
Geothlypis trichas	2	Island (N)	N	Seq
Parus hudsonicus	24	Island (N)	W	RFLP
Parus atricapillus	24	Island (N)	W	RFLP
Parus gambeli	24	Desert	S/S	RFLP
Dendroica nigrescens	17	Desert	S/S	RFLP
Strix occidentalis	18	Desert	S/S	Seq
Auriparus flaviceps	19	Desert	N	Seq, RFLP
Campylorhynchus brunneicapillus	19	Desert	N	Seq, RFLP
Polioptila melanura	19	Desert	N	Seq, RFLP
Toxostoma lecontei	19	Desert	S/S	Seq, RFLP
Toxostoma curvirostre	19	Desert	S/S	Seq, RFLP
Pipilo fuscus	19	Desert	S/S	Seq, RFLP
Passerculus sandwichensis	20	Desert (Baja)	W/S?	RFLP

[a]Abbreviations for level of phylogeographic variation: S, strongly differentiated; W, weak; N, none apparent. If S or W is indicated, the nature of variation may be coded as S (step clinal) or G (gradual).

[b]References: (1) Ball and Avise (1992); (2) J. T. Klicka and R. M. Zink (unpublished data); (3) Ball *et al.* (1988); (4) Wenink *et al.* (1993); (5) Johnson and Cicero (1991); (6) Zink *et al.* (1995); (7) Zink (1994); (8) Quinn (1992); (9) Klein and Brown (1995); (10) Seutin *et al.* (1994); (11) Zink and Dittmann (1993a); (12) Rising and Avise (1993); (13) Gill and Slikas (1992); (14) Zink and Dittmann (1993b); (15) Zink *et al.* (1991a); (16) Seutin *et al.* (1993); (17) Bermingham *et al.* (1992); (18) Barrowclough *et al.* (personal communication); (19) R. M. Zink and R. C. Blackwell (unpublished data); (20) Zink *et al.* (1991b); (21) Avise and Nelson (1989); (22) Fleischer *et al.* (1991); (23) Moore *et al.* (1991); (24) Gill *et al.* (1993); (25) Ellsworth *et al.* (1994); (26) Van Wagner and Baker (1990); (27) Shields (1990).

[c]Considering all four taxa conspecific.

1. *Effects of Distance*

Isolation by distance results in differences between populations at opposite ends of a more or less continuous group of populations if dispersal distances are relatively low (Wright, 1978). It is difficult to test for this effect because "distance" is often confounded by barriers, either current or ancient, that could be associated with phylogeographic divisions in modern-day species. I identified studies in which samples were taken from at least 500 km apart, and for which the species' habitat is more or less continuous without obvious current barriers to gene flow or major range disjunctions. Where phylogeographic differences were observed, I determined whether they occurred over very short distances (e.g., step clines) or whether the variation was gradual over the region compared (i.e., consistent with isolation by distance).

Most phylogeographic differences occurred over a geographically limited area (Table I). For example, in the fox sparrow (*Passerella iliaca*), there are four distinct groups of haplotypes that apparently are parapatrically distributed, and each group is relatively uniform *intra-se* (Zink, 1994). Overall data suggest that dispersal distances are too large, or population expansions too recent, for isolation by distance to be a major factor in structuring avian populations in North America, and that where differences do exist, there appears to be evidence of a barrier other than distance.

2. *Effects of Mountains*

Populations distributed on either side of mountain barriers could be expected to diverge because gene flow is limited, or the mountain ranges mark boundaries between different areas of endemism. My compilation suggests that mountain barriers could be a significant (6 of 13 species in Table I) cause of mtDNA differentiation. It is difficult to falsify the proposition that mountain ranges represent sites of secondary contact.

3. *Effects of Deserts*

Deserts can act as barriers to nondesert species, and intervening habitats can isolate taxa in different deserts (Hubbard, 1973). Species distributed in different deserts of southwestern North America show a variety of levels and patterns of differentiation. Canyon towhee (*Pipilo fuscus*) and curve-billed thrasher (*Toxostoma curvirostre*) show considerable mtDNA RFLP and sequence differentiation across the Sonoran and Chihuahuan deserts, whereas several other species [verdin (*Auriparus flaviceps*), cactus wren (*Campylorhynchus brunneicapillus*), black-tailed gnatcatcher (*Polioptila melanura*)] appear undifferentiated. However, in southern Baja California Sur, preliminary data suggest that the verdin and cactus wren are significantly differentiated (R. M. Zink, unpublished data), a pattern found in other vertebrates (Murphy, 1983). A subspecies of Le Conte's thrasher (*T. lecontei arenicola*) isolated along the west coast of central Baja California is strongly differentiated (Zink *et al.*, 1997) whereas populations of California gnatcatcher (*Polioptila californica*) are at most weakly differentiated throughout Baja California (R. M. Zink, G. F. Barrowclough, R. C. Blackwell, and J. L. Atwood, unpublished data). Thus, species sampled in the aridlands of North America exhibit a mixture of phylogeographic patterns.

4. *Effects of Islands*

Although the island of Newfoundland shows evidence of having populations genetically differentiated from adjoining continental ones, other island populations did not show differentiation. For example, neither fox nor song sparrows were differentiated in restriction sites on the Queen Charlotte Islands or Vancouver Island, relative to adjacent mainland populations. This suggests that these large islands recently were separated from the mainland, were colonized relatively recently, or receive immigrants at a rate preventing differentiation. Too few populations have been sampled to make firm conclusions about islands as isolating factors.

5. *Effects of Beringia*

Zink *et al.* (1995) compared small samples of species found on either side of Beringia. Of 13 species they studied, all but 2 showed evidence of mtDNA RFLP differentiation, with 4 species showing weak and 7 species strong differentiation. However, the relationship between morphological and mtDNA differentiation was inconsistent. Populations of three-toed woodpecker (*Picoides tridactylus*) exhibit little morphological difference between the two continents, yet they were very different in mtDNA, suggestive of species status. Other species, such as the marbled murrelet (*Brachyramphus marmoratus*), show both mtDNA and morphological differentiation. Overall, most species showed mtDNA differentiation to some degree, consistent with geographic isolation on different continents.

6. *Summary*

Contrary to allozyme studies, almost 40% of species examined exhibited geographic variation in mtDNA (Table I). Most occurrences of significant phylogeographic structure have been found in species that also exhibit morphological differentiation; often mtDNA and subspecies boundaries are congruent, whereas this was not the case with allozymic studies. The resultant question is whether there are consistent geographic correlates of structured haplotype trees. Although past barriers are difficult to judge from current conditions, most mtDNA phylogeographic structure seems associated with a barrier other than distance, although most general barriers appear associated with phylogeographic structure. However, the existence of an apparent barrier or a named subspecies is not a general predictor of mtDNA differentiation. That is, more subspecies lack mtDNA differentiation than exhibit it (Ball and Avise, 1992). Phylogenetic analysis of mtDNA haplotypes seems better able to document patterns of population differentiation than allozyme data. In summary, a new perspective on avian populations is emerging, recognizing more structure than suggested by allozyme analysis.

B. Haplotype Phylogenies and Directionality

If a haplotype phylogeny were rooted with an outgroup haplotype (e.g., Maddison *et al.*, 1984), or from coalescence theory (Crandall and Templeton, 1993), one might infer the historical direction of colonization by superimposing the haplotype tree on a map, starting from the basal haplotype and moving up the tree, revealing a directional geographic progression (Avise *et al.*, 1983; Lansman *et al.*, 1983). However, if a species has been evolving for a significant period, "dispersing" haplotypes will reach geographic range boundaries and be"deflected" backward, erasing the monotonic relationship between geographic and genetic distance. Equilibrium is often a signal of the latter phenomenon (Neigel and Avise, 1993).

Most phylogeographic studies did not include a sister species, resulting in an (unrooted) network. A rooted tree for the common grackle (Fig. 11.1) did not exhibit a directionality, potentially owing to high levels of gene flow after the range was fully occupied (Moore and Dolbeer, 1989). A rooted haplotype tree for the song sparrow (Zink and Dittmann, 1993a) suggested that basal haplotypes occurred in Newfoundland, and the more derived haplotypes were found in the northern part of the range. This is consistent with a relatively recent northward spread of song sparrows from an eastern refuge following retreat of glaciers. Several caveats render this only a hypothesis (Zink and Dittmann, 1993a). However, studies of other species point to basal haplotypes also occurring in Newfoundland (Zink, 1994; Gill *et al.*, 1993). Further research is required to determine whether Newfoundland, or nearby sites today submerged (see Pielou, 1991), were in fact a refuge for other species.

C. Estimates of Gene Flow

An expectation for molecular markers is that they will provide indirect measures of gene flow. Slatkin and Maddison (1989) suggest that coalescent events between haplotypes that are currently in different populations are evidence of recent gene flow. In the common grackle, sister haplotypes often occurred in different population samples (Fig. 11.1) suggesting gene flow. In general, RFLP data suggest considerable gene flow. However, in many RFLP data sets, Slatkin and Maddison's (1989) method seems inappropriate as individual haplotypes are often found in multiple populations, which could be evidence of gene flow or simply ancestral retentions and nonequilibrium conditions (see Edwards, 1993). Direct sequencing might resolve more haplotypes, which could be useful for gene flow calculations. Edwards (1993) sequenced individual babblers, and the haplotype phylogeny provided evidence of long-distance gene flow among populations.

Neigel *et al.* (1991) and Neigel and Avise (1993) use information on haplotype divergence, generation time, rate of molecular evolution, and the geographic occurrence of haplotypes to infer single generation dispersal distances. Importantly, their method does not require genetic or demographic equilibrium, conditions unlikely to hold for North American birds [and an assumption required for Slatkin and Maddison's (1989) method]. Dispersal distances for birds tend to be higher (>3.0 km/generation) than those for rodents (<0.5 km/generation) (Neigel and Avise, 1993; Zink, 1996a), consistent with interpretations derived from allozyme and RFLP data (Zink and Remsen, 1986). Estimates of dispersal distance differ from the typical measure of gene flow derived from allozymic data (or any frequency-based method), *Nm*, the average number of immigrants exchanged among demes per generation. *Nm* values are not easily compared to the single-generation dispersal distance. Because of the restrictive assumptions associated with calculations of *Nm* from DNA data, single-generation dispersal distances might be a better measure to compute and compare among avian populations. Such data also permit direct comparison with mark–recapture studies. Unfortunately, there are relatively few mtDNA dispersal distances estimated at this time. Nevertheless, mtDNA data do suggest that gene flow is high enough to prevent differentiation over distance.

D. Description and Significance of Genetic Variation among Populations

An important step in the evolutionary process is the conversion of genetic variation from within to among populations. With allozyme electrophoresis, measures such as F_{st} and G_{st} were computed, which estimate that proportion of genetic variance distributed among populations (Wright, 1978). The degree of population structure is taken to reflect degree of geographic isolation. Considering haplotypes as alleles at a locus (the mtDNA genome), one can compute F_{st} and G_{st} for mtDNA data. The analysis that seems best suited for this is N_{st} (Lynch and Crease, 1990); unfortunately, available computer programs require raw data that is often not included in published phylogeographic studies.

G_{st} or F_{st} values calculated from haplotype data without corrections for small sample size or sequence divergence between haplotypes, or consideration of phylogeography, can be misleading. For a distribution of haplotype frequencies among populations, the single possible G_{st}/F_{st} value is associated with multiple haplotype trees (see Felsenstein, 1978; for example, with only 9 haplotypes, there are 2,027,025 possible rooted trees). Thus, if a G_{st} value is significant, but the haplotype tree is

TABLE II G_{st} Values for Avian Species[a]

No clear phylogeographic structure			Clear phylogeographic structure		
Species	Ref.[b]	G_{st} value	Species	Ref.	G_{st} value
Song sparrow	1	0.09	Fox sparrow	6	0.26*
Chipping sparrow	2	0	Seaside sparrow	7	0.38*
Red-winged blackbird	3	0.11	Rufous-sided towhee	5	0.45*
Common grackle	4	0.06	Common yellowthroat	5	0.22
Downy woodpecker	5	0	Black-capped chickadee	8	0.19
Brown-headed cowbird	5	0.27	Boreal chickadee	8	0.57*
	Mean:	0.09	Carolina chickadee	8	0.25*
			Sharp-tailed sparrow	9	0.27*
				Mean:	0.32

[a] "No clear phylogeographic structure" means that haplotypes at a locality (or within a region) are not generally each others nearest relatives, whereas the alternative suggests the opposite. An asterisk indicates that the value was significant at the $p \leq 0.05$ level.

[b] References: (1) Zink and Dittmann (1993a); (2) Zink and Dittmann (1993b); (3) Ball *et al.* (1988); (4) Zink *et al.* (1991a); (5) Ball and Avise (1992); (6) Zink (1994); (7) Avise and Nelson (1989); (8) Gill *et al.* (1993); (9) Rising and Avise (1993).

unstructured, the investigator needs to consider which "message" to believe. I suggest that the haplotype tree be accorded primary significance if the two methods suggest different pictures of population history.

I calculated G_{st} as $(H_t - H_s)/H_t$, where H_t is the total gene diversity and H_s is the weighted average of within-population gene diversity; no correction was made for sequence divergence. H_t and H_s are calculated with a correction for sample size. The significance of the G_{st} is determined by calculating the probability that the observed G_{st} is significantly different from that obtained by randomly reallocating haplotypes among populations. The randomization is done by reallocating haplotypes among populations (keeping the original sample sizes) and recalculating G_{st}. The P value is obtained by dividing the number of times that the recalculated G_{st} is equal to or larger than the observed one by the total number of permutations. A P value of less than 0.05 indicates significant genetic structure.

I found 14 avian phylogeographic studies in North America from which G_{st} could be computed (Table II), and I divided these studies into those with and without phylogeographic structure. For the former, the average G_{st} was 0.32 ± 0.13 (SD; $n = 8$) and for the latter it was 0.09 ± 0.11 (SD; $n = 6$). Clearly, there can be biases introduced by overly rapid mutation rates (nonequilibrium), and differences in effective population size between organellar (e.g., mitochondrial) and nuclear genes, but I feel that comparisons of the distribution of G_{st} and F_{st} values derived from mtDNA and allozyme data are of interest. Although there is variation caused by differing sample sizes, population samples, and geographic area covered, the

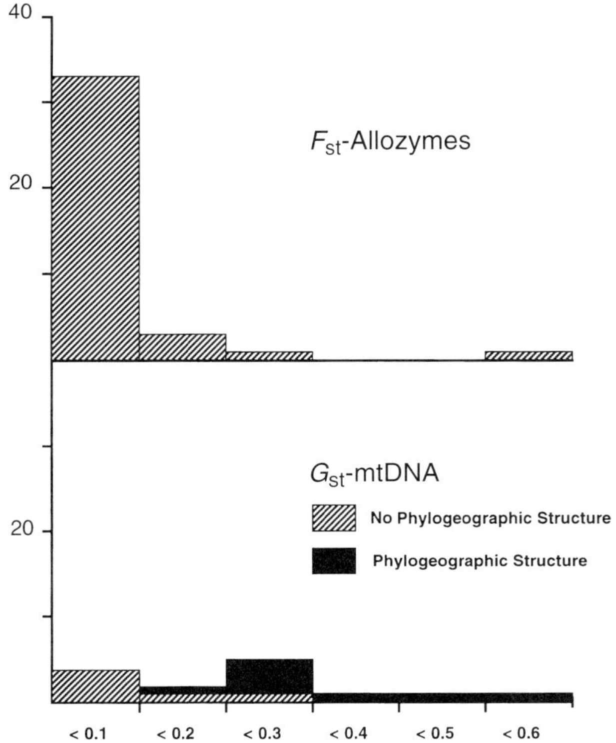

FIGURE 11.3 Distribution of G_{st} values from mtDNA data and F_{st} values from allozyme data. The single F_{st} value greater than 0.6 was based on a single locus (Barrowclough and Gutiérrez, 1990). Data used in computation of F_{st} distribution are available from the author.

mtDNA data suggest a broader range of population differentiation relative to that for allozyme studies (Fig. 11.3). Thirty-three of 38 (87%) allozyme F_{st} values are less than 0.10, whereas the comparable figure for mtDNA surveys was 4 of 14 (29%).

One might ask, therefore, what is to be made of the many allozymic studies of avian population structure in North America (Fig. 11.3). Clearly, those reporting no structure need to be corroborated by other molecular markers, because it seems likely that allozyme variation might not record recent population fragmentation events (Zink, 1991). Conclusions about high rates of gene flow, natural selection, or large effective population sizes also need to be interpreted cautiously. Degree of population structuring might indicate speciation potential, with a highly structured species being closer to the speciation boundary (Templeton, 1980). If true, allozyme data suggest relatively low speciation potential in birds, relative to groups such as rodents and salamanders. However, there are more species of birds in the world than rodents or salamanders, which further casts doubt on the validity of population structure inferences from allozyme data. Speciation in birds might be rapid owing to sexual selection (Zink, 1996b).

V. COMPARATIVE PHYLOGEOGRAPHY

A. Method

Haplotype phylogenies have a variety of uses, such as in conservation biology and mating system studies (Avise, 1994). Zink (1996a) reviewed a potential parallel between historical (vicariance) biogeography (Cracraft, 1982; Wiley, 1988) and phylogeography. In vicariance biogeography one examines the phylogenetic patterns among species in lineages distributed over the same areas of endemism. The most parsimonious explanation for congruent patterns is that the component lineages were historically widespread and codistributed (or "broadly sympatric"), and that they responded to the same set of vicariance events. Lack of congruence can result from dispersal across barriers, differential response to barriers, or lack of long-term sympatry in ancestral biotas (Zink and Hackett, 1988). On a more recent time scale, one can ask whether phylogeographic patterns among currently codistributed species are congruent, an endeavor that could be termed *comparative phylogeography*. If so, it would suggest that species' phylogeographic patterns were shaped by common responses to unique historical events. This interpretation assumes that the variation is selectively neutral, and that gene flow, mutation, and genetic drift produce phylogeographic patterns. Congruent phylogeographies suggest that the species composition of communities has been reasonably stable. If species turnover in communities is relatively frequent, congruent phylogeographic patterns would not be pervasive because, presumably, species would not be broadly sympatric for sufficient periods of time. Some paleoecologists predict that "communities have broken up and reformed in different configurations repeatedly and regularly on time scales of a few thousand years" (Bennett, 1990), suggesting that phylogeographic congruence might be the exception, not the rule. Thus, comparative phylogeography offers a perspective on factors producing geographic variation, and on the stability of particular associations of species.

Avise (1992) found that a diverse group of species, including freshwater fish, marine fish, marine invertebrates, and a terrestrial bird (seaside sparrow) showed a significant phylogeographic break in northeastern Florida. Because species in different taxonomic classes showed evidence of this division, it seems clear that ancestral species were historically codistributed and subsequently isolated in common; it is unlikely that such diverse species responded in common to a selective gradient. Comparison of taxonomically diverse species adds strength to the historical isolation interpretation. Thus, if one is inclined not to compare a marine fish and a terrestrial bird because they are "not comparable" the strength and validity of the comparison is tenuous. Choice of species compared should not be constrained *a priori* (Simberloff, 1987). If only species in a single genus were compared, one might suspect that the species were prone to respond to some selective gradient because of common phylogenetic background. Thus, the protocol for comparative phylogeography is to falsify the hypothesis that species currently codistributed (sympatric but not necessarily syntopic) over a broad area show congruent haplotype phylogenies.

Outcomes of phylogeographic comparisons include congruent patterns, lack of congruence, markedly incongruent phylogeographic patterns, or mixture of pattern(s) and no pattern. Lack of congruence and its causes are just as instructive as congruent patterns (Lamb *et al.*, 1989, 1992). If two species that are today broadly sympatric have phylogeographic patterns that differ, one might infer that they responded to different historical events (i.e., what was a barrier to one species was not a barrier to another). Two currently codistributed species that differ because one has discernable phylogeographic structure and the other does not could be a result of not sharing a long history of coassociation. The nature of mtDNA haplotype data suggests ways to distinguish alternative hypotheses for lack of phylogeographic congruence (Zink, 1996a).

B. Some Avian Results

The data in Table I serve as a basis for asking whether currently codistributed species of North American birds have similar phylogeographic patterns. Unfortunately, only species that have been surveyed over the same broad area with similar molecular methods yield useful comparisons, and there are relatively few such studies. Zink (1996a) compared five species that are currently widespread over the same continental area and were subjected to similar mtDNA RFLP analyses (Fig. 11.4). The fox sparrow has four geographically structured groups of haplotypes that can be traced to four common ancestors (Zink, 1994). The phenotypically variable song sparrow (*Melospiza melodia*) exhibits considerable haplotype variation, which surprisingly is geographically unstructured (Zink and Dittmann, 1993a). The chipping sparrow (*Spizella passerina*; Zink and Dittmann, 1993b) and red-winged blackbird (Ball *et al.*, 1988) exhibit little if any phylogeographic structure, yet the latter species has many morphological subspecies. Last, the Canada goose (*Branta canadensis*) exhibits a relatively deep division into two haplotype groups (Van Wagner and Baker, 1990).

The five species do not have congruent phylogeographies (Fig. 11.4). The fox sparrow and Canada goose have relatively deeply structured phylogeographic trees that are not congruent with each other, whereas the other species show no clear pattern of differentiation. One can, therefore, falsify the hypothesis that each species was historically codistributed and responded similarly to historical isolating events. The question becomes, why do these five species have different phylogeographic structures? The species might have had different dispersal characteristics that resulted in different responses to common barriers. The species might differ in levels of genetic variability, such that species without variation could not show differentiation because they lack the "raw materials." Species without phylogeographic patterns might be recently evolved, such that insufficient time has elapsed for

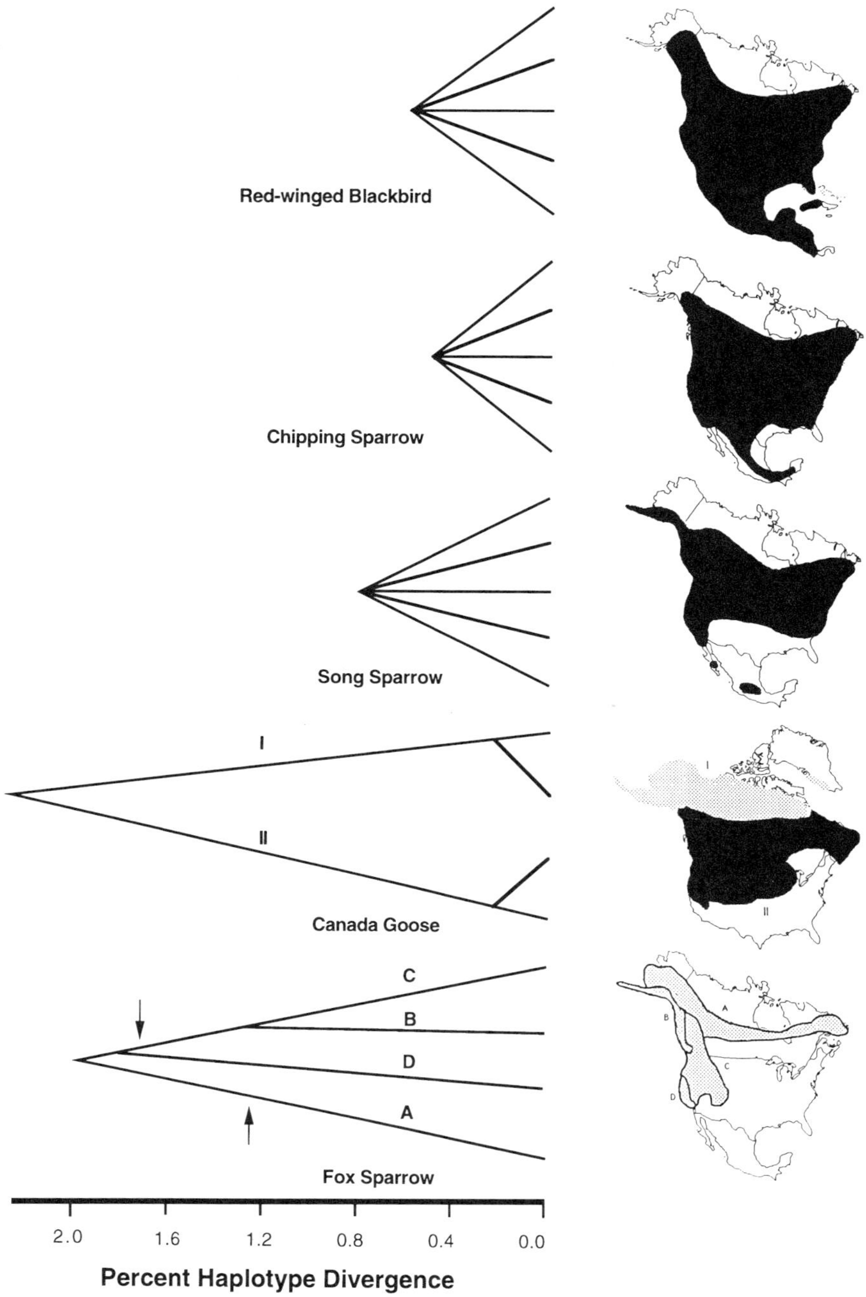

FIGURE 11.4 Approximate breeding distributions and diagrammatic haplotype phylogenies for five species of North American birds (from Zink, 1996a). If mtDNA evolution proceeds at a reasonably uniform rate, it is significant that the splits in the Canada goose and fox sparrow occurred at 2.5% sequence divergence or less, whereas each of the species compared is more than 2.5% distant from its nearest congener; hence, each species was extant during the times when isolating events fragmented populations of fox sparrow and Canada goose. Reproduced with permission of the publisher.

differentiation. Lack of pattern might occur in species with high levels of gene flow. Perhaps the five species were simply not historically codistributed.

Zink (1996a) discussed these alternatives. Each species was relatively well sampled with similar numbers of restriction endonucleases, which revealed similar amounts of mtDNA variability. Hence, incomplete sampling and lack of genetic variation were deemed implausible reasons for differing phylogeographic patterns. Mitochondrial DNA data suggest significant gene flow within species that are unstructured, or within the phylogeographic units of the fox sparrow and Canada goose. Thus, gene flow is probably too high to allow differentiation in the absence of geographic isolating barriers.

The mtDNA genetic distance of a species from its most closely related extant congener can be used as a relative measure of the length of time a species has been evolving independently. We might predict that species isolated for the longest time would show mtDNA differentiation because of a clocklike accumulation of genetic differences. The fox sparrow and Canada goose are the most differentiated from their extant sister taxa (Zink and Blackwell, 1996; Shields and Wilson, 1987), and these show phylogeographic structure. Zink (1996a) concluded that there was a relationship between elapsed time since sharing a common ancestor and phylogeographic structure, which could result because species that have been evolving independently longer have had a heightened probability of vicariant fragmentation. However, this does not mean that relatively old species were historically codistributed, as suggested by lack of congruence between the fox sparrow and Canada goose. Also, this relationship is not straightforward in explaining levels of phenotypic differentiation. For example, the chipping sparrow and song sparrow are each similarly distant from their sister taxa, yet the former is divided into 7 subspecies and the latter into 34. Even allowing for idiosyncracies in the way in which taxonomists delimit subspecies, the amount of phenotypic variation in species of the same relative "molecular age" is strikingly different. Therefore, one is left with the possibilities that phenotypic or mtDNA evolution is rate variable, or much of subspecific differentiation is ecophenotypic.

The mtDNA phylogeographic structures evident in these five species led to several conclusions. First, historical isolating events, rather than isolation by distance, are most consistent with the phylogeographic structure in the fox sparrow and Canada goose; in both, phylogeographic breaks occurred over short distances. The reason for the lack of phylogeographic structure in the chipping sparrow, song sparrow, and red-winged blackbird likely is that they have only recently colonized the current range (see below). The relatively deep splits in the haplotype trees of the fox sparrow and Canada goose probably occurred before the three currently undifferentiated species (in mtDNA) achieved their present ranges and high degree of sympatry. Obviously a larger sample of codistributed species is needed. Data suggest that one might not find much phylogeographic congruence among North American birds; perhaps this is because of the recent deglaciations and habitat displacements (Pielou, 1991) and their effects on bird distributions.

C. Mismatch Distributions and Population Histories

Central to my argument is the notion that present distributions are poor indicators of historical ones, especially over the times required for phylogeographic congruence to evolve (Chesser and Zink, 1994). Past population increases are probably likely to accompany major range expansions, and the former can be inferred from haplotype data (Avise *et al.*, 1988; Rogers and Harpending, 1992; Rogers, 1995, 1996). For example, a dramatic (500-fold) and sudden increase in population size results in a haplotype tree that is relatively unstructured, with most haplotypes tracing to a common ancestor just prior to the population expansion (a "bush" topology for the haplotype tree). Also, the distribution of pairwise restriction site differences between haplotypes, termed by Rogers and Harpending (1992) the *mismatch distribution*, exhibits a "wave" if there was a significant population expansion in an unstructured population. The wave is centered near the point of population expansion on the mutation distance scale used (Fig. 11.5). The method provides information about upper (θ_1) and lower (θ_0) bounds on population increases, and the relative timing of the increases in units of mutational distance (τ). If the above inferences about phylogeographic histories were correct, such waves should be evident in the data for the red-winged blackbird, chipping sparrow, and song sparrow.

The mismatch distributions (Fig. 11.5) show waves suggestive of population increases for the song sparrow and red-winged blackbird, but not for the chipping sparrow or fox sparrow (data for Canada goose unavailable). Avise *et al.* (1988) used a different method to suggest an increase in population size of the red-winged blackbird. Population increases in the song sparrow and red-winged blackbird are consistent with recent range expansions. The lack of such an expansion in the chipping sparrow could mean that although the species might have recently expanded its range, it did not have a low population size wherever it resided during Pleistocene glacial maxima. Because the fox sparrow is highly structured, the mismatch distribution shows two peaks (corresponding to within and among-group differences). The mismatch distribution (not shown) for the common grackle similarly indicated a population increase, a conclusion also reached by considering the haplotype tree (Fig. 11.1) and pairwise distribution of interindividual haplotype distances (Avise *et al.*, 1988).

The use of the mismatch distribution aids phylogeographic studies in revealing population expansions. Such inferences, not without caveats (Rogers, 1996), can help determine whether species that lack phylogeographic patterns only recently expanded their ranges. As one might predict for North American birds that colonized recently deglaciated areas, waves in mismatch distributions are apparent.

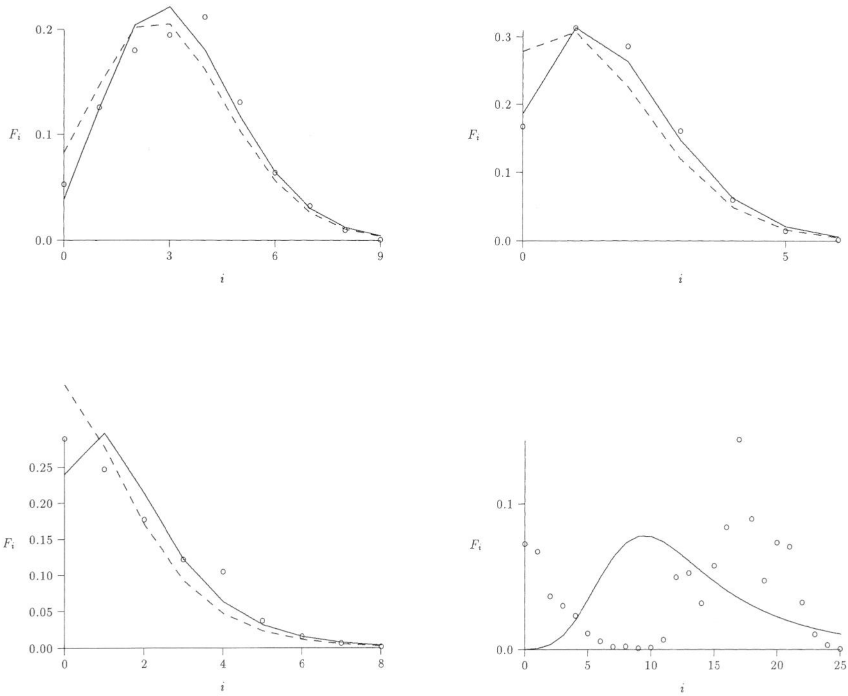

FIGURE 11.5 Mismatch distributions for four avian species. For each species, open circles indicate the mismatch distribution derived from the original matrix of restriction sites. F_i indicates the frequency of each class, and i represents the number of restriction sites separating pairs of haplotypes. The solid line is the theoretical distribution fit using Eqs. (2) and (3) of Rogers (1996), and the dashed line shows the fit of the three-parameter model (see Rogers, 1996). For the song sparrow (upper left), the mismatch distribution shows a peak at $3/(2u)$ generations ago, where u is the aggregate mutation rate of the region of DNA under study. The parameter estimates suggest that the population was initially very small ($\theta_0 = 0$), that the postexpansion population was moderately large ($\theta_1 = 18.13$), and that the expansion occurred $3.25/(2u)$ generations ago. The mismatch distribution for the red-winged blackbird (upper right) is consistent with a recent population expansion within 4 units of mutational time from present, starting from a relatively small population [alternatively, the mutation rate could be low (Rogers, 1996), but there is no evidence of this]. Mismatch distribution for the chipping sparrow (bottom left) does not have a clear wave, and is not consistent with a population expansion; the mismatch distribution is similar to that expected for the theoretical distribution of a population at equilibrium between mutation and drift. The mismatch distribution for the fox sparrow (bottom right) shows two distinct peaks, one at zero and one at 17. Such a distribution can reflect (1) a history with two bottlenecks in population size, or (2) geographic structure (A. R. Rogers, personal communication); Zink (1994) found clear evidence of geographic structure.

D. Comparative Phylogeography and the History of Communities

Comparative phylogeography provides a means to evaluate the stability of community structure and coevolutionary models. An implication of phylogeographic incongruence is lack of historical continuity in community membership (Bennett, 1990), which could influence the likelihood of certain types of coevolution. For example, Rothstein (1990) suggests that over long periods of time hosts of parasitic cowbirds evolve defenses and force the parasites to become specialists (see Lanyon, 1992, for an alternative view). If phylogeographic studies suggest lack of significant historical association of cowbirds and their parasites, host-rejection behavior might be inferred to be very rapid, a plausible hypothesis if parasitism exerts a strong selective force (Rothstein, 1990). Hypotheses of coevolution that assume sympatry (and syntopy) on an evolutionary timetable could be coupled with comparative phylogeographic studies. Congruent phylogeographies provide evidence for species' long-term associations, and using molecular characters these might be placed in a temporal scale.

VI. PROSPECTUS

Detecting genetic variation, documenting its geographic deployment, and inferring evolutionary processes are central themes of evolutionary analysis of populations. New methods of molecular analysis successively provide greater resolving power. Direct sequencing of DNA stands to be the next most influential method to provide insight into the nature of avian population structure. Although patterns of genetic variation at microsatellite loci (e.g., McDonald and Potts, 1994) can provide fine-scale resolution, the difficulty of phylogenetic interpretation of the data could limit use of the technique for some evolutionary analyses. Inferences from haplotype phylogenies based on coalescence theory provide powerful analytical tools (e.g., Slatkin and Maddison, 1989; Crandall and Templeton, 1993)

Although the need is often mentioned for nuclear gene assays, problems accompany nuclear gene analysis. Because of the four times greater effective population size of nuclear genes, the time to coalescence of nuclear gene alleles greatly exceeds that for mitochondrial genes (therefore, rapid, nonrecombining nuclear genes are needed). Although population structure can mitigate this relationship, nuclear genes on average will have a reduced probability of capturing population-level fragmentation relative to a well-resolved mtDNA gene tree (Moore, 1995). Nonetheless, at this writing, it is difficult to predict the future of mtDNA and nuclear DNA analyses; selective sweeps in mtDNA (Rand *et al.*, 1994) could hinder population inferences drawn from mtDNA data.

I suggest that studies of geographic variation will have greatest value when they encompass a large area, permitting comparisons of codistributed species. The numbers of individuals per locale need not be as high as they were for allozymes because the estimation procedures are haplotype and phylogeny based, not allelic frequency based. Roots for intraspecific haplotype phylogenies will also be required. These aspects will further the cross-enlightenment of population genetics and systematics (Avise *et al.*, 1987).

ACKNOWLEDGMENTS

I thank A. Rogers for the mismatch distribution analyses, and J. Klicka, who compiled Table I. Others who provided field or laboratory assistance are acknowledged in previous papers, but I am especially grateful to D. Dittmann for laboratory assistance, and to G. F. Barrowclough, S. J. Hackett, and K. V. Burns for field help. I thank the various state and federal agencies for granting collecting permits. The NSF funded many of the studies discussed herein. A. Lynch provided the G_{st} program. F. B. Gill, S. V. Edwards, D. P. Mindell, J. T. Klicka, A. J Fry, and J. D. Weckstein provided helpful comments on the manuscript.

REFERENCES

Aldrich, J., and James, F. C. (1991). Ecogeographic variation in the American robin (*Turdus migratorius*). *Auk* **108**, 230–249.

Avise, J. C. (1983). Commentary. *In* "Perspectives in Ornithology" (A. H. Brush and G. A. Clark, Jr., Eds.), pp. 262–270. Cambridge University Press, New York.

Avise, J. C. (1992). Molecular population structure and the biogeographic history of a regional fauna: A case history with lessons for conservation biology. *Oikos* **63**, 62–76.

Avise, J. C., and Nelson, W. S. (1989). Molecular genetic relationships of the extinct dusky seaside sparrow. *Science* **243**, 646–648.

Avise, J. C. (1994). "Molecular Markers, Natural History, and Evolution." Chapman & Hall, New York.

Avise, J. C., Shapira, J. F., Daniel, S. W., Aquadro, C. F., and Lansman, R. A. (1983). Mitochondrial DNA differentiation during the speciation process in *Peromyscus*. *Mol. Biol. Evol.* **1**, 38–56.

Avise, J. C., Arnold, J., Ball, R. Martin, Bermingham, E., Lamb, T., Neigel, J. E., Reeb, C. A., and Saunders, N. C. (1987). Intraspecific phylogeography: The mitochondrial DNA bridge between population genetics and systematics. *Annu. Rev. Ecol. Syst.* **18**, 489–522.

Avise, J. C., Ball, R. M., and Arnold, J. (1988). Current versus historical population sizes in vertebrate species with high gene flow: A comparison based on mitochondrial DNA lineages and inbreeding theory for neutral mutations. *Mol. Biol. Evol.* **5**, 331–344.

Avise, J. C., Ankney, C. D., and Nelson, W. S. (1990). Mitochondrial gene trees and the evolutionary relationship of mallard and black ducks. *Evolution* **44**, 1109–1119.

Ball, R. M., Jr., and Avise, J. C. (1992). Mitochondrial DNA phylogeographic differentiation among avian populations and the evolutionary significance of subspecies. *Auk* **109**, 626–636.

Ball, R. M., Jr., James, F. C., Freeman, S., Bermingham, E., and Avise, J. C. (1988). Phylogeographic population structure of red-winged blackbirds assessed by mitochondrial DNA. *Proc. Natl. Acad. Sci. U.S.A.* **85**, 1558–1562.

Barrowclough, G. F. (1980a). Genetic and phenotypic differentiation in a wood warbler (genus *Dendroica*) hybrid zone. *Auk* **97**, 655–668.

Barrowclough, G. F. (1980b). Gene flow, effective population sizes, and genetic variance components in birds. *Evolution* **34**, 789–798.

Barrowclough, G. F. (1983). Biochemical studies of microevolutionary processes. *In* "Perspectives in Ornithology" (A. H. Brush and G. A. Clark, Jr., Eds.), pp. 223–261. Cambridge University Press, New York.

Barrowclough, G. F., and Gutiérrez, R. J. (1990). Genetic variation and differentiation in the spotted owl (*Strix occidentalis*). *Auk* **107**, 737–744.

Barrowclough, G. F., and Johnson, N. K. (1988). Genetic structure of North American birds. *Acta XIX Congr. Int. Ornithol.* 1620–1638.

Barrowclough, G. F., Johnson, N. K., and Zink, R. M. (1985). On the nature of genic variation in birds. *In* "Current Ornithology" (R. F. Johnston, Ed.), Vol. 2, pp. 135–154. Plenum Press, New York.

Bennett, K. D. (1990). Milankovitch cycles and their effects on species in ecological and evolutionary time. *Paleobiology* **16**, 11–21.

Bermingham, E., Rohwer, S., Freeman, S., and Wood, C. (1992). Vicariance biogeography in the Pleistocene and speciation in North American wood warblers: A test of Mengel's model. *Proc. Natl. Acad. Sci. U.S.A.* **89**, 6624–6628.

Capparella, A. P. (1991). Neotropical avian diversity and riverine barriers. *Acta XX Congr. Int. Ornithol.* 307–316.

Chesser, R. T., and Zink, R. M. (1994). Modes of speciation in birds: An application of Lynch's (1989) method. *Evolution* **48**, 490–497.

Corbin, K. C. (1987). Geographic variation and speciation. *In* "Avian Genetics" (F. Cooke and P. A. Buckley, Eds.), pp. 321–353. Academic Press, New York.

Cracraft, J. (1982). Geographic differentiation, cladistics and vicariance biogeography: Reconstructing the tempo and mode of evolution. *Am. Zool.* **22**, 411–424.

Crandall, K. A., and Templeton, A. R.. (1993). Empirical tests of some predictions from coalescent theory with applications to intraspecific phylogeny reconstruction. *Genetics* **134**, 959–969.

Edwards, S. V. (1993). Long distance gene flow in a cooperative breeder detected in genealogies of mitochondrial DNA sequences. *Proc. R. Soc. Lond.* **252**, 177–185.

Ellsworth, D. L., Honeycutt, R. L., Silvy, N. J., Rittenhouse, K. D., and Smith, M. H. (1994). Mitochondrial DNA and nuclear gene differentiation in North American prairie grouse (genus *Tympanuchus*). *Auk* **111**, 661–671.

Evans, P. G. H. (1987). Electrophoretic variability of gene products. *In* "Avian genetics" (F. Cooke and P. A. Buckley, Eds.), pp. 105–162. Academic Press, New York.

Felsenstein, J. (1978). The number of evolutionary trees. *Syst. Zool.* **27**, 27–33.

Fleischer, R. C., Rothstein, S. I., and Miller, L. S. (1991). Mitochondrial DNA variation indicates gene flow across a zone of known secondary contact between two subspecies of the brown-headed cowbird. *Condor* **93**, 185–189.

Gill, F. B., and Slikas, B. (1992). Patterns of mitochondrial DNA divergence in North American crested titmice. *Condor* **94**, 20–28.

Gill, F. B., Mostrom, A. M., and Mack, A. L. (1993). Speciation in North American chickadees. I. Patterns of mtDNA genetic divergence. *Evolution* **47**, 195–212.

Gould, S. J., and Johnston, R. F. (1972). Geographic variation. *Annu. Rev. Ecol. Syst.* **3**, 457–498.

Gyllensten, U., Reuterwall, C., and Rymann, N. (1979). Genetic variability in Scandinavian populations of willow grouse (*Lagopus lagopus* L.) and rock ptarmigan (*Lagopus mutus* L.). *Hereditas* **91**, 301.

Hubbard, J. P. (1973). Avian evolution in the aridlands of North America. *Living Bird* **12**, 155–196.

James, F. C. (1970). Geographic size variation in birds and its relationship to climate. *Ecology* **51**, 365–390.

James, F. C. (1983). Environmental component of morphological differences in birds. *Science* **221**, 184–186.

James, F. C. (1991). Complementary descriptive and experimental studies of clinal variation in birds. *Am. Zool.* **31**, 694–706.

Johnson, N. K., and Cicero, C. (1991). Mitochondrial DNA sequence variability in two species of sparrows of the genus *Amphispiza*. *Acta XX Congr. Int. Ornithol.* 600–610.

Johnson, N. K., and Marten, J. A. (1988). Evolutionary genetics of flycatchers. II. Differentiation in the *Empidonax difficilis* complex. *Auk* **105**, 177–191.

Kessler, L. G., and Avise, J. C. (1985). A comparative description of mitochondrial DNA differentiation in selected avian and other vertebrate genera. *Mol. Biol. Evol.* **2**, 109–125.

Klein, N. K., and Brown, W. M. (1995). Intraspecific molecular phylogeny in the yellow warbler (*Dendroica petechia*), and implications for avian biogeography in the West Indies. *Evolution* **48**, 1914–1932.

Lamb, T., Avise, J. C., and Gibbons, J. W. (1989). Phylogeographic patterns in mitochondrial DNA of the desert tortoise (*Xerobates agassizi*), and evolutionary relationships among the North American gopher tortoises. *Evolution* **43**, 76–87.

Lamb, T., Jones, T. R., and Avise, J. C. (1992). Phylogeographic histories of representative herpetofauna of the southwestern U.S.: Mitochondrial DNA variation in the desert iguana (*Dipsosaurus dorsalis*) and the chuckwalla (*Sauromalus obesus*). *J. Evol. Biol.* **5**, 465–480.

Lansman, R. A., Avise, J. C., Aquadro, C. F., Shapira, J. F., and Daniel, S. W. (1983). Extensive genetic variation in mitochondrial DNA's among geographic populations of the deer mouse, *Peromyscus maniculatus*. *Evolution* **37**, 1–16.

Lanyon, S. M. (1992). Interspecific brood parasitism in blackbirds (Icterinae): A phylogenetic perspective. *Science* **255**, 77–79.

Lewontin, R. C. (1974). "The Genetic Basis of Evolutionary Change." Columbia Univ. Press, New York.

Lynch, M., and Crease, T. J. (1990). The analysis of population survey data on DNA sequence variation. *Mol. Biol. Evol.* **7**, 377–394.

McDonald, D. B., and Potts, W. K. (1994). Cooperative display and relatedness among males in a lek-mating bird. *Science* **266**, 1030–1032.

Maddison, W. P., Donoghue, M. J., and Maddison, D. R. (1984). Outgroup analysis and parsimony. *Syst. Zool.* **33**, 83–103.

Mindell, D. P., Knight, A., Baer, C., and Huddleston, C. J. (1996). Slow rates of molecular evolution in birds and the metabolic rate and body temperature hypotheses. *Mol. Biol. Evol.* **13**, 422–426.

Moore, W. S. (1995). Inferring phylogenies from mtDNA variation: Mitochondrial-gene trees versus nuclear-gene trees. *Evolution* **49**, 718–726.

Moore, W. S., and Dolbeer, R. A. (1989). The use of banding recovery data to estimate dispersal rates and gene flow in avian species: Case studies in the red-winged blackbird and common grackle. *Condor* **91**, 242–253.

Moore, W. S., Graham, J. H., and Price, J. T. (1991). Mitochondrial DNA variation in the northern flicker (*Colaptes auratus*). *Mol. Biol. Evol.* **8**, 327–344.

Murphy, R. W. (1983). Paleobiogeography and genetic differentiation of the Baja California herpetofauna. *Occ. Papers Cal. Acad. Sci.* **137**, 1–48.

Neigel, J. E., and Avise, J. C. (1993). Application of a random-walk model to geographic distributions in animal mitochondrial DNA variation. *Genetics* **135**, 1209–1220.

Neigel, J. E., Ball, Jr., R. M., and Avise, J. C. (1991). Estimation of single-generation migration distances from geographic variation in animal mitochondrial DNA. *Evolution* **45**, 423–432.

Pielou, E. C. (1991). "After the Ice Age." Univ. Chicago Press, Chicago.

Quinn, T. W. (1992). The genetic legacy of Mother Goose—phylogeographic patterns of lesser snow goose *Chen caerulescens caerulescens* maternal lineages. *Mol. Ecol.* **1**, 105–117.

Rand, D. M., Dorfsman, M., and Kann, L. M. (1994). Neutral and non-neutral evolution of *Drosophila* mitochondrial DNA. *Genetics* **138**, 741–756.

Redfield, J. A. (1974). Genetics and selection at the *Ng* locus in blue grouse (*Dendragapus obscurus*). *Heredity* **36**, 69–78.

Rising, J. D., and Avise, J. C. (1993). Application of genealogical-concordance principles to the taxonomy and evolutionary history of the sharp-tailed sparrow (*Ammodramus caudacutus*). *Auk* **110**, 844–856.

Rogers, A. R. (1995). Genetic evidence for a Pleistocene population expansion. *Evolution* **49**, 608–615.

Rogers, A. R. (1997). Population structure and modern human origins. *In* "Progress in Population Genetics and Human Evolution" (P. J. Donnelly and S. Tavaré, Eds.). Springer-Verlag, New York (in press).

Rogers, A. R., and Harpending, H. (1992). Population growth makes waves in the distribution of pairwise genetic differences. *Mol. Biol. Evol.* **9**, 552–569.

Rothstein, S. (1990). A model system for coevolution: Avian brood parasitism. *Annu. Rev. Ecol. Syst.* **21**, 481–508.

Seutin, G., Brawn, J., Ricklefs, R. E., and Bermingham, E. (1993). Genetic divergence among populations of a tropical passerine, the streaked saltator (*Saltator albicollis*). *Auk* **110**, 117–126

Seutin, G., Klein, N. K., Ricklefs, R. E., and Bermingham, E. (1994). Historical biogeography of the bananquit (*Coereba flaveola*) in the Caribbean region: A mitochondrial DNA assessment. *Evolution* **48**, 1041–1061.

Shields, G. F. (1990). Analysis of mitochondrial DNA of Pacific black brant (*Branta bernicla nigricans*). *Auk* **107**, 620–623.

Shields, G. F., and Wilson, A. C. (1987). Calibration of mitochondrial DNA evolution in geese. *J. Mol. Evol.* **24**, 212–217.

Simberloff, D. (1987). Calculating probabilities that cladograms match: A method of biogeographic inference. *Syst. Zool.* **36**, 175–195.

Slatkin, M., and Maddison, W. P. (1989). A cladistic measure of gene flow inferred from the phylogenies of alleles. *Genetics* **123**, 603–613.

Slatkin, M., and Maddison, W. P. (1990). Detecting isolation by distance using phylogenies of genes. *Genetics* **126**, 249–260.

Templeton, A. R. (1980). Modes of speciation and inferences based on genetic distances. *Evolution* **34**, 719–729.

Van Wagner, C. E., and Baker, A. J. (1990). Association between mitochondrial DNA and morphological evolution in Canada geese. *J. Mol. Evol.* **31**, 373–382.

Wenink, P. W., Baker, A. J., and Tilanus, M. G. J. (1993). Hypervariable control-region sequences reveal global population structuring in a long-distance migrant shorebird, the dunlin. *Proc. Natl. Acad. Sci. U.S.A.* **90**, 94–98.

Wenink, P. W., Baker, A. J., and Tilanus, M. G. J. (1994). Mitochondrial control-region sequences in two shorebird species, the turnstone and dunlin, and their utility in population genetic studies. *Mol. Biol. Evol.* **11**, 22–31.

Wiley, E. O. (1988). Vicariance biogeography. *Annu. Rev. Ecol. Syst.* **19**, 513–542.

Wright, S. (1978). "Evolution and the Genetics of Natural Populations. IV. Variability within and among Natural Populations." Univ. Chicago Press, Chicago.

Zink, R. M. (1991). The geography of mitochondrial DNA variation in two sympatric sparrows. *Evolution* **45**, 329–339.

Zink, R. M. (1994). The geography of mitochondrial DNA variation, population structure, hybridization, and species limits in the fox sparrow (*Passerella iliaca*). *Evolution* **48**, 96–111.

Zink, R. M. (1996a). Comparative phylogeography in North American birds. *Evolution* **50**, 308–317.

Zink, R. M. (1996b). Species concepts, speciation, and sexual selection. *J. Avian Biol.* **27**, 1–6.

Zink, R. M., and Blackwell, R. C. (1996). Comparison of evolutionary patterns in multiple data sets for four sparrow genera. *Auk* **113**, 59–67.

Zink, R. M., and Dittmann, D. L. (1993a). Gene flow, refugia, and evolution of geographic variation in the song sparrow (*Melospiza melodia*). *Evolution* **47**, 717–729.

Zink, R. M., and Dittmann, D. L. (1993b). Population structure and gene flow in the chipping sparrow and a hypothesis for evolution in the genus *Spizella*. *Wilson Bull.* **105**, 399–413.

Zink, R. M., and Hackett, S. J. (1988). Historical biogeography of the North American avifauna. *Acta XIX Congr. Int. Ornithol.* 2574–2580.

Zink, R. M., and Remsen, J. V. (1986). Evolutionary processes and patterns of geographic variation in birds. *Curr. Ornithol.* **4**, 1–69.

Zink, R. M., Rootes, W. L., and Dittmann, D. L. (1991a). Mitochondrial DNA variation, population structure, and evolution of the common grackle (*Quiscalus quiscala*). *Condor* **93**, 318–329.

Zink, R. M., Dittmann, D. L., Cardiff, S. W., and Rising, J. W. (1991b). Mitochondrial DNA variation and taxonomic status of the large-billed savannah sparrow. *Condor* **93**, 1016–1019.

Zink, R. M. Rohwer, S., Andreev, A. V., and Dittmann, D. L. (1995). Trans-Beringia comparisons of mitochondrial DNA differentiation in birds. *Condor* **97**, 639–649.

Zink, R. M., Blackwell, R. C., and Rojas-Soto, O. (1997). Species limits in the Le Conte's thrasher. *Condor* **99**, 132–138.

CHAPTER 12

The Speciation of South American and African Birds in Montane Regions

MICHAEL S. ROY,* JOSÉ MARIA CARDOSO DA SILVA,† PETER ARCTANDER,* JAIME GARCÍA-MORENO,* AND JON FJELDSÅ†

*Zoological Institute and †Zoological Museum, University of Copenhagen, DK 2100 Denmark

I. INTRODUCTION

Tropical forest ecosystems are credited with containing unparalleled biodiversity, but unfortunately are increasingly threatened by human activities. Up until now most conservation priorities have been modeled around present patterns of species richness and/or endemism. However, from an evolutionary perspective, only by understanding the processes that cause these patterns can informed and comprehensive conservation policies be formulated. Many theories have been postulated that link historic processes to the ecological patterns evident today (*paleogeography hypothesis*, *gradient hypothesis*, *river hypothesis*, *river refuge hypothesis*, and *refuge hypothesis*),

Avian Molecular Evolution and Systematics

but rather than clarify, these theories have tended to increase the controversiality of the subject (see reviews by Haffer, 1997, and Tuomisto and Ruokalainen, 1997).

The difficulty lies in disentangling today's patterns of species distributions from the historic processes that caused them. For example, replacement of closely related species across a geographic barrier or adjacent ecoregions may lead one to conclude that the present physical or environmental barrier caused the initial speciation event (Endler, 1977). However, such replacements may be secondary since sharp sutures between previously reproductively incompatible species are most easily maintained where there is a physical barrier or gradient. This has been exemplified by DNA studies of Andean rodents (Patton and Smith, 1992) and birds (Arctander and Fjeldså, 1994). These authors showed that species that display present-day parapatric altitudinal replacement on mountain slopes actually originated from different mountain ranges, suggesting that parapatry is secondary. In addition, Salo (1988) had associated habitat complexity in fluvial plains in the upper Amazon area with speciation, but later suggested (J. Salo, personal communication) that dynamic habitat mosaicism is more likely to be a diversity-maintaining process and not the cause of the initial speciation process.

To separate initial speciation events from the process of species redistribution and accumulation in areas of high current carrying capacity, we examined macroscale patterns of avian species richness and related this to macroscale patterns of diversification in tropical forests. Our aim was to find out where recent speciation has been most intensive, and where old lineages (which only show phyletic speciation) predominate. We then used this pattern to formulate a model hypothesis for continental-scale speciation and redistribution, which was then used as a guideline for identifying case studies for detailed evaluation using DNA sequence data.

This study compared data from South America and Africa, first because the marked difference in species richness indicates more intensive diversification in the former, and second because speciation patterns can be compared between the almost uninterrupted band of montane forest along the eastern slope of the tropical Andes region with the chains of mutually isolated "montane forest islands" in Africa.

We review here the development of the model hypothesis, where biogeographic information was combined with the DNA–DNA hybridization data of Sibley and Ahlquist (1990) (see Fjeldså, 1994, for details). Thereafter we present preliminary results of ongoing studies of smaller species groups, designed to evaluate different predictions of the model hypothesis.

II. THE MODEL HYPOTHESIS

A. Basic Assumptions

Regions of intensive speciation are likely to be characterized by a high proportion of species representing recent phylogenetic events relative to old lineages with little or no recent speciation. In addition, local aggregates of relictual species, which have undergone a severe contraction from large parts of the initial range of its lineage, indicate places that have remained ecologically stable during climatic fluctuations throughout the Quaternary (Fjeldså, 1995; Fjeldså and Lovett, 1997). To enter a phylogenetic framework and an approximate time dimension into our study, we used the results of DNA–DNA hybridizations by Sibley and Ahlquist (1990) (Fjeldså 1992, 1994). These assumptions provide us with the possibility of temporal and spatial systematic comparisons, of regions of species richness with regions of paleoecological stability.

B. Method

One advantage of Sibley–Ahlquist phylogenetic data for this study is that a uniform technology was used to measure the time dimension across all groups of birds. The data set covers the global avifauna remarkably well, with altogether 1700 species studied, the data gaps consisting mostly of terminal branches of the phylogeny. Certain taxa are underrepresented but the data set does not appear to be biased toward any particular geographical region.

The molecular technique is based on thermostability of heteroduplex DNA. All single-copy, nuclear genes of two species are hybridized, after one of the single-copy, nuclear genomes has been radio labeled to serve as a tracer. The method measures the melting curve of the single-stranded tracer DNA, taking the midpoint ($T_{50}H$) as a measure of the overall genetic divergence. Attempts to calibrate the molecular clock suggest a tentative rate $\Delta 1.0 = 2.3$ million years (MY) (Sibley and Ahlquist, 1990, p. 703), but the relationship between DNA distance values and time of divergence is not simple (Sibley and Ahlquist, 1990; see, e.g., p. 400 and Fig. 99) and the divergence may be retarded considerably in large birds with delayed sexual maturation. The relationships among species are determined by UPGMA clustering.

Despite comprehensive methodological control, including studies of generation time effects, a multitude of technical and molecular aspects cause doubts about the power of resolution claimed by the authors (see O'Hara, 1991; Mindell, 1992; Sibley *et al.*, 1993; Harshmann, 1994). Sibley and Ahlquist have taken the unusual approach to "correct" for different rates of DNA evolution rather than using an algorithm that does not assume rate constancy. It is also possible that the UPGMA clustering compresses the deep parts of the phylogeny, and that lineage-specific features such as biased base composition in some lineages could influence the apparent depth of nodes and their sequence, for example in the "explosive" radiation of passerine birds families in the mid-Tertiary.

However, since our only use of the Sibley–Ahlquist phylogenies was to define two markedly different categories of species (deep branches and recent radiations with many species), the effect of the above-mentioned inaccuracies is likely to be small. For this reason and despite reservations about the precise sequence and depths of nodes, we accepted the trees given by Sibley and Ahlquist as being representative of avian diversification.

The Sibley–Ahlquist data provide relative timings of early radiations of most groups, although there is a lack of detailed reconstruction of top branches causing unresolved polytomies. This was corrected by rigorously determining which species are included within specific nodes on the Sibley–Ahlquist trees and which are not, using all published systematic revisions and phylogenetic hypotheses (the source list is too large to be included here) and personal judgments, where the Sibley–Ahlquist phylogenies provide a basis for polarizing morphological characters. Thus, voucher specimens were used extensively. For the sake of methodological homogeneity, the Sibley–Ahlquist sequences of nodes were used wherever viewpoints were contradictory. For all parts of the phylogenies that were not supported by DNA–DNA hybridizations the nodes (equating N species $- 1$) were spaced regularly above the "baseline" $T_{50}H$ value. Several branches were discarded from the analysis because baseline dating was not possible owing to lack of original data. To make the choice of species as unbiased as possible, acceptable phylogenies that met our criteria and identified recent radiations and old lineages were reconstructed (Fig. 12.1), before the geographical analysis was initiated.

Phylogenetic lineages were identified and mapped. Groupings were assigned according to the age of the lineage: (1) "new species" for radiations with at least 10 species emerging after $T_{50}H$ 2.5 (i.e., during the last 6-MY period with increasing climatic fluctuations of the Croll–Milankovitch type), and (2) "old species" (mostly monotypic genera) representing a single lineage from before $T_{50}H$ 2.5 (in the original analysis by Fjeldså, 1994, this category was divided in two subcategories). As the categories are separated by a major evolutionary dichotomy, the errors will at most consist of inclusion (in group 1) of a few radiations that started slightly earlier than $T_{50}H$ 2.5, or the inclusion (in group 2) of a few monophyletic lineages that are slightly too young. For the African tropics 233 species were identified as recent radiations and 82 species as older branches; for South America the corresponding samples comprised 648 and 107 species (see Fjeldså, 1994, for taxa used). The ratio of new and old species was calculated for each 200×200 km geographic grid cell over both continents, using distributional data contained in standard reference books for the two continental avifaunas (Fig. 12.2).

III. OLD AND NEW SPECIES IN SOUTH AMERICA

The intensive folding in the tropical Andes region since the Miocene blocked the earlier outlet of the Amazon into the Pacific Ocean, thereby altering the patchwork of different habitats, and potential biogeographic barriers, over large parts of the continent. A number of new habitats were formed in the Andes, and a geological subsidence, creating a hydrologically unstable zone in the Chaco, effectively iso-

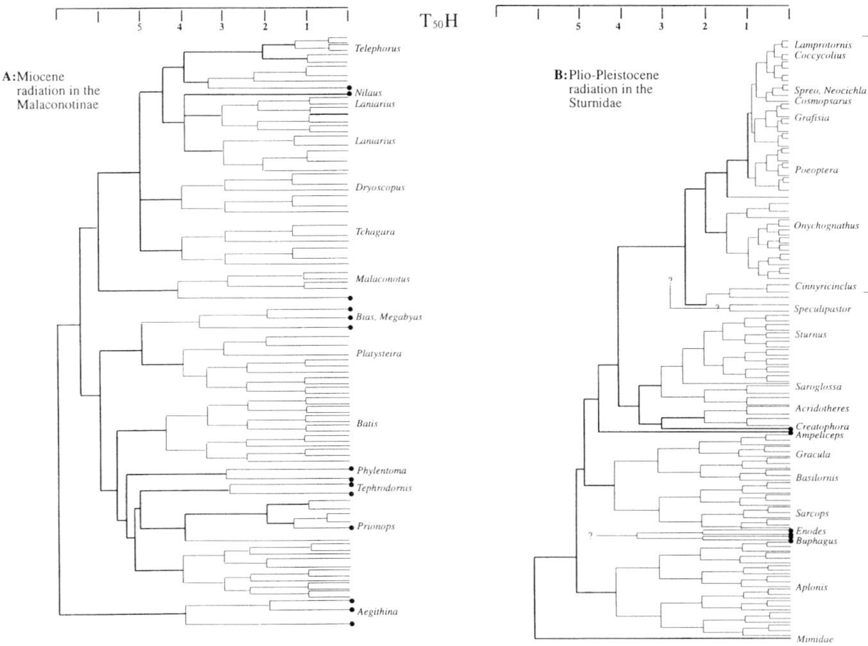

FIGURE 12.1 An example of the two categories of phylogenetic reconstructions representing (A) a rapid early radiation (probably in the early Miocene) and slow rates of later diversification (bush and helmet shrikes and vangas, Malaconotinae; for earlier discussions see Meise, 1968; Benson *et al.*, 1971; Traylor, 1970; Sibley and Ahlquist, 1990) and (B) a group that first radiated mainly in the Pacific and Indo-Malayan areas and showed a strong recent (Plio-Pleistocene) radiation in Africa (starlings, Sturnidae; last revised by Amadon, 1956; Beecher, 1978). Small terminal dots identify lineages identified as "old species" while bracketed groups identify radiations of "new species" (see text for explanation). Heavy lines indicate branching order as determined by Sibley and Ahlquist (1990), the rest were reconstructed according to our criteria (see text).

lated the Andean biota from that of the Brazilian Highland (Hanagarth, 1993; Silva, 1995). Today, humid forests form a continuous band along the eastern slope of the Andes, while the upper cloudforest zone is dissected by deep valleys with arid climates on the bottom.

The distribution of nodes connecting South American taxa on the Sibley–Ahlquist phylogenies shows an explosive burst of differentiation starting from $T_{50}H5$ (Fjeldså, 1994, Fig. 1). Figure 12.2A clearly demonstrates that this differentiation was most intensive in those parts of the tropical zone that were affected by mountain folding and tectonic changes. By far the most important area is the tropical Andes region and its transition toward the Amazon lowlands. Conversely, the proportion of young species is moderate to low in much of the Amazon lowland (Fig. 12.2A), being highest in regions with tectonically active crystalline arcs and lowest in those

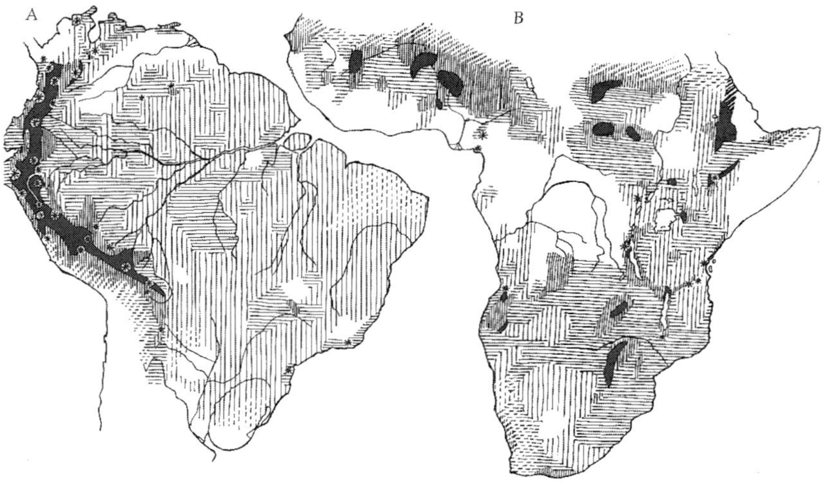

FIGURE 12.2 Geographical variation in (A) South America and (B) Africa, of the ratio between "new species" representing recent radiations and "old species" (modified from Fjeldså, 1994). Blackened areas indicate a ratio of more than 3 new species to old species; narrow spaced vertical lines, more than 2.5; horizontal lines, more than 2; and widely spaced vertical lines, more than 1.5. Asterisks mark an aggregate of biogeographic relicts (*sensu* Cronk, 1992), which may indicate long-term ecoclimatic stability. In Africa these regions are the Cameroon Mountains, Angola Scarp, Rwenzori Mountains/Itombwe Forest in eastern Zaire, Uluguru, Udzungwa and east Usumbara Mountains in Tanzania. Interrupted signatures indicate that the species density is too low for calculating reliable ratios.

parts that are characterized by high fluvial disturbance leading to complex and dynamic habitat mosaics of "fossil" and active floodplains (see Kalliola *et al.*, 1993). It is noteworthy that more than 80% of "new species" occupying the Amazon lowlands also inhabit regions lying outside of the lowland forest biome, thereby obscuring their origins. Also other hydrologically unstable areas (Chaco, Llanos, and northwestern Colombia) are characterized by old species, with little recent differentiation.

IV. OLD AND NEW SPECIES IN AFRICA

Unlike South America, Africa has been little affected by mountain folding, the most significant geological changes affecting the forest biota being (1) a general drying of northern Africa after the continent "collided" with Asia in the early Miocene and the Tethys Sea was closed (see Axelrod and Raven, 1978) and (2) uplifting and rifting isolating the eastern lowland forests from the main Guinea–Congolian rainforest block during the Miocene (Lovett, 1993; Coppes, 1994). The montane forests are continuous on the transition between the Congo Basin and the Albertine

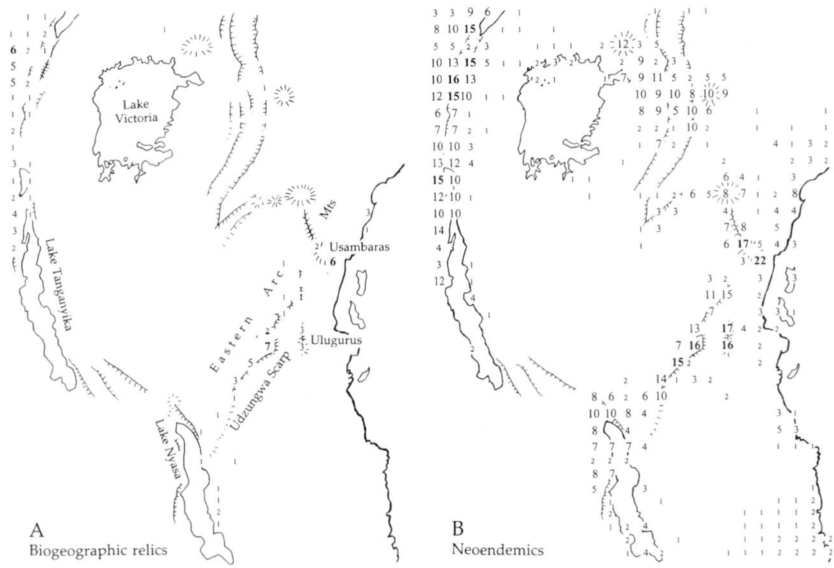

FIGURE 12.3 Geographical patterns of the density of highly distinctive forest bird species that can be regarded as (A) biogeographical relicts and (B) aggregates of neoendemics ["new species" with restricted range (<50,000 km^2) representing vicariance patterns]. Number of species of each category per grid square (as described in text) is indicated; large numbers in boldface represent highest densities of species in each category. Note a correlation in regions that contain both high densities of biogeographical relics and neoendemics (see text). Species defined as relictual comprise *Afropavo congensis*, *"Francolinus" nahani*, *Xenoperdix udzungwensis*, *Himantornis haematopus*, *Canirallus oculeus*, *Phodilus prigoginei*, *Otus irenae*, *Pseudocalyptomena graueri*, *Malaconotus alius*, *Prionops alberti*, *Arcanator orostruthus*, *Swynnertonia swynnertoni*, *Hemitesia neumanni*, *Graueria vittata*, *Orthtomus metopias*, *Apalis moreaui*, *Bathmocercus winnifredae*, *Anthreptes pallidigaster*, *Nectarinnia rufipennis*, *Ploceus golandi*.

Rift, but otherwise are discontinuous (Cameroon, Angola, Ethiopia highland) with a large but punctuated "montane circle" following the Albertine and Malawi rifts, the Eastern Arc crystalline fault-blocks in Tanzania and the Kenya highlands. Along this "circle," it is assumed that the ecoclimatically most stable parts are in upper Zaire and on east-facing escarpments of the Eastern Arc mountains, which are under direct climatic influence from the Indian Ocean (Lovett, 1993). The distribution of relictual species of forest birds seems to locate this stability quite precisely to west of the Rwenzori Slope to Itombwe forest in upper Zaire and the Udzungwa Scarp and Uluguru and East Usambara Mountains in Tanzania (Fig. 12.3).

The most recent diversification in Africa is associated with geological swells in the savanna zone, and mountains, while old species dominate in the extensive lowland forests and in hydrologically unstable lowlands (Chad and Sud Basins; Fig. 12.2B). A study of only forest birds (and forest plants, Fjeldså and Lovett, 1997) further emphasizes that the diversification in the Pleistocene was associated with highlands outside the Congo Basin, and the "montane circle" in eastern Africa in particular. "New species" occurring inside the Congolian rainforest are generally widespread or disjunct, and as with South America more than 80% of them also live elsewhere on the continent, obscuring their origins (Fig. 12.2).

V. IMPLICATIONS FOR THE MODEL

Species arise by dynamic processes acting in montane regions, but persist in floodplains. The species-rich tropical lowland forests were often regarded as "centers of origin" or "dispersal centers" for tropical biodiversity. Over the last three decades, the prevailing explanation for the origin of the extraordinary biodiversity of tropical lowland forests was the *refuge theory*. Proposed initially for South American birds by Haffer (1969, 1974), this theory was soon applied to the African avifauna (Diamond and Hamilton, 1980; Mayr and O'Hara, 1986; Crowe and Crowe, 1982) as well as other groups, such as plants, and insects in different tropical regions (reviews in Prance, 1982; Whitmore and Prance, 1987). The theory assumes that species evolved by isolation in forest areas that remained stable despite global ecoclimatic changes. The impact of these changes, forced by cyclical changes in the earth's orbit (Milankovitch cycles), has been accentuated by a general global cooling during the last few million years, with large glacial peaks (or arid periods in the tropics) during the last 0.9 MY (see Bartlein and Prentice, 1989; Bennett, 1990; Hooghiemstra *et al.*, 1993).

Lowland floodplains are dominated by species of pre-Pleistocene age, with no particular concentration of younger species in the postulated refuge areas (Fig. 12.2; see Amorim, 1991, for similar arguments). Instead, much more speciation takes place in areas with a distinctive topographic structure (see also Vrba, 1993). Since a large proportion of young species in lowland rainforests are widespread, often extending to forested escarpments or gallery forests in adjacent biomes, alternative explanations of the initial speciation events are indeed possible (notably for Africa).

Undoubtedly, a great deal of differentiation is due to vicariance caused by tectonism and erosion creating isolating barriers. However, a marked correlation between "hotspots" for species that are part of vicariance patterns (Fig. 12.3B) and peak concentrations of relictual forms (Fig. 12.3A) suggests an association between speciation and intrinsic properties of specific mountain scarps (see Fjeldså, 1995, for the Andes; see Stebbins and Major, 1965, for the Californian flora). Where prevalent atmospheric flows interact with topography, climate can be moderate, creating stable cloudforest conditions locally (Fjeldså *et al.*, 1997). In this case, the initial isolating mechanism is not necessarily a physical barrier between hotspots but could be their intrinsic high spatiotemporal heterogeneity that produces high species turnover and robust communities, where specialist species cannot easily remain established. This would form the basis for a highly dynamic process of isolation and opportunities for short-term dispersal between hotspots.

As a working hypothesis we postulate that the evolution of tropical forest birds is driven by a dynamic process of local isolation in stable montane forests with occasional dispersal between them, and that new species may gradually expand into other habitats, and in the end accumulate in the extensive tracts of lowland forest or woodland savannas. In the Amazon basin, the amount of tectonically induced flooding, and the dynamics of meandering rivers in the Amazon basin, make it evident that biodiversity is redistributed (Kalliola *et al.*, 1993). Tropical lowland forests are highly unstable on the local scale, and we suggest that this high level of spatiotemporal heterogeneity makes them act as "museums" where large numbers of species (of potentially diverse origins) have accumulated over long periods of time.

Critical evaluation of whether hotspots are centers of origin or reflect subsequent redistribution, and whether a dynamic speciation process in montane areas can deliver recruits to lowland biota, is needed. In the following section we describe three studies that assess specific questions of patterns and processes of distribution of montane avifauna in both Africa and South America.

VI. CASE STUDIES OF BIOGEOGRAPHIC PATTERNS IN TROPICAL MOUNTAINS

A. The Andes

1. Flycatchers of the genus *Leptopogon* are represented by four species that occur in the South American lowlands and highlands. All four species are forest dwelling, whose distribution near the Andes can be summarized as follows: (a) *L. amaurocephalus*, tropical lowland (to 600 m); (b) *L. superciliaris*, upper-tropical (600 to 2100 m); (c) *L. taczanowskii*, upper subtropical (1600 to 2700 m); and (d) *L. rufipectus*, upper subtropical (1600 to 2700 m). *Leptopogon taczanowskii* and *L. rufipectus* are allopatric taxa that are separated by the River Marañon valley, in northern Peru (Traylor, 1979; Fjeldså and Krabbe, 1990; Bates and Zink, 1994).

Bates and Zink (1994) studied the phylogenetic relationships of these four species, using both allozymes and mitochondrial DNA (mtDNA). They found that (1) *Leptopogon* is indeed a monophyletic genus; (2) *L. amaurocephalus* is the basal member of the genus; and (3) *L. superciliaris* is the sister-group of the clade formed by *L. taczanowskii* and *L. rufipectus*. This pattern supports the hypothesis that the evolution of *Leptopogon* species was driven by vicariance events that basically followed the uplift of the Andes during the late Tertiary. In fact, by using published estimates of molecular-clock calibrations for both allozymes and mtDNA, Bates and Zink (1994) found a good correlation between genetic differentiation within *Leptopogon* and the timing of uplift of the Bolivian Andes.

2. Tapaculos of the genus *Scytalopus* (family Rhinocryptidae) are small (20–40 g), sooty-gray-colored birds that inhabit the dense understory of humid forest

and scrub in the Andes, Central America, and eastern and central Brazil (Sibley and Monroe, 1990; Ridgeley and Tudor, 1994). Traditional systematics using plumage characters have recognized 11 species in this genus, but detailed studies including vocal characters suggest that many more species are involved (Krabbe and Schulenberg, 1997).

Most species of *Scytalopus*, as they are currently defined, have complex patterns of distribution and geographic variation along the Andes. For instance, species that are similar in plumage tend to replace each other in different altitudinal zones (Fjeldså and Krabbe, 1990). The simplest model for explaining this pattern could be one based on parapatric speciation along an environmental gradient, such as suggested by Endler (1977). In this case, species along a gradient are predicted to be monophyletic and no dispersal events are needed to explain this distribution pattern. Alternatively, one can propose that the pattern of altitudinal replacement of species along a mountain slope is a secondary event, caused by dispersal between different slopes. In this model, species found on the same slope are not predicted to be monophyletic. These species could have evolved by an allopatric (vicariant) mode of speciation due to isolation in different mountain ranges and thereafter dispersed between slopes, establishing the distribution pattern currently observed. As these two models suggest different sets of phylogenetic relationships (Patton and Smith, 1992), they can be tested by examining the phylogenetic relationships of populations.

Arctander and Fjeldså (1994) evaluated these models by studying 14 tapaculo taxa, some of which have adjacent distributions along the same mountain slope in Ecuador. They compared 285 bp of mitochondrial cytochrome *b* (cyt *b*) from each species in order to elucidate their phylogenetic relationships. Although they used only a small fragment of cyt *b*, pairs of taxa that have distinctive songs and live essentially in sympatry (i.e., those that are distinct biological species) differed from each other by 23–42 transitions (ts; average 28.3) and 1–13 transversions (tv; average 4.7), with a ts-to-tv ratio of 6:1.

On the basis of the preliminary phylogeny of the 14 taxa evaluated so far and 4 taxon tests used to evaluate specific hypotheses, Arctander and Fjeldså (1994) showed that in the 2 areas that have the highest number of parapatric species (5 species on one slope) every one of these was more closely related to a taxon inhabiting another mountain range than to its nearest neighbor on the same slope. Thus the parapatric model was falsified, and it is suggested that divergence in the Andean *Scytalopus* was allopatric, in small disjunct isolates in different parts of the Andes, and that the currently observed pattern of replacement in different altitudinal belts is a secondary event.

B. East African Mountains

The circle of mountain islands of East Africa are composed of the Albertine and Malawi Rift Mountains, the Tanzanian Eastern Arc Mountains, and the Kenya

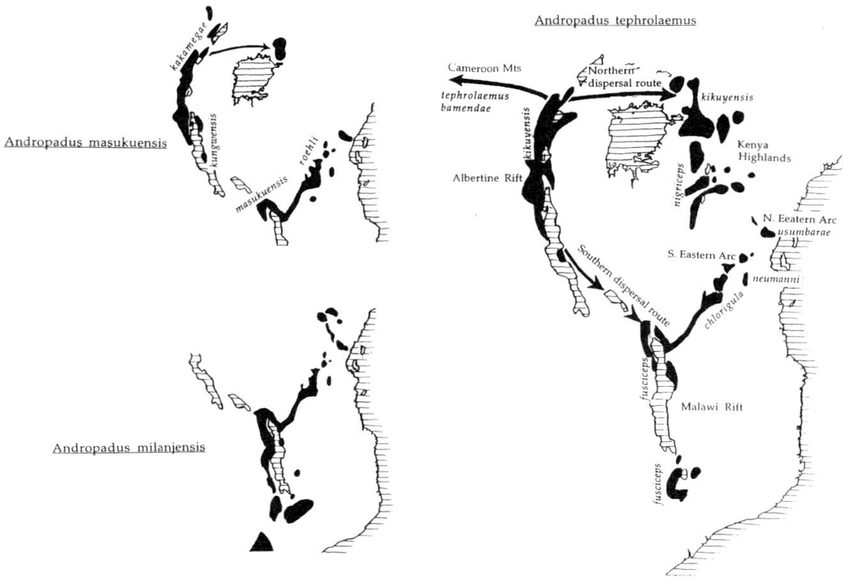

FIGURE 12.4 Distribution of the three montane *Andropadus* species of East Africa. Blackened areas indicate known species range with approximate distribution of relevant subspecies (in italic) as discussed in text. In addition, suggested dispersal routes are indicated by arrows and highland regions referred to in text are shown on the *A. tephrolaemus* distribution map.

Highlands (Figs. 12.2 and 12.4). Biogeographically linked to these mountains are the Cameroon Mountains and Angola Scarp. To elucidate the complex patterns of vicariance and dispersal between the mountain islands of Africa, we focused our attention on the strongly polytypic greenbuls of the genus *Andropadus*.

The genus *Andropadus* is represented by 11 obligate forest-dwelling species. Once united with *Pycnonotus* (bulbuls) (Delacour, 1943; Rand, 1958), they were separated by White (1962) and Hall and Moreau (1970). They differ in plumage and ecology, and possibly origins, with *Pycnonotus* being a primarily Asian genus. Four species (*A. tephrolaemus*, *A. masukuensis*, *A. milanjensis*, and *A. montanus*) are strictly montane and represent (according to our current evaluation) a recently radiated, monophyletic group. All except for *A. montanus* are widely sympatric in eastern Africa (Fig. 12.4), suggesting that the initial vicariance events leading to speciation were followed by periods of range expansion. Taxonomically this group of mountain greenbuls remains obscure and efforts to identify evolutionary relationships between species and taxonomic status of isolated populations within each species remain inconclusive, having been largely influenced by geographic rather than sound cladistic analysis (see Dowsett and Dowsett-Lemaire, 1993).

Most intriguing is the relationship between *A. masukuensis* and *A. tephrolaemus*, which are morphologically similar and show broadly overlapping distributions around the "montane circle." *Andropadus masukuensis* is subdivided into four relatively indistinct gray-headed (*kakamegae/kungwensis* along the Albertine Rift and Kenya) and green-headed forms (*masukuensis/roehli* along the Eastern Arc; Fig. 12.4). Conversely, *A. tephrolaemus* is subdivided into six distinctive forms within the montane circle (Fig. 12.4), replacing each other on different mountains (*kikuyensis* along the Albertine Rift into Kenya, *nigriceps* in southern Kenya/northern Tanzania, *usambarae* in Pare and Usambara mountains of the northern Eastern Arc, *neumanni* in Usambara mountains, *chlorigula* along the southern Eastern Arc mountains, *fusciceps* along the Malawi Rift). *Andropadus masukuensis* was once treated as a subspecies of *A. montanus* (White, 1962; separation suggested by Stuart, 1986; Dowsett-Lemaire, 1989), which is sympatric with *A. t. tephrolaemus* and *bamendae* in the Cameroon highlands. Interestingly, however, Keith *et al.* (1992) suggest that *A. masukuensis* is in fact more related to *A. t. tephrolaemus*. *Andropadus milanjensis* has a "southerly" distribution extending from the Malawi Mountains of Mozambique and up through Malawi to Tanzania and the northern Eastern Arc Mountains (Fig. 12.4). Southern populations resemble *A. tephrolaemus* whereas those of the north are more distinct.

In this analysis we compared 597 bp of the cyt *b* gene sequence of the mtDNA from individuals of each species (Fig. 12.5). Both 5′ (297 bp) and 3′ (300 bp) ends of the gene were amplified and sequenced in both directions using universal primers (Kocher *et al.*, 1989; Edwards *et al.*, 1991) (see caption to Fig. 12.5). Sequences were aligned by eye, and no insertions or deletions were found, as expected from a coding region. Aligned sequences were then analyzed using maximum parsimony (PAUP; Swofford, 1991). The samples included six subspecies of *A. tephrolaemus* (*chlorigula* and *neumanni* from the southern Eastern Arc, *usambarae* and *nigriceps* from the northern Eastern Arc and Kenya Highlands, *kikuyensis* from the northern Albertine Rift, and *tephrolaemus* from Cameroon), six samples of *A. masukuensis roehli* from comparable regions of the entire Eastern Arc, two *A. milanjensis striifacies* from different regions within the Eastern Arc, and *Phyllastrephus flavostriatus* as an outgroup. Our objectives were to investigate the evolutionary histories within and between these species, specifically to assess (1) the degree of genetic divergence and phylogenetic relationships within and between *A. masukuensis* and *A. tephrolaemus*, (2) dispersal and vicariance patterns between Albertine Rift and Eastern Arc Mountains, and (3) the potential interchange between the Cameroon Mountains and the "montane circle" of eastern Africa.

Phylogenetic analyses of these populations suggest complex historical interchanges between different montane areas, including an early vicariance and divergence of *A. milanjensis* (Fig. 12.5). However, the most interesting aspect is that *A. tephrolaemus* is paraphyletic in relation to the largely sympatric *A. masukuensis*. The genetic divergence among Tanzanian forms of *A. tephrolaemus* (N. Arc and S. Arc in Table I) is much larger than between *A. masukuensis* from the same geographical area (Table I), corresponding well with the difference in morphological divergence, and suggesting long isolation of different lineages of *A. tephrolaemus*. *Andropadus masukuensis* shows little genetic variation between populations separated by considerable geographical distances, indicating that its differentiation may have occurred only recently.

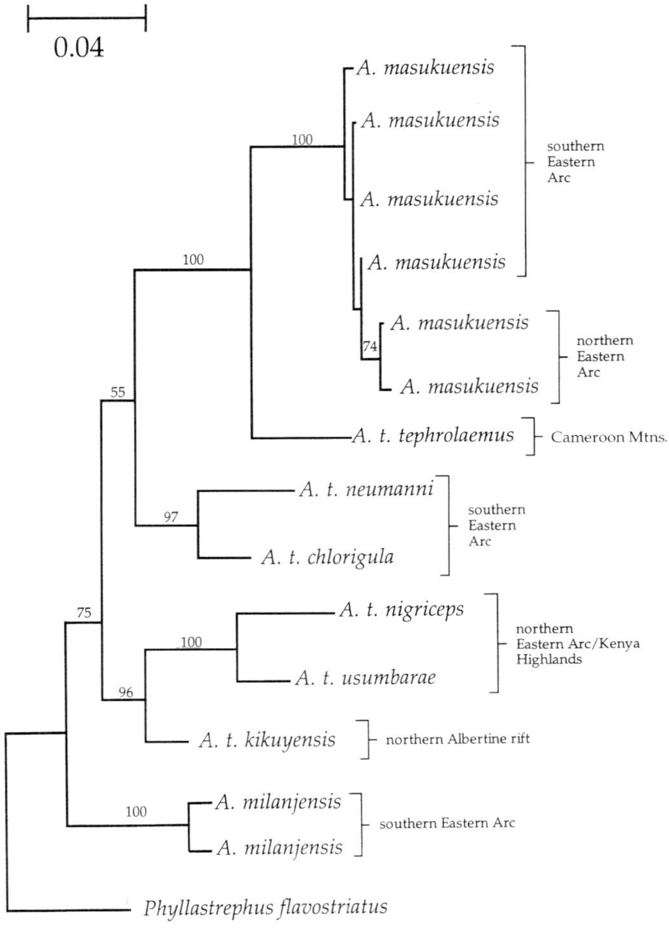

FIGURE 12.5 Phylogeny of *Andropadus* species based on 597 bp of mitochondrial cytochrome *b*. Species and regional distribution are indicated. One of two equally parsimonious trees is shown; the other tree differed only in the branching order of *A. masukuensis* from the southern Eastern Arc. Sequence data were analyzed by the exhaustive search option of PAUP 3.1.1 (Swofford, 1991), giving all positions equal weight. A mean distance measure is indicated as is percentage bootstrap support for nodes from 1000 replicates. The tree has a consistency index of 0.606 excluding uninformative characters and a tree length of 387 steps. PCR and sequencing were performed using mtDNA primers 14841 with 15149 (Kocher *et al.*, 1989), and 15564 with 15915 (Edwards *et al.*, 1991).

Andropadus tephrolaemus is divided into three geographical groups consisting of (1) the southern Eastern Arc, (2) the northern Eastern Arc/Kenya/northern Albertine Rift, and (3) the Cameroon Mountains. The nodes between these branches are all well supported and it is surprising that *A. t. tephrolaemus* is not an immediate sister-taxon to the geographically closest *A. t. kikuyensis* (see Fig. 12.4). The geographic distance between the nearest points of the Udzungwa Scarp (*chlorigula*) and

TABLE I Genetic Distances Averaged over Samples, Corrected for Missing Data

	Andropadus masukuensis	*Andropadus tephrolaemus*			
		N. Arc	N. Albert	S. Arc	Cam
Andropadus masukuensis					
Andropadus tephrolaemus	0.011[a]				
N. Arc	0.132	0.044[a]			
N. Albert	0.116	0.069	0		
S. Arc	0.13	0.109	0.098	0.076[a]	
Cam	0.082	0.124	0.108	0.119	0

[a]Average intrapopulation distances are also given where appropriate (from PAUP 3.1.1; Swofford 1991).

Abbreviations: N. Arc, northern Eastern Arc; S. Arc, southern Eastern Arc; N. Albert, northern Albertine Rift; Cam, Cameroon Mountains.

Uluguru Mountains (*neumanni*) is 75 km and the genetic distance is 0. 076. The distance between ranges inhabited by northern and southern groups (Mount Kanga with *chlorigula* to East Usambaras with *usambarae*) is 135 km and yet the average genetic distance between the southern and northern groups is 0.112 (Table I). This can be compared to a geographic distance of 950 km and genetic distance of 0.069 between northern Albertine Rift and northern Eastern Arc (Table I). This result may indicate that dispersal of an *A. tephrolaemus* ancestor to the northern and southern Eastern Arc may have been from different directions around the "montane circle."

During moist parts of interglacial periods, forest formed intermittently along the "northern dispersal corridor," permitting expansion through the Kenya Highlands to the humid Usambara Mountains (see Hamilton, 1982). The Udzungwa and Uluguru Mountains are assumed to have been permanently affected by humidity from the Indian Ocean, permitting persistence of isolated *A. t. chlorigula* and *A. t. neumanni* populations here (see Fig. 12.3).

Interestingly, in contrast to *A. tephrolaemus*, *A. masukuensis* displays little genetic variation between populations, over a large geographical range, indicating that although closely related, these two species have had different evolutionary histories. The resulting phylogenetic reconstruction indicates periods of vicariance through isolation accompanied by intermittent dispersal. Such a scenario could explain the origin and distribution of *A. masukuensis*, which appears to have originated from an *A. tephrolaemus*-like ancestor, perhaps after dispersal and isolation on Mount Cameroon. Subsequent reinvasion of the Albertine Rift Mountains then allowed it to confront its "parental species" as an independent species. The close association between *A. t. tephrolaemus* and *A. masukuensis* revealed by this analysis had been suspected by Keith *et al.* (1992). Interestingly, *A. masukuensis* shows a south to north cline (although weakly supported by the phylogenetic reconstruction; Fig. 12.5), indicating that invasion of the Eastern Arc from the Albertine Rift could have been from a "southern dispersal corridor" (Fig. 12.4).

VII. DISCUSSION AND SUMMARY

The case studies described above illustrate some aspects of the biogeographic dynamics that must be considered in order to understand the processes involved in the evolution of the rich avifauna of tropical forests. While *Leptopogon* shows a simple pattern of differentiation that correlated with the uplift of the Andes, *Scytalopus* illustrates a more complex case in which ancient differentiation caused by geological factors has been covered by new cycles of dispersal and vicariance that may be tentatively associated with Quaternary climatic–vegetational cycles. These two cases are possibly extremes of a continuum of patterns that will eventually include patterns of phylogenetic relationships showing a series of multiple biotic interchanges between mountain and lowland avifaunas. With regard to Africa, relationships within *Andropadus* show that a recent dynamic process of vicariance-induced speciation followed by dispersal has occurred within the mountains of East Africa. In addition, a complex interchange between East Africa and the Cameroon Mountains has also been revealed.

It is evident that a great deal of avian diversity has been generated in recent times within tropical montane regions and that further studies are needed to understand its magnitude. In particular, the prediction in the model hypothesis that lowland rainforest biota are recruited from radiations in montane regions needs to be fully tested using phylogenetic studies. Nothing can be concluded from species that now live only in lowlands, but a well-resolved population phylogeny of species that extend across Africa in lowlands as well as montane regions (*A. virens*, *curvirostris*, and *gracilirostris*) could provide more conclusive evidence. *Leptopogon* speciation could be interpreted in two ways, and a comprehensive study of population structure is needed to evaluate the possible interpretation that old species move into the lowlands as new species arise by vicariance in the mountains. A preliminary phylogenetic study of spinetails (genus *Cranioleuca*) in our laboratory, indicates that one species inhabiting lowland Amazonia comes out within the Andean species group, suggesting an interchange between these two regions (J. García-Moreno, unpublished data).

Several theories have been proposed to explain the geographic patterns of species richness and endemism in tropical forest biomes. None has received more attention than the refuge theory, which has become the predominant model for explaining the origin and biogeography of tropical forest organisms for the last three decades (Prance, 1982; Simpson and Haffer, 1978; Whitmore and Prance, 1987). We showed that by using a simple test that combines DNA–DNA hybridization with distribution data the refuge theory, in its original form (Haffer, 1969, 1974), cannot account for the most recent bursts of speciation of South American and African lowland avifaunas. In fact, molecular studies in lowland groups of birds (Capparella, 1988; Gerwin and Zink, 1989; Gill and Gerwin, 1989; Hackett, 1993; Hackett and Rosenberg, 1990), mammals (Patton *et al.*, 1994), and frogs (Heyer and Maxson, 1982) indicate high genetic divergence between pairs of closely related species. On the basis of different molecular clock calibrations, these levels of divergence among species suggest that most of the speciation in tropical lowland biotas occurred before the Quaternary.

This analysis provides evidence for bursts of speciation in montane regions during the Quaternary climatic–vegetational fluctuations and does not support the hypothesis that avian diversification was intensive in the lowland regions during this period. We suggest that since montane regions are highly heterogeneous with regard to vegetation, climate, and topography, there is a good chance that areas of paleoecological stability may exist as small pockets within them. These stable areas would be consistent in terms of climate and vegetational cover throughout periods of shifting global climate and would, in a sense, act as small refuges (*sensu* Brown and Ab'Saber, 1979; Vrba, 1993). It is not necessary to invoke barriers of open-vegetation habitats to explain the isolation of populations of forest birds in tropical mountains. From a metapopulation perspective, range disjunction can also arise as a consequence of ecoclimatic instability affecting community composition (Gilpin and Hanski, 1991). Because the assumed stable areas are small, the populations of animals and plants that would live in them would themselves be of small size. This could lead to rapid divergence from parent populations owing to rapid fixation of alleles and founder effect (Avise, 1994).

In our case studies, we have illustrated that forest montane avifaunas may have been assembled by a combination of factors. They include local to regional ecological changes caused by geological events (e.g., tectonism) to continental to global ecological changes caused by Croll–Milankovitch climatic cycles. In some groups of birds (e.g., *Leptopogon*), ancient processes of speciation caused mostly by geological changes are still conspicuous and recoverable. In other groups (*Scytalopus* and *Andropadus*), ancient processes of speciation have been hidden by new cycles of vicariance and dispersal associated with Quaternary climatic–vegetational cycles. In these groups, ancient processes of speciation are not recoverable, but the most recent ones are.

Molecular studies are a valuable part of any modern biogeographic analysis since they open the possibility for the development of rigorous protocols that can be used to untangle the patterns of species distributions from the processes that caused them. These protocols can be applied to a wide range of biotas, including complex ones such as tropical montane avifaunas.

ACKNOWLEDGMENTS

This study was supported in part by the Danish Center for Tropical Biodiversity. Studies by J.M.C.S. are supported by the scholarship of the Conselho Brasileiro de Desenvolvimento Cientifico e Technológico (CNPq), Brasilia, Brazil. We thank Santiago Aragón and Casper Paludan for helpful comments and discussions.

REFERENCES

Amadon, D. (1956). Remarks on the starlings, family Sturnidae. *Am. Mus. Novitates* **1803**, 1–14.

Amorim, D. S. (1991). Refuge model simulations: Testing the theory. *Revia Bras. Ent.* **35**, 803–812.

Arctander, P., and Fjeldså, J. (1994). Andean tapaculos of the genus *Scytalopus* (Aves, Rhinocryptidae): A study of modes of speciation, using DNA sequence data. *In* "Conservation Genetics" (V. Loeschcke, J. Tomiuk and S. K. Jain, Eds.), pp. 205–225. Birkhäuser Verlag, Basel.

Avise, J. C. (1994). "Molecular Markers, Natural History and Evolution." Chapman & Hall, New York.

Axelrod, D. I., and Raven, P. H. (1978). Late Cretaceous and Tertiary vegetation history of Africa. *In* "Biogeography and Ecology of Southern Africa" (M. J. A. Werger, Ed.), pp. 77–130. Dr W. Junk Publ., The Hague.

Bartlein, P. J., and Prentice, I. C. (1989). Orbital variations, climate and paleoecology. *Tree* **4**, 195–199.

Bates, J. M., and Zink, R. M. (1994). Evolution into the Andes: Molecular evidence for species relationships in the genus *Leptopogon*. *Auk* **111**, 507–515.

Beecher, W. J. (1978). Feeding adaptations and evolution in the starlings. *Bull. Chicago Acad. Sci.* **11**, 269–298.

Bennet, K. D. (1990). Milankovitch cycles and their effects on species in ecological and evolutionary time. *Paleobiology* **16**, 11–21.

Benson, C. W., Brooke, R. K., Dowsett, R. J., and Irwin, M. P. S. (1971). "The Birds of Zambia." Collins, London.

Brown, K. S., Jr., and Ab'Saber, A. N. (1979). Ice-age forest refuges and evolution in the neotropics: Correlation of paleoclimatological, geomorphological and pedological data with modern biological endemism. *Paleoclimas* **5**, 1–30.

Bush, M. B. (1994). Amazonian speciation: A necessarily complex model. *J. Biogeogr.* **21**, 5–17.

Capparella, A. P. (1988). Genetic variation in neotropical birds: Implications for the speciation process. *Acta XIX Congr. Int. Ornithol.* **1658**, 1973.

Coppes, Y. (1994). East side story: The origin of humankind. *Sci. Am.* **270**(5), 62–69.

Cronk, Q. C. B. (1992). Relict floras of Atlantic islands: Patterns assessed. *Biol. J. Linn. Soc.* **46**, 91–103.

Crowe, T. M., and Crowe, A. A. (1982). Patterns of distribution, diversity and endemism in Afrotropical birds. *J. Zool. (Lond.)* **198**, 417–442.

Delacour, J. (1943). A revision of the genera and species of the family Pycnonotidae (bulbuls). *Zool. N.Y.* **28**, 17–28.

Diamond, A. W., and Hamilton, A. C. (1980). The distribution of forest passerine birds and Quaternary climatic change in Africa. *J. Zool. (Lond.)* **191**, 379–402.

Dowsett-Lemaire, F. (1989). Ecological and biogeographical aspects of forest bird communities of Malawi. *Scopus* **13**, 1–80.

Dowsett, R. and Dowsett-Lemaire, F. (1993). "A Contribution to the Distribution and Taxonomy of Afrotropical and Malagasy Birds." Tauraco Press, Jupille, Liége, Belgium.

Edwards, S. V., Arctander, P., and Wilson, A. C. (1991). Mitochondrial resolution of a deep branch in the genealogical tree for perching birds. *Proc. R. Soc. Lond. B* **243**, 99–107.

Endler, J. A. (1977). "Geographic Variation, Speciation and Clines." Princeton Univ. Press, Princeton, New Jersey.

Fjeldså, J. (1992). Biogeographic patterns and evolution of the avifauna of the relict high-altitude woodlands of the Andes. *Steenstrupia* **18**, 9–62.

Fjeldså, J. (1994). Geographical patterns of relict and young species of birds in Africa and South America and implications for conservation priorities. *Biodivers. Conserv.* **3**, 107–226.

Fjeldså, J. (1995). Geographical patterns of neoendemic and relict species of Andean forest birds: The significance of ecologically stable areas. *In* "Biodiversity and Conservation of Neotropical Montane Forests" (S. P. Churchill, H. Balsu, E. Forero, and J. L. Luteyn, Eds.), pp. 89–102. The New York Botanical Garden, New York.

Fjeldså, J., and Krabbe, N. (1990). "Birds of the High Andes." Zoological Museum, Copenhagen, and Apollo Books, Svendborg, Denmark.

Fjeldså, J., and Lovett, J. (1997). Geographical patterns of phylogenetic relicts and phylogenetically subordinate species in tropical African forest. *Biodivers. Conserv.* (in press).

Fjeldså, J., Ehrlich, D., Lambin, E., and Prins, E. (1997). Are biodiversity "hotspots" correlated with ecoclimatic stability? A pilot study using the NOAA-AVHRR remote sensing data. *Biodivers. Conserv.* (in press).

Gerwin, J. A., and Zink, R. M. (1989). Phylogenetic patterns in the genus *Heliodoxa* (Aves: Trochilidae): An allozymic perspective. *Wilson Bull.* **101**, 525–544.

Gill, F. B., and Gerwin, J. A. (1989). Protein relationships among hermit hummingbirds. *Proc. Natl. Acad. Sci. U.S.A.* **141**, 409–421.

Gilpin, M. E., and Hanski, I. (1991). "Metapopulation Dynamics: Empirical and Theoretical Investigations." Academic Press, London.

Hackett, S. J. (1993). Phylogenetic and biogeographic relationships in the neotropical genus *Gymnopithys* (Formicariidae). *Wilson Bull.* **105**, 301–315

Hackett, S. J., and Rosenberg, K. V. (1990). Evolution of South American antwrens (Formicariidae): Comparisons of phenotypic and genetic differentiation. *Auk* **107**, 473–489.

Haffer, J. (1969). Speciation in Amazonian forest birds. *Science* **165**, 131–137.

Haffer, J. (1974). Avian speciation in tropical South America. *Publ. Nuttal ornithol. Club* **14**, 1–390.

Haffer, J. (1997). Alternative models of vertebrate speciation in Amazonia: An overview. *Biodivers. Conserv.* (in press).

Hall, B. P., and Moreau, R. E. (1970). "An Atlas of Speciation in African Passerine Birds." British Museum of Natural History, London.

Hamilton, A. C. (1982). "Environmental History of East Africa: A Study of the Quaternary." Academic Press, London and New York.

Hanagarth, W. (1993). "Acerca de la Geoecologia de las Sabanas del Beni en el Noroeste de Bolivia." Instituto de Ecologia, La Paz, Bolivia.

Harshmann, J. (1994). Reweaving the tapestry: What can we learn from Sibley and Ahlquist (1990)? *Auk* **11**, 377–388.

Heyer, W. R., and Maxson, L. R. (1982). Distribution, relationships, and zoogeography of lowland frogs. *In* "The Leptodactylus Complex in South America, with Special Reference to Amazonia" (G. T. Prance, Ed.), pp. 375–388. Columbia Univ. Press, New York.

Hooghiemstra, H., Milica, J. L., Berger, A., and Schackelton, N. J. (1993). Frequency spectra and paleoclimatic variability of the high-resolution 30-1450 ka Funza I pollen record (Eastern Cordillera, Colombia). *Quaternary Sci. Rev.* **12**, 141–156.

Kalliola, R., Puhakka, M., and Danjoy, W. (1993). "Amazonia Peruana. Vegetación Húmeda Tropical en el Llano Subandino." Oficina Nacional de Evaluación de Recursos Naturales, Lima, Peru.

Keith, S., Urban, E. K., and Fry, C. H. (1992). "The Birds of Africa," Vol. IV. Academic Press, London.

Kocher, T. D., Thomas, W. K., Meyer, A., Edwards, S. V., Pääbo, S., Villablanca, F. X., and Wilson, A. C. (1989). Dynamics of mitochondrial DNA evolution in animals: Amplification and sequencing with conserved primers. *Proc. Natl. Acad. Sci. U.S.A.* **86**, 6196–6200.

Krabbe, N., and Schulenberg, T. S. (1997). Species limits and natural history of *Scytalopus tapaculos* (Rhinocryptidae) of Ecuador, with descriptions of three new species and notes on extralimital forms. *Auk* (in press).

Lovett, J. C. (1993). Climatic history and forest distribution in eastern Africa. Eastern arc moist flora. *In* "Biogeography and Ecology of the Rain Forests of Eastern Africa" (J. C. Lovett and S. K. Wasser, Eds.), pp. 23–29 and 33–55. Cambridge Univ. Press, London.

Mayr, E., and O'Hara, R. J. (1986). The biogeographic evidence supporting the Pleistocene forest refuge hypothesis. *Evolution* **40**, 55–67.

Meise, W. (1968). Zur Speciation afrikanischer besonderer angolischer Singvögel der Gattungen Tersiphone, Dicrurus und Malaconotus. *Zool. Beitr. (Berlin)* **14**, 1–60.

Mindell, D. P. (1992). [Review of] Phylogeny and classification of birds: A study in molecular evolution, by C. G. Sibley and J. Ahlquist. *Syst. Biol.* **41**, 126–134.

O'Hara, R. J. (1991). [Review of] Phylogeny and classification of birds: A study of molecular evolution, by C. G. Sibley and J. E. Ahlquist. *Auk* **108**, 990–994.

Patton, J. L., and Smith, M. F. (1992). mtDNA phylogeny of Andean mice: A test of diversification across ecological gradients. *Evolution* **46**, 174–183.

Patton, J. L., DaSilva, M. N. F., and Malcolm, J. R. (1994). Gene genealogy and differentiation among arboreal spiny rats (Rodentia, Echimyidae) of the Amazon basin—a test of the riverine barrier hypothesis. *Evolution*, **48**(4), 1314–1323.

Prance, G. T. (Ed.) (1982). "Biological Diversification in the Tropics." Columbia Univ. Press, New York.

Rand, A. L. (1958). Notes on African bulbuls. Family Pycnonotidae: Class Aves. *Fieldiana Zool.* **35**, No. 6.

Ridgeley, R. S., and Tudor, G. (1988). "The Birds of South America: The Suboscine Passerines." Univ. of Texas Press, Austin, Texas.

Salo, J. (1988). Rainforest diversification in the western Amazon basin: The role of river dynamics. Ph.D. thesis, Department of Biology, Univ. Turku, Turku, Finland.

Sibley, C. G., and Ahlquist, J. E. (1990). "Phylogeny and Classification of Birds: A Study in Molecular Evolution." Yale Univ. Press., New Haven, Connecticut.

Sibley, C. G., and Monroe, B. L. (1990). "Distribution and Taxonomy of Birds of the World." Yale Univ. Press, New Haven, Connecticut.

Sibley, C. G., Ahlquist, J. E., and DeBenedictis, P. (1993). The phylogenetic relationships of the rails, based on DNA comparisons. *J. Yamashina Inst. Ornithol.* **25**, 1–11.

Silva, J. M. C. (1995). Biogeographic analysis of the South American cerrado avifauna. *Steenstrupia* **21**, 49–67.

Simpson, B. B., and Haffer, J. (1978). Speciation patterns in the Amazonian forest biota. *Annu. Rev. Ecol. Syst.* **9**, 497–518.

Stebbins, G. L., and Major, J. (1965). Endemism and speciation in the California flora. *Ecol. Monogr.* **35**, 1–35.

Stuart, S. N. (1986). "Conservation of Cameroon Montane Forests." I.C.B.P., Cambridge.

Swofford, D. L. (1991). "PAUP: Phylogenetic Analysis Using Parsimony," version 3.1. Center for Biodiversity of the Illinois Natural History Survey, Champaign, Illinois.

Traylor, M. A., Jr. (1970). Notes on African Muscicapidae. *Ibis* **112**, 395–397.

Traylor, M. A., Jr. (Ed.). (1979). "Check-List of Birds of the World." Vol. 8. Museum of Comparative Zoology, Cambridge, Massachusetts.

Tuomisto, H., and Ruokolainen, K. (1997). The role of ecological knowledge in explaining biogeography and biodiversity in Amazonia. *Biodivers. Conserv.* (in press).

Vrba, E. S. (1993). Mammal evolution in the African Neogene and a new look at the Great American Interchange. *In* "Biological Relationships between Africa and South America" (P. Goldblatt, Ed.), pp. 393–432. Yale Univ. Press, New Haven, Connecticut.

White, C. M. N. (1962). "A Revised Check List of African Shrikes, Orioles, Drongos, etc." Government Printer, Lusaka, Zambia.

Whitmore, T. C., and Prance, G. T. (1987). "Biogeography and Quaternary History in Tropical America." Clarendon Press, Oxford.

CHAPTER 13

Studies of Avian Ancient DNA: From Jurassic Park to Modern Island Extinctions

ALAN COOPER
Department of Biological Anthropology, Oxford University, Oxford OX2 6QS, United Kingdom

I. INTRODUCTION

During the first decade of ancient DNA analysis the concept of analyzing prehistoric genetic information has emerged from obscurity to become a relatively well-known and mainstream pursuit (Higuchi *et al.*, 1984; Crichton, 1991; Pääbo, 1993). Ancient DNA techniques are particularly suited to the study of avian evolution since birds make up a disproportionate number of the world's recently extinct and currently threatened taxa. Extensive museum collections of avian skins and skeletons, often significant contributors to this situation, are now a considerable resource for systematics research. Many extinct avian taxa were the results of evolution within ecosystems that have since disappeared, and represent unique unrepeatable experiments. In these situations ancient DNA techniques may allow "lost" genetic information to be used to reconstruct these evolutionary paths and assist in the conservation of remaining biota.

The study of preserved macromolecules has provided access to prehistoric genetic information and allowed molecular evolutionary change to be examined in real time, rather than through extrapolation from DNA of living organisms. Ancient DNA has been used in subjects as diverse as systematics, population genetics, paleoecology, archaeology, conservation biology, and forensics (Pääbo, 1993) creating new importance for museum specimens, and changing the role of museums themselves (Houde and Braun, 1988).

Since its discovery, the polymerase chain reaction (PCR) (Saiki *et al.*, 1988) has formed the basis of nearly all ancient DNA research because it permits a designated sequence to be amplified from a small number of damaged templates amidst a background of nonspecific DNA (Pääbo and Wilson, 1988). From 1985 through to the mid-1990s the template of choice for most vertebrate ancient DNA studies has been mitochondrial DNA (mtDNA), primarily because of the high number of mitochondrial genomes per cell and the prevalence of its use in other systematic and population studies. Because ancient DNA is invariably damaged, studies have generally been restricted to sequences of less than 500 base pairs (bp), but until recently this has been sufficient for phylogenetic comparisons with extant taxa.

Nuclear DNA sequences are a powerful complement to mtDNA data because they provide an independently inherited set of molecular characters. Rapidly evolving nuclear microsatellites are particularly useful for studies of population genetics, and avian microsatellites have already been successfully amplified from museum specimens (Ellegren, 1991, 1993; Roy *et al.*, 1994). Unfortunately, the comparatively low ratio of single-copy nuclear to mitochondrial genes in most cells means that the range of ancient samples likely to yield single-copy nuclear sequences is far less than that for mtDNA. Nevertheless, the amount of information accessible through nuclear sequence data will ensure that both genomes feature in future ancient DNA studies.

Ancient DNA research has obvious utility in determining the systematic relationships of extinct taxa, but may also clarify the evolutionary relationships of extant relatives by balancing taxon sampling across parts of the tree (Cooper *et al.*, 1992; Höss *et al.*, 1995). However, more substantial evolutionary questions can be addressed if the data are combined with information from other fields such as paleontology and ecology.

II. REVIEW OF ANCIENT DNA RESEARCH

The majority of ancient DNA publications have concerned the systematic position of extinct taxa (Thomas *et al.*, 1989; Cooper *et al.*, 1992; Höss *et al.*, 1995; Christidis *et al.*, 1996; Houde *et al.*, Chapter 5 in this volume), although studies involving taxon identification (Hagelberg *et al.*, 1989, 1991; Höss *et al.*, 1992; Sallares *et al.*, 1995; Cooper *et al.*, 1996) and population genetics (Thomas *et al.*, 1990; Wayne and Jenks, 1991; Stone and Stoneking, 1993; Roy *et al.*, 1994) have become more prevalent. Although there has been a surge in the number of evolutionary questions addressed using preserved macromolecules, appreciation of the practical difficulties inherent in ancient DNA research has lagged behind.

Mammoths, mummies, and dinosaurs are the stuff of legends, and it is the responsibility of all those involved in ancient DNA research that scientific publications do not also fall into this category. The majority of published ancient DNA studies involve late Pleistocene or Holocene specimens (50,000 years old to present) and therefore do not conflict with observed DNA decay rates (Lindahl, 1993a,b). In contrast, reports of DNA preservation over many millions of years (Golenberg *et al.*, 1990; Soltis *et al.*, 1992; Cano *et al.*, 1993; Poinar *et al.*, 1993; Woodward *et al.*, 1994) have attracted considerable publicity and scrutiny, yet are still to be independently verified (Sidow *et al.*, 1991). Even if such studies are legitimate, they are valueless without independent replication. Perhaps the most important point these studies demonstrate is the central role of authentication procedures in ancient DNA research (Pääbo *et al.*, 1989; Lindahl, 1993b; Handt *et al.*, 1994a, 1996).

Difficulties in demonstrating the authenticity of ancient DNA sequences vary considerably, and in some cases only circumstantial supporting evidence exists (Handt *et al.*, 1994a). Cryptic contamination is a significant problem when contaminating DNA, often from within the laboratory, is similar or identical to real ancient sequences. Systematic studies of ancient DNA are somewhat less susceptible to this problem than population studies, because sequences can be phylogenetically contrasted with extant taxa. However, cloning artifacts (Higuchi *et al.*, 1987; Pääbo and Wilson, 1988), damaged modern contaminant DNA (Collura and Stewart, 1995), and chimeric sequences (DeSalle *et al.*, 1993; Hackett *et al.*, 1995) can still confuse analysis. A further complication is nuclear copies of mtDNA genes, which may appear functional (Quinn and White, 1987; Arctander, 1995; Collura and Stewart, 1995; Zischler *et al.*, 1995; Sorenson and Fleischer, 1996) and can act as default ancestral sequences. Nuclear pseudogenes will be more problematical if samples have a high ratio of nuclear DNA to mtDNA, such as nucleated avian erythrocytes.

As the difficulties involved in preventing contamination became appreciated, new criteria have been adopted. Authentication techniques currently in use emphasize the independent replication of results (Handt *et al.*, 1994b; Taylor, 1996) and reporting of failed attempts (Hänni *et al.*, 1994). Recent developments include cloning PCR products to examine individual sequences when ambiguities exist in direct sequences (Handt *et al.*, 1996), and examining the sample for evidence of suitable preservation by measuring amino acid racemization (Poinar *et al.*, 1996) or histological preservation and nitrogen content (Hedges *et al.*, 1995; Colson *et al.*, 1997).

A. Techniques

Because a variety of source materials are used in ancient avian DNA studies, some of the commonly used extraction procedures are briefly reviewed below. A broader discussion of extraction and amplification techniques can be found in *Ancient DNA* (Herrmann and Hummel, 1994).

Preserved bone is often a better source of DNA than surrounding tissue (Cooper *et al.*, 1992) and this is also apparent in the macroscopic preservation of many remains. Compact bone from weight-bearing limbs appears to be a reliable source of well-preserved DNA, whereas yields from cancellous (trabeculae or marrow) bone are poor and the risk of environmental contamination is increased. Desirable bones for sampling exhibit few external signs of diagenesis, such as cancellous bone showing through damaged epiphyses, cracked surfaces, and bleached or discolored sections. To avoid ingrained human DNA from handling of the bone, the surface of the sample area should be mechanically removed to the practical maximum. However, sweat and dust may penetrate deeply below the surface of samples and the potential contribution from this source of DNA should not be underestimated (Richards *et al.*, 1995; Handt *et al.*, 1996).

The speed at which preserved specimens are dehydrated appears to be a significant factor controlling the size of amplifiable DNA fragments, and this is presumably related to the period during which endogenous endonucleases remain activated (Pääbo, 1993). Accordingly, DNA from museum specimens that are prepared quickly may permit relatively long PCR amplifications (<1000 bp), in contrast to naturally preserved mummies or bones, from which smaller amplifications (<400 bp) are normal. Tissue remains are often the most accessible in museum specimens, but bone samples are advisable when there is a need for long sequences or protection from some external treatment (such as alum, varnish, arsenic, or shellac) that may be detrimental to enzyme activity. It is helpful that bones such as phalanges and sections of humerus are often left in prepared skins, but if they are not available then tissue samples of thick skin from the extremities of the specimen (such as toe pads) may do. As a general rule for museum specimens, 0.1–1.0 g of bone or 2–5 mm^3 of tissue is normally sufficient for analysis, although this is dependent on specimen preservation. Formalin-fixed samples vary in DNA content, and the processes involved are often complex (Grody, 1994).

There are two broad categories of extraction technique presently in use with ancient specimens. The traditional technique involves digestion of the sample with a proteinase (after, or simultaneously with, decalcification of bone) followed by extraction of DNA using organic solvents (Hagelberg *et al.*, 1991; Cooper *et al.*, 1992). Disadvantages of this technique include the number of procedural steps, and the possibility of copurifying inhibitors. The second category uses silica to bind DNA in the presence of chaotropic agents (Boom *et al.*, 1990; Höss and Pääbo, 1993). Silica-based techniques are relatively simple and remove inhibitors efficiently, although it is possible that they are less efficient at recovering DNA than organic extraction. It is also possible to "mix and match" the extraction and purification steps from the different techniques.

The laboratory situation used for ancient DNA research is one of the most important aspects of any study. DNA extraction and PCR setup should be conducted in a location physically well separated from PCR products, and a site in a different building is strongly recommended. Protective clothing (disposable paper coveralls, footwear, and breathing masks) is needed to prevent contamination of samples and reagents, especially for population studies, where cryptic contamination is always a concern. To fully comprehend these precautions it is necessary to appreciate that aerosol droplets from successful PCR reactions can contain enormous amounts of amplified DNA, perhaps up to 10,000 copies of a sequence. As a result of this, shoes/clothing or drafts can easily transmit amplified DNA fragments as dust between work areas via laboratory floor surfaces and air conditioning. Consequently, a sensible precaution is to complete research on ancient specimens before working on modern relatives that might become cryptic contaminants.

III. SYSTEMATICS AND PALEOECOLOGICAL APPLICATIONS

The following three projects demonstrate practical applications of the above-described techniques and range in time from the Cretaceous [145–65 million years ago (MYA)] to the Holocene (10,000 YA to present). Each study uses DNA sequence information from modern and extinct taxa to augment evidence of temporal change from traditional fields such as paleontology, geology, ecology, and biology.

The first project concerns the evolution of the ratite birds, and illustrates how ancient DNA can contribute to systematic studies. New sequence data are presented that support the conclusions of an earlier study (Cooper *et al.*, 1992), and ancient DNA sequences are shown to be essential for the evaluation of alternative phylogenetic hypotheses.

The remaining two projects use ancient DNA to investigate Pacific paleoecosystems that have since been drastically altered. The first concerns the effects of a paleoecological catastrophe on three endemic New Zealand avian taxa and illustrates how ancient DNA can provide important information about geological events in the absence of a fossil record. The second involves an extinct Hawaiian duck population, and shows how ancient DNA can reveal information needed for current conservation attempts.

A. Ratite Systematics

The living ratite birds are the ostrich (*Struthio*), emu (*Dromaius*), cassowaries (*Casuaris*), kiwis (*Apteryx*), and rheas (*Rhea*) of the southern continents. They are linked by several morphological characters including the paleognathous palate and rhamphothecal grooves, and share these with the flighted tinamous of South America, which are commonly believed to be their closest living relatives. Despite the seemingly inordinate amount of research that has followed their scientific recognition (reviewed in Sibley and Ahlquist, 1981; Houde, 1988) the phylogenetic relationships of the ratite birds are still not fully resolved.

Most recent research supports ratite monophyly (Cracraft, 1974; Sibley and Ahlquist, 1981, 1990; Caspers *et al.*, 1994; Cooper and Penny, 1997), although Houde and Olson (1981) and Houde (1986) have suggested that fossil paleognathous birds from the late Paleocene/Eocene of the Northern Hemisphere indicate flighted polyphyletic origins. The relationship of these fossil paleognathes to ratites is uncertain and they could be ancestral tinamous, sister taxa to either tinamous or ratites, or unrelated (Houde, 1988). In addition, the Northern Hemisphere fossil taxa are younger than mid-Paleocene rhea fossils (Tambussi *et al.*, 1994; Tambussi, 1995), so it seems unlikely that they represent ancestral ratites unless the latter evolved more than once (Houde, 1988).

Two of the largest avian species known were members of recently extinct ratite groups, namely the Madagascan elephant birds (*Aepyornis maximus* was approximately 500 kg and 2.5 m tall) and the New Zealand moas (*Dinornis giganteus* could reach 3 m and weighed approximately 250 kg; Cooper *et al.*, 1993). Ratite fossils of the recently extinct and still living taxa are found on all of the southern continents created by the break-up of the Cretaceous supercontinent Gondwana, including India (Olson, 1985). Cracraft (1974) used this geographical distribution, and morphological data, to suggest a vicariant biogeographic origin of the ratites, with a basal divergence of the New Zealand kiwi and moa (see Cracraft, Chapter 7 in this volume). Subsequent morphological, DNA–DNA hybridization, and mtDNA sequence analyses (Sibley and Ahlquist, 1981, 1990; Bledsoe, 1988; Cooper *et al.*, 1992) corroborated the vicariant origin hypothesis, but converged on a strikingly different phylogeny. Instead, all three studies place the kiwi, emu, and cassowary in a derived clade, a situation supported by cytogenetic studies (De Boer, 1980). In fact, the three studies differ essentially only in the relative position of the rhea and ostrich near the base of the tree, an encouraging degree of concordance (Sheldon and Bledsoe, 1993).

The position of the moa, as well as that of the kiwi, are the central differences between the conflicting phylogenetic hypotheses, because otherwise they differ only in the position of the root (see Cracraft, Chapter 7 in this volume). Because Sibley and Ahlquist (1981) could not analyze the extinct moa using DNA–DNA hybridization techniques (although see Houde *et al.*, 1995), the phylogenetic position of the moa lineage had been determined only by morphology (Cracraft, 1974; Bledsoe, 1988). The advent of ancient DNA techniques allowed Cooper *et al.* (1992) to include moa mitochondrial sequences in a ratite molecular phylogeny and in contrast to Cracraft (1974) the tree placed the moa as a basal lineage, and the kiwi as a derived taxa, suggesting New Zealand was invaded twice by ratites.

Consequently, two areas of disagreement remain about the ratite phylogeny. First, the basal and monophyletic New Zealand ratite clade of Cracraft (1974) contrasts with the derived position of the kiwi in the other three studies. Second, the other studies disagree on whether the rhea, ostrich, or both are the extant basal lineage (Sibley and Ahlquist, 1981, 1990; Bledsoe, 1988; Cooper *et al.*, 1992). To resolve these issues further, the 12S data set of Cooper *et al.* (1992) is reanalyzed using the computer packages PAUP* 4.0d 44–51 (Swofford, 1996) and Spectrum 1.0.5 (Charleston, 1996) and new phylogenetic techniques. In addition, new data sets from the mitochondrial NADH subunit 6 (ND6)/transfer RNA-proline intragenic region and the nuclear protooncogene c-*mos* are presented.

1. New Phylogenetic Analyses of the 12S Data Set

The ratite 12S data set of Cooper *et al.* (1992) consisted of an approximately 390-bp region of domain III, one of the most conserved areas of the mitochondrial genome (Mindell and Honeycutt, 1990). Sequences from four (later expanded to five, Cooper, 1993) of the six moa genera were presented along with all of the extant ratites except for two of the three cassowary species. Phylogenetic analyses of the data set with parsimony, distance, and maximum-likelihood methods produced the same highly supported tree (Cooper *et al.*, 1992), in which the kiwi and moa were not each other's closest relatives as suggested by Cracraft (1974), but assumed phylogenetic positions similar to those described by Bledsoe (1988). A relative rate test of the data using the tinamous as an outgroup found no significant variation within the ratites (Steel *et al.*, 1996).

The sequence data of Cooper *et al.* (1992) were aligned with limited reference to a secondary structure model of the mitochondrial 12S gene because no appropriate avian model existed. Such a model has become available (Hickson *et al.*, 1996) and a revised alignment of the data with additional sequences of the tataupa and spotted tinamous (*Crypturellus tataupa* and *Northura maculosa*, respectively) is presented in Fig. 13.1. The new alignment identifies 366 homologous positions, of which 105 are variable and 75 are parsimony sites. Parsimony, distance, and maximum-likelihood analyses of the revised ratite data set strongly support the phylogeny of Cooper *et al.* (1992), as shown in Fig. 13.2. LogDet analysis (Lockhart *et al.*, 1994) produces the same phylogeny, demonstrating that the tree topology is independent of base composition and lineage evolution rates.

Two approaches are used to investigate the discrepancies between the 12S phylogeny and those of Cracraft (1974), Bledsoe (1988), and Sibley and Ahlquist (1990). First, optimality criteria such as parsimony and maximum likelihood are used with the 12S data set to determine how optimal (or suboptimal) the various phylogenetic hypotheses are. The length of the most parsimonious trees, and maximum-likelihood values of the various hypotheses, are shown in Table I for two data sets. The alternative phylogenies are clearly not supported by either criteria, with the phylogeny of Cracraft (1974) being significantly worse in maximum-likelihood analyses of both data sets. Importantly, the ability of the optimality criteria to distinguish between the various phylogenetic hypotheses is severely com-

```
                 25        26         26' | 20 |29     29      29'   29'  2'      32          33    34     35     35'
Brown KIWI       GCTTAGCCCTAAATCCTGGTGCTTGCA-TTACCCAAGTACCCGCCCGAGAACTACGAGCACAAACGCTTAAAACTCTAAGGACTTGGCGGTGCCCTAAACCCACCTAGAGGAGCCTGTTCTATAATCGATAACCCACGAT
L. spotted KIWI  ........................A..-.....T..........................................................................................................
Roa KIWI         ....................A...A.G-.....T..........................................................................................................
EMU              ...............T..A.A..CA.C-.....AG..C.T.....T..............................................................................................
CASSOWARY        ...............T..A.A..C..T-A....TG....T....................................................................................................
OSTRICH          ........T......A..A.A...A.C-CC...T.....T..............................................................................................T.......
Emeus MOA        ................A.A.A...A.C-C...A......T.......................................................C....................................T.......
Pachyorn. MOA    ................A.A.A...A.C-C...A......T.......................................................C....................................T.......
Anomalop. MOA    ................A.A.A...A.C-C...A......T.......................................................C....................................T.......
Dinornis MOA     ....G..........TA.A.A...A.C-C...A......T.....T.................................................C....................................T.......
Megalap. MOA     ................A.A.A...A.C-CC..A......T.......................................................C..........................................T.
Common RHEA      ..................A.A...A.C-CC...T.....T....................................................................................................
Lesser RHEA      ................C.A.A...A.C-CC.........T....................................................................................................
E. Cr TINAMOU    ...............TA.A.A...AA.-C..........T.....T.....T.........C.....................................................................CT.......
Spot. TINAMOU    ...............TAAA.....A.C-....TT...C.TT..........................................................................................CT.......
Tat. TINAMOU     ................AAA.....AAC-.....T...C.TT..................G...................................C...................................CT.......
CHICKEN          ..C............TA.A.A.C.C.CA.C..A..T...T.....T.................................................C....................................T.......

                          36              38       38        39 | 40   40'        39'      42          42'           38'            36'   34'
Brown KIWI       ACACCCAACCATCTCTTGCCAAC-ACAGCCTACATACCGCCGTCGCCAGCCCGCCTAAAATGAAAG-AACAGCAGCGAGCACAACAGCTACA-CCAGCTAACAAGACAGGTCAAGGTATAGCCCACGAGATGGGAGAAAT
L. spotted KIWI  .......................-........T.................T............G..-....A......A...........TC-..............................T.T.......A......
Roa KIWI         .......................-........T.................T............G..-....A......A...........TC-..............................T.T.......A......
EMU              .......G..............T-................................--........-....AT...........T...C..--TTC..........................GT.T.......A......
CASSOWARY        ......................TT................................--.G...G..-....AT........T......C..--T.C..........................GT.T.......A......
OSTRICH          T..........C.C.......-T-G...............................C--....G..-....AT...........T...CCA--..C..........................AT.T.GAG...A......
Emeus MOA        ......G......C.C....C.T-G...................C...........--........-....AT....G..........CG.CC..C..........................AT.T.......A......
Pachyorn. MOA    ......G......C.C....C.T-G...................C...........--........-....AT...............CT..-..C..........................AT.T.......A......
Anomalop. MOA    ......G......C.C....CGT-G...................C...........--........-....AT...............CCTCC..C..........................AT.T.......A......
Dinornis MOA     ......G.............C.T-G...................C...........--........-....AT.................T--.TC..........................AT.T.......A......
Megalap. MOA     ......G.............C.T-G...................C...........--........-....AT...........T...C..--..C..........................AT.T.......A......
Common RHEA      ......G.............C.T-G...................C...........T..G......-....A....AG...T..T.....---..C..........................AT.T.G.....A......
Lesser RHEA      ......G.............C.T-G...................C...........T......G..-....A....A....T..T.....---..C....G.....................AT.T.......A......
E. Cr TINAMOU    .T....G.........A...CC.-T...............................--........CC..-A.........T......CC---..C.....T....................A..T.......C..C...
Spot. TINAMOU    ...........C........C..-........T.......................--........-G...A............T...CC---..C.....T.....................T.T.......A......
Tat. TINAMOU     .....T..............CCA-............................A...--.T......-T...A.....G...T..T..T.C.--..CA...G.....................T..T.......A......
CHICKEN          T..........C.C.......G.-............................A...CT........-....A...T....T...T...CC---.TC.....T.....................T.T.G.G..........

                  34'    45        45'               47           47'           33'    48                48'            32'
Brown KIWI       GGGCTACATTTTCTAAA-ATAGAATACT--------------ACGAAAAAGGGTATGAAA-TCCCCCTTAGAAGGAGGATTTAGCAGTAAAACAGAATAAGA------AAGTCTATTTTAAG-CCGGCCCTAGGGC
L. spotted KIWI  .................-..........--------------............G.....-..............C..........................------G.............-.T...........
Roa KIWI         .................-..........--------------............G.....--T............C..........................------G.............-.T...........
EMU              ................C-........TCC-------------......G.A.A.G.....-..A..T.C......C..........................------G............A-.T...T.......
CASSOWARY        ................C-.........CC-------------..........A.G.....-..AT..........C.....................C....------G............A-.............
OSTRICH          ................C-.........AC-------------......G...A.......-..AT...C......C................T.........------G.............-T....T.......
Emeus MOA        ................C-......C..CC-------------......G..AAGG.....-C..T...C......C................T....C....------.T.C..........C.T......G....
Pachyorn. MOA    ................C-......C..CC-------------......G..AAG......-CT.T...C......C................T....C....------.T.C..........C.T......G....
Anomalop. MOA    ................C-......C..CC-------------......G..AAGG.....-C..T...C.A....C................T....C....------.T.C..........C........G....
Dinornis MOA     ................C-......C..CT-------------......G..AAGG.....-C..T...C......C................T..G.C....------.C.C...C......C.T......G....
Megalap. MOA     ................C-......C..CC-------------......G..AAG......-C..T...C......C................T..G.C....------.C.C.C........C.T......G....
Common RHEA      ................T-C.....C.ACC-------------...G..G....C......--.A....CC.....C...............GT..G..C...------...C.C.C......-.............
Lesser RHEA      ...............GT-C.....C.GCC-------------......G....C......--.A....CC.....C...............GTG.G..C...------...C.C.C......-.............
E. Cr TINAMOU    ................C-C.....C.ACCTTTACTTATCTTT...G..G...TA......--......CG.....C...............CTG...CT...------.......C.....A-.T...T.......
Spot. TINAMOU    ................CTT.....C.A---TTATCCACCCA-...G..G...AA......--.A....C......C...............CT.........------.......A.....A-.T...T.......
Tat. TINAMOU     ................C-......C..-CCCTTTTAACTC--......G....A......--.A....CT.....T........T......GTG...CC...------..A..C.C.....A-.....T..G.A..
CHICKEN          ................C-......C.A---------------..........A.G.....CC.G...........................GTGAG..C.T.CCCCCT...CTC.C......-A....T..GA...
```

FIGURE 13.1 Approximately 390 nucleotide bases from domain III of the mitochondrial 12S gene of eight living, and five extinct, ratites as well as three tinamous and a chicken. The sequences correspond to positions 1754–2147 of the published chicken sequence (Desjardins and Morais, 1990), and the stem numbers given above the sequences follow the secondary structure model of Hickson *et al.* (1996). The 366 homologous positions are underlined and dots indicate a base identical to the brown kiwi sequence given above. Taxon names: Brown KIWI (*Apteryx australis*), L. spotted KIWI (*A. haastii*), Roa KIWI (*A. owenii*), CASSOWARY (*Casuarius casuarius*), Emeus MOA (*Emeus crassus*), Pachyorn. MOA (*Pachyornis elephantopus*), Anomalop. MOA (*Anomalopteryx didiformis*), Dinornis MOA (*Dinornis novaezealandiae*), Megalap. MOA (*Megalapteryx didinus*), Common RHEA (*Rhea americana*), Lesser RHEA (*Rhea pennata*),

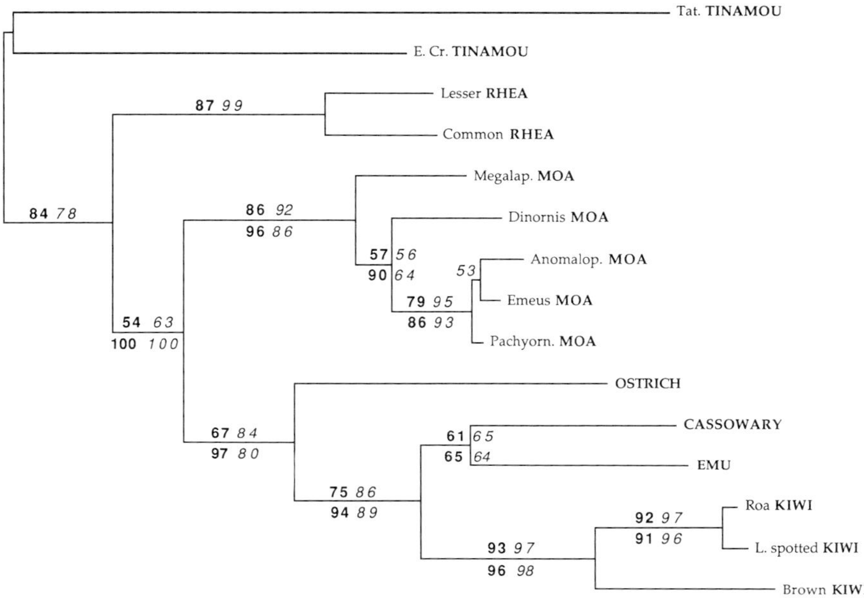

FIGURE 13.2 Phylogenetic tree of the ratite and tinamou 12S sequence data in Fig. 13.1, obtained from parsimony, neighbor-joining (two parameter with 0.5 gamma distribution, and LogDet corrections), and maximum-likelihood methods (Swofford *et al.*, 1996) using PAUP* 4.0d 44–51 (Swofford, 1996). To break the long branch between the ratite and tinamou taxa, the two most divergent tinamou taxa (*Eudromia* and *Crypturellus*) are used in rooted analyses. The branch lengths are calculated using neighbor joining with the LogDet correction and are drawn proportional to evolutionary distance. The proportion of invariant sites is estimated to be 0.4 from comparisons of more than 100 divergent avian 12S sequences (Cooper and Penny, 1997). Maximum-likelihood analyses were used to estimate transversion-to-transition ratios of 6.67 : 1 and 4.67 : 1 for the ratite, and ratite plus tinamou, data sets, respectively. Bootstrap values from 1000 unweighted parsimony heuristic replications (in boldface), and 1000 neighbor-joining (LogDet correction) replications (in italics) of the ratite data set are given under the branches, whereas values for the same data set plus the two tinamous are given above the lines. Unweighted parsimony analysis produces 2 shortest trees (differing only in the resolution of a trichotomy among the moa taxa) of 123 steps, and the same 2 shortest topologies are obtained if transversions are weighted between 2 and 25 times that of transitions.

promised if the moa taxa are excluded from the analysis. Without the moa taxa the only difference between the hypotheses becomes the position of the root (see Cracraft, Chapter 7 in this volume), so all are equally likely in the unrooted data set (Table I). When the tinamou outgroups are included, the position of the root provides limited resolving power between the hypotheses, but none is significantly worse. Consequently, the moa taxa are essential to evaluate the various hypotheses and fully resolve the ratite phylogeny.

A more comprehensive approach to measuring suboptimal signals in a data set is to avoid using any optimality criterion. Spectral analyses (Hendy and Penny, 1993;

TABLE I Phylogenetic Analyses of the Alternative Hypotheses Using Optimality Criteria[a]

Ratite phylogenetic hypothesis	Ratite data set			Ratite data set plus two tinamous			Ratite data set minus moa taxa			Ratite data set plus two tinamous minus moa taxa		
	−ln L.	Parsimony		−ln L.	Parsimony		−ln L.	Parsimony		−ln L.	Parsimony	
	(Tv = 6.67)	Tv = 1	Tv = 6.67	(Tv = 4.67)	Tv = 1	Tv = 4.67	(Tv = 6.67)	Tv = 1	Tv = 6.67	(Tv = 4.67)	Tv = 1	Tv = 4.67
Cooper *et al.* (1992)	1159.1	123 (2)	219.4 (2)	1434.9	181 (3)	305.8 (3)	977.99	94	162	1246.9	150 (3)	252.8 (3)
Cracraft (1974)	**1178.7** (3)	132 (5)	234.1 (5)	**1463.0** (3)	192 (3)	320.5 (3)	977.99	94	162	1262.9	155	261.4
Bledsoe (1988)	1170.6	129 (5)	231.1 (5)	1443.4	186	307.1	977.99	94	162	1259.9	154	260.4
Sibley and Ahlquist (1990)												
b	1170.6	129 (5)	231.1 (5)	1443.4	186	307.1	977.99	94	162	1259.9	154	260.4
c	1170.6	129 (2)	231.1 (2)	1444.2	187 (3)	308.1 (3)	977.99	94	162	1260.2	154	260.4

[a]Parsimony tree lengths and maximum-likelihood values (−ln L.) for alternative phylogenetic hypotheses using 12S sequences of the ratites, and ratites plus two tinamous. If more than one topology is equally optimal the number is given in parentheses. Transversions (Tv) are either unweighted (Tv = 1) or weighted 6.67 or 4.67 times (see Fig. 13.2). Kashino–Hasegawa tests (Swofford *et al.*, 1996) of the various trees show that the phylogeny of Cracraft (1974) is significantly worse than that of Fig. 13.2 using either the ratite, or ratite plus tinamou, data sets (see entries in boldface, $p < 0.05$ and $p < 0.01$, respectively). In contrast, the phylogenies of Sibley and Ahlquist (1990) and Bledsoe (1988) are not significantly worse for either data set. If the moa taxa are excluded from the ratite data set, the alternative hypotheses are all equally likely, as the only difference between the trees is the position of the root. When the moa taxa are excluded from the ratite plus tinamou data set there is limited resolution, but no significant differences between the hypotheses. This demonstrates that the moa taxa are essential to test the various hypotheses properly.

[b]Figure 354 of Sibley and Ahlquist (1990); rhea and ostrich as a basal monophyletic clade.

[c]Figure 326 of Sibley and Ahlquist (1990); ostrich as the basal divergence within ratites.

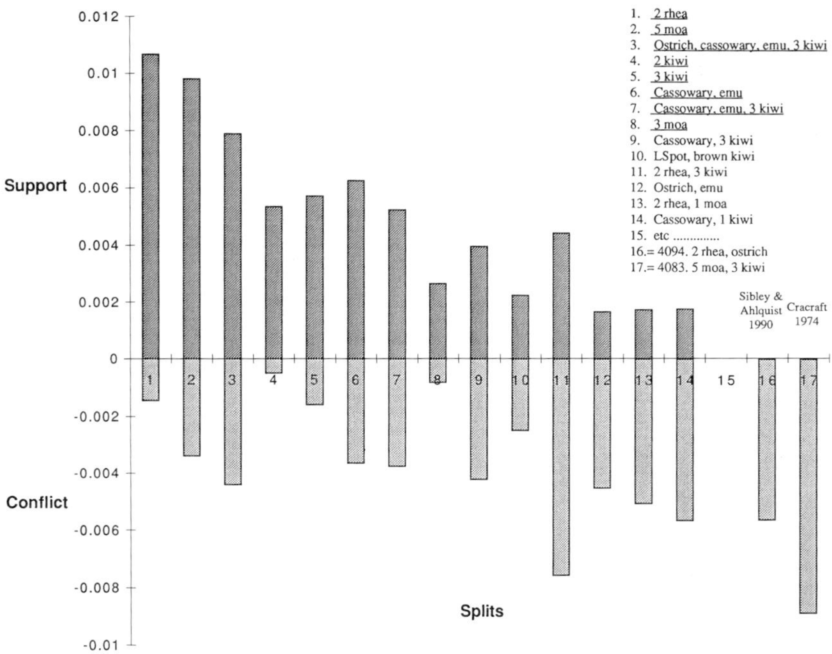

FIGURE 13.3 Spectral analysis of the 366-bp 12S ratite data set, with taxonomic groupings (splits) ranked by the frequency of signals in the data set that support them (above axis, in black), minus those that conflict (below axis, in gray). The frequency of a split is the sum of occurrences of that split in the data divided by the total number of nucleotides (Lento *et al.*, 1995). Because there are many possible patterns in which the data could conflict with a given split, conflict values are normalized so that they sum to the same value as the support signals. The taxonomic groupings corresponding to splits are listed at top right, with clades in the phylogeny identified by the optimality criteria (Fig. 13.2) underlined. There are 4096 possible splits in this data set, and the vast majority have little or no support, so only the strongest 14 are shown. Phylogenetic groupings found by the optimality criteria (Fig. 13.2) have signals with high support and low conflict values (splits 1–8), in contrast to the taxon groupings in splits 9 and above. The signals for taxon groupings from the alternative phylogenetic hypotheses are shown as splits 16 and 17 (actually 4094 and 4083 by support values, respectively), and have no support and large conflict values.

Hendy *et al.*, 1994; Lento *et al.*, 1995) measure the direct support for every taxon grouping (split) within a data set, and present these data independently of any phylogenetic tree. Therefore, this method can measure the amount of support for, and conflict against, any given phylogenetic arrangement and is particularly useful when several alternative phylogenetic hypotheses are to be compared. Spectral analyses of the ratite 12S sequences were performed using Spectrum 1.0.5 (Charleston, 1996) on data sets with, and without, two tinamou outgroups (Figs. 13.3 and 13.4 respectively). The resulting spectrums identify a phylogeny identical to that presented in Fig. 13.2 and indicate that the strongest signals (i.e., highest ratios of support to

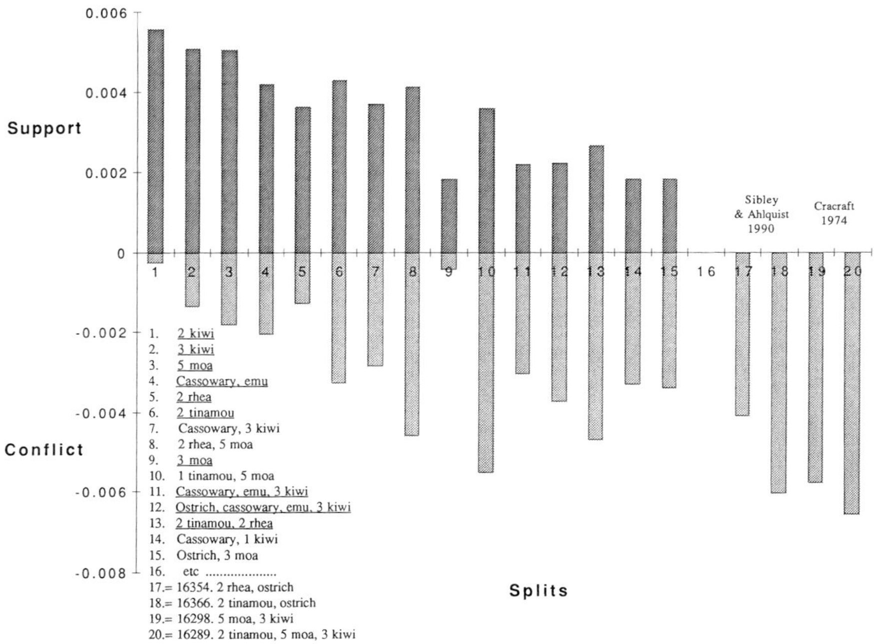

FIGURE 13.4 Spectral analysis of the ratite and tinamou data set with normalized conflict values. Only the 15 strongest signals out of the possible 16,384 are given. While the addition of the distant outgroup decreases the signal within the ingroups, and increases noise, the clades found by the optimality criteria (underlined) are still well supported. Splits 7, 8, and 10 are not seen in Fig. 13.2 but have slightly stronger signals than splits 9, 11, and 12, which were identified by the optimality criteria. The anomalous taxon groupings are not biologically sensible and presumably occur because of the reduced level of resolution. No support can be found for the alternative phylogenetic hypotheses in the data set (splits 17–20, representing rhea/ostrich monophyly, basal ostrich lineage, and New Zealand ratite monophyly, respectively) and they are ranked between 16,289 and 16,366 by support values.

conflict) are for clades identified by the optimality criteria. As the ratio of support to conflict values decrease, other, less likely taxon groupings appear. Because the spectrum contains 2^{n-1} splits (where n is the number of taxa) and the vast majority have low support-to-conflict ratios, only the 14 strongest signals are represented. When the tinamou outgroups are included in the analysis (Fig. 13.4) the average ratio of support to conflict decreases, because the long tinamou branch allows many opportunities for convergent (homoplasious) substitutions with the ingroup taxa, reducing resolution.

Spectral analyses permit the direct measurement of the signals for kiwi/moa monophyly (Cracraft, 1974), rhea/ostrich monophyly (Bledsoe, 1988; Sibley and Ahlquist, 1990), and the arrangement in which ostrich is the sister taxa to other ratites (Sibley and Ahlquist, 1990). As Figs. 13.3 and 13.4 show, these signals have almost no support, and high conflict values. To investigate the amount of support for these taxon groupings in the data set, all possible splits were ranked by support values. Of the 4096 splits possible when only the ratite taxa are considered, the signals for kiwi/moa and rhea/ostrich monophyly are ranked 4083 and 4094 respectively (Fig. 13.3). When the tinamous are included, 16,384 splits are possible and the monophyletic groupings, as well as that for the ostrich as a basal lineage, are ranked between 16,289 and 16,354 (Fig. 13.4). Therefore, spectral analyses demonstrate both that the phylogeny in Fig. 13.2 is well supported, and that there is no direct support in the data for any of the alternative phylogenetic hypotheses.

2. New Ratite Molecular Datasets

Ratite sequence data were obtained from the mitochondrial ND6/transfer RNA-proline (ND6/tRNA^{Pro}) intragenic region and the nuclear c-*mos* protooncogene (c-*mos*), using the same techniques and DNA samples described above. Polymerase chain reaction primers and conditions are given in Cooper and Cooper (1995) and Cooper and Penny (1997).

The ND6/tRNA^{Pro} intragenic regions of the extant ratite genera and three moas are shown in Fig. 13.5. The basal state observed in the outgroup galliform and passerine is an intragenic spacer sequence of 6–7 bp, typical of avian mitochondrial intragenic regions (Desjardins and Morais, 1990). The intragenic regions of the ostrich, moas, and rhea are similar, or slightly bigger, while the kiwi, emu, and cassowary intragenic regions all have a large insertion, ranging from 20 to 30 bp. The intragenic region may still be expanding in the kiwi species; the brown kiwi (*Apteryx australis*) has a 3-bp insertion relative to the other two kiwi species. The insert is strongly biased toward G and against C, with the 30-bp brown kiwi insertion being 57% G and 0% C. The ratite intragenic regions correlate well with

```
Translation        Y  G  T  I  R  A  V  *
Brown KIWI         CGAGGGACAATTCGGGCAGTGTAA  GGGGGAAAGGAGGGAGAGGATTGGAGTTGT  CAGAAGATAGTTTAATGGAGA
L. spotted KIWI    ..G.....G..............G  ....---.........G.AG...T..C.A.  .....................
Roa KIWI           ..G.....G..............G  ....---.........G.A....TT...A.  .....................
EMU                ..G...G..........T..A..G  ....--------.TG..A.G...T--..A.  ...G.............AA...
CASSOWARY          ..G...G..........T..T..G  ...A--------.TG.GT.G...T--..AC  ...G..G...........A...
OSTRICH            ..G...G.............T..G  .AAT----------------.A..----..  ......G.......G.A....
Anomalop. MOA      ..G...G.......A..G.....G  ....----------------...---G.A.  .....A........G..A...
Dinornis MOA       ..G...G.......A..G.....G  .T..----------------...---G...  .....A........G..A...
Megalap. MOA       ..G...G.......A..G.....G  .T..----------------...---A.A.  .....A........G..A...
Lesser RHEA        ..G...G.G........T..T..G  ..AT----------------.A..T.A..G  ..............GCA....
CHICKEN            ..G...G..........G..T...  ...T-----------------------T.  ....GAG.......T.T..A.
Rock WREN          TAT..T..G...........T..G  ...-----------------------...  .....AG..........TGA.

                   <-  NADH subunit 6 gene <-         Intragenic region         ->  tRNApro  ->
```

FIGURE 13.5 Sequences between the NADH subunit 6 (ND6) and transfer RNA proline (tRNA^{Pro}) genes in the extant ratite genera, three moa taxa, and the chicken and wren outgroups. Mitochondrial heavy-strand sequences (corresponding to positions 16,151–16,207 of the published chicken sequence; Desjardins and Morais, 1990) are shown. The translated brown kiwi ND6 amino acid sequence is given ("Translation") with* as the termination codon. Taxon names are given in Fig. 13.1, except for rock wren (*Xenicus gilviventris*).

```
Translation       R  L  C  L  L  Q  P  L  G  S  G  G  F  G  A  V  Y  K  A  T  Y  H  G  V  T  V  A  V  K  Q  V  K  K  S  S  K  N  R  L  A  S  R  Q  S
CHICKEN          CGGCTCTGCCTCCTGCAGCCCCTGGGCTCTGGGGGCTTTGGGGCTGTCTACAAGGCCACCTACCACGGTGTGACTGTGGCTGTGAAGCAGGTGAAGAAGAGCAGCAAAAACCGGCTGGCATCCCGACAGAGC
Lesser RHEA      .A.............A.......A.................CT..........A..T........T....CA..............A..............T.........A..........A.....A...
OSTRICH          .A......T..............A.................CT..........A..T........T....CTGT............A........................A..........A.....A...
CASSOWARY        .A.....................A.................CT..........A..T........T....CA..............A...........C............A..........G.........
EMU              .A.....................A.................CT..........A..T........T....CA..............A...........C............A..........G.........
Brown KIWI       .A.....................A..............C..CT..........A..T........T....CA..............A........................A..........G.........
Roa KIWI         .A.....................A........A.....C..CT..........A..T........T....CA..............A........................A..........G.........
L. spotted KIWI  .A.....................A........A.....C..CT..........A..T........T....CA..............A........................A..........G.........

Translation       F  W  A  E  L  N  V  A  R  L  Q  H  D  N  V  V  R  V  V  A  A  S  T  C  A  P  A  S  Q  N  S  L  G  T  I  I  M  E  Y  V  G  N  V  T
CHICKEN          TTCTGGGCTGAGCTGAACGTAGCCCGACTGCAGCATGATAATGTGGTGCGTGTGGTGGCTGCTAGCACGTGTGCCCCTGCCAGCCAGAACAGCCTGGGCACCATCATCATGGAGTATGTTGGCAATGTCACC
Lesser RHEA      ..T..............T........T..........C...C.....A..............C.....C.....T....................C.....T.................G.....CA.....
OSTRICH          ..T.......................T..........C.........A..............C.....A..CA........................................A.....G.....CA.....
CASSOWARY        ..T..............T........T..........C...C.....A..............C..........................................G.......A.....G.....CA.....
EMU              ..T..............T........T..........C.........A..............C..................................................A.....G.....CA.....
Brown KIWI       ..T.....C........T........T..........C.........A..............C.....A............................................A.....G.....CA.T...
Roa KIWI         ..T.....C........T........T..........C.........A..............C.....A............................................A.....G.....CA.....
L. spotted KIWI  ..T.....C........T......T.T..........C.........A..............C.....A............................................A.....G.....CA.....

Translation       L  H  H  V  I  Y  G  T  R  D  A  W  R  Q  G  E  E  E  E  G  G  C  G  R  K  A  L  S  M  A  E  A  V  C  Y  S  C  D  I  V  T  G  L  A
CHICKEN          CTGCACCATGTCATCTACGGCACTAGAGATGCGTGGAGGCAGGGCGAGGAGGAGGAAGGAGGATGTGGGAGGAAGGCTCTGAGCATGGCGGAGGCTGTGTGCTACTCGTGTGACATCGTGACTGGCTTAGCC
Lesser RHEA      ...........A.....T..T.....T....TA...........T...---..T..............C.AA.....C.........AAA...A.......T..T..T........T...............
OSTRICH          ...........A.....T........T...A.A...........T.....T..T..............A.AA.....C..........AA...A.......T..T..T........TA..............
CASSOWARY        ...........A.....T........T..CA.A....A......T.....T..T..............A.AA.....C..........AA...A.......T..T..T........TA..............
EMU              ...........A.....T........T...A.A....A......T.....T..T..............A.AA.....C..........AA...A.......T..T..T........TA..............
Brown KIWI       ...........A.....T........T.....A...........T...A.T..T..............A.AA.....C..........AA...A.......T..T..T........TA..............
Roa KIWI         ...........A.....T........T.....A...........T.....T..T..............A.AA.....C..........AA...A.......T..T..T........TA..............
L. spotted KIWI  ...........A.....T........T.....A...........T.....T..T..............A.AA.....C..........AA...A.......T..T..T........TA..............

Translation       F  L  H  S  Q  G  I  V  H  L  D  L  K  P  A  N  I  L  I  T  E  H  G  A  C  K  I  G  D  F  G  C  S  Q  R  L  E  E  G  L  S  Q  S  H
CHICKEN          TTCCTTCACTCGCAGGGCATCGTGCACCTCGACCTGAAGCCTGCCAATATCCTCATCACTGAGCACGGAGCGTGCAAGATCGGAGACTTCGGCTGCTCCCAGAGACTGGAGGAGGGCTTGTCCCAGAGCCAC
Lesser RHEA      ..T........A.....T..T......T.G..............A...G..T..........A..G....T.....................G........A.A.........T..............T.C.
OSTRICH          ..T........A........T......T.G..............A...G..T..........A..G....T.........T........T..G........A.A.........T..............T.C.
CASSOWARY        ..T........A........T......T.G..............A...G..T..........A..G....T.........T........T..G........A.A.........T..............T.C.
EMU              ..T........A........T......T.G..............A...G..T..........A..G....T.........T........T..G........A.A.........T..............T.C.
Brown KIWI       ..T........A........T......T.G..............A...G..T..........A..G....T.........T........T..G........A.A.........T..............T.C.
Roa KIWI         ..T........A........T......T.G..............A...G..T..........A..G....T.........T........T..G........A.A.........T..............T.C.
L. spotted KIWI  ..T........A........T......T.G..............A...G..T..........A..G....T.........T........T..G........A.A.........T..............T.C.

Translation       H  V  C  Q  Q  G  G  T  Y  T  H  R  A  P  E  L  L  K  G  E  R  V  T  A  K  A  D  I  Y  S  F  A  I  T  L  W  Q  I  V  M  R  E  Q  P
CHICKEN          CATGTTTGCCAGCAAGGGGGCACGTACACCCACCGCGCTCCTGAGCTCCTCAAGGGCGAGAGGGTCACTGCCAAAGCAGACATCTACTCCTTTGCCATCACCCTCTGGCAGATCGTCATGCGGGAGCAGCCC
Lesser RHEA      ..A........A...........A.....G..T..T..C..............A..G..A..A.................T........A.....T..............A..T.......AA.........
OSTRICH          ..G....................A.....A..T..T..C.................A................................A.....T.....T........A..T........A.........
CASSOWARY        ..G....................A.....G..T.....C.................G.......................T........A.....T.....T........A..T........A.........
EMU              ..G....................A.....A..T..T..C.................A................................A.....T.....T........A..T........A.........
Brown KIWI       ..G....................A.....G..T..T..C.................G................................A.....T.....T........A..T........A.........
Roa KIWI         ..G....................A.....G..T..T..C.................G................................A.....T.....T........A..T........A.........
L. spotted KIWI  ..G....................A.....G..T..T..C.................G................................A.....T.....T........A..T........A.........
```

FIGURE 13.6 Aligned sequences of a 660-bp region of the protooncogene c-*mos* for the extant ratites and a chicken. The chicken amino acid sequence ("Translation") is shown above the DNA sequences. The underlined 657 homologous positions are used in the phylogenetic analysis. There are 105 variable, and 20 parsimony, sites in the data. Taxon names are given in Fig. 13.1. The proportion of invariant sequence positions was estimated to be 0.4, on the basis of the number of leucine and third codon positions, and variation observed in 10 avian orders (Cooper and Penny, 1997).

the 12S phylogeny, since the derived ratites (kiwi, emu, and cassowary) all possess a derived insert relative to the other ratites and outgroups. However, the nature and distribution of this character state contrast strongly with a monophyletic kiwi/moa clade.

The c-*mos* protooncogene is a single-copy intronless nuclear gene that encodes Mos, a serine/threonine kinase with important oocyte maturation-controlling functions (Sagata *et al.*, 1988). Sequences of a 657-bp fragment of the ratite c-*mos* are given in Fig. 13.6. It was difficult to obtain sequences for the tinamou taxa and consequently the chicken sequence is used as an outgroup for the analysis. Unfortunately, it was also not possible to amplify a moa c-*mos* sequence using PCR, presumably owing to the low concentration of any surviving single-copy c-*mos* sequences. In contrast, relatively large amplifications (600–800 bp) were obtainable from museum specimens up to 20 years old (data not shown). The ratite nuclear c-*mos* sequences were used for phylogenetic analyses separately, and in combination with the mitochondrial 12S data.

Phylogenetic analysis of the c-*mos* data (Fig. 13.7) reveals that the rhea is the basal divergence within ratites, whereas the kiwi, ostrich, and emu/cassowary lineages form an unresolved trichotomy. The topology is consistent with Fig. 13.2,

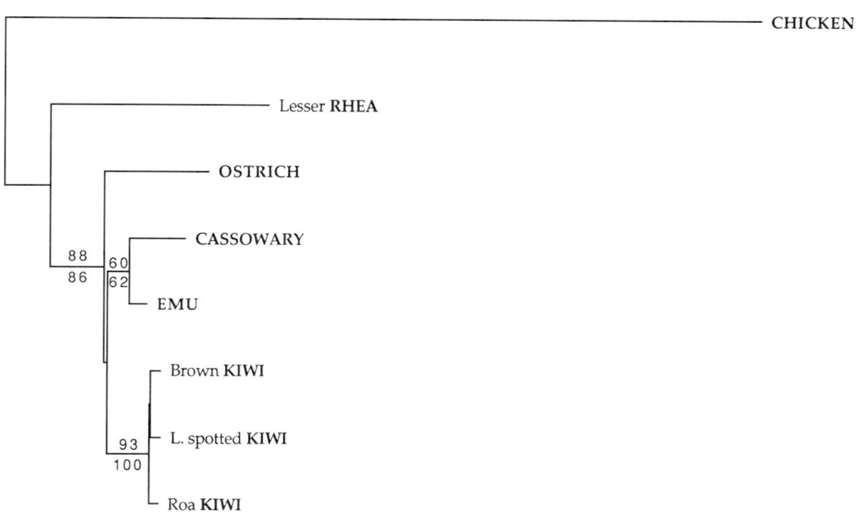

FIGURE 13.7 Unweighted LogDet corrected distance tree of the c-*mos* data (Fig. 13.6). The tree topology is consistent with Fig. 13.2, suggesting the rhea is the basal ratite lineage. The long branch joining the rhea to the ancestor of the remaining ratites suggests that a long period of time existed between the initial split within ratites, and a subsequent radiation. The most parsimonious tree (not shown) joins the ostrich, kiwi, and emu/cassowary lineages as an unresolved trichotomy. Bootstrap values from 1000 heuristic replications using parsimony (above line) and LogDet corrected neighbor joining (below line) are shown.

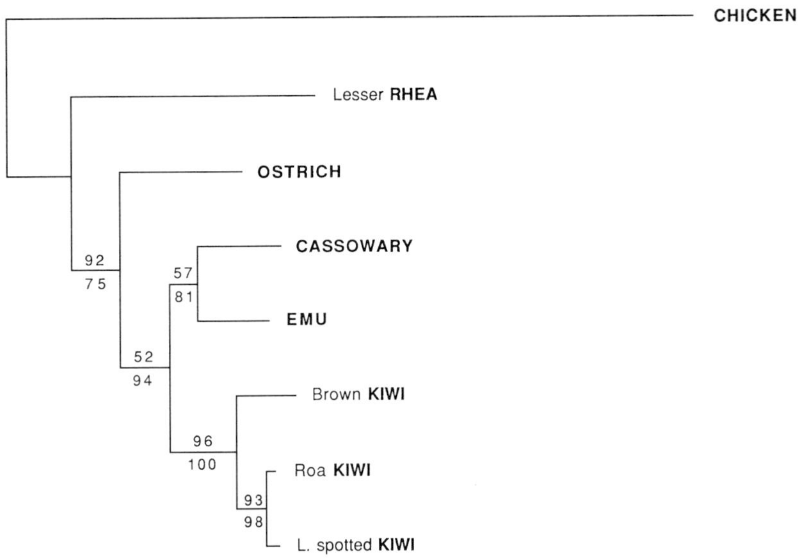

FIGURE 13.8 Unweighted LogDet corrected distance tree of the combined nuclear and mitochondrial data sets (1023 bp). The same tree topology is produced by weighted and unweighted parsimony and maximum-likelihood analyses, and is identical to Fig. 13.2. Bootstrap values from 1000 heuristic replications using parsimony and LogDet corrected neighbor joining are given above and below branches, respectively.

although there is less phylogenetic resolution, consistent with the slower evolutionary rate of the nuclear gene and the use of a distant outgroup. The long branch separating the rhea from the ancestor of the other ratites indicates that a considerable period of time elapsed between the first divergence within ratites and a subsequent radiation. It would be interesting to include the moa in this phylogeny to determine whether the early separation of South America and New Zealand from the other Gondwanic land masses might correspond to early rhea and moa divergences relative to the other taxa.

A combined nuclear and mitochondrial data set of 1023 bp was created for the rhea, ostrich, cassowary, emu, three kiwis, and chicken taxa for which both 12S and c-*mos* sequences had been obtained. Parsimony, distance, and maximum-likelihood analyses of the data set (Fig. 13.8) identify the same phylogeny as in Fig. 13.2, with similar levels of support. None of the alternative phylogenies are statistically worse than this topology. This is not surprising because the moa taxa are missing from the phylogeny and a distant outgroup is used, which severely reduces resolution with respect to the 12S data set. Spectral analyses (data not shown) demonstrate that of the 128 possible splits, a basal kiwi lineage is ranked 40 by support values, while a basal ostrich lineage and rhea/ostrich monophyly are ranked 13 and

23, respectively. Consequently, even analyses with reduced levels of resolution demonstrate the phylogeny of Cracraft is poorly supported.

3. *Why Do the Phylogenies Differ?*

Phylogenetic analyses of the mitochondrial and nuclear data sets strongly support the topology presented in Fig. 13.2, and indicate that not only are the phylogenies of Cracraft (1974), Sibley and Ahlquist (1981, 1990), and Bledsoe (1988) suboptimal, but that there is absolutely no support in the data set for the taxon groupings in which they differ from Fig. 13.2. It is important to examine the sequence data for systematic biases that could obscure signals for alternative phylogenies, although in this case the signals would have to be removed, rather than just obscured. No significant evolutionary rate variation was observed in the data, and LogDet analyses indicate that the topology in Figs. 13.2, 13.7, and 13.8 is independent of base composition. To detect biases due to long-branch attraction (Hendy and Penny, 1989), analyses were carried out in which ingroup taxa were excluded in turn from the data set to see if the tree topology changed. For example, if the emu and cassowary are left out of the 12S data set, the kiwi branch becomes longer than those of the rhea, moa, or ostrich but still the kiwi does not shift from its derived position in the phylogeny to move closer to the tinamou outgroup. No long-branch attractions were detected in the mitochondrial or combined mitochondrial and nuclear data sets, using this method.

Because no shortcomings are apparent in the sequence data or analyses, it is necessary to examine the discrepancies between the alternative phylogenies and Figs. 13.2 and 13.8. Sibley and Ahlquist (1981, 1990) and Bledsoe (1988) identified the kiwi, emu, and cassowary as a derived clade, but differ from Fig. 13.2 in how well they resolve the position of the rhea lineage. Sibley and Ahlquist (1981, 1990; see Cracraft, Chapter 7 in this volume) use the same data to obtain three different outgroup combinations of the rhea and ostrich lineages, suggesting the data lack resolving power among the deeper divergences. Two of the phylogenies are identical to Fig. 13.2 if basal branches are collapsed (e.g., owing to a lack of resolution), while the conflicting phylogeny weakly places the ostrich as the basal lineage. Bledsoe (1988) tentatively placed the moa taxa as the basal lineage, but suggested this could be an artifact. Otherwise the phylogeny of Bledsoe is identical to Fig. 13.2, except that once again the long rhea and ostrich branches are joined, perhaps owing to a long-branch attraction. Consequently, the discrepancy between the DNA–DNA hybridization, morphological data, and Fig. 13.2 is easily explained as a difference in resolution of the basal rhea and ostrich lineages. The resolving power of the 12S data, provided by the moa taxa splitting the long rhea and ostrich branches, is a clear example of the value of ancient DNA sequences to systematic research.

The phylogeny of Cracraft (1974) is widely divergent from those of all the other studies, including Fig. 13.2. Analysis of the 12S sequence data shows the phylogeny to be significantly worse than that of in Fig. 13.2, and none of the data sets have any support for a basal kiwi, or monophyletic New Zealand ratite clade. Furthermore, the ND6/tRNA^{Pro} insertion is difficult to reconcile with the topology of Cracraft. Subsequent publications (Sibley and Ahlquist, 1981; Bledsoe, 1988) have criticized Cracraft's character measurements, and the unusually small amount of homoplasy of the data set, especially given the convergence expected in a group of large flightless birds, many of whom exploit similar habitats. Bledsoe (1988) used most of the morphological characters measured by Cracraft and produced a widely divergent phylogeny that is similar to those of the genetic studies. The two different interpretations of what is essentially the same morphological data strongly suggest that subjective decisions about character states and polarity are influencing the morphological phylogenies. In contrast, the DNA–DNA hybridization and sequence studies utilize completely different, and objective, data to yield closely matched topologies. As further DNA studies support the same topology, the confidence in its accuracy should increase correspondingly.

B. Paleoecological Studies

Owing to the repetitive destructive tendencies of humans, the Pacific Islands are an area in which ancient DNA research is particularly suited. Conservation management is critical in attempts to halt the decline of biodiversity and ecosystem health in these environments. Unfortunately, many of these islands have insufficient paleontological information to form accurate views of the evolution and processes of past paleoecosystems. When bones and pollen are preserved in deposition sites, caves, or lava tubes, the picture that is produced is often far more complex than expected (James *et al.*, 1987; Diamond, 1990; Olson and James, 1991; James and Olson, 1991; Cooper and Millener, 1993; Worthy and Holdaway, 1993; Steadman, 1995; Cooper *et al.*, 1996). In these situations, ancient DNA sequences can provide important information about extinct and endangered taxa, including phylogenetic relationships and changes in genetic diversity and gene flow between populations through time. The following two projects demonstrate how ancient DNA data can have quite different practical applications for surviving island endemics.

1. The Oligocene Drowning of New Zealand

New Zealand separated from the remnant Gondwana land mass around 80 MYA and carried a range of Gondwanic biota, augmented by wind-blown additions, to a position 1500 km from the nearest land mass (Cooper and Millener, 1993). Paleoecological "ghost" signals abound in New Zealand (Diamond, 1990) but the lack of a vertebrate terrestrial fossil record prior to the Pliocene (Fordyce, 1991) has constrained temporal aspects of evolutionary research. Ancient DNA studies are particularly useful in island situations such as this, in which a variety of morphologically unique taxa lack paleontological records and therefore have poorly known evolutionary histories.

As discussed previously, New Zealand holds a central position in the evolution of several avian groups. In turn, the ecology of New Zealand has been shaped to a large degree by avian evolution. Because New Zealand lacked endemic terrestrial mammals, many typical "mammalian" niches were filled by birds, reptiles, and insects (Daugherty *et al.*, 1993). The dominant herbivore in New Zealand was undoubtedly the moa, as this giant nonruminant required large amounts of vegetation in the temperate climate. Moa browsing has been suggested to have exerted large selective effects on the growth patterns of New Zealand flora (Atkinson and Greenwood, 1989; Cooper *et al.*, 1993). The extinction of the moa has even been hypothesized to have changed the forest ecology of New Zealand from one of gymnosperm, to angiosperm, dominance (Wellman, 1994; Cooper, 1994). Consequently, the population history of the moa, and possibly also that of the kiwi, are important factors in evaluating the paleoecological interactions of New Zealand biota.

a. Molecular Studies

The ratite 12S sequences in Fig. 13.1 indicate that the five moa genera, which are morphologically quite diverse, possess a surprisingly limited amount of genetic diversity. Furthermore, the three kiwi species show a similarly limited amount of diversity. To contrast this pattern with that from a more rapidly evolving sequence, a 244-bp region of the ND6 gene was sequenced for the kiwi and moa taxa as well as three New Zealand acanthisittid wrens, one of which is extinct (Cooper and Cooper, 1995). Surprisingly, the maximum genetic diversity within ND6 sequences of each group (moa, kiwi, and wren) was similar (range, 0.254–0.377; Cooper and Cooper, 1995), and repeated the pattern observed in the 12S sequences (range, 0.084–0.106). This situation is difficult to reconcile with hypotheses suggesting the three groups arrived and radiated in New Zealand at different times, with the kiwi a recent, perhaps early Tertiary, dispersal from Australia (Sibley and Ahlquist, 1981; Cooper *et al.*, 1992) while the moa and wren are ancient, possibly Gondwanic, lineages (Fleming, 1979). A further problem is that there are several deep splits within each group and the observed 12S and ND6 sequence diversity in the three groups is similar to that of avian taxa thought to have radiated as recently as mid-Oligocene to mid-Miocene times (Moum *et al.*, 1994). Therefore the sequence data appear to indicate that each of these ecologically diverse bird groups radiated at a common, possibly Oligocene to Miocene, point in time.

Because these results are unexpected, it is important to reevaluate the authenticity of the ancient moa and wren sequences. The six moa taxa have different, but closely related, sequences and several individuals of most taxa have been sequenced (Cooper *et al.*, 1992; Cooper and Cooper, 1995). Although no intraspecific 12S variation is seen, small amounts of ND6 sequence variation corresponding to geographic patterns are seen within species (data not shown). Moa and wren interspecific variation occurs at positions known to be variable in 12S and ND6 sequences, and the ratio of transversion to transition substitutions is consistent with mitochondrial patterns. The sequences have been replicated in four physically separate laboratories, and also match the pattern observed in the living kiwis. Last, phylogenetic analyses cluster the moa sequences, and the wren sequences, in 2 clades when compared to sequences from 15 other avian orders (data not shown).

Since the ancient sequences appear authentic, other explanations are needed. The similar, and limited, amounts of genetic diversity could result from independent Oligo–Miocene arrivals of the three groups in New Zealand, presumably from Australia, but this is unlikely for several reasons. The ratites are flightless (the moa has totally lost its wings) and the wrens are barely flighted, and no closely related Australian ancestral population exists for the moa or wren, while the kiwi appears to have diverged from the emu/cassowary lineage in the Eocene (Sibley and Ahlquist, 1981). This scenario also demands that all three groups would have to arrive independently, but almost simultaneously.

b. The Oligocene Marine Transgression

Another explanation of the data is that while the moa, kiwi, and wren arrived and radiated independently in New Zealand, some relatively recent event reduced diversity in each group to a single mitochondrial lineage, from which there has been a subsequent radiation. This would conceal any previous diversity, and the groups would appear to have simultaneously radiated. Mitochondrial DNA is particularly sensitive to population size fluctuations and quickly loses diversity during periods of constant, or decreasing population size (Wilson *et al.*, 1985). Consequently, the sequence data are compatible with a widespread ecological event that could simultaneously constrain, or drastically reduce, the population size of the giant herbivorous moas, tiny insectivorous wrens, and nocturnal omnivorous kiwis.

The Tertiary geological record of New Zealand is briefly reviewed in Cooper and Cooper (1995) and reveals that a lack of tectonic activity in the Paleogene (65–23 MYA) reduced New Zealand to a broad lowland. During the large sea level changes of the late Oligocene (29–23 MYA), New Zealand was inundated by a marine transgression that is thought to have lasted some 6 million years, reducing it to a string of low-lying islands. The peak of this drowning event is estimated to have reduced the land area to about 18% of the current size (Cooper and Cooper, 1995). In the absence of a fossil record, the transgression has been hypothesized to have stimulated speciation (Stevens, 1985), although the exact effects of the large reduction in niche diversity and population sizes were unknown. The sequence data appear to remedy this situation, as it correlates perfectly with this geological catastrophe, both in timing and because mitochondrial diversity would be severely reduced by a prolonged period (up to 6 million years) of limited population sizes and local extinctions. Furthermore, the subsequent star-like radiation of mtDNA diversity is consistent with a Miocene increase in land area and niche diversity. Consequently, the combined genetic and geological data suggest a severe loss of endemic taxonomic diversity in the mid-Tertiary, providing an important new view of a catastrophic period in the history of New Zealand (Cooper and Cooper, 1995).

Importantly, the Oligocene drowning hypothesis is testable because it predicts which habitats and taxa would have been most adversely affected during the marine transgression. Studies of nuclear sequence data in New Zealand endemics will enhance the view of this event, but if it was as ecologically widespread as the mtDNA data suggest then conservation studies will need to accommodate the model when interpreting the genetic diversity of endemics.

2. Conservation of the Endangered Laysan Duck

Ancient DNA data have considerable potential to provide information about the evolutionary history of extant, as well as extinct, taxa. Genetic data from subfossils (preserved nonmineralized bone) can be used to analyze the prior range and habitat of taxa with recently restricted distributions. This is important in areas like the Pacific, where many flighted taxa currently endemic to islands are relics of formerly widespread populations (Steadman, 1995).

Laysan Island is one of the many eroded islands that has formed as the Pacific plate moves across the Hawaiian hotspot, and currently lies some 600 km to the northwest of the main Hawaiian islands. It is only 370 ha in size with a maximum altitude of 12 m, and is dominated by a large central hypersalinic lagoon. The brine flies that live on the lagoon are the main food source of the last remaining population of Laysan ducks (*Anas laysanensis*). This endangered population has varied in size from 500 to less than 20 individuals during this century and is highly vulnerable to disease or climatic disruptions. Because the Laysan duck is historically known only from Laysan Island, it has been difficult to gain permission to establish a second population elsewhere, given the historically negative effects of introduced taxa in the Hawaiian islands. In addition, the Laysan duck and the Hawaiian duck, or koloa (*Anas wyvilliana*), an inhabitant of wetlands on most of the main Hawaiian islands, are thought to have evolved from stray migratory mallards (*Anas platyrhynchos*), which has greatly influenced recovery programs (Moulton and Weller, 1984).

Paleontological studies have shown that fossil and subfossil bones of small duck species occur in late Pleistocene and Holocene deposits on the main Hawaiian islands (Olson and James, 1991; Giffin, 1993). Interestingly, the bones are found in association with a variety of paleontological habitats and indicate that the fossil species was widely adaptable. The bones are intermediate in size between the koloa and Laysan duck and do not provide sufficient information for identification purposes (Cooper *et al.*, 1996), because morphology and body size are not particularly diagnostic in dabbling ducks (Worthy, 1988; Livezey, 1991).

If the paleontological duck bones on the main Hawaiian islands are part of the former range for the Laysan duck, then considerable data about the paleoecology and evolution of the species might be gained. The information would be important in potential relocation plans because it might identify whether the habitat of Laysan Island was optimal for the species. Consequently, the identity of the bones was crucial, and in the absence of definitive morphological data it appeared that ancient DNA techniques might resolve the issue.

a. Phylogenetic Analysis of Subfossil Ducks

DNA was extracted from femurs and tibiotarsi of subfossils from lava tubes on the island of Hawaii. Two variable regions of the control region, spanning 312 and 133 bp, respectively (positions 78–390 and 1117–1251 in the published chicken sequence; Desjardins and Morais, 1990), were amplified and sequenced using PCR. Interestingly, DNA could not be amplified from subfossils found in low-altitude sites (0–500 m) while those at high-altitude sites (around 2300 m) worked well. This correlation was also found in studies of other Hawaiian avian subfossils, even in high-altitude sites that were regularly wet, and is thought to relate to cold temperature.

The mt control region sequences of three subfossil bones, three Laysan ducks, three koloas, two genetically diverse mallards, and an outgroup (African black duck, *Anas sparsa*) were aligned and 366 homologous positions identified. Of these, 60 were variable and 36 were informative among the ingroup taxa (Fig. 13.9a). The results of phylogenetic analyses of the data are shown in Fig. 13.9b. The sequences clearly demonstrate the subfossil taxa are closely related to the extant Laysan ducks, differing only by one transition. The long branch between the Laysan duck/subfossil clade and any other taxa indicates they are not genetically closely related to either mallard sequence. Conversely, the koloa taxa form a clade with mallard haplotype 2, indicating that the migratory mallard ancestry hypothesis may be correct, or that some degree of hybridization has taken place. Interestingly, mallard haplotype 1 does not group strongly with the mallard 2/koloa clade, indicating that considerable mtDNA genetic diversity exists in the mallard, as previously noted (Avise *et al.*, 1992).

The analysis shows that the Laysan Island population is a relict of a formerly widespread distribution, and provides justification for reestablishing populations of Laysan duck on the main Hawaiian islands. The ecological situations that currently exist on the main Hawaiian islands are obviously vastly altered from that of the former population, and this must be taken into account when reintroducing a species. However, the widespread distribution of paleontological remains indicates that the species was surprisingly adaptable, inhabitating high-altitude (up to 1800 m) forested sites far from water as well as sites near sea level (Cooper *et al.*, 1996). Furthermore, the considerable numbers of seabirds that frequent Laysan Island have probably introduced, and stimulated immunity to, many avian diseases that have decimated other Hawaiian endemics. Whether sufficient genetic variability remains in the population after the many bottlenecks experienced on Laysan Island is currently unknown. Nevertheless, the establishment of any significant breeding a)

```
                              5'  variable  region                    3'  v.r.
                         11111111111111111111111111112222222222222222222222
                      2255990111112222333455666667888999011112344445555556668124457
                      2301567234563470592691378995680173123467046801267812406408 64

Mallard 1             TACTTCCC-TATACTCGCATCACTCAACCC-AAGATGACTCATCAATTCTACCACCCCAC
Mallard 2             .....T..C.G.........TG.C.....T-....-.GTCT...G..CT.G.TAT.T.G.
Koloa 1               .....T..C.G.........TG.C...T.T-....-.GTCT...G..CTC..TAT.T.G.
Koloa 2               ....CT..C.G......T..TG.C...T.T-....-.GTC....G..CT...TAT...GG
Koloa 3               ....CT..C.G......T..TG.C...T.T-....-.GTC....G..CT...TAT...GG
Laysan duck1          C....T..CC...T.T..GC.G.......TCG..G-..T.....G..C....TGTTT...
Laysan duck 2         C....T..CC...T.T..GC.G.......TCG..G-..T.....G..C....TGTTT...
Laysan duck 3         C....T..CC...T.T..GC.G.......TCG..G-..T.....G..C....TGTTT...
Subfossil bone 1      C....T..CC...T.T..GC.G.......TCG..G-..T.....G..C....TGTTT...
Subfossil bone 2      C....T..CC...T....GC.G.......TCG..G-..T.....G..C....TGTTT...
Subfossil bone 3      NNNNNNNNNNNNNNNNNNNNNNNNNNNNNNNNNNNNNNNNNNNNNNNNNNNNNNNNNTTT...
African black duck    .GTC.TATCC.AG.C.A..C.GTCTGG.T.-.GAG-A...TGCAGTC-...T.GT..T..
```

b)

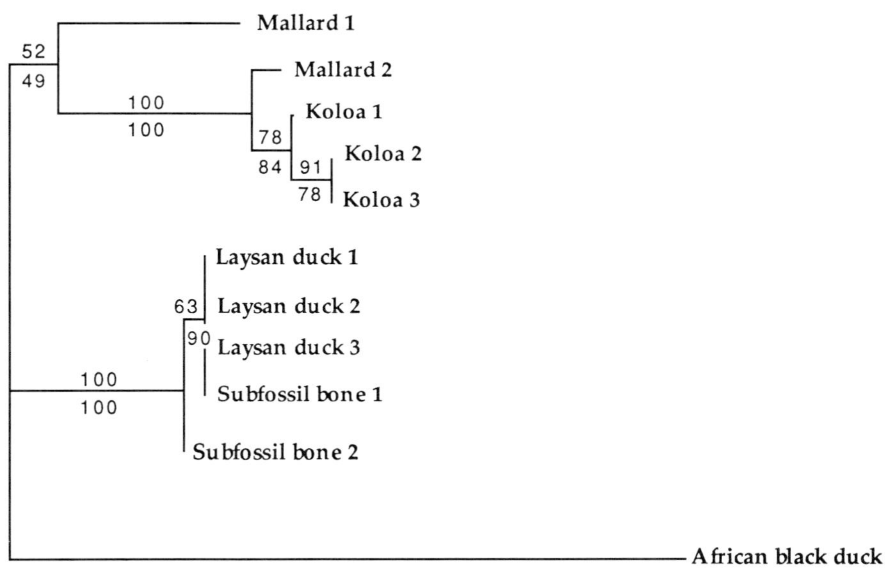

FIGURE 13.9 (a) Variable positions from two portions of the mt control region of three Laysan ducks, three subfossil bones, three koloas, two divergent mallard haplotypes, and an outgroup African black duck (*Anas sparsa*) from Cooper *et al.* (1996). A total of 366 bp was obtained from the 5′ and 3′ variable regions of the control region. The numbering system uses the 3′ base of the light-strand primers as 0 for each fragment. Taxon names are given in text. (b) Unweighted LogDet corrected neighbor-joining tree using gaps as characters with branch lengths drawn proportional to distance. The partial sequence from subfossil bone 3 was excluded from analysis. Bootstrap values from 1000 replications are given for both parsimony and neighbor joining, above and below branches, respectively. When subfossil bone 3 was included in the analysis, it joined the Laysan duck/subfossil bone clade immediately prior to the divergence of subfossil bone 2.

population outside Laysan Island must represent a considerable reduction of the threat of immediate extinction.

IV. SUMMARY

The three projects described above demonstrate how genetic information from the past can be incorporated into modern evolutionary studies. In phylogenetic studies of the ratite birds, ancient DNA sequences from moa taxa drastically increased resolution among the basal lineages by splitting a long branch and improving taxon distribution across the tree. The moa sequences were also essential in demonstrating that several alternative phylogenetic hypotheses are not supported by the sequence data. Without the moa data the alternative hypotheses differed only in the position of a root, and a monophyletic moa and kiwi clade could not be tested.

The ratite data led to a further discovery in New Zealand, where the combination of geological data and ancient DNA sequences revealed the extent of a mid-Tertiary ecological disaster, previously concealed by a missing paleontological record. In both the New Zealand and Hawaiian studies, genetic data from recently exterminated taxa provided temporal information about paleoecosystems that have been severely disrupted. In so doing, the studies allowed the ecology of modern taxa to be reinterpreted in the light of paleoecological data, providing information for the conservation of surviving taxa. While both studies focused on avian taxa they serve as models for the investigation of many other island endemics or isolated populations.

V. FUTURE RESEARCH

The future of ancient DNA in systematic research is currently difficult to predict, as the increasing availability of automated sequencers means that studies involving many taxa can now realistically use sequences of thousands of base pairs. Whole mitochondrial genomes are now routinely used in vertebrate systematic studies (e.g., Horai *et al.*, 1995; Xu *et al.*, 1996) and sequences of this length hold considerable advantages for phylogenetic reconstruction (Charleston *et al.*, 1994). As increasingly long DNA sequences become standard in systematic studies of extant taxa it will become correspondingly difficult to obtain ancient DNA sequences of a similar size. The use of DNA repair systems to increase the length of ancient DNA amplifications (Lindahl, 1993a) may improve the situation slightly, but this problem is only likely to grow.

In contrast, ancient DNA studies that investigate the identity, or genetic diversity, of extinct or preserved taxa are likely to become more prevalent. The potential of ancient microsatellite data is still being explored, but this new area will undoubtedly increase the scope of genetic studies of extinct populations in the next 10 years of ancient DNA research.

ACKNOWLEDGMENTS

I thank David Penny, Jeremy Austin, Roger Cooper, and Rod Page for fruitful discussion and comments. I also thank members of the Wilson, Chambers, Fleischer, and Harvey groups for assistance, and particularly the many museum staff and individuals who have cooperated with sampling requests and allowed these studies to proceed. Phylogenetic software was kindly provided by D. Swofford and M. Charleston. The Laysan duck research was supported in part by the U.S. Fish and Wildlife Service and the Smithsonian Institution Scholarly Studies Program. This work was supported in part by the New Zealand Marsden Fund and the U.K. NERC.

REFERENCES

Arctander, P. (1995). Comparison of a mitochondrial gene and a corresponding nuclear pseudogene. *Proc. R. Soc. Lond. B* **262**, 13–19.

Atkinson, I. A. E., and Greenwood, R. M. (1989). Relationships between moas and plants. *N. Z. J. Ecol.* **12**(Suppl.), 67–96.

Avise, J. C., Alisauskas, R. T., Nelson, W., and Ankney, C. D. (1992). Matriarchal population genetic structure in an avian species with female natal philopatry. *Evolution* **46**, 1084–1096.

Bledsoe, A. H. (1988). A phylogenetic analysis of postcranial skeletal characters of the ratite birds. *Ann. Carnegie Mus.* **57**, 73–90.

Boom, R., Sol, C. J. A., Salimans, M. M. M., Jansen, C. L., Wertheim-van Dillen, P. M. E., and Van Der Noordaa, J. (1990). Rapid and simple method for purification of nucleic acids. *J. Clin. Microbiol.* **28**, 495–503.

Cano, R. J., Poinar, H. N., Pieniazek, N. J., Acra, A., and Poinar, G. O., Jr. (1993). Amplification and sequencing of DNA from a 120–135 million year old weevil. *Nature (London)* **363**, 536–538.

Caspers, G.-J., Wattel, J., and de Jong, W. W. (1994). αA-crystallin sequences group tinamou with ratites. *Mol. Biol. Evol.* **11**, 711–713.

Charleston, M. A. (1996). "Spectrum," version 1.0.5. University of Glasgow, Glasgow, Scotland, U.K.

Charleston, M. A., Hendy, M. D., and Penny, D. (1994). The effects of sequence length, tree topology and number on the performance of phylogenetic methods. *J. Comput. Biol.* **1**, 133–151.

Christidis, L., Leeton, P. R., and Westerman, M. (1996). Were bowerbirds part of the New Zealand fauna? *Proc. Natl. Acad. Sci. U.S.A.* **93**, 3898–3901.

Collura, R. V., and Stewart, C.-B. (1995). Insertions and deletions of mtDNA in the nuclear genomes of Old World monkeys and hominoids. *Nature (London)* **378**, 485–489.

Colson, I. B., Bailey, J. F., Varcauteren, M., Sykes, B. C., and Hedges, R. E. M. (1997). Bone diagenesis and the preservation of ancient DNA. *Ancient Biomol.* (in press).

Cooper, A. (1993). DNA from museum specimens. *In* "Ancient DNA" (B. Herrmann and S. Hummel, Eds.), pp. 149–165. Springer-Verlag, New York.

Cooper, A. (1994). Moas, gizzard stones, and New Zealand plants: Further commentary. *N. Z. Sci. Rev.* **51**, 60–63.

Cooper, A., and Cooper, R. A. (1995). The Oligocene bottleneck and New Zealand biota; genetic record of a past environmental crisis. *Proc. R. Soc. Lond. B* **261**, 293–302.

Cooper, A., and Penny, D. (1997). Mass survival of birds across the K-T boundary: Molecular evidence. *Science* (in press).

Cooper, A., Mourer-Chauviré, C., Chambers, G. K., von Haeseler, A., Wilson, A. C., and Pääbo, S. (1992). Independent origins of the New Zealand moas and kiwis. *Proc. Natl. Acad. Sci. U.S.A.* **89**, 8741–8744.

Cooper, A., Atkinson, I. A. E., Lee, W., and Worthy, T. H. (1993). The evolution of the moa and their effect on the flora. *Trends Ecol. Evol.* **8**, 433–437.

Cooper, A., Rhymer, J., James, H. F., Olson, S. L., McIntosh, C. E., Sorenson, M. D., and Fleischer, R. C. (1996). Ancient DNA and island endemics. *Nature (London)* **381**, 484.

Cooper, R. A., and Millener, P. R. (1993). The New Zealand biota: Historical background and new research. *Trends Ecol. Evol.* **8**, 429–433.

Cracraft, J. (1974). Phylogeny and evolution of the ratite birds. *Ibis* **116**, 494–521.

Crichton, M. (1991). "Jurassic Park," p. 399. Century, London.

Daugherty, C. H., Gibbs, G. W., and Hitchmough, R. A. (1993). Mega-island or micro-continent? New Zealand and its fauna. *Trends Ecol. Evol.* **8**, 437–442.

De Boer, L. E. M. (1980). Do the chromosomes of the kiwi provide evidence for a monophyletic origin of the ratites? *Nature (London)* **287**, 84–85.

DeSalle, R., Barcia, M., and Wray, C. (1993). PCR jumping in clones of 30-million-year-old DNA fragments from amber preserved termites (*Mastotermes electrodominicus*). *Experientia* **49**, 906–909.

Desjardins, P., and Morais, R. (1990). Sequence and gene organization of the chicken mitochondrial genome: A novel gene order in higher vertebrates. *J. Mol. Biol.* **212**, 599–634.

Diamond, J. M. (1990). Biological effects of ghosts. *Nature (London)* **345**, 769–770.

Ellegren, H. (1991). DNA typing of museum birds. *Nature (London)* **354**, 113.

Ellegren, H. (1993). Genomic DNA from museum bird feathers. *In* "Ancient DNA" (B. Herrmann and S. Hummel, Eds.), pp. 211–217. Springer-Verlag, New York.

Fleming, C. A. (1979). "The Geological History of New Zealand and Its Life." Auckland University, Auckland, New Zealand.

Fordyce, R. E. (1991). A new look at the vertebrate fossil record of New Zealand. *In* "Vertebrate Palaeontology of Australasia" (P. V. Rich, Ed.), pp. 1191–1317. Pioneer Design Studio Monash University, Victoria, Australia.

Giffin, J. G. (1993). New species of fossil birds found at Pu'u Wa'awa'a. *'Elepaio* **35**, 1–3.

Golenberg, E. M., Giannasi, D. E., Clegg, M. T., Smiley, C. J., Durbin, M., Henderson, D., and Zurawski, G. (1990). Chloroplast DNA sequence from a Miocene *Magnolia* species. *Nature (London)* **344**, 656–658.

Grody, W. W. (1994). Screening for pathogenic DNA sequences in clinically collected human tissues. *In* "Ancient DNA" (B. Herrmann and S. Hummel, Eds.), pp. 69–91. Springer-Verlag, New York.

Hackett, S. J., Griffiths, C. S., Bates, J. M., and Klein, N. K. (1995). Re: A commentary on the use of sequence data for phylogeny reconstruction. *Mol. Phylogenet. Evol.* **4**, 350–356.

Hagelberg, E., Sykes, B., and Hedges, R. (1989). Ancient bone DNA amplified. *Nature (London)* **342**, 485.

Hagelberg, E., Gray, I. C., and Jeffreys, A. J. (1991). Identification of the skeletal remains of a murder victim by DNA analysis. *Nature (London)* **352**, 427–429.

Handt, O., Höss, M., Krings, M., and Pääbo, S. (1994a). Ancient DNA: Methodological challenges. *Experientia* **50**, 524–529.

Handt, O., Richards, M., Trommsdorff, M., Kilger, C., Simanainen, J., Georgiev, O., Bauer, K., Stone, A., Hedges, R., Schaffner, W., Utermann, G., Sykes, B., Pääbo, S. (1994b). Molecular genetic analyses of the Tyrolean ice man. *Science* **264**, 1775–1778.

Handt, O., Krings, M., Ward, R. H., and Pääbo, S. (1996). The retrieval of ancient human DNA sequences. *Am. J. Hum. Genet.* **59**, 368–376.

Hänni, C., Laudet, V., Stehelin, D., and Taberlet, P. (1994). Tracking the origins of the cave bear (*Ursus spelaeus*) by mitochondrial DNA sequencing. *Proc. Natl. Acad. Sci. U.S.A.* **91**, 12336–12340.

Hedges, R. E. M., Millard, A. R., and Pike, A. W. G. (1995). Measurements and relationships of diagenetic alteration of bones from three archaeological sites. *J. Archeol. Sci.* **22**, 201–211.

Hendy, M. D., and Penny, D. (1989). A framework for the quantitative study of evolutionary trees. *Syst. Zool.* **38**, 297–309.

Hendy, M. D., and Penny, D. (1993). Spectral analysis of phylogenetic data. *J. Classification* **10**, 5–24.

Hendy, M. D., Penny, D., and Steel, M. (1994). A discrete Fourier analysis for evolutionary trees. *Proc. Natl. Acad. Sci. U.S.A.* **91**, 3339–3343.

Herrmann, B., and Hummel, S. (1994). "Ancient DNA," p. 263. Springer-Verlag, New York.

Hickson, R. E., Simon, C., Cooper, A., Spicer, G. S., Sullivan, J., and Penny, D. (1996). Conserved motifs, alignment, and secondary structure for the third domain of animal 12S rRNA. *Mol. Biol. Evol.* **13**, 150–169.

Higuchi, R. G., Wrischnik, L. A., Oakes, E., George, M., Tong, B., and Wilson, A. C. (1987). Mitochondrial DNA of the extinct quagga: Relatedness and extent of postmortem changes. *J. Mol. Evol.* **25**, 283–287.

Higuchi, R., Bowman, B., Freiberger, M., Ryder, O. A., and Wilson, A. C. (1984). DNA sequence from the quagga, an extinct member of the horse family. *Nature (London)* **312**, 282–284.

Horai, S., Hayasaka, K., Kondo, R., Tsugane, K., and Takahata, N. (1995). Recent African origin of modern humans revealed by complete sequences of hominoid mitochondrial DNAs. *Proc. Natl. Acad. Sci. U.S.A.* **92**, 532–536.

Höss, M., and Pääbo, S. (1993). DNA extraction from Pleistocene bones by a silica-based extraction method. *Nucleic Acids Res.* **21**, 3913–3914.

Höss, M., Kohn, M., Pääbo, S., Knauer, F., and Schröder, W. (1992). Excrement analysis by PCR. *Nature (London)* **359**, 199.

Höss, M., Dilling, A., Currant, A., and Pääbo, S. (1995). Molecular phylogeny of the extinct ground sloth *Mylodon darwinii*. *Proc. Natl. Acad. Sci. U.S.A.* **93**, 181–185.

Houde, P. (1986). Ostrich ancestors found in the Northern hemisphere suggest new hypothesis of ratite origins. *Nature (London)* 324, 563–565.

Houde, P. (1988). Paleognathous birds from the early Tertiary of the Northern Hemisphere. *Publ. Nuttall Ornithol. Club* **22**, 1–148.

Houde, P., and Braun, M. J. (1988). Museum collections as a source of DNA for studies of avian phylogeny. *Auk* **105**, 773–776.

Houde, P., and Olson, S. L. (1981). Paleognathous carinate birds from the Early Tertiary of North America. *Science* **214**, 1236–1237.

Houde, P., Sheldon, F. H., and Kreitman, M. (1995). A comparison of solution and membrane-bound DNA × DNA hybridization, as used to infer phylogeny. *J. Mol. Evol.* **40**, 678–688.

James, H. F., and Olson, S. L. (1991). "Descriptions of 32 New Species of Birds from the Hawaiian Islands, Part II: Passeriformes." Ornithological Monographs, Vol. 46, pp. 1–88. American Ornithologists' Union, Washington, D.C.

James, H. F., Stafford, T. W., Jr., Steadman, D. W., Olson, S. L., Martin, P. S., Tull, A. J. T., and McCoy, P. (1987). Radiocarbon dates on bones of extinct birds from Hawaii. *Proc. Natl. Acad. Sci. U.S.A.* **84**, 2350–2354.

Lento, G. M., Hickson, R. E., Chambers, G. K., and Penny, D. (1995). Use of spectral analysis to test hypotheses on the origin of pinnipeds. *Mol. Biol. Evol.* 12, 28–52.

Lindahl, T. (1993a). Instability and decay of the primary structure of DNA. *Nature (London)* **362**, 709–715.

Lindahl, T. (1993b). Recovery of antediluvian DNA. *Nature (London)* **365**, 700.

Livezey, B. C. (1991). A phylogenetic analysis and classification of recent dabbling ducks (tribe Anatini) based on comparative morphology. *Auk* **108**, 471–507.

Lockhart, P. J., Steel, M. A., Hendy, M. D., and Penny, D. (1994). Recovering trees under a more general model of sequence evolution. *Mol. Biol. Evol.* **11**, 605–612.

Mindell, D. P., and Honeycutt, R. L. (1990). Ribosomal RNA in vertebrates: Evolution and phylogenetic applications. *Annu. Rev. Ecol. Syst.* **21**, 541–566.

Moulton, D. W., and Weller, M. W. (1984). Biology and conservation of the Laysan duck (*Anas laysanensis*). *Condor* **86**, 105–117.

Moum, T., Johansen, S., Erikstad, K. E., and Piatt, J. F. (1994). Phylogeny and evolution of the auks (subfamily Alcinae) based on mitochondrial DNA sequences. *Proc. Natl. Acad. Sci. U.S.A.* **91**, 7912–7916.

Olson, S. L. (1985). The fossil record of birds. *In* "Avian Biology" (D. S. Farner, J. R. King, and K. C. Parkes, Eds.), Vol. VIII, pp. 79–238. Academic Press, Orlando, Florida.

Olson, S. L., and James, H. F. (1991). "Descriptions of 32 New Species of Birds from the Hawaiian

Islands, Part I: Non-Passeriformes." Ornithological Monographs, Vol. 45, pp. 1–88. American Ornithologists' Union, Washington, D.C.

Pääbo, S. (1993). Ancient DNA. *Sci. Am.* **269**, 86–92.

Pääbo, S., and Wilson, A. C. (1988). Polymerase chain reaction reveals cloning artefacts. *Nature (London)* **334**, 387–388.

Pääbo, S., Higuchi, R. G., and Wilson, A. C. (1989). Ancient DNA and the polymerase chain reaction. *J. Biol. Chem.* **264**, 9709–9712.

Poinar, H., Cano, R. J., and Poinar, G. O., Jr. (1993). DNA from an extinct plant. *Nature (London)* **363**, 677.

Poinar, H. N., Höss, M., Bada, J. L., and Pääbo, S. (1996). Amino acid racemization and the preservation of ancient DNA. *Science* **272**, 864–866.

Quinn, T. W., and White, B. N. (1987). Analysis of DNA sequence variation. *In* "Avian Genetics" (F. Cooke and P. A. Buckley, Eds.), pp. 163–198. Academic Press, London.

Richards, M. B., Sykes, B. C., and Hedges, R. E. M. (1995). Authenticating DNA extracted from ancient skeletal remains. *J. Archeol. Sci.* **22**, 291–299.

Roy, M. S., Girman, D. J., Taylor, A. C., and Wayne, R. K. (1994). The use of museum specimens to reconstruct the genetic variability and relationships of extinct populations. *Experientia* **50**, 551–557.

Sagata, N., Oskarsson, M., Copeland, T., Brumbaugh, J., and Vande Woude, G. F. (1988). Function of *c-mos* proto-oncogene product in meiotic maturation in *Xenopus* oocytes. *Nature (London)* **335**, 519–525.

Saiki, R. K., Gelfand, D. H., Stoffel, S., Scharf, S. J., Higuchi, R., Horn, G. T., Mullis, K. B., and Erhlich, H. A. (1988). Primer-directed enzymatic amplification of DNA with a thermostable DNA polymerase. *Science* **239**, 487–491.

Sallares, R., Allaby, R. G., and Brown, T. A. (1995). PCR-based identification of wheat genomes. *Mol. Ecol.* **4**, 509–514.

Sheldon, F. H., and Bledsoe, A. H. (1993). Avian molecular systematics, 1970s to 1990s. *Annu. Rev. Ecol. Syst.* **24**, 243–278.

Sibley, C. G., and Ahlquist, J. E. (1981). The phylogeny and relationships of the ratite birds as indicated by DNA–DNA hybridization. *In* "Evolution Today: Proceedings of the 2nd International Congress of Systematic and Evolutionary Biology" (G. G. E. Scudder and J. L. Reveal, Eds.), pp. 301–335. Carnegie-Mellon University, Pittsburgh, Pennsylvania.

Sibley, C. G., and Ahlquist, J. E. (1990). "Phylogeny and Classification of Birds." Yale University Press, London.

Sidow, A., Wilson, A. C., and Pääbo, S. (1991). Bacterial DNA in *Clarkia* fossils. *Philos. Trans. R. Soc. Lond. B* **333**, 429–433.

Soltis, P. S., Soltis, D. E., and Smiley, C. J. (1992). An *rbc*L sequence from a Miocene *Taxodium* (bald cypress). *Proc. Natl. Acad. Sci. U.S.A.* **89**, 449–451.

Sorenson, M. D., and Fleischer, R. C. (1996). Multiple independent transpositions of mitochondrial DNA control region sequences of the nucleus. *Proc. Natl. Acad. Sci. U.S.A.* **93**, 15239–15243.

Steadman, D. (1995). Prehistoric extinctions of Pacific island birds: Biodiversity meets zooarchaeology. *Science* **267**, 1123–1131.

Steel, M., Cooper, A., and Penny, D. (1996). Confidence intervals for the divergence time of two clades. *Syst. Biol.* **45**, 127–134.

Stevens, G. R. (1985). "Lands in Collision." Information Series 161. New Zealand Department of Scientific and Industrial Research, Wellington, New Zealand.

Stone, A. C., and Stoneking, M. (1993). Ancient DNA from a pre-Columbian Amerindian population. *Am. J. Phys. Anthropol.* **92**, 463–471.

Swofford, D. (1996). "PAUP*," version 4.0d 44–51. Sinauer, Sunderland, Massachusetts (in press).

Swofford, D. L., Olsen, G. J., Waddell, P. J., and Hillis, D. M. (1996). *In* "Molecular Systematics" (D. M. Hillis, C. Moritz, and M. K. Mable, Eds.), 2nd ed., pp. 407–514. Sinauer, Sunderland, Massachusetts.

Tambussi, C. P. (1995). The fossil Rheiformes from Argentina. *Courier Forsch. Inst. Senckenberg* **181**, 121–129.

Tambussi, C. P., Noriega, J., Gazdzicki, A., Tatur, A., Reguero, M. A., and Vizcaino, S. F. (1994). The first occurrence of ratite birds in the Paleocene of Antarctica. *In* "XXI Polar Symposium" (S. M. Zalewki, Ed.), pp. 45–58. Polish Academy of Sciences, Warsaw.

Taylor, P. G. (1996). Reproducibility of ancient DNA sequences from extinct Pleistocene fauna. *Mol. Biol. Evol.* **13**, 283–285.

Thomas, R. H., Schaffner, W., Wilson, A. C., and Pääbo, S. (1989). DNA phylogeny of the extinct marsupial wolf. *Nature* (*London*) **340**, 465–467.

Thomas, W. K., Pääbo, S., Villablanca, F. X., and Wilson, A. C. (1990). Spatial and temporal continuity of kangaroo rat populations shown by sequencing mitochondrial DNA from museum specimens. *J. Mol. Evol.* **31**, 101–112.

Wayne, R. K., and Jenks, S. M. (1991). Mitochondrial DNA analysis implies extensive hybridization of the endangered red wolf *Canis rufus*. *Nature* (*London*) **351**, 565–568.

Wellman, H. (1994). Moas, gizzard stones, and New Zealand plants. *N. Z. Sci. Rev.* **51**, 57–60.

Wilson, A. C., Cann, R. L., Carr, S. M., George, M., Gyllensten, U. B., Helm-Bychowski, K. M., Higuchi, R. G., Palumbi, S. R., Prager, E. M., Sage, S. D., and Stoneking, M. (1985). Mitochondrial DNA and two perspectives on evolutionary genetics. *Biol. J. Linn. Soc.* **26**, 375–400.

Woodward, S. R., Weyand, N. J., and Bunnell, M. (1994). DNA sequence from Cretaceous period bone fragments. *Science* **266**, 1229–1232.

Worthy, T. H. (1988). Loss of flight ability of the New Zealand duck *Euryanas finschi*. *J. Zool. Lond.* **215**, 619–628.

Worthy, T. H., and Holdaway, R. N. (1993). Quaternary fossil faunas from Westland caves. *J. R. Soc. N.Z.* **23**, 147–254.

Xu, X., Janke, A., and Arnason, U. (1996). The complete mitochondrial DNA sequence of the greater Indian rhinoceros, *Rhinoceros unicornis*, and the phylogenetic relationship among Carnivora, Perissodactyla, and Artiodactyla (+ Cetacea). *Mol. Biol. Evol.* **13**, 1167–1173.

Zischler, H., Höss, M., Handt, O., von Haeseler, A., van der Kuyl, A. C., Goudsmit, J., and Pääbo, S. (1995). Detecting dinosaur DNA. *Science* **268**, 1192–1193.

Index